AURORA

DOVER MODERN MATH ORIGINALS

Dover Publications is pleased to announce the publication of the first volumes in our new Aurora Series of original books in mathematics. In this series we plan to make available exciting new and original works in the same kind of well-produced and affordable editions for which Dover has always been known.

Aurora titles currently in the process of publication are:

Optimization in Function Spaces by Amol Sasane. (978-0-486-78945-3)

The Theory and Practice of Conformal Geometry by Steven G. Krantz. (978-0-486-79344-3)

Numbers: Histories, Mysteries, Theories by Albrecht Beutelspacher. (978-0-486-80348-7)

Elementary Point-Set Topology: A Transition to Advanced Mathematics by André L. Yandl and Adam Bowers. (978-0-486-80349-4)

Additional volumes will be announced periodically.

AN INTRODUCTORY COURSE ON
DIFFERENTIABLE MANIFOLDS

SIAVASH SHAHSHAHANI

DOVER PUBLICATIONS, INC.
Mineola, New York

Bibliographical Note

An Introductory Course on Differentiable Manifolds is a new work, first published by Dover Publications, Inc., in 2016, as part of the "Aurora Dover Modern Math Originals" series.

International Standard Book Number

ISBN-13: 978-0-486-80706-5
ISBN-10: 0-486-80706-1

Manufactured in the United States by RR Donnelley
80706101 2016
www.doverpublications.com

in memory of my parents

Contents

Preface

This book is the outgrowth of a course on differentiable manifolds that the author has taught many times over the years. The audience has mostly consisted of advanced undergraduate and first-year graduate students in mathematics. A course on manifolds has now become a staple of solid education in mathematics not only for mathematicians, but also for many theoretical physicists and others. One could argue that such a course is the natural modern day setting for doing advanced calculus on a curved space. The material in this book is slightly more than the author has been able to cover in a one-semester course for the intended audience, but a two-quarter course should allow a leisurely paced coverage. Alternatively, various choices of omission are available for a one-semester course after covering most of the first seven chapters.

The prerequisite for this course is solid grounding in undergraduate mathematics, including rigorous multivariable calculus, linear algebra, elementary abstract algebra and point set topology. In practice, we have found that the ideal background is often lacking, so we have not shied away from "recalling" some of the material either in the text or in the appendices. We have also strived to make the treatment complete, providing sufficient detail for the novice so that they will be able to gain confident mastery of the material. In particular, we have made a point of not leaving any details other than the completely routine or repetitive to the reader, nor have we relegated any crucial argument in the text material to the exercises. In the same vein, we have avoided the use of time-worn phrases such as "it is obvious...," believing that the obvious should best be passed by in silence. On the other hand, over 250 exercises, tuned to the text material, should offer students sufficient opportunity to gauge their skills and to gain additional depth and insight.

An explanation of the arrangement of material in this book is in order. In the earliest lecture note versions of the book, we followed the familiar sequence of manifolds-vector bundles-vector fields and tensor fields, as a starter. In the present work, based on our recent teaching practice, we have divided the book into four parts. Part I, called **Pointwise**, consists of a single chapter devoted to the tensor algebra of linear spaces and their mappings. This may be construed as what happens at a single tangent space on a manifold, indifferent to neighboring points. Part II, **Local**, brings in neighboring points, essentially through the role of the derivative, to develop such genuinely local matters as integrating vector fields, Lie bracket, exterior derivative and Lie derivative. In Part III, we pass on to **Global** through the introduction of manifolds and vector bundles and develop the main body of the course. Vector bundle theory is developed only to the extent needed for later chapters. Finally, in Part IV, **Geometric Structures**, consisting of the single final chapter of the book, we provide a glimpse into geometric structures through the introduction of connections on the tangent bundle as a tool to implant the second derivative and the derivative of vector fields on the base manifold. In the early versions of the work, we concluded with a chapter on symplectic geometry and Hamiltonian mechanics. But symplectic geometry is now witnessing such rapid and broad development that we thought it wise to forego such an introduction and replace Hamiltonian mechanics with a purely "Newtonian" account tied closely to the notions of connection and acceleration vector as developed in the final chapter.

One may cite logical, pedagogical and even historical justification for this choice of presentation. While the appearance of the word "manifold" (*Mannigfaltigkeit*) in mathematics literature predates the introduction of topological spaces by at least half a century, the formal definition of a manifold had to await the development of point set theory. Before that happened, many tools of differential geometry were developed locally in tensor language. The pioneers used coordinates and local computations while differentiating between pure analysis and geometry by requiring that geomeric entities be held to certain rigid transformations under coordinate change. The student who is initially presented with the whole apparatus of manifolds and vector bundles will often need plenty of time to sort out the logical niche of various notions and frequently experiences difficulty working out examples and making computations. Our experience has been that by first developing the algebraic and local computational skills of students, and by treating what is genuinely local as local, we can better prepare the student to appreciate the necessity and usefulness of global notions and the facility they provide in dealing directly and intrinsically with geometric notions. Even the integration of plane fields (Frobenius theorem) and the notion of connection could have been included in the early chapters on local theory, but the significant global aspects of these notions make it more expedient to present them after global tools have been developed.

The editorial organization of the book is as follows. Each chapter is divided into sections serialized alphabetically by capital letters A, B, C,... Each section is divided into subsections identified by numbers. These numbers run consecutively not by section, but by the chapter they are in. A subsection may be a significant theorem, an important definition, a propostion describing elementary consequences of a definition, a set of examples or even a heuristic discussion. Further, some subsections are subdivided into sub-subsections serially identified by small letters a, b, c, ... Not many statements are honored as "theorems" throughout the book. Most assertions requiring proofs are identified by a subsection number or a small letter indicating a sub-subsection. The dearth of serious theorems is natural as this is only an introductory text aimed at providing necessary language skills.

Acknowledgments

I have had the fortune of being associated with highly talented students during repeated presentations of the course on which this book is based. Their spirited interaction has sharpened and reshaped the original notes. Many of these former students are now accomplished mathematicians. I will refrain from naming many, lest I inadvertently leave out some equally deserving. But I will mention a few, in chronological order, who gave me written suggestions I happen to have kept: F. Rezakhanloo, A.S. Tahvildarzadeh, H. Torabi Tehrani, B. Khanedani, A. Taghavi and S. Zakeri. The latter has also been, more recently, my ever generous TeX advisor. The finished manuscript was improved by comments and suggestions from A. Taghavi, S. Habibi-Esfahani, M. Shahriari, M. Khoshnevis and M. R. Koushesh. Other mathematicians whose contributions I should recognize include A. Shafii-Dehabad, M. Ardeshir, A. Jafari, P. Safari, M. Zeinalian and Sohrab Shahshahani. Most of all, Bahman Khanedani has been a constant mathematical companion and teaching colleague from whom I have learned much; his footprint is all over the book.

Special thanks are due to A. Kamalinejad who initiated the TeX version of the book. I have benefited in this and earlier work from the expert graphical skill of Amin Sadeghi whose tutelage also launched M. Mazaheri into generously giving many hours of his time to preparing and importing the diagrams. The final English version of the book was prepared in New York where I had access to the facilities of CUNY Graduate Center. I thank Dennis Sullivan for the opportunity.

S. Shahshahani
December 2015

Siavash Shahshahani received his university education at Berkeley in the 1960s, getting a PhD in mathematics under Steve Smale in 1969. Subsequently he held positions at Northwestern and the University of Wisconsin, Madison. From 1974 to his retirement in 2012, he was mainly at Sharif University of Technology, Tehran, Iran, and helped develop a strong mathematics program there.

Part I

Pointwise

CHAPTER 1

Multilinear Algebra

In this chapter, we discuss various structures and mappings that involve one or several vector spaces over a fixed field. It will always be assumed that the field characteristic is zero; in fact, the reader may assume that the underlying field is $\mathbb{R}$ or $\mathbb{C}$. In Section D, we will consider special features of vector spaces over $\mathbb{R}$.

A. Dual Space

Let V be a finite dimensional vector space over a field F. The set of linear mappings $V \to F$ will be denoted by V^*. This set will be endowed with the structure of a vector space over F. Let α and β be elements of V^* and r an element of F, then we define $\alpha+\beta$ and $r\alpha$ by

$$(\alpha+\beta)(x) = \alpha(x) + \beta(x) \tag{1.1}$$

$$(r\alpha)(x) = r\alpha(x) \tag{1.2}$$

where x is an arbitrary element of V. With these operations, V^* becomes a vector space over F and will be called the ***dual space*** to V. Suppose $(e_1, \ldots, e_n)$ is a basis for V. We define elements e^i of V^* by their value on e_j, $j = 1, \ldots, n$, as follows:

$$e^i(e_j) = \delta^i_j \tag{1.3}$$

where δ^i_j denotes the value 1 or 0 depending on whether $i{=}j$ or $i{\neq}j$. Note that any element α of V^* can be written as a linear combination of $e^1, \ldots, e^n$. In fact,

$$\alpha = \sum_{i=1}^{n} \alpha(e_i)e^i$$

since the value of both sides on an arbitrary basis element e_j is the same. Further, $\{e^1, \ldots, e^n\}$ is a linearly independent set, for if $\sum_i r_i e^i{=}0$, evaluating both sides on the basis element e_j yields $r_j{=}0$. Therefore, the ordered set $(e^1, \ldots, e^n)$ is a basis for V^*, called the ***dual basis*** for V^* relative to $(e_1, \ldots, e_n)$. Thus V^* has the same dimension as V.

By repeating the operation of dual making, one can look at $(V^*)^*$, the so-called ***double dual*** of V, usually denoted by V^{**}. The double dual will then have the same dimension as the original space V, and since all linear spaces of the same dimension over a given field are isomorphic, there are isomorphisms between V, V^* and V^{**}. But in the case of V and V^{**}, there is a distinguished ***natural isomorphism***, denoted by $I_V : V \to V^{**}$, which is given as follows. For each $v \in V$, the element $I_V(v)$ is defined by

$$(I_V(v))(\alpha) = \alpha(v) \tag{1.4}$$

It follows from (1.1) and (1.2) that $I_V(v)$ is indeed linear, i.e., it is a member of V^{**}. That I_V is linear follows from the linearity of α. To show that I_V is an isomorphism, it suffices to show that its kernel is $\{0\}$ since the domain and target linear spaces are finite dimensional of the same dimension. But $\alpha(v)=0$ for all α in V^* implies that $v=0$, and the isomorphism is established. Note that the definition of I_V was independent of the specific nature of the linear space V or the choice of basis for it. In fact, one can state the following general assertion.

1. Theorem *For any basis $(e_1, \ldots, e_n)$ of V, $(I_V(e_1), \ldots, I_V(e_n))$ is the dual basis in V^{**} relative to the basis $(e^1, \ldots, e^n)$ for V^*.*

Proof. We must show

$$(I_V(e_i))(e^j) = \delta_i^j$$

This is a consequence of (1.4) and (1.3). □

By virtue of the natural isomorphism I_V, the space V^{**} is often identified with V. Under this identification, $I_V(e_i)$ is identified with e_i, so that $(e_1, \ldots, e_n)$ becomes the dual basis for V^{**} relative to $(e^1, \ldots, e^n)$.

B. Tensors

Let $V_1, \ldots, V_p$ and W be vector spaces over a field. A map $\alpha : V_1 \times \cdots \times V_p \to W$ is called ***p*-linear** provided that by fixing any $p-1$ components of $(v_1, \ldots, v_p) \in V_1 \times \cdots \times V_p$, α is linear with respect to the remaining component. As we shall see in some of the following examples, operations generally known as "products" in elementary mathematics are of this nature.

2. Examples

(a) Let V be a vector space over a field F. Regard F as a one-dimensional vector space over F. Then the product $F \times V \to V$ given by

$$(r, v) \mapsto rv$$

is 2-linear (***bilinear***).

(b) Let F be a field. Then the p-fold product $F \times \cdots \times F \to F$ given by

$$(r_1, \ldots, r_p) \mapsto r_1 \cdots r_p$$

is p-linear.

(c) Let V be a vector space over $\mathbb{R}$. Then any inner product $V \times V \to \mathbb{R}$ is bilinear. The vector product $\mathbb{R}^3 \times \mathbb{R}^3 \to \mathbb{R}^3$ is another example of a bilinear mapping. In general, let $\beta : V{\times}V{\to}F$ be bilinear and consider a basis $(e_1, \ldots, e_n)$ for V. The $n{\times}n$ matrix B=$[\beta_{ij}]$, where $\beta_{ij} = \beta(e_i, e_j)$, determines β completely as

$$\beta(\sum_i u^i e_i, \sum_j v^j e_j) = \sum_{i,j} \beta_{ij} u^i v^j \tag{1.5}$$

If B is a symmetric matrix with positive eigenvalues, then β is an inner product. Conversely, any inner product on V is obtained in this manner.

(d) For a vector space V over a field F, the ***evaluation pairing*** $V{\times}V^*{\to}F$, given by $(v, \alpha){\mapsto}\alpha(v)$, is bilinear.

(e) Let F be a field and $V_1, \ldots, V_p; W_1, \ldots, W_q$ be vector spaces over F. Suppose p-linear and q-linear maps $\alpha{:}V_1{\times}\cdots{\times}V_p{\to}F$ and $\beta{:}W_1{\times}\cdots{\times}W_q{\to}F$ are given. Then the ***tensor product***

$$\alpha \otimes \beta : V_1 \times \cdots \times V_p \times W_1 \times \cdots \times W_q \to F$$

is defined by

$$(\alpha \otimes \beta)(v_1, \ldots, v_p, w_1, \ldots, w_q) = \alpha(v_1, \ldots, v_p)\beta(w_1, \ldots, w_q) \tag{1.6}$$

Note that $\alpha{\otimes}\beta$ is a $(p{+}q)$-linear mapping. Further, it follows from the associativity of the product operation in the field F that $\otimes$ is associative, hence the product $\alpha_1{\otimes}\cdots{\otimes}\alpha_k$ is unambiguously defined by induction.

In what follows, V will be a finite dimensional vector space over a field F. The n-fold product $V{\times}\cdots{\times}V$ will be denoted by V^n.

3. Definition

(a) A p-linear map $V^p{\to}F$ will be called a ***covariant p-tensor*** , or a ***tensor of type (p,0)***, on V.

(b) A q-linear map $(V^*)^q{\to}F$ will be called a ***contravariant q-tensor***, or a ***tensor of type (0,q)***, on V.

(c) A $(p + q)$-linear map $V^p{\times}(V^*)^q{\to}F$ will be called a ***mixed (p,q)-tensor***, or a ***tensor of type (p,q)***, on V.

4. Examples An element of V^* is a covariant 1-tensor on V. In view of the natural isomorphism I_V, any member of V may be regarded as a contravariant tensor on V. The evaluation pairing (Example 2d) is a $(1, 1)$-tensor on V. Inner products are examples of covariant 2-tensors.

We use the symbols $L^p(V)$, $L_q(V)$ and $L^p_q(V)$, respectively, to denote the sets of $(p, 0)$-, $(0, q)$- and (p, q)-tensors on V. Under functional addition, and multiplication by elements of the field F, each of these becomes a vector space over F. The dimensions of these spaces are, respectively, n^p, n^q and n^{p+q}, as the following will imply.

5. Basis for the Space of Tensors *Let* $(e_1, \ldots, e_n)$ *be a basis for* V. *Then the following are basis elements for the spaces of tensors.*

(a) *For* $L^p(V)$:

(1.7) $$e^{i_1}\otimes\cdots\otimes e^{i_p}, \quad 1 \le i_\mu \le n$$

(b) *For* $L_q(V)$:

(1.8) $$e_{j_1}\otimes\cdots\otimes e_{j_q}, \quad 1 \le j_\nu \le n$$

(c) *For* $L_q^p(V)$:

(1.9) $$e^{i_1}\otimes\cdots\otimes e^{i_p}\otimes e_{j_1}\otimes\cdots\otimes e_{j_q}, \quad 1 \le i_\mu, j_\nu \le n$$

Proof. Note that by virtue of Example 2e, the displayed tensors are actually elements of the stated spaces. We prove the third case which includes the other two. To show linear independence, suppose that

$$\sum_{i_\mu, j_\nu} c^{j_1\ldots j_q}_{i_1\ldots i_p}\, e^{i_1}\otimes\cdots\otimes e^{i_p}\otimes e_{j_1}\otimes\cdots\otimes e_{j_q} = 0$$

By applying the two sides to $(e_{i_1}, \ldots, e_{i_p}, e^{j_1}, \ldots, e^{j_q})$, we see that the coefficients are zero, and linear independence is established. On the other hand, note that any $\alpha \in L_q^p$ can be written as

(1.10) $$\alpha = \sum_{i_\mu, j_\nu} \alpha(e_{i_1}, \ldots, e_{i_p}, e^{j_1}, \ldots, e^{j_q}) e^{i_1}\otimes\cdots\otimes e^{i_p}\otimes e_{j_1}\otimes\cdots\otimes e_{j_q}$$

which can be verified by applying both sides to $(e_{i_1}, \ldots, e_{i_p}, e^{j_1}, \ldots, e^{j_q})$. □

By convention, we let $L^0V = L_0V = F$.

6. Change of Basis

The bases introduced above for the spaces of tensors as well as the resulting components of the tensors depend on the original choice of basis for the linear space. We are now going to investigate how a linear change of basis for the space affects the value of tensor components. We take V to be an n-dimensional vector space over F. It will be convenient to write $n\times n$ matrices with entries from F as $A=[a^i_j]$, where the superscript denotes the row index and the subscript indicates the column of the matrix entry. Suppose two bases $\mathcal{B}=(e_1, \ldots, e_n)$ and $\bar{\mathcal{B}} = (\bar{e}_1, \ldots, \bar{e}_n)$ are given for V, related linearly by matrix $A=[a^i_j]$ as

(1.11) $$\bar{e}_j = \sum_\mu a^\mu_j e_\mu$$

Thus the components of $\bar{e}_j$ with respect to the basis $\mathcal{B}$ are the entries of the jth column of matrix A. Corresponing to $\mathcal{B}$ and $\bar{\mathcal{B}}$, we have the dual bases $\mathcal{B}^*=(e^1, \ldots, e^n)$ and

$\bar{\mathcal{B}}^*=(\bar{e}^1,\ldots,\bar{e}^n)$. We will first investigate the linear relationship between these two bases. We write

$$\bar{e}^i = \sum_{\nu} b^i_{\nu} e^{\nu} \tag{1.12}$$

Therefore, the components of $\bar{e}^i$ with respect to the basis $\mathcal{B}^*$ are the entries of the ith row of matrix B=$[b^i_j]$. To identify B, we note that

$$\delta^i_j = \bar{e}^i(\bar{e}_j) = \left(\sum_{\nu} b^i_{\nu} e^{\nu}\right)\left(\sum_{\mu} a^{\mu}_j e_{\mu}\right) = \sum_k b^i_k a^k_j$$

Therefore, the matrix B is the inverse of the transpose of the matrix A:

$$\mathrm{B}^{-1} = \mathrm{A}^{\mathrm{T}}$$

Now let α be a (p,q)-tensor on V. With respect to the above bases, the following two representations for α are obtained.

$$\alpha = \sum_{\substack{i_1,\ldots,i_p \\ j_1,\ldots,j_q}} \alpha^{j_1,\ldots,j_q}_{i_1,\ldots,i_p}\, e^{i_1} \otimes \cdots \otimes e^{i_p} \otimes e_{j_1} \otimes \cdots \otimes e_{j_q}$$

$$\alpha = \sum_{\substack{i_1,\ldots,i_p \\ j_1,\ldots,j_q}} \bar{\alpha}^{j_1,\ldots,j_q}_{i_1,\ldots,i_p}\, \bar{e}^{i_1} \otimes \cdots \otimes \bar{e}^{i_p} \otimes \bar{e}_{j_1} \otimes \cdots \otimes \bar{e}_{j_q}$$

We wish to express the components $\bar{\alpha}^{j_1,\ldots,j_q}_{i_1,\ldots,i_p}$ in terms of $\alpha^{j_1,\ldots,j_q}_{i_1,\ldots,i_p}$. Using (1.10), we have

$$\bar{\alpha}^{j_1,\ldots,j_q}_{i_1,\ldots,i_p} = \alpha(\bar{e}_{i_1},\ldots,\bar{e}_{i_p};\bar{e}^{j_1},\ldots,\bar{e}^{j_q})$$

This is equal to

$$\sum_{\substack{k_1,\ldots,k_p \\ l_1,\ldots,l_q}} \alpha^{l_1,\ldots,l_q}_{k_1,\ldots,k_p} e^{k_1}\otimes\cdots\otimes e^{k_p}\otimes e_{l_1}\otimes\cdots\otimes e_{l_q}\Big(\sum_{\mu} a^{\mu}_{i_1} e_{\mu},\ldots,\sum_{\mu} a^{\mu}_{i_p} e_{\mu};\sum_{\nu} b^{j_1}_{\nu} e^{\nu},\ldots,\sum_{\nu} b^{j_q}_{\nu} e^{\nu}\Big)$$

$$= \sum_{\substack{k_1,\ldots,k_p \\ l_1,\ldots,l_q}} \alpha^{l_1,\ldots,l_q}_{k_1,\ldots,k_p}\, a^{k_1,\ldots,k_p}_{i_1,\ldots,i_p}\, b^{j_1,\ldots,j_q}_{l_1,\ldots,l_q}$$

Thus we have obtained the desired formula for the change of tensor components under a linear change of variables

$$\bar{\alpha}^{j_1,\ldots,j_q}_{i_1,\ldots,i_p} = \sum_{k_1,\ldots,k_p=1}^{n} \sum_{l_1,\ldots,l_q=1}^{n} a^{k_1,\ldots,k_p}_{i_1,\ldots,i_p}\, \alpha^{l_1,\ldots,l_q}_{k_1,\ldots,k_p}\, b^{j_1,\ldots,j_q}_{l_1,\ldots,l_q} \tag{1.13}$$

Classically, a tensor is defined as a collection of numerical quantities $\alpha^{j_1,\cdots,j_q}_{i_1,\cdots,i_p}$ which transform under a linear change of variables according to formula (1.13).

(a) **Special case (p=1,q=0)** For a covariant 1-tensor

$$\alpha = \sum_{i=1}^{n} \alpha_i e^i = \sum_{i=1}^{n} \bar{\alpha}_i \bar{e}^i$$

we obtain

$$\bar{\alpha}_i = \sum_{k=1}^{n} a_i^k \alpha_k \tag{1.14}$$

(b) **Special case (p=0,q=1)** Consider a contravariant 1-tensor, or by virtue of the natural isomorphism I_V, an element x of V

$$x = \sum_{j=1}^{n} x^j e_j = \sum_{j=1}^{n} \bar{x}^j \bar{e}_j$$

In this case, we have

$$\bar{x}^j = \sum_{l=1}^{n} b_l^j x^l \tag{1.15}$$

7. Functoriality

Let V and W be vector spaces over a field F, and suppose $f{:}V \to W$ is a linear map. For each non-negative integer p, a map $L^p f{:}L^p W \to L^p V$ is defined as follows. If $p = 0$, $L^0 f = \mathbb{1}_F$. For $p > 0$, suppose $\alpha \in L^p W$ and $v_1, \ldots, v_p \in V$, then

$$(L^p f(\alpha))\,(v_1, \ldots, v_p) = \alpha(f(v_1), \ldots, f(v_p)) \tag{1.16}$$

That $(L^p f)(\alpha) \in L^p V$ follows from the linearity of f and the fact that $\alpha \in L^p W$. The linearity of $L^p f$ follows from the definition of linear space operations in the space of tensors. The following two properties are straightforward consequences of definition and establish L^p as a ***contravariant functor*** .

(a) *For any vector space V and any non-negative integer p,*

$$L^p \mathbb{1}_V = \mathbb{1}_{L^p V} \tag{1.17}$$

(b) *For linear maps $f : V \to W$ and $g : U \to V$, and any non-negative integer p,*

$$L^p(f \circ g) = L^p g \circ L^p f \tag{1.18}$$

Of course, $L^1 V = V^*$. The induced linear map $L^1 f$ is denoted by f^*. Note that by definition, $L^q V^* = L_q V$. For a linear map $f : V \to W$, we denote $L^q f^*$ by $L_q f$. The following properties follow from (a) and (b) above and are summarized by saying that L_q is a ***covariant functor***.

(c) *For any vector space V and non-negative integer q,*

$$L_q \mathbb{1}_V = \mathbb{1}_{L_q V} \tag{1.19}$$

(d) *For linear maps $f : V \to W$ and $g : U \to V$, and any non-negative integer q,*

$$L_q(f \circ g) = L_q f \circ L_q g \tag{1.20}$$

C. Anti-symmetric Tensors

The so-called *anti-symmetric tensors* are among the most powerful tools in the study of geometric structures. As we shall see in the following section, these are closely related to the concepts of volume and orientation in the case of real vector spaces.

We recall some elementary facts about the group S_n of permutations on n symbols $\{1, \ldots, n\}$. A ***transposition*** is a permutation that exchanges two symbols and leaves the other symbols fixed. Any permutation $\sigma \in S_n$ can be written as a composition of transpositions, $\sigma = \tau_1 \circ \cdots \circ \tau_k$, where k is not unique but its parity (even- or odd-ness) is determined by σ. Thus a permutation σ is called ***even*** or ***odd*** depending on whether k is even or odd. We write $\varepsilon(\sigma) = +1$ or $\varepsilon(\sigma) = -1$, respectively, if σ is even or odd. The map ε, called the ***sign***, is a homomorphism from S_n onto the two-element multiplicative group $\{+1, -1\}$; thus $\varepsilon(\sigma_1 \circ \sigma_2) = \varepsilon(\sigma_1) \circ \varepsilon(\sigma_2)$ and $\varepsilon(\mathbb{1}_{S_n}) = +1$. The set of even permutations form a subgroup of index 2 in S_n.

8. Definition Let V be a finite-dimensional linear space over a field F and let p be a natural number. An element $\alpha \in L^p V$ is called ***anti-symmetric*** (or ***alternating***) if for every $\sigma \in S_p$ and any $u_1, \ldots, u_p$ in V

$$\alpha(u_{\sigma(1)}, \ldots, u_{\sigma(p)}) = \varepsilon(\sigma)\alpha(u_1, \ldots, u_p) \tag{1.21}$$

We denote the set of anti-symmetric elements of $L^p V$ by $\Lambda^p V$. By convention, $\Lambda^0 V = L^0 V = F$.

9. Elementary Properties of Anti-symmetric Tensors

(a) A tensor $\alpha \in L^p V$ *is anti-symmetric if and only if for each transposition τ and any $u_1, \ldots, u_p$ in V,*

$$\alpha(u_{\tau(1)}, \ldots, u_{\tau(p)}) = -\alpha(u_1, \ldots, u_p)$$

Proof. The statement follows from the facts that $\varepsilon(\tau) = -1$, any permutation is a composition of transpositions and that ε is a homomorphism. □

(b) A tensor $\alpha \in L^p V$ *is anti-symmetric if and only if it has the property that $\alpha(u_1, \ldots, u_p) = 0$ whenever $u_i = u_j$ for $i \neq j$.*

Proof. Suppose the property holds and $u_1, \ldots, u_p$ are elements of V. Taking $i < j$, and expanding $\alpha(u_1, \ldots, u_i + u_j, \ldots, u_i + u_j, \ldots, u_p)$ we obtain

$$\begin{aligned} 0 = &\alpha(u_1, \ldots, u_i, \ldots, u_i, \ldots, u_p) + \alpha(u_1, \ldots, u_i, \ldots, u_j, \ldots, u_p) \\ &+ \alpha(u_1, \ldots, u_j, \ldots, u_i, \ldots, u_p) + \alpha(u_1, \ldots, u_j, \ldots, u_j, \ldots, u_p) \end{aligned}$$

The first and the last term above vanish by the property, and the result follows.

Conversely, suppose that for $i<j$, we have $u_i=u_j=u$ and consider the transposition that switches i and j. Applying (a) we obtain

$$\alpha(u_1,\ldots,u_i,\ldots,u_j,\ldots,u_p) = -\alpha(u_1,\ldots,u_j,\ldots,u_i,\ldots,u_p)$$

Since the field characteristic was assumed to be 0, we have $1\neq -1$, and

$$\alpha(u_1,\ldots,u,\ldots,u,\ldots,u_p) = 0$$

as claimed. □

(c) *Let* $\alpha\in\Lambda^p V$. *If* $\{u_1,\ldots,u_p\}\subset V$ *is linearly dependent, then* $\alpha(u_1,\ldots,u_p)=0$.

Proof. We write one u_i as a linear combination of the rest and expand by p-linearity. The result follows from (b). □

10. Basis for $\Lambda^p V$

Let $(e_1,\ldots,e_n)$ be a basis for V and consider an element $\alpha\in\Lambda^p V$. If $p>n$, then $\alpha=0$ by 9c above. For $0<p\leq n$, anti-symmetry implies that α is completely determined by its value on p-tuples $(e_{i_1},\ldots,e_{i_p})$, where $i_1<\cdots<i_p$. Therefore, to define an element of $\Lambda^p V$, it suffices to specify its values on p-tuples $(e_{i_1},\ldots,e_{i_p})$, where $i_1<\cdots<i_p$. Thus we introduce $e^{i_1\cdots i_p}\in\Lambda^p V$, where $i_1<\cdots<i_p$, by giving its value as

$$e^{i_1\cdots i_p}(e_{j_1},\ldots,e_{j_p}) = \delta^{i_1}_{j_1}\cdots\delta^{i_p}_{j_p}, \text{ where } j_1<\cdots<j_p \tag{1.22}$$

There are $\binom{n}{p}$ such elements in $\Lambda^p V$. One may extend the definition to an arbitrary multi-superscript $(i_1\cdots i_p)$ by stipulating that $e^{i_1\cdots i_p}=0$ if there is repetition in superscripts, and by multiplying by $\varepsilon(\sigma)$, where σ is the permutation that arranges $i_1,\ldots,i_p$ in increasing order.

We can now state and prove a couple of very useful propositions.

(a) *Let* $(e_1,\ldots,e_n)$ *be a basis for* V. *Then for* $0<p\leq n$, *a basis for* $\Lambda^p V$ *is given by* $e^{i_1\cdots i_p}$, *where* $i_1<\cdots<i_p$. *Further,* $\dim\Lambda^0 V=1$ *and* $\dim\Lambda^p V=0$ *for* $p>n$.

Proof. By earlier convention, $\Lambda^0 V=L^0 V$ is the underlying field. The case $p>n$ was treated at the beginning of the previous paragraph. For $0<p\leq n$, suppose that

$$\sum_{i_1<\cdots<i_p} c_{i_1\cdots i_p} e^{i_1\cdots i_p} = 0$$

Applying both sides to $(e_{j_1},\ldots,e_{j_p})$, where $j_1<\cdots<j_p$, yields $c_{j_1\cdots j_p}=0$, and linear independence is established. Further, we have the representation

$$\alpha = \sum_{i_1<\cdots<i_p} \alpha(e_{i_1},\ldots,e_{i_p})e^{i_1\cdots i_p} \tag{1.23}$$

for $\alpha\in\Lambda^p V$, as can be verified by applying the two sides to a p-tuple $(e_{j_1},\ldots,e_{j_p})$, $j_1<\cdots<j_p$. □

Let $\dim V{=}n$. Then $\dim \Lambda^n V{=}\binom{n}{n}{=}1$. Any non-zero element of $\Lambda^n V$ is called a ***volume element*** for V and serves as a basis for this one-dimensional linear space. The following amplifies 9c in the case $p{=}n$.

(b) *Let* $\dim V{=}n$ *and* ω *be a volume element for* V. *Then a subset* $\{u_1, \ldots, u_n\}{\subset}V$ *is linearly dependent if and only if* $\omega(u_1, \ldots, u_n){=}0$.

PROOF. As shown in 9c, linear dependence of $\{u_1, \ldots, u_n\}$ implies that $\omega(u_1, \ldots, u_n){=}0$. Conversely, if $\{u_1, \ldots, u_n\}$ is linearly independent, then it is a basis for the n-dimensional space V. Therefore, $\omega(u_1, \ldots, u_n){=}0$ would imply that ω vanishes on any n-tuple of elements of V, i.e., $\omega{=}0$. □

11. Functoriality

Let $f{:}V{\to}W$ be a linear map. We recall the definition of $L^p f{:}L^p W{\to}L^p V$ in (1.16). If $\alpha{\in}\Lambda^p W$, it follows that $L^p f(\alpha){\in}\Lambda^p V$. Thus denoting the restriction of $L^p f$ to $\Lambda^p W$ by $\Lambda^p f$, we obtain a linear map

$$\Lambda^p f : \Lambda^p W \to \Lambda^p V$$

by

$$(\Lambda^p f(\alpha))\,(v_1, \ldots, v_p) = \alpha(f(v_1), \ldots, f(v_p)) \tag{1.24}$$

For $p{>}n{=}min\{\dim V, \dim W\}$, $\Lambda^p f$ is the zero map, and for $p{=}0$, we have the convention $\Lambda^0 f{=}L^0 f{=}\mathbb{1}_F$. From Subsection 7, we obtain the following by restriction.
(a) *For any linear space* V *and any non-negative integer* p,

$$\Lambda^p \mathbb{1}_V = \mathbb{1}_{\Lambda^p V} \tag{1.25}$$

(b) *For linear maps* $f{:}V{\to}W$ *and* $g{:}U{\to}V$, *and any non-negative integer* p,

$$\Lambda^p(f{\circ}g) = \Lambda^p g{\circ}\Lambda^p f \tag{1.26}$$

Note that $\Lambda^1 V{=}L^1 V$ and $\Lambda^1 f{=}L^1 f{=}f^*$.

12. Determinants

An important consequence of the one-dimensionality of $\Lambda^n V$ is that for a linear map $f{:}V{\to}V$, the induced linear map $\Lambda^n f{:}\Lambda^n V{\to}\Lambda^n V$ is multiplication by a (fixed) element of the field. This element we call the **determinant** of f and denote it by $\det f$. Thus,

$$(\Lambda^n f)(\omega) = (\det f)\omega, \ \forall\omega{\in}\Lambda^n V \tag{1.27}$$

In the next section on real vector spaces we will give an incisive geometric interpretation of the determinant, but for now we concentrate on developing the formal algebraic properties of the concept.

13. Elementary Properties of the Determinant

(a) We have

$$\det(\mathbb{1}_V) = 1$$
$$\det(f\circ g) = (\det f)(\det g)$$

These are consequences of (1.25) and (1.26).

(b) *A linear map $f{:}V\to V$ is invertible if and only if* $\det f\neq 0$. *In this case,* $\det f^{-1}=(\det f)^{-1}$.

Proof. The second statement is a consequence of 13a. For the first, let $(e_1,\ldots,e_n)$ be a basis and ω a volume element for V. By the definition of determinant

$$\omega(f(e_1),\ldots,f(e_n)) = (\det f)\,\omega(e_1,\ldots,e_n) \tag{1.28}$$

By 10b, we have $\omega(e_1,\ldots,e_n)\neq 0$, therefore it follows that $\det f\neq 0$ if and only if the set $\{f(e_1),\ldots,f(e_n)\}$ is linearly independent, i.e., f is invertible. □

(c) **Expansion of the Determinant**

Suppose that the matrix of a linear map $f{:}V\to V$ relative to a basis for V is $[a^i_j]$, then

$$\det f=\sum_{\sigma\in S_n}\varepsilon(\sigma)\,a_1^{\sigma(1)}\cdots a_n^{\sigma(n)} \tag{1.29}$$

Proof. Let $(e_1,\ldots,e_n)$ be a basis for V, and consider the volume element $e^{1\cdots n}$ for V (see (1.22) for the definition of $e^{1\cdots n}$). Thus $e^{1\cdots n}(e_1,\ldots,e_n)=1$. Now using (1.28),

$$\begin{aligned}\det f &= e^{1\cdots n}(f(e_1),\ldots,f(e_n))\\ &= e^{1\cdots n}\Big(\sum_{i_1=1}^{n} a_1^{i_1}e_{i_1},\ldots,\sum_{i_n=1}^{n} a_n^{i_n}e_{i_n}\Big)\\ &= \sum_{i_1=1}^{n}\cdots\sum_{i_n=1}^{n} a_1^{i_1}\cdots a_n^{i_n}\, e^{1\cdots n}(e_{i_1},\ldots,e_{i_n})\end{aligned}$$

If there is any repetition among $i_1,\ldots,i_n$, we get $e^{1\cdots n}(e_{i_1},\ldots,e_{i_n})=0$, otherwise $(e_{i_1},\ldots,e_{i_n})$ represents a permutation σ of $(e_1,\ldots,e_n)$, and $e^{1\cdots n}(e_{i_1},\ldots,e_{i_n})=\varepsilon(\sigma)$. The desired result is thus obtained. □

Note that as σ ranges over S_n in the sum (1.29), σ^{-1} also ranges over S_n. Moreover, $\varepsilon(\sigma)=\varepsilon(\sigma^{-1})$, therefore one may also write

$$\det f = \sum_{\sigma\in S_n}\varepsilon(\sigma)\,a^1_{\sigma(1)}\cdots a^n_{\sigma(n)} \tag{1.30}$$

An interpretation of this result is that the determinant of the transpose of a matrix is equal to the determinant of the original matrix. Equivalently, in (1.29), the products of matrix entries are picked consecutively from columns 1 to n, while in (1.30), the products are taken consecutively from rows 1 to n. All familiar formulas about the expansion of the determinant according to column or row follow from (1.29) and (1.30). For these facts and a generalization, see Exercise 1.6 at the end of the chapter.

D. Real Linear Spaces

We now wish to specialize the considerations of the previous section to the case where the ground field is $\mathbb{R}$. We look at $\mathbb{R}^n$ with the standard basis $(e_1, \ldots, e_n)$. Thus, points of $\mathbb{R}^n$ will be represented as $x=(x^1, \ldots, x^n)$ or $x=\sum_{i=1}^n x^i e_i$. We will first try to find an interpretation for elements $e^{i_1 \cdots i_p}$ of $\Lambda^p \mathbb{R}^n$. Let us look at the cases $p = 1, 2, 3$, where x, y and z are elements of $\mathbb{R}^n$.

$$e^i(x) = x^i \tag{1.31}$$

$$e^{ij}(x, y) = \det \begin{bmatrix} x^i & y^i \\ x^j & y^j \end{bmatrix} \tag{1.32}$$

$$e^{ijk}(x, y, z) = \det \begin{bmatrix} x^i & y^i & z^i \\ x^j & y^j & z^j \\ x^k & y^k & z^k \end{bmatrix} \tag{1.33}$$

From elementary analytic geometry, we know that the absolute values of the right-hand sides of the above have, respectively, the following interpretations: the length of the projection of x on the i-axis, the area of the projection of the parallelogram determined by x and y on the (i, j)-plane, and the volume of the projection of the parallelepiped determined by x, y and z on the (i, j, k)-space. Further, the signs of the above have the following meaning. In (1.31), x^i is positive or negative depending on whether the projection of x points in the same or against the direction of e_i. In (1.32), the determinant is positive or negative depending on whether the projection of the ordered pair (x, y) is right-handed or left-handed relative to the ordered pair (e_i, e_j). Likewise, the sign of the determinant in (1.33) signifies whether the projection of the ordered triple (x, y, z) on (i, j, k)-space has the same or opposite handedness as the ordered triple (e_i, e_j, e_k). (See Figure 1.)

Based on the above intuition, we generalize the notions of volume and orientation to arbitrary real linear spaces. Let V be a linear space of dimension n on $\mathbb{R}$, and consider two ordered bases $\mathcal{B} = (e_1, \ldots, e_n)$ and $\bar{\mathcal{B}}=(\bar{e}_1, \ldots, \bar{e}_n)$ for this space. There is a unique linear map $f{:}V \to V$ with $f(e_j)=\bar{e}_j$, $j = 1, \ldots, n$. f is invertible as it carries basis to basis, so $\det f \neq 0$. We say that $\mathcal{B}$ has the *same orientation* as $\bar{\mathcal{B}}$ if and only if $\det f > 0$. This is an equivalence relation and breaks up the set of ordered bases for V into two classes, each called an ***orientation*** for V. An equivalent approach is the following. For each ordered basis $\mathcal{B} = (e_1, \ldots, e_n)$, consider the corresponding volume element $e^{1 \cdots n}$ as in (1.22). Now for any basis $\bar{\mathcal{B}} = (\bar{e}_1, \ldots, \bar{e}_n)$, it follows from (1.28) that $\bar{\mathcal{B}}$ has the same orientation as $\mathcal{B}$ if and only if the determinant of the relating linear map is positive. Thus, the non-zero elements of the one-dimensional space $\Lambda^n V$ (i.e., the volume elements) break up into two classes, each signifying one of the two orientations of V.

Continuing as above with the linear space V of dimension n on $\mathbb{R}$, we consider a volume element ω on V. Let $(a_1, \ldots, a_n)$ be an ordered n-tuple of elements of V. We

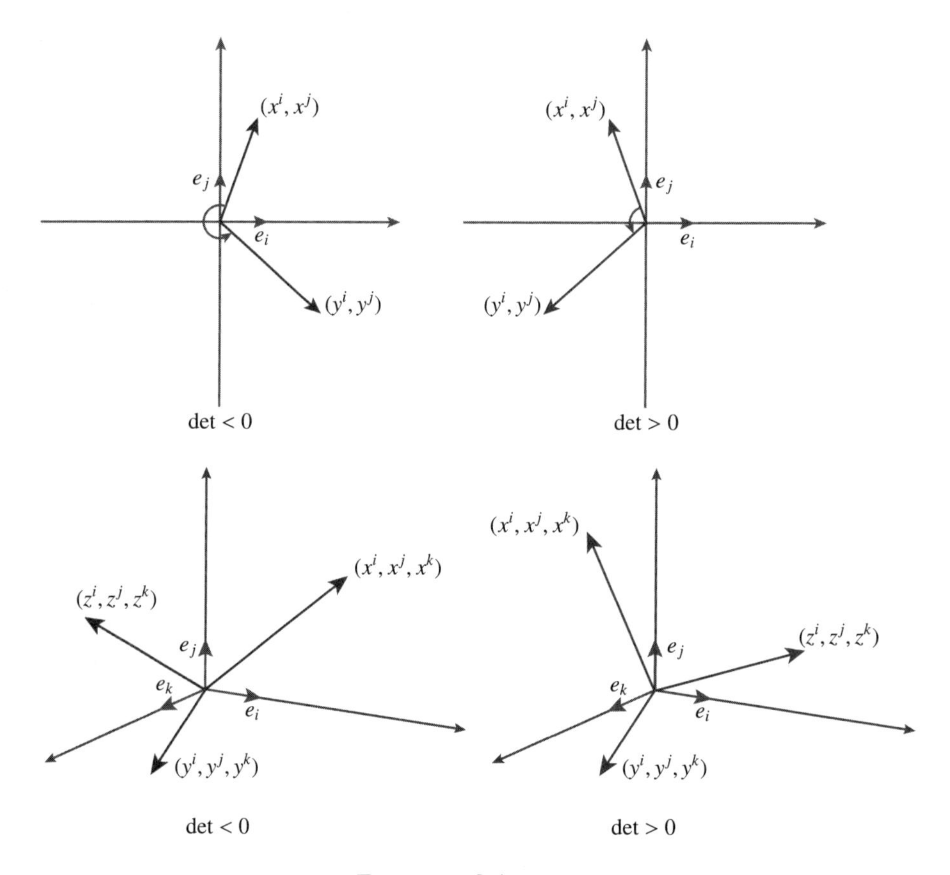

FIGURE 1. Orientation

define the ***n-dimensional parallelepiped*** determined by $(a_1, \ldots, a_n)$ to be the set

$$P(a_1, \ldots, a_n) = \Big\{ \sum_{i=1}^{n} t^i a_i : 0 \leq t^i \leq 1, i = 1, \ldots, n \Big\} \tag{1.34}$$

The ***volume*** (relative to ω) of $P(a_1, \ldots, a_n)$ is defined as follows:

$$Vol_\omega P(a_1, \ldots, a_n) = |\omega(a_1, \ldots, a_n)| \tag{1.35}$$

This definition is compatible with the discussion at the beginning of the section. We let $(u_1, \ldots, u_n)$ be a basis for V with $\omega(u_1, \ldots, u_n) = 1$, i.e., we take $P(u_1, \ldots, u_n)$ to be a "unit parallelepiped" relative to ω. Consider the linear map $f : V \to V$ that sends u_j to a_j for each $j = 1, \ldots, n$. Then

$$\begin{aligned} Vol_\omega P(a_1, \ldots, a_n) &= |\omega(f(u_1), \ldots, f(u_n))| \\ &= |\det f||\omega(u_1, \ldots, u_n)| \\ &= |\det f| \end{aligned}$$

Note that the matrix of f relative to the basis $(u_1, \ldots, u_n)$ has $a_1, \ldots, a_n$ as columns.

E. Product Structure

In Example 2e, we defined the tensor product $\alpha \otimes \beta$ of $\alpha \in L^p V$ and $\beta \in L^q V$ to be an element of $L^{p+q} V$. It is not the case, however, that if $\alpha \in \Lambda^p V$ and $\beta \in \Lambda^q V$, then $\alpha \otimes \beta$ is anti-symmetric, i.e., it is an element of $\Lambda^{p+q} V$. Here we modify $\otimes$ by a sort of "anti-symmetric averaging" to obtain an anti-symmetric tensor.

Let $\alpha \in \Lambda^p V$ and $\beta \in \Lambda^q V$. Then the ***wedge product*** $p \wedge q$ will be a $(p+q)$-covariant tensor defined by giving its value on an arbitrary $(p+q)$-tuple $(u_1, \ldots, u_{p+q})$ of elements of V as follows:

(1.36)

$$(\alpha \wedge \beta)(u_1, \ldots, u_{p+q}) = \frac{1}{p!q!} \sum_{\sigma \in S_{p+q}} \varepsilon(\sigma) \alpha(u_{\sigma(1)}, \ldots, u_{\sigma(p)}) \beta(u_{\sigma(p+1)}, \ldots, u_{\sigma(p+q)})$$

As we shall soon see, the choice of coefficient before the summation above makes the associative law come out true for the wedge product and allows for computational simplifications.

14. Elementary Properties of the Wedge Product

(a) *If $\alpha \in \Lambda^p V$ and $\beta \in \Lambda^q V$, then $\alpha \wedge \beta \in \Lambda^{p+q} V$.*

Proof. Let $\rho \in S_{p+q}$. Then

$$\begin{aligned}(\alpha \wedge \beta)(u_{\rho(1)}, \ldots, u_{\rho(p+q)}) &= \frac{1}{p!q!} \sum_{\sigma \in S_{p+q}} \varepsilon(\sigma) \alpha(u_{\sigma\rho(1)}, \ldots, u_{\sigma\rho(p)}) \beta(u_{\sigma\rho(p+1)}, \ldots, u_{\sigma\rho(p+q)}) \\ &= \frac{\varepsilon(\rho)}{p!q!} \sum_{\sigma \in S_{p+q}} \varepsilon(\sigma\rho) \alpha(u_{\sigma\rho(1)}, \ldots, u_{\sigma\rho(p)}) \beta(u_{\sigma\rho(p+1)}, \ldots, u_{\sigma\rho(p+q)})\end{aligned}$$

For a fixed $\rho \in S_{p+q}$, the element $\sigma\rho$ takes on all the values in the group S_{p+q} as σ does, so we may consider the above summation as running over $\sigma\rho$. It follows that

$$(\alpha \wedge \beta)(u_{\rho(1)}, \ldots, u_{\rho(p+q)}) = \varepsilon(\rho)(\alpha \wedge \beta)(u_1, \ldots, u_{p+q})$$

and the proof is complete. □

(b) *The wedge product is a bilinear map $\Lambda^p V \times \Lambda^q V \to \Lambda^{p+q} V$.*

Proof. This is an immediate consequence of the definition (1.36). □

(c) *The wedge product is associative.*

Proof. We take $\alpha \in \Lambda^p V$, $\beta \in \Lambda^q V$, $\gamma \in \Lambda^r V$ and $u_1, \ldots, u_{p+q+r}$ in V. Then the value of $((\alpha \wedge \beta) \wedge \gamma)(u_1, \ldots, u_{p+q+r})$ is equal to

$$\frac{1}{(p+q)!r!} \sum_{\sigma \in S_{p+q+r}} \varepsilon(\sigma)(\alpha \wedge \beta)(u_{\sigma(1)}, \ldots, u_{\sigma(p+q)}) \gamma(u_{\sigma(p+q+1)}, \ldots, u_{\sigma(p+q+r)})$$

For each $\sigma\in S_{p+q+r}$, we let S_σ be the subgroup of S_{p+q+r} consisting of permutations that leave each of $\sigma(p+q+1)$ to $\sigma(p+q+r)$ fixed. This is isomorphic to S_{p+q}. The expression inside the summation above is then equal to

$$\varepsilon(\sigma)\Big\{\frac{1}{p!q!}\sum_{\rho\in S_\sigma}\varepsilon(\rho)\alpha(u_{\rho\sigma(1)},\ldots,u_{\rho\sigma(p)})\beta(u_{\rho\sigma(p+1)},\ldots,u_{\rho\sigma(p+q)})\Big\}\gamma(u_{\sigma(p+q+1)},\ldots,u_{\sigma(p+q+r)})$$

$$=\frac{1}{p!q!}\sum_{\sigma\in S_{p+q+r}}\sum_{\rho\in S_\sigma}\varepsilon(\rho\sigma)\alpha(u_{\rho\sigma(1)},\ldots,u_{\rho\sigma(p)})\beta(u_{\rho\sigma(p+1)},\ldots,u_{\rho\sigma(p+q)})\gamma(u_{\rho\sigma(p+q+1)},\ldots,u_{\rho\sigma(p+q+r)})$$

Now for given $\sigma,\tau\in S_{p+q+r}$, we consider the equation $\rho\sigma=\tau$, where $\rho\in S_\sigma$. For $i>(p+q)$, the definition of S_σ implies that $\tau(i)=\sigma(i)$, so for a given τ, the number of σ's that can satisfy this equation is $(p+q)!$ On the other hand, for given τ and σ, a unique ρ satisfies this equation, therefore the value of the double summation above is

$$(p+q)!\sum_{\tau\in S_{p+q+r}}\varepsilon(\tau)\alpha(u_{\tau(1)},\ldots,u_{\tau(p)})\beta(u_{\tau(p+1)},\ldots,u_{\tau(p+q)})\gamma(u_{\tau(p+q+1)},\ldots,u_{\tau(p+q+r)})$$

It follows then that $((\alpha\wedge\beta)\wedge\gamma)(u_1,\ldots,u_{p+q+r})$ is equal to

$$\frac{1}{p!q!r!}\sum_{\tau\in S_{p+q+r}}\varepsilon(\tau)\alpha(u_{\tau(1)},\ldots,u_{\tau(p)})\beta(u_{\tau(p+1)},\ldots,u_{\tau(p+q)})\gamma(u_{\tau(p+q+1)},\ldots,u_{\tau(p+q+r)})$$

The associativity of multiplication in the field then shows that the alternative $(\alpha\wedge(\beta\wedge\gamma))(u_1,\ldots,u_{p+q+r})$ will also come out to the same expression above, and the proof is complete. □

As an outcome of the above calculation, the expression $\alpha\wedge\beta\wedge\gamma$ is meaningful and the following formula holds:

$$(\alpha\wedge\beta\wedge\gamma)(u_1,\ldots,u_{p+q+r})=$$
$$\frac{1}{p!q!r!}\sum_{\sigma\in S_{p+q+r}}\varepsilon(\sigma)\alpha(u_{\sigma(1)},\ldots,u_{\sigma(p)})\beta(u_{\sigma(p+1)},\ldots,u_{\sigma(p+q)})\gamma(u_{\sigma(p+q+1)},\ldots,u_{\sigma(p+q+r)})$$

Inductively, the expression $\alpha_1\wedge\cdots\wedge\alpha_k$ is unambiguously defined for $\alpha_i\in\Lambda^{p_i}V$, $i=1,\ldots,k$, and

(d) *Let* $\alpha_i\in\Lambda^{p_i}V$, $i=1,\ldots,k$, *and* $u_j\in V$, $j=1,\ldots,p_1+\cdots+p_k$, *then*

$$(\alpha_1\wedge\cdots\wedge\alpha_k)(u_1,\ldots,u_{p_1+\cdots+p_k})=$$
$$\frac{1}{(p_1)!\cdots(p_k)!}\sum_{\sigma\in S_{p_1+\cdots+p_k}}\varepsilon(\sigma)\alpha_1(u_{\sigma(1)},\ldots,u_{\sigma(p_1)})\cdots\alpha_k(u_{\sigma(p_1+\cdots+p_{k-1}+1)},\ldots,u_{\sigma(p_1+\cdots+p_k)})$$

(e) *Let* $\alpha^i\in V^*$ *and* $u_i\in V$, $i=1,\ldots,k$, *then*

$$(\alpha^1\wedge\cdots\wedge\alpha^k)(u_1,\ldots,u_k)=\det\left[\alpha^i(u_j)\right]$$

Proof. This is the special case of (d) for $p_1=\cdots=p_k=1$. □

(f) *For* $\alpha\in\Lambda^pV$ *and* $\beta\in\Lambda^qV$, *one has*

$$\alpha\wedge\beta=(-1)^{pq}\beta\wedge\alpha \tag{1.37}$$

Proof. In view of bilinearity, it suffices to consider the case of basis elements $\alpha = e^{i_1}\wedge\cdots\wedge e^{i_p}$ and $\beta = e^{j_1}\wedge\cdots\wedge e^{j_q}$. Note also from the definition of the wedge product that $e^i\wedge e^j = -e^j\wedge e^i$. Therefore in order to transform $(e^{i_1}\wedge\cdots\wedge e^{i_p})\wedge(e^{j_1}\wedge\cdots\wedge e^{j_q})$ into $(e^{j_1}\wedge\cdots\wedge e^{j_q})\wedge(e^{i_1}\wedge\cdots\wedge e^{i_p})$, using associativity, one has to move each of $e^{j_1},\ldots,e^{j_q}$, in order, p places to the left by making consecutive transpositions. Since there are q such e^{j_ν}, it takes pq transpositions to effect this transformation, and the claim is proved. □

(g) *If p is odd and $\alpha\in\Lambda^p V$, then $\alpha\wedge\alpha=0$.*

Proof. This is an immediate consequence of (f). □

(h) *Let $(e_1,\ldots,e_n)$ be a basis for V with dual basis $(e^1,\ldots,e^n)$. Then*

$$e^{i_1\cdots i_p} = e^{i_1}\wedge\cdots\wedge e^{i_p}$$

Proof. The two sides have the same effect on any p-tuple $(e_{j_1},\ldots,e_{j_p})$. □

Let $f{:}V\to W$ be a linear map. Recalling the induced linear maps $\Lambda^p f{:}\Lambda^p W\to\Lambda^p V$, it is a routine matter to check that the induced maps preserve the wedge product.
(i) *Let $f{:}V\to W$ be a linear map, $\alpha\in\Lambda^p W$ and $\beta\in\Lambda^q W$. Then*

$$\Lambda^{p+q}f\,(\alpha\wedge\beta) = \Lambda^p f\,(\alpha)\wedge\Lambda^q f\,(\beta)$$

Proof. Let $u_1,\ldots,u_{p+q}$ be elements of V. Then by definition of the induced map, $(\Lambda^{p+q}f\,(\alpha\wedge\beta))(u_1,\ldots,u_{p+q})$ is equal to $(\alpha\wedge\beta)(f(u_1),\ldots,f(u_{p+q}))$, which in turn is equal to

$$\begin{aligned}
&\frac{1}{p!q!}\sum_{\sigma\in S_{p+q}}\varepsilon(\sigma)\,\alpha(f(u_{\sigma(1)}),\ldots,f(u_{\sigma(p)}))\,\beta(f(u_{\sigma(p+1)}),\ldots,f(u_{\sigma(p+q)}))\\
&=\frac{1}{p!q!}\sum_{\sigma\in S_{p+q}}\varepsilon(\sigma)\,((\Lambda^p f)\alpha)(u_{\sigma(1)},\ldots,u_{\sigma(p)})\,((\Lambda^q f)\beta)(u_{\sigma(p+1)},\ldots,u_{\sigma(p+q)})\\
&=((\Lambda^p f)\alpha\wedge(\Lambda^q f)\beta)(u_1,\ldots,u_{p+q})
\end{aligned}$$

as desired. □

(j) **Graded Exterior Algebra** $\Lambda^* V$
Let $\dim V = n$. It is convenient to consider the direct sum $\bigoplus\Lambda^p V$ as one linear space. Then

$$\dim\Big(\bigoplus_{p=0}^{n}\Lambda^p V\Big) = \sum_{p=0}^{n}\binom{n}{p} = 2^n$$

Thus an element of $\Lambda^* V$ is an $(n+1)$-tuple $(\alpha_0,\alpha_1,\ldots,\alpha_n)$, or a formal sum $\alpha_0+\alpha_1+\cdots+\alpha_n$, where α_i is an element of $\Lambda^i V$. α_i is called an element of degree i in $\Lambda^* V$. Now the wedge product is extended to a product

$$\wedge : \Lambda^* V\times\Lambda^* V\to\Lambda^* V$$

by stipulating that the distributive laws

$$\alpha \wedge (\beta + \gamma) = (\alpha \wedge \beta) + (\alpha \wedge \gamma)$$
$$(\alpha + \beta) \wedge \gamma = (\alpha \wedge \gamma) + (\beta \wedge \gamma)$$

hold. These are of course consistent with the bilinearity of $\wedge$ as in (b). With the operations + and $\wedge$, $\Lambda^* V$ is known as the ***(graded) exterior algebra*** of V.

For a linear map $f{:}V{\to}W$, the induced linear maps $\Lambda^p f$ give rise to a linear and $\wedge$-preserving linear map $\Lambda^* W{\to}\Lambda^* V$, which is denoted by f^*. It is also customary to denote all $\Lambda^p f$ by f^*, a convention we will henceforth adopt, unless there is danger of confusion.

15. Interior Product or Contraction

Let V be a linear space over a field F and let $x{\in}V$. As the final topic in this section, we consider a method for reducing the degree of an anti-symmetric tensor, known as ***contraction by x*** or ***interior product with x***. We define an operator i_x or $x\lrcorner$ that will map each $\Lambda^p V$ to $\Lambda^{p-1} V$. For $p{=}0$, we let i_x be the zero map, adding the convention that $\Lambda^p V{=}\{0\}$ for $p{<}0$. For $p{\geq}1$, $\alpha{\in}\Lambda^p V$, and $u_2, \ldots, u_p$ in V, we define

$$(i_x\alpha)(u_2, \ldots, u_p) = \alpha(x, u_2, \ldots, u_p) \tag{1.38}$$

We will now state and prove the main properties of contraction.

(a) *If $\alpha{\in}\Lambda^p V$, then $i_x\alpha{\in}\Lambda^{p-1} V$.*

PROOF. $(p-1)$-linearity and anti-symmetry of $i_x\alpha$ follow from the corresponding properties of α. □

(b) *$i_x\alpha$ is linear with respect to x, i.e., $i_{x+y}{=}i_x{+}i_y$ for x and y in V, and $i_{rx}{=}ri_x$ for x in V and r in F.*

PROOF. This is linearity with respect to the first component of the argument of α. □

(c) *For x and y in V, $i_x{\circ}i_y{=}-i_y{\circ}i_x$, therefore $i_x{\circ}i_x{=}0$.*

PROOF. For $p{<}2$, both sides are zero. Otherwise,

$$\begin{aligned}((i_x \circ i_y)(\alpha))(u, \ldots, z) &= (i_x(i_y\alpha))(u, \ldots, z)\\ &= (i_y\alpha)(x, u, \ldots, z)\\ &= \alpha(y, x, u, \ldots, z)\end{aligned}$$

The exchange of x and y is a transposition and will change the sign. □

(d) **Basic Example**

Let $(e_1, \ldots, e_n)$ be a basis for V. Then

$$i_{e_k}(e^{i_1}\wedge\cdots\wedge e^{i_p}) = \begin{cases} 0 & k{\notin}\{i_1, \ldots, i_p\} \\ (-1)^{\nu-1} e^{i_1}\wedge\cdots\wedge\widehat{e^{i_\nu}}\wedge\cdots e^{i_p} & k{=}i_\nu \end{cases} \tag{1.39}$$

where the symbol $\widehat{*}$ always indicates the deletion of $*$.

PROOF. We check that the two sides have the same effect on an arbitrary ordered $(p-1)$-tuple $(e_{j_2}, \ldots, e_{j_p})$. If k is not one of $\{i_1, \ldots, i_p\}$, then using 14e we see that the first column of the matrix consists of zeros, so the determinant is zero. Now suppose $k = i_\nu$, then

$$i_{e_k}(e^{i_1} \wedge \cdots \wedge e^{i_p})(e_{j_2}, \ldots, e_{j_p}) = (e^{i_1} \wedge \cdots \wedge e^{i_p})(e_k, e_{j_2}, \ldots, e_{j_p})$$

If the subscript set $\{j_2, \ldots, j_p\}$ is not the same as $\{i_1, \ldots, i_{\nu-1}, i_{\nu+1}, \ldots, i_p\}$, then both sides are zero. If they are the same, reordering so that $e^k = e^{i_\nu}$ moves to the first position in $e^{i_1} \wedge \cdots \wedge e^{i_p}$ involves $\nu-1$ transpositions, and the claim is proved. □

(e) *Let* $\alpha \in \Lambda^p V$, $\beta \in \Lambda^q V$ *and* $x \in V$. *Then*

$$i_x(\alpha \wedge \beta) = i_x\alpha \wedge \beta + (-1)^p \alpha \wedge i_x\beta \tag{1.40}$$

PROOF. We take an ordered basis $(e_1, \ldots, e_n)$ for V. Because of the bilinearity of the wedge product and the linearity of i_x with respect to x, it suffices to consider the case where $x = e_k$, $\alpha = e^{i_1} \wedge \cdots \wedge e^{i_p}$ and $\beta = e^{j_1} \wedge \cdots \wedge e^{j_q}$, where $i_1 < \cdots < i_p$ and $j_1 < \cdots < j_q$. We consider four cases:

Case 1: $k \notin \{i_1, \ldots, i_p\} \cup \{j_1, \ldots, j_q\}$.

In this case, $i_x\alpha = 0$, $i_x\beta = 0$ and $i_x(\alpha \wedge \beta) = 0$, all by the first case of (1.39).

Case 2: $k = i_\mu \in \{i_1, \ldots, i_p\}$, but $k \notin \{j_1, \ldots, j_q\}$.

Here again by (1.39), $i_x\beta = 0$, and

$$\begin{aligned} i_{e_k}(e^{i_1} \wedge \cdots \wedge e^{i_p} \wedge e^{j_1} \wedge \cdots \wedge e^{j_q}) &= (-1)^{(\mu-1)} e^{i_1} \wedge \cdots \widehat{e^{i_\mu}} \cdots \wedge e^{i_p} \wedge e^{j_1} \wedge \cdots \wedge e^{j_q} \\ &= i_{e_{i_\mu}}(e^{i_1} \wedge \cdots \wedge e^{i_p}) \wedge (e^{j_1} \wedge \cdots \wedge e^{j_q}) \end{aligned}$$

Case 3: $k \notin \{i_1, \ldots, i_p\}$, but $k = j_\nu \in \{j_1, \ldots, j_q\}$.

This is similar to Case 2, except that an extra $(-1)^p$ appears.

Case 4: $k = i_\mu \in \{i_1, \ldots, i_p\}$, and $k = j_\nu \in \{j_1, \ldots, j_q\}$.

Here we have $\alpha \wedge \beta = 0$, so the left-hand side of (1.40) is zero. On the other hand,

$$\begin{aligned} i_x(\alpha) \wedge \beta &= (-1)^{(\mu-1)} e^{i_1} \wedge \cdots \widehat{e^{i_\mu}} \cdots \wedge e^{i_p} \wedge e^{j_1} \wedge \cdots \wedge e^{j_q} \\ (-1)^p \alpha \wedge i_x\beta &= (-1)^{p+\nu-1} e^{i_1} \wedge \cdots \wedge e^{i_p} \wedge e^{j_1} \wedge \cdots \widehat{e^{j_\nu}} \cdots \wedge e^{j_q} \end{aligned}$$

It takes $(p-\mu+\nu-1)$ transpositions to move e^k from its position on the right-hand side of the second line to where it is in the first, therefore the sum of the two terms is zero, and the proof is complete. □

EXERCISES

1.1 Let V be a vector space and V^* its dual. Show there is no isomorphism $V \to V^*$ that maps every basis to its dual.

1.2 Let V_1 and V_2 be vector spaces with ordered bases $\mathcal{B}_1$ and $\mathcal{B}_2$, respectively, and let A be the matrix of a linear map $f: V_1 \to V_2$ with respect to these bases. Show that the matrix of the induced linear map $f^*: V_2{}^* \to V_1{}^*$ with respect to the dual bases is the transpose of A.

1.3 Let V be a vector space and $\alpha^1, \ldots, \alpha^k \in V^*$. Show that $\{\alpha^1, \ldots, \alpha^k\}$ is linearly dependent if and only if $\alpha^1 \wedge \cdots \wedge \alpha^k = 0$.

1.4 For anti-symmetric tensors α and β on a vector space V, we define

$$[\alpha, \beta] = \alpha \wedge \beta - \beta \wedge \alpha$$

If α, β and γ are anti-symmetric tensors on V, show that $[[\alpha, \beta], \gamma] = 0$.

1.5 Let $\mathcal{B}$ be an ordered basis for n-dimensional vector space V. Denote the basis constructed in 10a for $\Lambda^p V$ by $\mathcal{B}^p$ (you may choose any fixed order for the basis elements of $\mathcal{B}^p$).
(a) For $n = 3$, let $A = [a^i_j]$ be the matrix of a linear map $f: V \to V$ with respect to $\mathcal{B}$. Describe the matrices of $\Lambda^p f$ with respect to $\mathcal{B}^p$ for $p = 0, 1, 2, 3$.
(b) Do the same for arbitrary n and p.

1.6 (Laplace Expansion) Let V be a vector space with basis $(e_1, \ldots, e_n)$ and suppose $f: V \to V$ is a linear map.
(a) Using the formula 14i for $p = 1$ and $q = n - 1$, obtain the expansion formula for determinant in terms of the first row (column).
(b) For arbitrary p and q with $p+q=n$, obtain a more general formula.

1.7 Denote the vector space of 2×2 matrices with entries from the field F by $M_2(F)$. Consider a fixed element $\mathrm{M} \in M_2(F)$ and denote the linear map $\mathrm{X} \mapsto \mathrm{MX}$ from $M_2(F)$ to itself by f.
(a) Show that $\det f = (\det \mathrm{M})^2$.
(b) Compute the matrices of $\Lambda^p f$ relative to the bases described in 10a for all p.

1.8 Let $\dim V = n$, $0 \le p \le n$ and suppose that $f: V \to V$ is a linear map. Show that

$$(\det \Lambda^p f)(\det \Lambda^{n-p} f) = (\det f)^{\binom{n}{p}}$$

.

1.9 We denote by τ_p the trace of the linear map $\Lambda^p f$, where $f: V \to V$ is a linear map, and $\dim V = n$. Identify τ_0, τ_1 and τ_n. Show that

$$\det(\mathbb{1}_V + f) = \sum_{p=0}^{n} \tau_p$$

1.10 Let V be a finite-dimensional vector space over a field F and suppose that $\beta{:}V\times V\to F$ is a bilinear map. We define $\beta^\flat{:}V\to V^*$ by

$$(\beta^\flat(u))(v) = \beta(v, u), \quad u, v\in V$$

(a) Prove that in fact $\beta^\flat(u)\in V^*$ and that $\beta^\flat$ is linear.
(b) Show that $\beta^\flat$ is an isomorphism if β is an inner product.
(c) Let $\mathcal{B}=(e_1, \ldots, e_n)$ be a basis for V. Prove that the rank of $\beta^\flat$ (also called the *rank of* β) is equal to the rank of the matrix $[\beta_{ij}]$, where $\beta_{ij} = (\beta^\flat(e_j))(e_i) = \beta(e_i, e_j)$.
(d) If $\mathcal{B}^*$ is the dual basis for V^*, identify the matrix of $\beta^\flat$ relative to $\mathcal{B}$ and $\mathcal{B}^*$.

1.11 Let dim $V{=}n$. We use $\Lambda_p V$ as an alternative notation for $\Lambda^p V^*$. Suppose $(e_1, \ldots, e_n)$ is a basis for V. Using the natural identification $I_V{:}V\to V^{**}$, we may consider objects of the form $e_{i_1}\wedge\cdots\wedge e_{i_p}$, $i_1<\cdots<i_p$ to form a basis for $\Lambda_p V$. Give a geometric interpretation for such objects when the base field is $\mathbb{R}$. In general, if $u_1, \ldots, u_p\in V$, explain why the appellation ***oriented parallelepiped*** is an apt description for $u_1\wedge\cdots\wedge u_p$.

1.12 Using the notation of the previous exercise, define a bilinear map

$$B : \Lambda_p V \times \Lambda^p V \to F,$$

where F is the ground field, in the following way. For $u_1, \ldots, u_p\in V$ and $\alpha^1, \ldots, \alpha^p\in V^*$, let

$$B(u_1 \wedge \cdots \wedge u_p, \alpha^1 \wedge \cdots \wedge \alpha^p) = \det[\alpha^i(u_j)]$$

Show that B is well-defined and of rank $\binom{n}{p}$ (for the definition of rank, see Exercise 1.10 above). Thus by Exercise 1.10, a linear isomorphism $B^\flat{:}\Lambda_p V\to\Lambda^p V$ is obtained. Deduce that for each $x\in V$ and each $p\geq 1$, a linear map $i^x{:}\Lambda_{p-1}V\to\Lambda_p V$ exists that satisfies $i_x\circ B^\flat\circ i^x{=}B^\flat$. Show that

$$i^x(u_2 \wedge \cdots \wedge u_p) = x \wedge u_2 \wedge \cdots \wedge u_p$$

What is the geometric interpretation of i^x in the case $F{=}\mathbb{R}$?

1.13 Consider $\mathbb{R}^m$ with the standard inner product, $m\geq 2$. The goal here is to generalize the concept of cross-product in $\mathbb{R}^3$. Show there exists an $(m-1)$-linear map $\mathbb{R}^m\times\cdots\times\mathbb{R}^m \to \mathbb{R}^m$ that assigns to each $(m-1)$-tuple $(u_1, \ldots, u_{m-1})$ of elements of $\mathbb{R}^m$ an element $v\in\mathbb{R}^m$ with the following properties:
(a) v is perpendicular to each u_i,
(b) $|u|$ has the numerical value of the $(m-1)$-dimensional volume of the parallelepiped $P(u_1, \ldots, u_{p-1})$, and
(c) the m-tuple $(u_1, \ldots, u_{m-1}, v)$ is positively oriented if the set $\{u_1, \ldots, u_{m-1}\}$ is linearly independent.

Derive an explicit formula for this product. Let V be the $(m-1)$-dimensional volume of the parallelepiped $P(u_1, \ldots, u_{p-1})$ and V_i be the $(m-1)$-dimensional volume of the orthogonal projection of this parallelepiped on the hyperplane $x_i{=}0$.

Prove the "Pythagorean theorem"

$$V^2 = V_1^2 + \cdots + V_n^2$$

1.14 Consider a set of k vectors $\{u_1, \ldots, u_k\}$ in $\mathbb{R}^m$, $k<n$. There are $\binom{n}{k}$ coordinate subspaces of dimension k in $\mathbb{R}^m$. Denote the k-dimensional volumes of the orthogonal projections of the parallepiped $P(u_1, \ldots, u_k)$ on these subspaces by V_i and the volume of the parallepiped $P(u_1, \ldots, u_k)$ by V. Generalize the Pythagorean theorem of the previous exercise as $V^2 = \sum_i V_i^2$. (This is harder to prove!)

1.15 (Hodge star operator) Let V be a vector space of dimension n over $\mathbb{R}$ and $\sigma: V \times V \to \mathbb{R}$ an inner product. We use the isomorphism $\sigma^\flat$ described in Exercise 1.10. Consider an orthonormal basis $(e_1, \ldots, e_n)$ and denote the volume element $e^1 \wedge \cdots \wedge e^n$ by ω.
(a) Show that for $u_i, v_i \in V$, $i=1, \ldots, n$,

$$\omega(u_1, \ldots, u_n)\omega(v_1, \ldots, v_n) = \det[\sigma(u_i, v_j)]$$

(b) Denote $\sigma^\flat(u)$ by $\tilde{u}$. For $\alpha \in \Lambda^p V$, prove there exists a unique $\beta \in \Lambda^{n-p} V$ such that

$$\beta(u_1, \ldots, u_{n-p})\omega = \alpha \wedge \tilde{u}_1 \cdots \wedge \tilde{u}_{n-p}$$

for all $u_1, \ldots, u_{n-p} \in V$. We will denote β by $*\alpha$. Thus $*: \Lambda^p V \to \Lambda^{n-p} V$.
(c) Show that $**\alpha = (-1)^{p(n-p)}\alpha$ and that $*$ is a linear isomorphism.
(d) For $\alpha \in \Lambda^p V$ and $\beta \in \Lambda^q V$, show that $\alpha \wedge (*\beta) = \beta \wedge (*\alpha)$, and $\alpha \wedge (*\alpha) = a\omega$, where $a > 0$.

1.16 Let V be of dimension n over the field F and $\alpha \in \Lambda^2 V$.
(a) Show that there exists a basis $(e_1, \ldots, e_n)$ for V and elements $a_{ij} \in F$, with $a_{ij} + a_{ji} = 0$, such that $\alpha = \sum_{i,j} a_{ij}\, e^i \wedge e^j$.
(b) Prove that in fact the basis $(e_1, \ldots, e_n)$ can in fact be so chosen that $\alpha = \sum_{i=1}^{s} e^i \wedge e^{s+i}$. Show that here the rank of α is $2s$, demonstrating that the rank of an anti-symmetric 2-tensor is necessarily even. (For the definition of rank see Exercise 1.10.)

1.17 Let V be a linear space over F, and $\alpha^i, \beta^i \in V^*$ for $i = 1, \ldots, s$. Suppose that $\{\alpha^1, \ldots, \alpha^s\}$ is linearly independent and $\alpha^1 \wedge \beta^1 + \cdots + \alpha^s \wedge \beta^s = 0$. Prove that there exists a symmetric matrix $[a^i_j]$ with entries from F so that

$$\beta^i = \sum_{j=1}^{s} a^i_j \alpha^j$$

1.18 Let $\dim V = n$ and $\alpha \in \Lambda^{n-1} V$. Show that there exist $\alpha^1, \ldots, \alpha^{n-1} \in V^*$ so that

$$\alpha = \alpha^1 \wedge \cdots \wedge \alpha^{n-1}$$

For $n = 4$, give an example of $\alpha \in \Lambda^2 V$ that cannot be decomposed as $\alpha^1 \wedge \alpha^2$.

Part II

Local

CHAPTER 2

Vector Fields: Local Theory

In the present and following chapter we will introduce and develop the local version of some of our basic tools. One says that a property or a function is *locally determined* provided that its value at a given point depends on the values of the function on an arbitrary small neighborhood of that point. Properties arising from the derivative are usually of this nature, and the natural habitat for the study of these concepts is an open subset of $\mathbb{R}^n$.

Given an open subset U of $\mathbb{R}^n$, we will "install" a copy of the vector space $\mathbb{R}^n$ at each point x of U, and regard the elements of this vector space as vectors tangent to U and emanating from the point x. We can then perform constructions of Chapter 1 on this vector space to obtain spaces of various tensor types. A *field* will be a function that selects an element of the constructed space at each point of U.

A. Tangent Space

There is often some confusion in elementary calculus and physics courses about whether "vectors" all originate at a single fixed "origin" or their initial point may move around. We will formalize this distinction to begin our discussion. Consider an open subset U of $\mathbb{R}^n$. Regarding $\mathbb{R}^n$ as a vector space of dimension n over $\mathbb{R}$, points v of $\mathbb{R}^n$ may be alternatively thought of as directed segments (vectors) with initial point $\mathbf{0}$ and terminal point v. Now imagine shifting the origin of the vector space to an arbitrary point x of U so that the directed segments originate from x. We will use the set-theoretic artifact of ordered pair (x, v) to indicate a directed segment obtained from v above by parallel translating it so that it originates at x (see Figure 1).

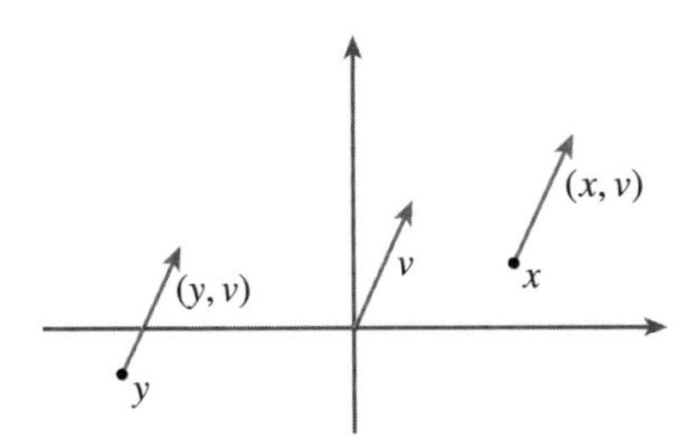

FIGURE 1. Free vector

Formally, we define

$$T_xU = \{(x, v) : v \in \mathbb{R}^n\} \tag{2.1}$$

as the ***tangent space to U at x***. The intuitive explanation for the word *tangent* is that this set is obtained by parallel-translating or sliding the *flat* $\mathbb{R}^n$ along itself so that the origin moves to x. We will not make any unjustified use of the word "tangent". In fact, as we shall soon see, the union of T_xU's, for $x \in U$, is best represented as a Cartesian product $U \times \mathbb{R}^n$, where the tangent visualization is not appropriate. One could usefully imagine the T_xU as infinitely thin non-intersecting copies of $\mathbb{R}^n$ placed along U, one at each $x \in U$, and each tangent to U (see Figure 2).

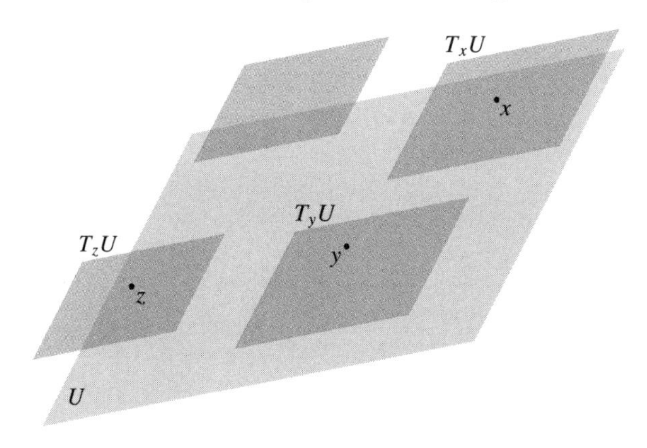

FIGURE 2. Tangent spaces along U

Next we shall endow each T_xU with the structure of a vector space over $\mathbb{R}$ by defining

$$(x, v) + (x, w) = (x, v + w) \tag{2.2}$$

$$r(x, v) = (x, rv) \quad \text{for } r \in \mathbb{R} \tag{2.3}$$

With these operations, T_xU becomes a vector space isomorphic to $\mathbb{R}^n$ via the ***natural isomorphism*** $(x, v) \mapsto v$. This is consistent with imagining the origin of the vector space T_xU as placed at x. We denote the zero vector $(x, \mathbf{0})$ of T_xU by $\mathbf{0}_x$.

Now consider the union

$$TU = \bigcup_{x \in U} T_xU = U \times \mathbb{R}^n \tag{2.4}$$

This is indeed the *disjoint union* of T_xU's since the first components are distinct. TU is called the ***tangent bundle of U***. We endow TU with the product topology, or equivalently, the topology of the open subset $U \times \mathbb{R}^n$ of $\mathbb{R}^n \times \mathbb{R}^n \cong \mathbb{R}^{2n}$. The ***tangent bundle projection***, $\tau_U : TU \to U$, is defined as projection on the first component, called the ***base point*** of the tangent vector.

$$\tau_U(x, u) = x \tag{2.5}$$

Being the restriction of a linear map, τ_U is C^∞ (in fact real-analytic).

Consistent with our venture of moving the origin of vector space around, we re-interpret the notion of derivative as the "tangent map". Let U and V be open subsets, respectively, of $\mathbb{R}^n$ and $\mathbb{R}^m$, and consider a mapping $f{:}U{\to}V$. If f is differentiable at a point x of U, the derivative of f at x, denoted by $Df(x)$, is a linear mapping $\mathbb{R}^n{\to}\mathbb{R}^m$. Moving the origins of vector spaces to x and $f(x)$, respectively, we define the ***tangent map*** of f at x by

$$T_xf : T_xU \to T_{f(x)}V$$

$$T_xf(x,v) = \big(f(x), (Df(x))(v)\big) \tag{2.6}$$

Thus, T_xf carries the information both of where the point x is moving under f, and what its derivative at x is. In view of the linear structure defined for tangent spaces, T_xf is a linear map.

All the rules of differentiation can now be recast in the new framework. Most notably, we mention the chain rule. Let U, V and W be open subsets, respectively, of $\mathbb{R}^n$, $\mathbb{R}^m$ and $\mathbb{R}^p$, and consider mappings $f{:}U{\to}V$ and $g{:}V{\to}W$. Suppose that f is differentiable at x and g is differentiable at $f(x)$, then $g{\circ}f$ is differentiable at x and the chain rule can be stated as

$$T_x(g \circ f) = T_{f(x)}g \circ T_xf \tag{2.7}$$

Now if the map $f{:}U{\to}V$ is differentiable at every point of U, the ***tangent (bundle) map*** $Tf{:}TU{\to}TV$ is defined by putting together all T_xf, i.e.,

$$Tf\,|_{T_xU} = T_xf$$

If both $f{:}U{\to}V$, and $g{:}V{\to}W$ are differentiable throughout their domains, the chain rule can be stated as

$$T(g \circ f) = Tg \circ Tf \tag{2.8}$$

Together with

$$T\mathbb{1}_U = \mathbb{1}_{TU} \tag{2.9}$$

T is established as a covariant functor.

1. Vector Fields: Notation and Representation

We are now in a position to consider our first example of a field. Let U be an open subset of $\mathbb{R}^n$. By a ***(tangent) vector field*** on U, we mean a map $X{:}U{\to}TU$ such that

$$\tau_U{\circ}X = \mathbb{1}_U$$

which means that $X(x){\in}T_xU$ for all $x{\in}U$. Thus X assigns to each $x{\in}U$, an element $X(x)$ from T_xU (see Figure 3).

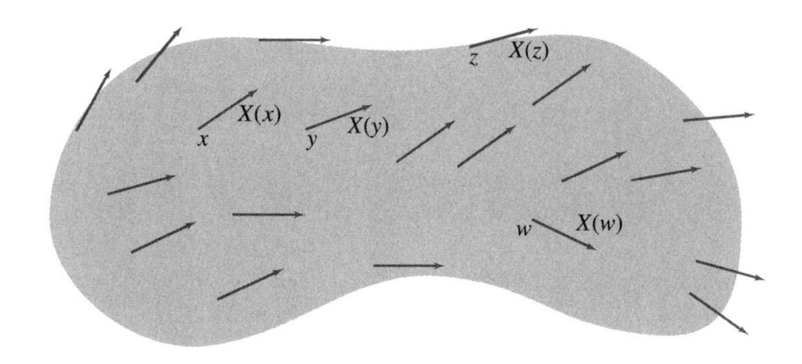

FIGURE 3. A vector field

Since U is an open subset of $\mathbb{R}^n$ and X a map of U to $TU \subset \mathbb{R}^{2n}$, it makes sense to talk about X being continuous, differentiable, C^r, etc. In fact, it follows from the definition of vector field that

$$X(x) = (x, f(x)) \tag{2.10}$$

where $f: U \to \mathbb{R}^n$. Since the first component of the right-hand side of (2.10) is the value of the identity function, adjectives such as continuous and C^r merely refer to the function f. Thus for a natural number r, a C^r vector field is one for which f has continuous partial derivatives through order r, C^0 means continuous, C^∞ or *smooth* means having partial derivatives of all orders, and C^ω is used for real analytic functions.

The points of $\mathbb{R}^n$ are denoted by $x=(x^1, \ldots, x^n)$ or $x = \sum_{i=1}^n x^i e_i$, where $(e_1, \ldots, e_n)$ is the standard basis of $\mathbb{R}^n$. For $i=1, \ldots, n$, the vector field $\frac{\partial}{\partial x^i}$ is defined by

$$\frac{\partial}{\partial x^i}(x) = (x, e_i) \tag{2.11}$$

We visualize $\frac{\partial}{\partial x^i}$ as arrows of length 1 pointing parallel to the positive direction of the ith coordinate axis emanating from every point of $\mathbb{R}^n$ (see Figure 4). The justification for using the partial derivative notation will appear in Section C.

For an open subset U of $\mathbb{R}^n$, and $a \in U$, a natural vector space basis for T_aU is $(\frac{\partial}{\partial x^1}(a), \ldots, \frac{\partial}{\partial x^n}(a))$. Consider a mapping $f: U \to V$, where U and V are open subsets, respectively, of $\mathbb{R}^n$ and $\mathbb{R}^m$. We denote the points of U by $x=(x^1, \ldots, x^n)$, points of V by $y = (y^1, \ldots, y^m)$, and we describe f explicitly as $y = f(x)$, or

$$\begin{cases} y^1 = f^1(x^1, \ldots, x^n) \\ \quad\vdots \\ y^m = f^m(x^1, \ldots, x^n) \end{cases}$$

If f is differentiable at $a \in U$, then the derivative $Df(a)$ is given as the $m \times n$ matrix $[\frac{\partial y^i}{\partial x^j}(a)]$ relative to the standard basis. In view of (2.11), the tangent map at a takes

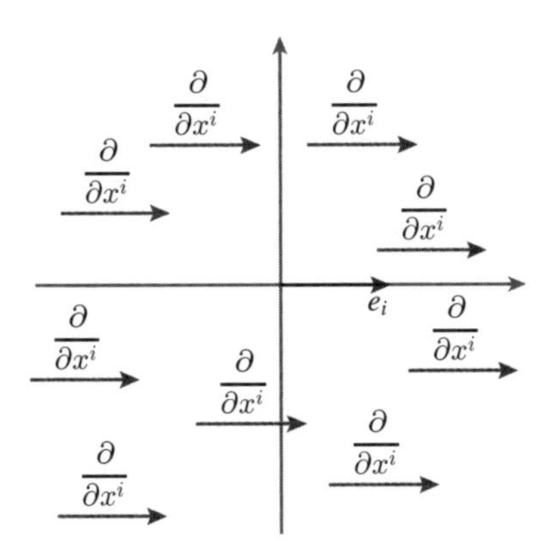

FIGURE 4. $\frac{\partial}{\partial x^i}$

the form

$$T_a f\Big(\sum_{j=1}^{n} v^j \frac{\partial}{\partial x^j}(a)\Big) = \sum_{i=1}^{m}\Big(\sum_{j=1}^{n} \frac{\partial y^i}{\partial x^j}(a) v^j\Big)\frac{\partial}{\partial y^i}(f(a)) \tag{2.12}$$

For a general vector field X on U, writing $f(x)$ in (2.10) as $f(x)=\sum_{i=1}^{n} f^i(x)e_i$, we obtain the representation

$$X = \sum_{i=1}^{n} f^i \frac{\partial}{\partial x^i} \tag{2.13}$$

where $f^i:U\to\mathbb{R}$. Thus the vector field X is C^r if and only if all f^i are C^r.

B. Vector Fields and Differential Equations

Let us begin by fixing some notation. Consider an open subset U of $\mathbb{R}^n$, and let $\gamma:I\to U$ be a differentiable curve, where I is an interval in $\mathbb{R}$. In case the interval I is closed at one or both endpoints, we consider one-sided derivative at the included endpoint. It is customary to visualize the *velocity vector* $\gamma'(t)$ as a directed segment *tangent* to the image of γ, and having as initial point, the point $\gamma(t)=(\gamma^1(t),\ldots,\gamma^n(t))$. We therefore define the ***velocity vector*** by

$$\gamma'(t) = \Big(\gamma(t); \frac{d\gamma^1}{dt},\ldots,\frac{d\gamma^n}{dt}\Big) \tag{2.14}$$

emphasizing the presence of basepoint $\gamma(t)$. Thus $\gamma'(t)\in T_{\gamma(t)}U$. By contrast, we use the notation

$$\frac{d\gamma}{dt} = \Big(\frac{d\gamma^1}{dt},\ldots,\frac{d\gamma^n}{dt}\Big) \tag{2.15}$$

to indicate the n-tuple of the derivatives of components. The latter is an element of $\mathbb{R}^n$, i.e., it can be visualized as a vector with initial point at the origin in $\mathbb{R}^n$.

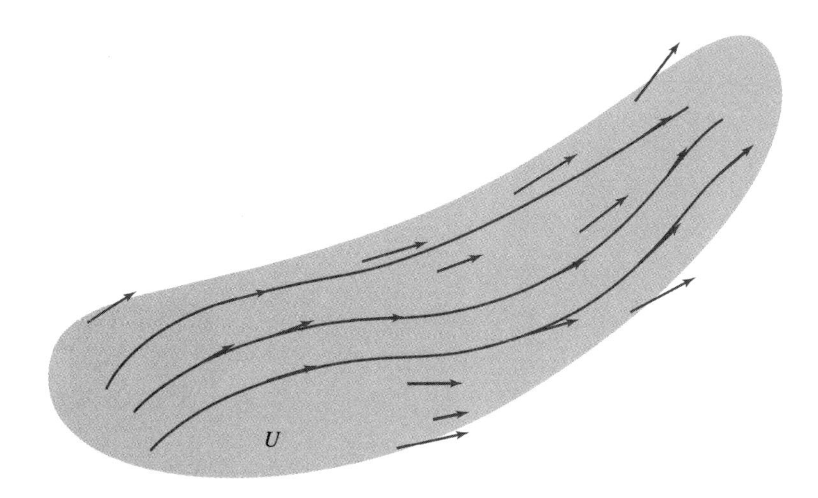

FIGURE 5. Integral curves

Now let X be a vector field on U. A differentiable curve $\gamma{:}I\to U$ is called an ***integral curve*** for X provided that

$$\gamma'(t) = X(\gamma(t)), \ \ \forall t\in I \tag{2.16}$$

Thus an integral curve is such that its velocity vectors coincide with the given vector field at points of the image curve (see Figure 5).

We turn to an alternative description. Writing $x{=}\gamma(t)$, or equivalently, $(x^1,\ldots,x^n){=}(\gamma^1(t),\ldots,\gamma^n(t))$, (2.16) can be written as

$$\frac{dx}{dt} = (f^1(x),\ldots,f^n(x)) \tag{2.17}$$

or, equivalently,

$$\begin{cases} \frac{dx^1}{dt} = f^1(x) \\ \vdots \\ \frac{dx^n}{dt} = f^n(x) \end{cases} \tag{2.18}$$

This is a system of ordinary differential equations on U. In this framework, an integral curve of X is a ***solution*** of (2.17) or (2.18). A solution (or integral curve) with ***initial condition*** $\bar{x}{=}\gamma(\bar{t})$ is a solution that satisfies $\gamma(\bar{t}){=}\bar{x}$, where $\bar{t}\in I$ and $\bar{x}\in U$ are given.

It would be more accurate to refer to (2.18) as an ***autonomous*** or ***time-independent*** (system of) differential equations, meaning that the variable t does not explicitly appear on the right-hand side of equations in (2.18). Later on we will briefly discuss *time-dependent* systems and note that their study can be reduced to the time-independent case. But for now, we have established a one-to-one correspondence between vector fields and autonomous differential equations. This correspondence enables us to recast the fundamental existence-uniqueness theorem for the solution of differential equations as a theorem about vector fields. A proof of the fundamental theorem may be found in any text on theory of ordinary differential equations (see, e.g., **[3]**, **[6]**, **[11]** or **[14]**).

2. Theorem *Let X be a C^1 vector field on an open subset U of $\mathbb{R}^n$. Then for any $\bar{x}\in U$ and any $\bar{t}\in\mathbb{R}$, the following hold:*
(i) (Existence) There exists an open interval I with $\bar{t}\in I$ and an integral curve $\gamma:I\to U$ of X with initial condition $\gamma(\bar{t})=\bar{x}$.
(ii) (Uniqueness) If I and J are two open intervals that contain $\bar{t}$, and $\alpha:I\to U$, $\beta:J\to U$ are integral curves with initial condition $\alpha(\bar{t})=\beta(\bar{t})=\bar{x}$, then $\alpha(t)=\beta(t)$ for all $t\in I\cap J$.
(iii) (Uniformity) If in (i), $\bar{t}\in[a,b]\subset I$, then an open neighborhood V of $\bar{x}$ in U exists so that for any $x\in V$, an integral curve λ with initial condition $\lambda(\bar{t})=x$ exists whose domain of definition includes $[a,b]$.

Thus for any initial condition, an integral curve satisfying that initial condition exists and is unique in the sense of (ii). However, uniqueness and uniformity deserve further elaboration, something that will occupy us in the next few subsections.

3. Stronger Forms of Uniqueness

The uniqueness statement in part (ii) of the theorem above does not, at face value, convey the full geometric uniqueness of solutions; the full version, however, can be easily derived from the above. Our first amendment will be to free (ii) from the initial value of the parameter of the integral curve. Let I_1 and I_2 be two open intervals, $t_1\in I_1$, $t_2\in I_2$, $\gamma_1:I_1\to U$ and $\gamma_2:I_2\to U$ two integral curves of X with initial conditions $\gamma_1(t_1)=\bar{x}$ and $\gamma_2(t_2)=\bar{x}$. We will show that the images of γ_1 and γ_2 coincide to the extent that they are both defined. Define a new curve $\gamma_3:I_3\to U$, where

$$I_3 = \{t+t_1-t_2 : t\in I_2\}$$

by

$$\gamma_3(t) = \gamma_2(t-t_1+t_2)$$

One has $\gamma_3(t_1)=\bar{x}$, and γ_3 is an integral curve for X since

$$\gamma_3{}'(t) = \gamma_2{}'(t-t_1+t_2) = X(\gamma_2(t-t_1+t_2)) = X(\gamma_3(t))$$

Therefore, by (ii), a piece of the image of γ_3 around $\gamma_3(t_1)$ (which is also a piece of the image of γ_2 around $\gamma_2(t_2)$) coincides with a piece of the image of γ_1 around $\gamma_1(t_1)$. Thus the image curves of solutions do not "cross".

Taking advantage of this fact, we can define the "maximal" solution curve that passes through a given point. Given $x\in U$, consider all integral curves $\gamma:I\to U$, where I is open interval, $0\in I$ and $\gamma(0)=x$. The union of all such I is an open interval that contains 0; we denote this union by I^x. Now an integral curve $\Phi^x:I^x\to U$ with $\Phi^x(0)=x$ is defined as follows. Since by the definition of I^x, each $t\in I^x$ belongs to some open interval I, which is the domain of an integral curve γ with $\gamma(0)=x$, we can define

$\Phi^x(t)=\gamma(t)$. This is independent of the particular (I,γ) by virtue of the discussion in the previous paragraph, hence Φ^x is well-defined. Moreover, by construction, I^x is the largest open interval on which an integral curve sending 0 to x can be defined. Thus it is natural to call Φ^x the ***maximal solution*** with initial point x. One often uses the notation $I^x =]\alpha_x, \omega_x[$. In case $I^x=\mathbb{R}$ for all $x\in U$, we say that the vector field X is ***complete***.

4. Examples

(a) Linear vector fields are examples of complete vector fields. Let A be an $n\times n$ real matrix and consider the system of differential equations

$$\frac{dx}{dt} = A\cdot x \tag{2.19}$$

on $\mathbb{R}^n$. It is shown in Appendix I that if M is any real or complex square matrix, then the series $\sum_{i=0}^{\infty} \frac{M^i}{i!}$ is always convergent. The sum will naturally be called as $\exp M$. It is also shown there that the unique solution of (2.19) with initial point $\bar{x}$ is $(\exp tA)\bar{x}$, which is defined over all of $\mathbb{R}$.

(b) We will now observe that the simple vector field $X=(x^2)\frac{d}{dx}$ on $\mathbb{R}$ is not complete. Here x is the variable describing the points of $\mathbb{R}$. Simple integration of the differential equation $\frac{dx}{dt}=x^2$ gives the solution with initial condition $\gamma(0)=a$ as

$$\gamma(t) = \frac{a}{1-at}$$

If $a = 0$, the constant solution $\gamma(t)=0$ is obtained, which is defined for all t. If $a>0$, the domain of the maximal solution is $]-\infty, \frac{1}{a}[$, and for $a<0$, one has $]\alpha_a, \omega_a[=]\frac{1}{a}, +\infty[$. Thus the vector field X is not complete.

(c) One can sometimes "complete" a vector field by either extending the domain of the vector field so that the solutions do not "run out" from the boundary of the domain, or modify the length of tangent vectors so as to "slow down" the vector field without affecting the geometric solution set. A very simple example is the vector field $X=\frac{d}{dx}$ on the interval $]0, +\infty[$. The solution with initial condition $\gamma(0)=a$ is $\gamma(t)=a+t$. For this solution we have $\alpha_a=-a$ and $\omega_a=+\infty$. All solutions share the same geometric image set, which is $]0, +\infty[$. One way to complete this vector field is to extend the domain from $]0, +\infty[$ to the entire real line $]-\infty, +\infty[$. Another method, keeping the same domain, is to multiply the vector field by the positive function $f(x)=x$, and look instead at the vector field $X = x\frac{d}{dx}$. For the new vector field, the solution with the initial condition $\gamma(0)=a$ is $\gamma(t)=a\exp t$. This solution is defined for all t. Moreover, the geometric solution set (image of the solution) is exactly the same as before, namely the interval $]0, +\infty[$.

For a generalization of the above ideas, the reader is referred to Exercise 2.3 at the end of the current chapter. Also, using the uniformity part of the fundamental existence-uniqueness theorem, we will derive in the next subsection a useful sufficient condition for $\omega_x = +\infty$ or $\alpha_x = -\infty$.

5. The Flow of a Vector Field

We will now discuss a method for describing and making use of the totality of maximal integral curves of a vector field. As usual, we consider an open subset U of $\mathbb{R}^n$ and a C^r, $r\geq 1$, vector field X on U. The maximal integral curve with initial condition $0\mapsto x$ will then be denoted by $\Phi^x: I^x \to U$. We define

(2.20)
$$\tilde{U} = \bigcup_{x\in U} I^x \times \{x\} \subset \mathbb{R} \times U$$

It is a consequence of the uniformity assertion, (iii), of the fundamental existence-uniqueness theorem that there exist $\varepsilon>0$ and open neighborhood V of x in U so that $]-\varepsilon, +\varepsilon[\times V \subset \tilde{U}$, hence $\tilde{U}$ is an open subset of $\mathbb{R}\times U$. We then define

(2.21)
$$\Phi : \tilde{U} \to U, \ \Phi(t, x) = \Phi^x(t)$$

Note that if $(t, x)\in\tilde{U}$, then by (2.20), $t \in I^x$, hence $\Phi^x(t)$ is defined. Φ is called the ***flow*** of the vector field X. The following important theorem from the theory of ordinary differential equations will be stated without proof. For a proof, see [**6**] or [**11**].

6. Theorem (**C^r Dependence on Initial Condition**) *If the vector field X is C^r on U with $r\geq 1$, then the associated flow $\Phi: \tilde{U} \to U$ is C^r.*

It is worthwhile to explain the choice of name for this theorem. $\Phi(t, x)$ depends on $n+1$ variables $(t, x^1, \ldots, x^n)$, of which the last n variables describe the initial point, i.e., $\Phi(0, x)$. We now show that if the vector field is C^r, then $\Phi(t, x)$ is automatically C^{r+1} with respect to the variable t, therefore the real content of the theorem is indeed C^r-ness with respect to the initial condition x. The assertion regarding the C^{r+1} dependence on t is proved by induction on r. Considering an arbitrary integral curve γ of the vector field, we have

$$\frac{d\gamma}{dt} = f(\gamma(t))$$

where $f: U \to \mathbb{R}^n$ is C^r. For $r=1$, since the solution γ must by its nature be at least differentiable, and f is assumed C^1, then the right-hand side of the equation is differentiable with respect to t. It follows that $\frac{d\gamma}{dt}$ is differentiable and *a fortiori* continuous. Therefore, γ is C^1 implying that the right-hand side of the equation above is C^1. It follows then that γ must be at least C^2! Continuing this argument, if f is assumed C^2, it would follow that γ is C^3, and so on, proving the assertion.

Before continuing on with the consequences of this theorem, we mention, also without proof, a complement to the theorem that we will later have occasion to use. For a proof, see the same references as for Theorem 6.

7. Theorem **(C^r Dependence on Parameter)** *Let M and U be, respectively, open subsets of $\mathbb{R}^m$ and $\mathbb{R}^n$, and suppose that $X{:}M\times U\to\mathbb{R}^n$ is C^r, $r\geq 1$. For each $\mu\in M$, the restriction of X to $\{\mu\}\times U$ is a C^r vector field on U, the flow of which we denote by Φ_μ. Then Φ_μ is a C^r function of μ.*

We now proceed to develop some basic properties of flows. Throughout, X will be a C^r vector field with $r\geq 1$, and Φ will be its associated flow. Depending on the point of view, each of the notations below for the flow may be used.

$$\Phi(t,x) = \Phi^x(t) = \Phi_t(x)$$

8. Elementary Properties of Flows
(a) *Let X be a C^r vector field, $r\geq 1$, on an open subset U of $\mathbb{R}^n$, with associated flow Φ. Then the following hold*

$$\Phi_0 = \mathbb{1}_U$$
$$\Phi_{s+t} = \Phi_s \circ \Phi_t, \quad s,t\in\mathbb{R}$$

in the sense that if two of $\Phi_t(x)$, $\Phi_s(\Phi_t(x))$ and $\Phi_{s+t}(x)$ are defined, then so is the third, and the equality $\Phi_{s+t}(x) = \Phi_s(\Phi_t(x))$ holds.

The above is referred to as the ***(local) group property*** of the flow.

Proof. The first identity is immediate from the definition of flow. For the second, we first assume that for a given x, all three of $\Phi_t(x)$, $\Phi_s(\Phi_t(x))$ and $\Phi_{s+t}(x)$ are defined, and proceed with proving the equality. Define two curves α and β by

$$\alpha(s) = \Phi_s(\Phi_t(x)) = \Phi^{\Phi_t(x)}(s)$$
$$\beta(s) = \Phi_{s+t}(x) = \Phi^x(s+t)$$

These are both integral curves for X, being defined by the flow, and they agree for $s=0$, so they agree for all s by uniqueness. Thus the equality is proved.

Now suppose that $\Phi_t(x)$ and $\Phi_s(\Phi_t(x))$ are defined; we must show that $\Phi_{s+t}(x)$ is defined. We assume $s>0$; the case $s<0$ is similarly handled. Let S be the set of all $r\geq 0$ with the property that $\Phi_{u+t}(x)$ is defined for all $0\leq u<r$. S is an interval. We wish to show that $s_0=\sup(S)>s$. Suppose $s_0\leq s$, and take a sequence (s_n) of points of S, $s_n<s_0$, that converge to s_0. By the definition of S, $\Phi_{s_n+t}(x)$ is defined and therefore $\Phi_{s_n+t}(x)=\Phi_{s_n}(\Phi_t(x))$. It follows from the continuity of the flow that $\Phi_{s_0+t}(x)=\Phi_{s_0}(\Phi_t(x))$. The solution curve $\Phi_{u+t}(x)$ will then be defined for $u>s_0$ if $u-s$ is sufficiently small, contradicting the assumption that s_0 is the supremum of S. This demonstrates that $s_0>s$, as desired. The other cases are similar or simpler and are left to the reader. □

In general, if W is an open subset of $\mathbb{R}^n$, a mapping $\phi{:}W\to\mathbb{R}^n$ is called a C^r ***diffeomorphism***, $r\geq 1$, if ϕ is C^r, one-to-one, and its inverse $\phi^{-1}{:}\phi(W)\to W$ is also C^r. In particular, such a map will be a homeomorphism onto its image and the image

$\phi(W)$ will be an open subset of $\mathbb{R}^n$ (see Appendix II).

(b) *Let X be a C^r vector field, $r \geq 1$, on an open subset U of $\mathbb{R}^n$, with associated flow Φ. Then the domain of Φ_t is open, and Φ_t is a C^r diffeomorphism onto its image provided that the domain of Φ_t is not empty. In this case, the inverse of Φ_t is Φ_{-t}.*

PROOF. Note that the domain of the mapping Φ_t consists of those $x \in U$ for which $t \in I^x$, i.e., those x for which the maximal solution with initial condition $0 \mapsto x$ includes $[0, t]$ or $[t, 0]$ depending on whether $t \geq 0$ or $t \leq 0$. It follows from the openness of $\tilde{U}$ (or ultimately from the uniformity assertion in the fundamental theorem) that the domain of Φ_t is also open. Further, since Φ_t is obtained as a restriction of Φ, and X is assumed C^r, Theorem 6 implies that Φ_t is C^r. Finally, if the domain W of Φ_t is not empty, then $\Phi_t(W)$ and $\Phi_{-t+t}(W) = \Phi_0 W = W$ are defined, and it follows from **(a)** that $\Phi_{-t}(\Phi_t(W))$ is defined and that Φ_{-t} is the inverse of Φ_t. □

The following gives a very useful sufficient condition for the extension of time parameter of solutions to infinity.

(c) *Let X be a C^r vector field, $r \geq 1$, on an open subset U of $\mathbb{R}^n$, with associated flow Φ. If for $x \in U$, the closure of the set $\Phi^x[0, \omega_x[$ (respectively, the closure of the set $\Phi^x]\alpha_x, 0]$) is bounded and contained in U, then $\omega_x = +\infty$ (respectively, $\alpha_x = -\infty$).*

PROOF. We prove the assertion for ω_x, the case of α_x being similar. By assumption, the closure of $\Phi^x[0, \omega_x[$ is compact and is contained in U, therefore an increasing sequence $t_1 < t_2 < \cdots$ exists with $t_n \to \omega_x$, and $\Phi_{t_n}(x) \to \bar{x}$, where $\bar{x} \in U$. By the uniformity assertion of the Fundamental Theorem, an open neighborhood V of $\bar{x}$ exists and $T > 0$ so that all solutions with initial point in V are defined throughout the interval $[0, T]$. We take n so large that $\omega_x - t_n < T$ and $\Phi_{t_n}(x) \in V$. Then $\Phi_{T+t_n}(x) = \Phi_T(\Phi_{t_n}(x))$ is defined. However, $T + t_n > \omega_x$, which contradicts the definition of ω_x. □

(d) **Classification of Orbits**

Let X be a C^1 vector field on an open set U of $\mathbb{R}^n$, and $x \in U$. The maximal integral curve whose image includes x is called the ***orbit*** or ***trajectory*** of x. Often the image curve itself is also called the orbit or trajectory through x. It turns out to be quite easy to give a geometric classification of orbits. There are three kinds.

(i) If the solution curve is constant, i.e., $\gamma(t) = a$ for all t in the domain interval I, then by c above, $I^a = \mathbb{R}$, $\frac{d\gamma}{dt} = \mathbf{0}$ for all t, and $X(a) = \mathbf{0}$. Such a point a is known by various names such as a ***zero*** (of the vector field), a ***singularity*** (or ***singular point***) of X, a ***rest point*** or an ***equilibrium point*** .

(ii) If γ is one-to-one, then $\gamma'(t) \neq \mathbf{0}$ for all t, for if $\gamma'(t_0) = \mathbf{0}$, then $X(\gamma(t_0)) = \mathbf{0}$, giving a constant solution with image $\gamma(t_0)$, and contradicting the uniqueness of solution through $\gamma(t_0)$. Thus if γ is one-to-one, γ is a *regular curve*, in the sense its velocity vector is everywhere non-zero.

(iii) Suppose γ is neither one-to-one nor a constant. We shall see that in this case the image curve is a simple closed curve representing a ***periodic solution***. More precisely, if x is a point on the orbit, we shall see that $I^x=\mathbb{R}$, and there is a smallest $T_0 > 0$ such that

$$\gamma(t+T_0) = \gamma(t), \ \forall t\in\mathbb{R}$$

Since γ is assumed not to be one-to-one, there exist t_1 and $_2$, $t_1<t_2$, so that $\gamma(t_1)=\gamma(t_2)$. Let $T=t_2-t_1$, and $\gamma(t_1)=x_1$. For any point x on the image of γ, there is $t\in I^x$ so that $x=\Phi_t(x_1)$. We thus obtain

$$\begin{aligned}\Phi_T(x) &= \Phi_T \circ \Phi_t(x_1)\\ &= \Phi_t \circ \Phi_T(x_1)\\ &= \Phi_t(x_1)\\ &= x\end{aligned}$$

Therefore, every point on the orbit will return to itself after time T. It follows from $x=\Phi_{-T}\circ\Phi_T(x)$ that $\Phi_{-T}(x)=x$. Simple induction shows that $\Phi_{nT}(x)=x$ for every integer n. As I^x is an interval, we must then have $I^x=\mathbb{R}$. Note now that the set of periods of the orbit form a subgroup of $(\mathbb{R},+)$, which is not the trivial subgroup $\{0\}$ because the existence of non-trivial periods was established. Such a subgroup must either have a smallest positive element or be dense in $\mathbb{R}$. If the subgroup were dense, then continuity of the flow would imply that the orbit is constant, contrary to the original assumption. It follows that there is a smallest positive element, T_0, of the subgroup. T_0 is called the ***period*** of the ***periodic orbit*** γ. The image $\gamma[0, T_0]$ is a simple closed curve.

We follow the above discussion on flows by showing how a family of diffeomorphisms parametrized by $\mathbb{R}$, satisfying the local group conditions 8a and a modicum of differentiability, necessarily arises from a vector field. This may be viewed as the source of ubiquity of ordinary differential equations in the study of evolutionary phenomena. Regard the subset U of $\mathbb{R}^n$ as the set of "states" of a system undergoing change, and Φ_t as the transformation law under the passage of time t. Thus the state x will be transformed to the state $\Phi_t(x)$ after the passage of time t. It is then natural to assume that 8a is satisfied, provided that the rule Φ_t itself does not change in time ("time independence"). Each state x then traverses a time-path $\Phi_t(x)=\Phi^x(t)$. It will be seen that under mild differentiability assumptions, these time paths are the integral curves of a vector field.

Let U be an open subset of $\mathbb{R}^n$, and let U' be an open neighborhood of $\{0\}\times U$ in $\mathbb{R}\times U$ with the property that for each $x\in U$, the set $U'\cap(\mathbb{R}\times\{x\})$ is connected, i.e., it is homeomorphic to an interval in $\mathbb{R}$. Then a smooth map $\Psi: U'\to U$ will be called a ***(smooth) local one-parameter group*** if the following conditions are satisfied:
(i) $\Psi(0, x)=x$ for all $x\in U$.
(ii) $\Psi(s+t, x)=\Psi(s, \Psi(t, x))$, in the sense that whenever two of (t, x), $(s, \Psi(t, x))$ and $(s+t, x)$ are in U', then the third is also in U', and the stated equality holds.

Given these data, we define a vector field X on U as follows. Let $U'\cap(\mathbb{R}\times\{x\})=J^x\times\{x\}$, and define the curve $\Psi^x: J^x \to U$ by

$$\Psi^x(t) = \Psi(t, x)$$

Then the vector field X is given by

$$X(x) = (\Psi^x)'(0) \tag{2.22}$$

Equivalently, the differential equation

$$\frac{dx}{dt} = \frac{\partial}{\partial t}(\Psi(t, x)) \,|_{t=0}$$

is considered. Note that X is smooth, being obtained from smooth Ψ by differentiation. The following describes the relationship between the original local one-parameter group Ψ and the flow of X.

9. Theorem *Let $\Psi: U' \to U$ be a local one-parameter group, and let the vector field X be defined by* (2.22). *If $\Phi: \tilde{U} \to U$ is the flow of X, then $U' \subset \tilde{U}$, and $\Psi = \Phi \,|_{U'}$.*

Proof. It suffices to show that for each $x \in U$, the curve $\Psi^x: J^x \to U$ is an integral curve for X. Note that in the definition of X, only the point where $t=0$ was considered. Thus we differentiate $\Psi^x(t)$ with respect to t at an arbitrary $t \in J^x$ and show that at $y = \Psi^x(t)$ the following equality holds.

$$(\Psi^x)'(t) = (\Psi^y)'(0) \tag{2.23}$$

We have

$$\begin{aligned}
\frac{d\Psi^x}{dt}(t) &= \lim_{h \to 0} \frac{\Psi^x(t+h) - \Psi^x(t)}{h} \\
&= \lim_{h \to 0} \frac{\Psi(t+h, x) - y}{h} \\
&= \lim_{h \to 0} \frac{\Psi(h, \Psi(t, x)) - y}{h} \\
&= \lim_{h \to 0} \frac{\Psi^y(h) - \Psi^y(0)}{h} \\
&= \frac{d\Psi^y}{dt}(0)
\end{aligned}$$

and the equality (2.23) is established. □

The vector field X defined above is called the ***generator*** for the local one-parameter group Ψ.

10. Effect of Diffeomorphisms

Let $h: U \to V$ be a smooth diffeomorphism of two open subsets of $\mathbb{R}^n$. We will investigate how a vector field on U is transferred to a vector field on V and what the relationship of the flows will be. Consider a C^r vector field X on U, $r \geq 1$, with associated flow $\Phi:\tilde{U} \to U$. We define a vector field h_*X on V by

$$h_*X = Th \circ X \circ h^{-1} \tag{2.24}$$

More explicitly, let $y=h(x)$ be an arbitrary point of V, then

$$h_*X(y) = T_x h\,(X(x)) \tag{2.25}$$

with h_*X sometimes referred to as the ***push-forward*** of X by h. Since h is assumed smooth (C^∞), and X is C^r, then h_*X is also C^r. Now let $\Psi:\tilde{V} \to V$ be the flow of h_*X. We claim that Φ^x and $\Psi^{h(x)}$ have the same interval in $\mathbb{R}$ as domain, and that

$$\Psi_t \circ h = h \circ \Phi_t,\ \forall t \in I^x \tag{2.26}$$

To see this, let $\gamma:I \to U$ be an integral curve for X. We show that $h \circ \gamma: I \to V$ is an integral curve for h_*X. This is an application of the chain rule:

$$\begin{aligned}(h \circ \gamma)'(t) &= Th(\gamma'(t)) \\ &= Th\big(X(\gamma(t))\big) \\ &= h_*X((h \circ \gamma)(t))\end{aligned}$$

Since the same argument holds for the inverse diffeomorphism h^{-1}, the correspondence between the maximal solutions is established, as well as the equality (2.26). One can summarize the above statements by saying that the following diagram is commutative:

$$\begin{array}{ccc}\tilde{U} & \xrightarrow{\tilde{h}} & \tilde{V} \\ \downarrow{\scriptstyle\Phi} & & \downarrow{\scriptstyle\Psi} \\ U & \xrightarrow{h} & V\end{array}$$

Here $\tilde{h}:\tilde{U} \to \tilde{V}$ is defined by $\tilde{h}(t,x)=(t,h(x))$.

Let X be a smooth vector field on U with associated flow Φ. We know from 8b, that if the domain of Φ_t is not empty, then Φ_t is a smooth diffeomorphism onto its image. The identity $\Phi_{-t} \circ \Phi_t \circ \Phi_t = \Phi_t$ then implies that

$$(\Phi_t)_*X = X \tag{2.27}$$

11. Note on Time-Dependent Vector Fields

There will be just a few occasions in the book where we will have to make use of time-dependent vector fields. We will now show how questions about time-dependent vector fields can be reduced to those of time-independent ones by adding an extra dimension. Regarding interpretation in terms of evolving phenomena, here the evolution of a state x depends not only on x and the elapsed time interval, but also on the time of the original "observation," since the law of evolution itself is undergoing change in time.

By a ***time-dependent vector field*** X on an open subset U of $\mathbb{R}^n$, we shall mean a map

$$X : \mathbb{R} \times U \to TU$$

such that

$$X(t, x) = (x, f(t, x)) \in T_x U, \quad \forall t \in \mathbb{R}, \ \forall x \in U$$

One could more generally replace $\mathbb{R}$ in the above by an open interval of $\mathbb{R}$ with obvious modifications in the rest of the discussion, but we shall work with $\mathbb{R}$ for the convenience of presentation. Geometrically, instead of being given a single vector $X(x)$ in each tangent space $T_x U$, one has a family $X_t(x)$ of tangent vectors at each point. Adjectives such as C^r will refer to the function f of $n + 1$ variables $(t, x^1, \ldots, x^n)$.

Given X above, and renaming $t{=}x^0$, $\tilde{x}{=}(x^0, x^1, \ldots, x^n)$, one obtains a time-independent vector field $\tilde{X}$ on $\mathbb{R}{\times}U$ as $\tilde{X}(\tilde{x}){=}(1, f(\tilde{x}))$. Written as a differential equation, this will look as

$$\text{(2.28)} \qquad \begin{cases} \frac{dx^0}{dt} = 1 \\ \frac{dx}{dt} = f(x^0, x^1, \ldots, x^n) \end{cases}$$

where x stands for $(x^1, \ldots, x^n)$. $\tilde{X}$ is a special kind of vector field on the open subset $\mathbb{R}{\times}U$ of $\mathbb{R}^{n+1}$, one where the first component of the assigned vector at each point is 1. The fundamental existence-uniqueness theorem and other considerations apply here as a special case. Although the solution curves of $\tilde{X}$ in $\mathbb{R}{\times}U$ do not cross, the projections of these curves on U can cross, and one has to be careful about the geometric interpretation of uniqueness of solutions in U itself.

Denoting the flow of $\tilde{X}$ in $\mathbb{R}{\times}U$ by $\tilde{\Phi}$, we define Φ_t as follows:

$$\text{(2.29)} \qquad \Phi_t(x) = \pi(\tilde{\Phi}_t(0, x))$$

where π denotes the projection on the second coordinate, $\mathbb{R}{\times}U{\to}U$. The domain of Φ_t consists of those x in U where $\tilde{\Phi}_t(0, x)$ is defined. Certainly Φ_0 is defined for all x, and $\Phi_0(x){=}x$. The following important fact is to be noted.

(a) **Lemma** *The domain of Φ_t is an open subset of U. If the domain is not empty, then Φ_t is a diffeomorphism onto its image.*

Proof. We know from 8b that the domain of $\tilde{\Phi}_t$ is open, but the domain of Φ_t is the intersection of this set with U, identified as the subset $\{0\}{\times}U$ of $\mathbb{R}{\times}U$, so it is open in U. Now suppose this domain is not empty. Φ_t is one-to-one on this domain since $\Phi_t(x){=}\Phi_t(y)$ implies that the second components of $\tilde{\Phi}_t(0, x)$ and $\tilde{\Phi}_t(0, y)$ are the same, but the first components of these are also equal (to t), therefore the one-to-one nature of $\tilde{\Phi}_t$ proves that Φ_t is one-to-one. Now $\tilde{\Phi}_t$ is a diffeomorphism, and the projection π has rank n, therefore the derivative of the composition $\pi{\circ}\tilde{\Phi}_t$ has rank n, and by the inverse function theorem (Appendix II), Φ_t will be a diffeomorphism onto its image. □

We should remark, however, that the local group property $\Phi_{s+t}=\Phi_s \circ \Phi_t$ is not generally valid in time-dependent context (see Exercise 2.2 at the end of this chapter).

C. Vector Fields as Operators

For smooth, i.e., C^∞, vector fields, there is a completely equivalent and highly useful approach to vector fields that we shall alternatively use. Here one identifies a tangent vector at a point with the operator that gives the value of the tangent maps of real smooth functions at that point on that vector, akin to the directional derivative. First we fix some notation. Let $C^\infty(U)$ denote the set of C^∞ real-valued functions on U. Under pointwise addition and multiplication of functions and multiplication by elements $\mathbb{R}$, $C^\infty(U)$ has an $\mathbb{R}$-algebra structure, i.e., it is vector space over $\mathbb{R}$ with an associative ring multiplication. For each $f \in C^\infty(U)$, we define the ***differential*** df as follows:

$$df : TU \to \mathbb{R}$$

$$df(x, v) = (Df(x))(v)$$

Thus the differential of f at x is nothing other than the derivative of f at x, except that it is regarded as a linear map from T_xU to $\mathbb{R}$, instead of a linear map from $\mathbb{R}^n$ to $\mathbb{R}$. To emphasize this point, we denote the restriction of df to T_xU by $df(x)$. Therefore, $df(x)$ is an element of the dual space $(T_xU)^*$ of T_xU. For $w=(x, v) \in T_xU$, the following are equivalent notations:

$$df(w) = (df(x))(w) = (Df(x))(v)$$

Note also that

$$df = \pi \circ Tf$$

where π is the projection on the second component. Then, for open subsets U and V, respectively, of $\mathbb{R}^n$ and $\mathbb{R}^m$, and smooth functions $f: U \to V$ and $g: V \to \mathbb{R}$, the chain rule can be written as

$$d(g \circ f) = dg \circ Tf$$

We now proceed to define the effect of tangent vectors as operators. For $w=(x, v) \in T_xU$ and $f \in C^\infty(U)$, we define

$$w \cdot f = df(w) = (df(x))(w) = (Df(x))(v) \tag{2.30}$$

Thus $w \cdot f$ is just the value of the derivative of f at x relative to v. This shows that $w \cdot f$ depends not only on $f(x)$, but also on the values of f in an arbitrarily small open set around x.

As a basic example, consider $w=\frac{\partial}{\partial x^j}(x)$. We have

$$\begin{aligned} \frac{\partial}{\partial x^j}(x) \cdot f &= (df(x))(\frac{\partial}{\partial x^j}(x)) \\ &= (Df(x))(e_j) \\ &= \frac{\partial f}{\partial x^j}(x) \end{aligned}$$

proving the identity

$$\frac{\partial}{\partial x^j}(x)\cdot f = \frac{\partial f}{\partial x^j}(x) \tag{2.31}$$

which justifies the use of the partial-derivative symbol for vector fields. By a slight abuse of notation, one traditionally uses the symbol x^i to denote projection on the ith coordinate, i.e., $x^i(a^1,\dots,a^n)=a^i$. This is a smooth function, being linear, and one obtains $Dx^i(a)=x^i$. Therefore, at every point $a\in U$

$$dx^i(a)(\frac{\partial}{\partial x^j}(a)) = \frac{\partial}{\partial x^j}(a)\cdot x^i = \delta^i_j$$

which shows that $(dx^1(a),\dots,dx^n(a))$ is the dual basis in $(T_aU)^*$ relative to the basis $(\frac{\partial}{\partial x^1}(a),\dots,\frac{\partial}{\partial x^n}(a))$ for T_aU. Now $df(a)\in(T_aU)^*$, so we write $df(a)=\sum_{i=1}^n f_i(a)dx^i(a)$. To identify the coefficients, we evaluate both sides on $\frac{\partial}{\partial x^j}(a)$ to obtain $\frac{\partial f}{\partial x^j}(a)=f_j(a)$. Therefore,

$$df(a) = \sum_{i=1}^n \frac{\partial f}{\partial x^i}(a)dx^i(a)$$

Or, in short-hand notation, the familiar

$$df = \sum_{i=1}^n \frac{\partial f}{\partial x^i}dx^i$$

Turning now to vector fields, let X be a smooth vector field on U. For $f\in C^\infty(U)$, we define $X\cdot f:U\to\mathbb{R}$ by

$$(X\cdot f)(x) = X(x)\cdot f \tag{2.32}$$

Since X and f are C^∞, the definition (2.32) shows that indeed $X\cdot f\in C^\infty(U)$. The previous discussion shows, for example, that

$$\frac{\partial}{\partial x^j}\cdot f = \frac{\partial f}{\partial x^j}$$

and, in general,

$$\Big(\sum_{j=1}^n g^j\frac{\partial}{\partial x^j}\Big)\cdot f = \sum_{j=1}^n g^j\frac{\partial f}{\partial x^j}$$

where g^j are smooth functions. For a given smooth vector field X, we define the operatorD_X by

$$D_X : C^\infty(U)\to C^\infty(U)$$
$$D_X(f) = X\cdot f$$

Algebraic properties of D_X are described below.

(i) D_X *is* $\mathbb{R}$*-linear, i.e.,*

$$D_X(f+g) = D_X(f)+D_X(g),\quad \forall f,g\in C^\infty(U)$$
$$D_X(rf) = rD_X(f),\quad \forall r\in\mathbb{R}, \forall f\in C^\infty(U)$$

(ii) D_X *is* ***Leibnizian****, i.e.,*

$$D_X(fg) = (D_X f)g + f(D_X g), \quad \forall f, g \in C^\infty(U)$$

Proof. In view of the definitions (2.32) and (2.30), these describe the corresponding properties of the derivative. □

Any operator $D: C^\infty(U) \to C^\infty(U)$ with the properties (i) and (ii) above is called a ***derivation*** of the $\mathbb{R}$-algebra $C^\infty(U)$. We have thus shown that any smooth vector field gives rise to a derivation of $C^\infty(U)$. Our next goal will be to show that all derivations of $C^\infty(U)$ arise this way from smooth vector fields. A few general preliminary facts will be needed.

12. Lemma (Smooth Bump Functions)
Let B_1 and B_2 be concentric open disks in $\mathbb{R}^n$ so that $B_1 \subset \overline{B_1} \subset B_2$. Then a smooth function $\theta: \mathbb{R}^n \to [0, 1]$ exists which takes value 1 on $\overline{B_1}$ and value 0 outside B_2.

Proof. The function θ will be constructed in several stages. The auxiliary functions constructed on the way will also be of use in the future. The basic ingredient is the following function $\alpha: \mathbb{R} \to \mathbb{R}$

$$\alpha(x) = \begin{cases} \exp(-\frac{1}{x}) & if\ x > 0 \\ 0 & if\ x \leq 0 \end{cases} \tag{2.33}$$

This is probably the best-known C^∞, non-analytic, function in elementary analysis. Derivatives of α to any order are zero at 0 (see Exercise 2.1 at the end of this chapter). For given $a, b \in \mathbb{R}$ with $a<b$, we define $\beta: \mathbb{R} \to \mathbb{R}$ by

$$\beta(x) = \alpha(x-a)\alpha(b-x) \tag{2.34}$$

Note that β is also C^∞, and it is non-zero precisely on $]a, b[$ (see Figure 6). Next we define $\gamma: \mathbb{R} \to \mathbb{R}$ by

$$\gamma(x) = \frac{\int_a^x \beta}{\int_a^b \beta} \tag{2.35}$$

Also γ is C^∞, vanishes on $]-\infty, a]$, takes values in $]0, 1[$ on $]a, b[$ and has value 1 on $[b, +\infty[$. Finally, we consider open disks, B_1 and B_2 of radius r_1 and r_2, $0<r_1<r_2$, both centered at a point c of $\mathbb{R}^n$. Letting $a=r_1$ and $b=r_2$, the function

$$\theta(x) = 1 - \gamma(|x-c|) \tag{2.36}$$

has the desired property. Here $|\cdot|$ is the Euclidean norm. Note that although the norm function is not differentiable at 0, no differentiability is lost since $\gamma(|x-c|)$ is constant in a neighborhood of $x=c$. Thus θ will be C^∞. □

13. Lemma *For a derivation $D: C^\infty(U) \to C^\infty(U)$, the following hold:*
(i) If f is a constant function, then $Df=0$.

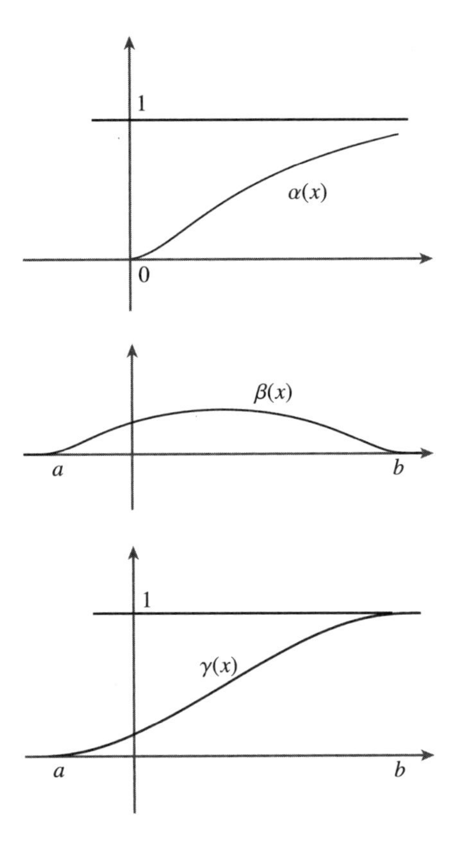

FIGURE 6. Auxiliary functions

(ii) For $f \in C^\infty(U)$ and $x \in U$, $Df(x)$ depends only on the values of f in an arbitrarily small neighborhood of x.

PROOF. (i) We denote by $\hat{c}$ the function with constant value c. Since $\hat{c}=c\hat{1}$, and D is $\mathbb{R}$-linear, it suffices to show that $D\hat{1}=\hat{0}$. We write $\hat{1}=\hat{1}\cdot\hat{1}$. From the Leibnizian property we obtain $D\hat{1}=2D\hat{1}$, hence $D\hat{1}=\hat{0}$.
(ii) Let V be an arbitrary open neighborhood of x with $V \subset U$, and suppose $g \in C^\infty(U)$ satisfy $g|_V = f|_V$. We wish to show that $Df(x)=Dg(x)$. Consider open balls B_1 of radius r_1 and B_2 of radius r_2, centered at x, so that $B_1 \subset \overline{B_1} \subset B_2 \subset V$, and take θ as in Lemma 12. We have

$$f-g = (\hat{1}-\theta)(f-g)$$

since both sides are zero on V, and θ is zero outside V. By Leibnizian property, we obtain

$$D(f-g) = D(\hat{1}-\theta)(f-g) + (\hat{1}-\theta)D(f-g)$$

At the point x we have $(f-g)(x)=0$ and $(\hat{1}-\theta)(x)=0$, so $(D(f-g))(x)=0$. It follows from the $\mathbb{R}$-linearity of D that $Df(x)=Dg(x)$. □

The following elementary fact of analysis will be of use on various occasions.

14. Lemma *Let U be an open and convex subset of $\mathbb{R}^n$ and $f:U\to\mathbb{R}$ a C^r function, $r\geq 1$. Then for every $a\in U$, there exist C^{r-1} functions, $f_i:U\to\mathbb{R}$, $i = 1,\dots,n$, so that $f_i(a) = \frac{\partial f}{\partial x^i}(a)$, $i = 1,\dots,n$, and for every $x\in U$, we have*

$$f(x) = f(a) + \sum_{i=1}^{n}(x^i - a^i)f_i(x)$$

Proof. U being convex, the straight segment joining a to any point x of U will remain in U. We can then define $\sigma:[0,1]\times U\to\mathbb{R}$ by

$$\sigma(t,x) = f((1-t)a + tx)$$

This is a C^r function. We have

$$\begin{aligned} f(x) - f(a) &= \sigma(1,x) - \sigma(0,x) \\ &= \int_0^1 \frac{\partial\sigma}{\partial t}(t,x)dt \\ &= \int_0^1 \Big(\sum_{i=1}^{n} \frac{\partial f}{\partial x^i}((1-t)a + tx)(x^i - a^i)\Big)dt \\ &= \sum_{i=1}^{n}(x^i - a^i)\Big(\int_0^1 \frac{\partial f}{\partial x^i}((1-t)a + tx)dt\Big) \end{aligned}$$

Letting $f_i(x)=\int_0^1 \frac{\partial f}{\partial x^i}((1-t)a+tx)dt$, we obtain $f_i(a)=\frac{\partial f}{\partial x^i}(a)$. Since f_i has been obtained from f by once differentiating with respect to x^i, f_i is C^{r-1}, and the proof is complete. □

Naturally, if f is C^∞, then f_i will also be C^∞. To show that every derivation arises from a smooth vector field, we begin by considering the case that the open set U is also convex. Let $f\in C^\infty(U)$ and represent f as in Lemma 14. We then obtain

$$Df(x) = D(\widehat{f(a)}) + \sum_{i=1}^{n}\Big(D(x^i - a^i)(x)f_i(x) + (x^i - a^i)Df_i(x)\Big)$$

Now using part (i) of Lemma 13, and letting $x=a$, we have

$$Df(a) = \sum_{i=1}^{n}(Dx^i)(a)\frac{\partial f}{\partial x^i}(a)$$

In view of this result and (2.31), we consider the vector field X on U defined by

$$X = \sum_{i=1}^{n} (Dx^i)\frac{\partial}{\partial x^i} \tag{2.37}$$

and conclude that $D{=}D_X$ on the open-convex set U.

For a general open subset U of $\mathbb{R}^n$, our strategy will be the following. We write U as a union of open convex subsets, e.g., the union of open balls contained in U. We show that for each open subset V of U, a derivation D on $C^\infty(U)$ induces a derivation D^V on $C^\infty(V)$, which is compatible with D (in a sense to be described). Taking open sets V to be convex, we construct vector fields X^V on them corresponding to D^V according to (2.37). Finally, we show that for two such sets V and W with non-empty intersection, $X^V|_{V\cap W} = X^W|_{V\cap W}$ and thus obtain a C^∞ vector field defined on the entire set U.

15. Lemma (Localization) *Let U and V be open subsets of $\mathbb{R}^n$ with $V{\subset}U$. If D is a derivation for $C^\infty(U)$, there is a unique derivation D^V for $C^\infty(V)$ with the property that for every $f{\in}C^\infty(U)$*

$$D^V(f|_V) = (Df)|_V \tag{2.38}$$

The process $D{\mapsto}D^V$ is called a ***localization***.

Proof. Given $g \in C^\infty(V)$, and $a{\in}V$, we wish to give a definition of $(D^V g)(a)$. If g were extendable to all of U as a smooth function $\tilde{g}$, one could define $(D^V g)(a)$ as $D\tilde{g}(a)$. However, there certainly exist $g{\in}C^\infty(V)$ that are not even continuously extendable to U. But since D^V is supposed to be a derivation, $(D^V g)(a)$ will depend only on the values of g in an arbitrarily small neighborhood of a, so we may try to modify g outside a small neighborhood of a so as to make it extendable to all of U. Consider a smooth $\theta{:}U{\to}\mathbb{R}$, as in Lemma 12, which has value 1 on a neighborhood W_1 of a, is 0 ouside a bigger neighborhood W_2, and $W_1{\subset}\overline{W_1}{\subset}W_2{\subset}\overline{W_2}{\subset}V$. Now the function θg can be extended smoothly outside V with value 0, a function we denote by $\tilde{g}$. We define

$$(D^V g)(a) = D\tilde{g}(a) \tag{2.39}$$

This is well-defined since any other smooth function on U that coincides with g in a neighborhood of a will yield the same value for the right-hand side. That D^V is a derivation follows from the fact that D is a derivation. If $f{\in}C^\infty(U)$, f will be a smooth extension of $f|_V$, and (2.38) follows from (2.39). Finally, the uniqueness of D^V follows from (2.38) in view of the fact that every element of $C^\infty(V)$ can be modified outside a neighborhood of a so as to extend smoothly to U. □

Now let D be a derivation for $C^\infty(U)$, and define a vector field X on U by the formula (2.37). With the above localization tool on hand, it is an easy matter to check that indeed $D_X = D$. For every $f \in C^\infty(U)$ and every $a \in U$, we must show that $Df(a) = \sum_{i=1}^n Dx^i(a) \frac{\partial f}{\partial x^i}(a)$. Take an open convex subset V of U with $a \in V$. By localization, $Df(a) = (D^V(f|_V))(a)$, but $D^V = D_{X|_V}$ is already known to be valid for convex V, so the verification is complete.

There is actually more to the (vector field) $\longleftrightarrow$ (derivation) correspondence than the bijection just established. Let us denote the set of smooth vector fields on U by $\mathcal{X}(U)$ and the set of derivations for $C^\infty(U)$ by $\mathcal{D}(U)$. If $D \in \mathcal{D}(U)$, and $g \in C^\infty(U)$, then gD defined by $(gD(f))(x) = g(x)(Df(x))$ is again a derivation, as are the sum of two derivations. We have observed the same closure properties for vector fields. Thus both $\mathcal{X}(U)$ and $\mathcal{D}(U)$ are modules over the ring $C^\infty(U)$. The following immediate consequence sums up what was established through the auxiliary lemmas and the discussion above.

16. Theorem *The correspondence $X \mapsto D_X$ and its inverse,* (2.37), *provide a module isomorphism between $\mathcal{X}(U)$ and $\mathcal{D}(U)$ over $C^\infty(U)$.*

We end this section with a generalization of the concept of "directional derivative", or "derivative with respect to a vector". Let U be an open subset of $\mathbb{R}^n$, $f: U \to \mathbb{R}$, $a \in U$, and $v \in \mathbb{R}^n$. The usual definition of the derivative of f at a with respect to v is

$$D_v f(a) = \lim_{t \to 0} \frac{f(a + tv) - f(a)}{t}$$

Here we are considering the rate of change of f along a straight line through a on which the variable moves with velocity v. The point of the following simple theorem is that we can replace the straight line, which is the integral curve of the constant vector field $X(x) = (x, v)$, with the integral curve of *any* vector field that has the same value at the point a.

17. Theorem *Let U be an open subset of $\mathbb{R}^n$, $a \in U$, $w \in T_a U$ and $f: U \to \mathbb{R}$ a C^1 function. Then for every smooth vector field X on U such that $X(a) = w$, we have*

$$w \cdot f = \lim_{t \to 0} \frac{f(\Phi^a(t)) - f(a)}{t} \tag{2.40}$$

where Φ is the flow of X.

Proof. Letting $w = (a, v)$, we have by definition

$$w \cdot f = (df(a))(w) = (Df(a))(v)$$

But $v = \frac{d\Phi^a}{dt}(0)$, so $(w \cdot f)(a) = (Df(a))(\frac{d\Phi^a}{dt}(0))$, which by the chain rule is equal to

$$
\begin{aligned}
&= (f \circ \Phi^a)'(0) \\
&= \lim_{t \to 0} \frac{(f \circ \Phi^a)(t) - (f \circ \Phi^a)(0)}{t} \\
&= \lim_{t \to 0} \frac{f(\Phi^a(t)) - f(a)}{t}
\end{aligned}
$$

as desired. □

D. Lie Bracket of Vector Fields

Let U be an open subset of $\mathbb{R}^n$ and let D_1 and D_2 be derivations of $C^\infty(U)$. We define $[D_1, D_2]$ as follows

$$[D_1, D_2] = D_1 \circ D_2 - D_2 \circ D_1$$

It is a straightforward matter to check that $[D_1, D_2]$ is also a derivation. Let now X and Y be smooth vector fields with corresponding derivations D_X and D_Y. Invoking the one-to-one correspondence between vector fields and derivations, we define the vector field $[X, Y]$, called the ***(Lie) bracket*** of X and Y, as the unique vector field corresponding to $[D_X, D_Y]$. Thus one could also use the following as definition:

$$[X, Y] \cdot f = X \cdot (Y \cdot f) - Y \cdot (X \cdot f), \quad \forall f \in C^\infty(U)$$

18. Elementary Properties of the Bracket

(a) $[,]$ *is additive in each component.*

(b) $[,]$ *is anti-symmetric, i.e.,* $[X, Y] = -[Y, X]$*, or equivalenly,* $[Z, Z] = \mathbf{0}$*, for all smooth vector fields X, Y and Z on U.*

(c) *For all smooth vector fields X, Y and Z, the* ***Jacobi identity*** *holds, namely,*

$$[X, [Y, Z]] + [Y, [Z, X]] + [Z, [X, Y]] = 0$$

(d) *For all* $X, Y \in \mathfrak{X}(U)$ *and all* $f \in C^\infty(U)$*,*

$$
\begin{aligned}
[X, fY] &= f[X, Y] + (X \cdot f)Y \\
[fX, Y] &= f[X, Y] - (Y \cdot f)X
\end{aligned}
$$

By considering constant functions f in (d) above and using (2.30), it follows that the bracket is $\mathbb{R}$-linear in each component. On the other hand, as (d) above shows, the bracket is not $C^\infty(U)$-linear, i.e., a non-constant function f cannot be simply taken out of the bracket. Thus $[X, Y](x)$ is NOT determined solely by $X(x)$ and $Y(x)$; it depends on the values of X and Y in a neighborhood of x. This will be further elaborated in 19b and Theorem 23 that will follow.

Proof. Properties (a) and (b) are immediate consequences of the definition. By the additivity condition (a), the identity

$$[X + Y, X + Y] = [X, X] + [Y, Y] + [X, Y] + [Y, X]$$

holds, which implies the equivalence of the two conditions in (b). For (c), let $f \in C^\infty(U)$, then

$$\begin{aligned}[X,[Y,Z]]\cdot f &= X\cdot([Y,Z]\cdot f) - [Y,Z]\cdot(X\cdot f)\\ &= X\cdot(Y\cdot(Z\cdot f) - Z\cdot(Y\cdot f)) - Y\cdot(Z\cdot(X\cdot f)) + Z\cdot(Y\cdot(X.f))\\ &= X\cdot(Y\cdot(Z\cdot f)) - X\cdot(Z\cdot(Y\cdot f)) - Y\cdot(Z\cdot(X\cdot f)) + Z\cdot(Y\cdot(X\cdot f))\end{aligned}$$

By permuting $X\to Y\to Z\to X$ and $X\to Z\to Y\to X$, we obtain two similar sets of equalities. By adding the three, we obtain (c). For (d), we compute the effect of the left-hand side on arbitrary $g\in C^\infty(U)$:

$$\begin{aligned}[X, fY]\cdot g &= X\cdot(fY\cdot g) - (fY)\cdot(X\cdot g)\\ &= (X\cdot f)(Y\cdot g) + fX\cdot(Y\cdot g) - fY\cdot(X\cdot g)\\ &= ((X\cdot f)Y + f[X,Y])\cdot g\end{aligned}$$

Note that we are using the Leibnizian property on the second line above. Finally, the second identity in (d) follows from the first together with (b). □

The appearance of Jacobi identity may seem strange and unmotivated. Note, however, that using anti-symmetry, we may rewrite Jacobi identity as

$$[[Y,Z],X] = [[Y,X],Z] + [Y,[Z,X]]$$

This is reminiscent of the Leibniz rule. We will indeed interpret the operation $[\cdot, X]$ as some kind of differentiation with respect to X in Theorem 23.

Any $\mathbb{R}$-vector space equipped with a product $[,]$ that satisfies conditions (a), (b) and (c) above is called a ***Lie algebra*** over $\mathbb{R}$. In Chapter 8, we will study Lie algebras in connection with Lie groups. Lie algebras over other fields are also extensively studied.

19. Example and Explicit Computation

(a) In the special case $X=\frac{\partial}{\partial x^i}$, $Y=\frac{\partial}{\partial x^j}$, one has

$$[\frac{\partial}{\partial x^i}, \frac{\partial}{\partial x^j}] = 0$$

This amounts to $\frac{\partial}{\partial x^i}(\frac{\partial}{\partial x^j}(f))=\frac{\partial}{\partial x^j}(\frac{\partial}{\partial x^i}(f))$ for arbitrary $f\in C^\infty(U)$, which is just a restatement of the equality of mixed partials for C^2 functions. In fact, the bracket measures the extent of deviation from commutativity of the effect two vector fields.

(b) For $X=\sum_{i=1}^n u^i\frac{\partial}{\partial x^i}$ and $Y=\sum_{j=1}^n v^j\frac{\partial}{\partial x^j}$, we claim that $[X,Y]=\sum_{k=1}^n w^k\frac{\partial}{\partial x^k}$, where

$$w^k = \sum_{l=1}^n (u^l\frac{\partial v^k}{\partial x^l} - v^l\frac{\partial u^k}{\partial x^l}) \tag{2.41}$$

Proof. To compute w^k, we use the elementary properties above:

$$\begin{aligned} w^k &= \Big(\sum_{k=1}^n w^k \frac{\partial}{\partial x^k}\Big) \cdot x^k \\ &= \Big[\sum_{i=1}^n u^i \frac{\partial}{\partial x^i}, \sum_{j=1}^n v^j \frac{\partial}{\partial x^j}\Big] \cdot x^k \\ &= \sum_i \sum_j [u^i \frac{\partial}{\partial x^i}, v^j \frac{\partial}{\partial x^j}] \cdot x^k \\ &= \sum_i u^i \frac{\partial}{\partial x^i} \cdot v^k - \sum_j v^j \frac{\partial}{\partial x^j} \cdot u^k \\ &= \sum_l (u^l \frac{\partial v^k}{\partial x^l} - v^l \frac{\partial u^k}{\partial x^l}) \end{aligned}$$

as claimed. □

Notice that the first order partial derivatives of u^i and v^j figure in the value of the bracket. It follows, incidentally, that the bracket of constant vector fields, i.e., vector fields with constant coefficients, vanishes. We will soon arrive at an important interpretation of the vanishing of the bracket. Our computation above shows that properties (a), (b) and one of (d) (the other following from anti-symmetry) determine the assignment $(X, Y) \mapsto [X, Y]$ uniquely, once the bracket of basis vector fields over $C^\infty(U)$ are known.

We will now attempt to expand the discussion of the Subsection 10 on the transformation of smooth vector fields under the effect of diffeomorphisms. As was seen there, a smooth diffeomorphism $h: U \to V$ gives rise to a one-to-one correspondence $X \mapsto h_* X$ between smooth vector fields on U and smooth vector fields on V. It follows from (2.25) that in fact the map $X \mapsto h_* X$ is $\mathbb{R}$-linear. We will show in fact that the push-forward map h_* preserves the bracket relationship as well; thus h_* is a Lie algebra isomorphism. We begin with a useful restatement of the Chain Rule as:

20. Lemma *Let $h: U \to V$ be a smooth map from an open subset U of $\mathbb{R}^n$ to an open subset V of $\mathbb{R}^m$, and suppose that $w \in T_a U$. Then for any $f \in C^\infty(V)$, one has*

$$Th(w) \cdot f = w \cdot (f \circ h)$$

Further, if h is a diffeomorphism and X is a smooth vector field on U, then

$$(h_* X \cdot f) \circ h = X \cdot (f \circ h)$$

PROOF. We use the definition of the effect of a tangent vector on a smooth function, and the chain rule.

$$\begin{aligned} Th(w)\cdot f &= df(h(a))(Th(w)) \\ &= (d(f\circ h)(a))(w) \\ &= w\cdot(f\circ h) \end{aligned}$$

as claimed. The case of the vector field follows by letting $X(a)=w$ and using the above. □

21. Theorem *If $h:U\to V$ is a smooth diffeomorphism of open subsets of $\mathbb{R}^n$, then h_* is a Lie algebra isomorphism, i.e., h_* is an isomorphism of vector spaces over $\mathbb{R}$ with the additional property that*

$$h_*[X,Y] = [h_*X, h_*Y]$$

PROOF. The fact that h_* is an isomorphism of vector spaces over $\mathbb{R}$ was already noted at the beginning of this subsection. The preservation of the bracket will be verified by repeated application of the previous Lemma. We check that the two sides have the same effect on an arbitrary element of $C^\infty(V)$. Let $f\in C^\infty(V)$. Then

$$\begin{aligned} (h_*[X,Y]\cdot f)\circ h &= [X,Y]\cdot(f\circ h) \\ &= X\cdot(Y\cdot(f\circ h)) - Y\cdot(X\cdot(f\circ h)) \\ &= X\cdot((h_*Y\cdot f)\circ h) - Y\cdot((h_*X\cdot f)\circ h) \\ &= (h_*X\cdot(h_*Y\cdot f))\circ h - (h_*Y\cdot(h_*X\cdot f))\circ h \\ &= ([h_*X, h_*Y]\cdot f)\circ h \end{aligned}$$

and the claim is proved.

□

22. Remark We call the attention of the reader to the fact that the only use made above of the assumption that h is a diffeomorphism was that h_* transfers smooth vector fields to well-defined and smooth vector fields. In later chapters, we will encounter situations where a smooth map h is not one-to-one, yet h_*Z is well-defined and smooth for vector fields Z in question. Under such circumstances, the above proof goes through verbatim and the conclusion of the theorem holds.

The following theorem expresses the bracket as a derivative-like concept. The bracket of two vector fields can be viewed as the rate of change of one vector field along the trajectories of the other.

23. Theorem *Let X and Y be smooth vector fields on the open subset U of $\mathbb{R}^n$ and suppose that $a \in U$. If Φ is the flow of X, then*

$$[Y, X](a) = \lim_{t \to 0} \frac{((\Phi_t)_* Y)(a) - Y(a)}{t} \tag{2.42}$$

This is sometimes written as

$$[Y, X] = \lim_{t \to 0} \frac{(\Phi_t)_* Y - Y}{t}$$

which should be viewed as just a short-hand representation of (2.42). The space of smooth vector fields on U is an infinite-dimensional vector space over $\mathbb{R}$ and allows different limit concepts, a subject into which we need not venture.

Proof. We have to show that the limit on the right-hand side exists and that the effect of both sides on an arbitrary $f \in C^\infty(U)$ is the same. Note that

$$\frac{((\Phi_t)_* Y)(a) - Y(a)}{t} \cdot f = df(a) \left(\frac{((\Phi_t)_* Y)(a) - Y(a)}{t} \right) \tag{2.43}$$

It follows that if we prove that the limit of the left-hand side of the above exists for every f as $t \to 0$, then the limit on the right-hand side of (2.42) exists as well, for if we take $f = x^i, i = 1, \ldots, n$, the right-hand side of (2.43) will yield the components of the limit in (2.42). So we examine the left-hand side of (2.43). Using Lemma 14, we can write

$$(f \circ \Phi_t)(a) = (f \circ \Phi^a)(t) = f(a) + tg(t, a)$$

where g is smooth and

$$\begin{aligned} g(0, a) &= \frac{d(f \circ \Phi_t)}{dt}(a)|_{t=0} \\ &= (X \cdot f)(a) \end{aligned}$$

the latter by Theorem 17. Thus we have

$$f \circ \Phi_t = f + t\rho(t)$$

where $\rho(0) = X \cdot f$. Substituting into the left-hand side of 2.43, we obtain

$$\begin{aligned} \frac{((\Phi_t)_* Y)(a) \cdot f - Y(a) \cdot f}{t} &= \frac{Y(\Phi_{-t}(a)) \cdot (f \circ \Phi_t) - Y(a) \cdot f}{t} \\ &= -\frac{Y(\Phi_{-t}(a)) \cdot f - Y(a) \cdot f}{-t} + Y(\Phi_{-t}(a)) \cdot \rho(t) \end{aligned}$$

The fraction on the right-hand side tends to $-(X \cdot (Y \cdot f))(a)$ as $t \to 0$, by Theorem 17, and the second term has limit $(Y \cdot (X \cdot f))(a)$ by the expression for $\rho(0)$ above. This concludes the proof. □

Replacing t by $(-t)$, one also derives

$$[Y, X] = \lim_{t \to 0} \frac{Y - (\Phi_{-t})_*(Y)}{t}$$

and using the anti-symmetry of the bracket,

$$[X, Y] = \lim_{t\to 0} \frac{(\Phi_{-t})_* Y - Y}{t}$$

24. Corollary *Let X and Y be smooth vector fields on U with flows Φ and Ψ, respectively. Then $[X,Y]=0$ if and only if $\Phi_s \circ \Psi_t = \Psi_t \circ \Phi_s$ whenever the two sides of this equation are defined.*

Proof. The commutativity relation of flows implies by (2.27) that $(\Phi_{-s})_* Y = Y$, hence $[X, Y] = 0$ by theorem above. Conversely, suppose that $[X,Y](a)=\mathbf{0}$ for every $a \in U$. We define a curve $\gamma(t)$ in $T_a U$ by

$$\gamma(t) = ((\Phi_t)_* Y)(a)$$

Differentiating with respect to t,

$$\begin{aligned}\frac{d}{dt}\gamma(t) &= \lim_{h\to 0} \frac{((\Phi_{t+h})_* Y)(a) - ((\Phi_t)_* Y)(a)}{h} \\ &= \lim_{h\to 0} \frac{(\Phi_t)_*((\Phi_h)_* Y)(a) - ((\Phi_t)_* Y)(a)}{h} \\ &= (\Phi_t)_* \lim_{h\to 0} \frac{(\Phi_h)_* Y(a) - Y(a)}{h}\end{aligned}$$

Now since the limit on the right-hand side vanishes by the theorem, we conclude that $\gamma(t)$ is constant, $\gamma(t) = \gamma(0) = Y(a)$, so $(\Phi_t)_* Y = Y$ for all t. It follows from (2.42) that the flows commute. □

25. Examples

(a) We saw earlier that the bracket of constant vector fields is zero. We now confirm this fact by looking at the flows. Let $X = \sum_{i=1}^n a^i \frac{\partial}{\partial x^i}$ and $Y = \sum_{j=1}^n b^j \frac{\partial}{\partial x^j}$ on $\mathbb{R}^n$, where the coefficients $a^i, b^j \in \mathbb{R}$. Representing the flows of X and Y, respectively, by Φ and Ψ, we have

$$\Phi_t(x) = x + ta$$
$$\Psi_s(x) = x + sb$$

where $a = (a^1, \ldots, a^n)$, $b = (b^1, \ldots, b^n)$ and $x = (x^1, \ldots, x^n)$. The commutativity relationship between the flows holds for all s and t (see Figure 7).

(b) Consider the following mutually perpendicular unit vector fields in $\mathbb{R}^2 - \{\mathbf{0}\}$.

$$X = \frac{x}{\sqrt{x^2+y^2}} \frac{\partial}{\partial x} + \frac{y}{\sqrt{x^2+y^2}} \frac{\partial}{\partial y}$$
$$Y = \frac{-y}{\sqrt{x^2+y^2}} \frac{\partial}{\partial x} + \frac{x}{\sqrt{x^2+y^2}} \frac{\partial}{\partial y}$$

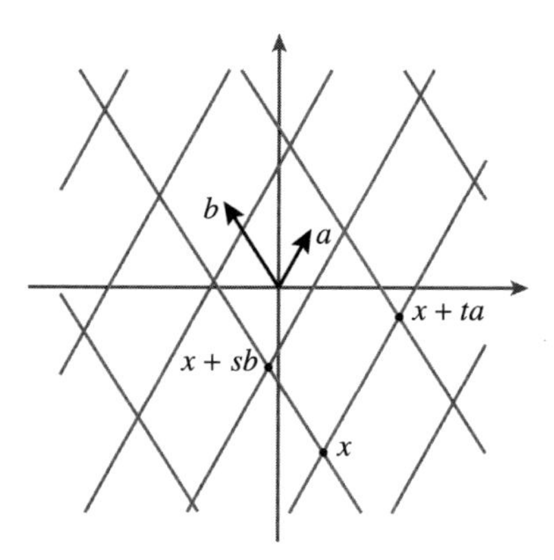

FIGURE 7. Commuting flows

The unit vector field X points in the radial direction away from the origin, and Y indicates counter-clockwise rotation around the origin (see Figure 8). We denote the flows of X and Y, respectively, by Φ and Ψ. Starting from a point P on the circle of radius R around the origin, we obtain $(\Psi_s \circ \Phi_t)(P)$ by first moving P in radial direction away from the origin a distance of t units (note that the velocity vector has length 1), and then rotating the resulting point around the circle of radius $R + t$ by s units of length in the positive direction. This will generally not give the same result as reversing the order of applying Φ_t and Ψ_s. Thus here the bracket $[X, Y]$ does not identically vanish.

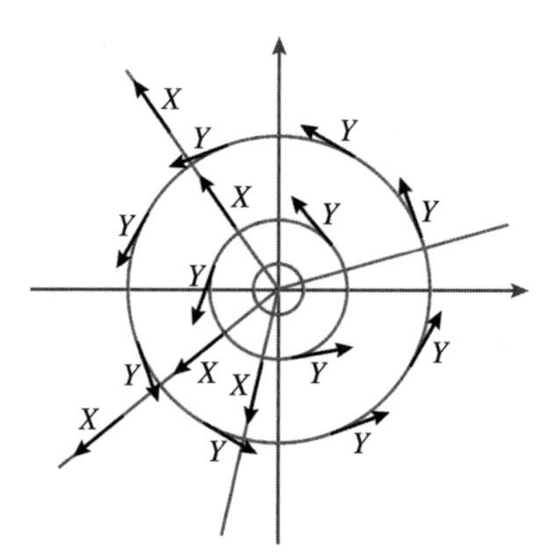

FIGURE 8. Two non-commuting vector fields

(c) By a modification of the length of Y in the above example, we can achieve commutativity of flows. Keeping X as above, define the vector field Z by

$$Z = -y \frac{\partial}{\partial x} + x \frac{\partial}{\partial y}$$

Let us denote the flow of Z by Θ. Representing P in polar coordinates as $P = (R\cos\alpha, R\sin\alpha)$, one checks that

$$\Theta_s(P) = (R\cos(\alpha + s), R\sin(\alpha + s))$$

which commutes with the flow of X.

As the last topic in this chapter, we will consider an application of the Lie bracket to the problem of local coordinate change. Let U and V be open subsets of $\mathbb{R}^n$ and suppose that $h{:}U\to V$ is a C^r diffeomorphism, $r\geq 1$. We can use two sets of coordinates to describe the points of V. First, there are the standard coordinates for a subset of $\mathbb{R}^n$, wherein the points of V are represented as n-tuples $x=(x^1,\ldots,x^n)$. Then, using the diffeomorphism h, we assign to each point x of V, the n-tuple $y=(y^1,\ldots,y^n)$, where $y=h^{-1}(x)$. In other words, $y^i=x^i\circ h^{-1}$, in which x^i denotes projection on the ith component. We call h, or h^{-1}, a C^r ***change of coordinates***. h^{-1} is a diffeomorphism, hence the $dy^i(a)=(dx^i\circ Th^{-1})(a)$, $i = 1,\ldots,n$, form a basis for the dual of T_aV. We denote the dual basis in T_aV by $(\frac{\partial}{\partial y^1}(a),\ldots,\frac{\partial}{\partial y^n}(a))$, and we claim that

$$\frac{\partial}{\partial y^i} = h_*(\frac{\partial}{\partial x^i}) \tag{2.44}$$

Using Lemma 20, we show that this is indeed the dual basis relative to $(dy^1,...,dy^n)$:

$$\begin{aligned}(dy^i(a))(h_*(\frac{\partial}{\partial x^j})(a)) &= h_*(\frac{\partial}{\partial x^j}(a)\cdot y^i\\ &= \frac{\partial}{\partial x^j}(h^{-1}(a))\cdot(y^i\circ h)\\ &= \frac{\partial}{\partial x^j}(h^{-1}(a))\cdot x^i\\ &= \delta^i_j\end{aligned}$$

proving (2.44).

We now pose a basic question. Suppose $X_1,\ldots,X_k$, $k\leq n$, are smooth vector fields on an open subset V of $\mathbb{R}^n$ so that at every point x of V, the set $\{X_1(x),\ldots,X_k(x)\}$ is linearly independent. Is there a smooth change of coordinates $h{:}U\to V$ such that $X_i=h_*(\frac{\partial}{\partial x^i})$, $i=1,\ldots,k$? In view of $[\frac{\partial}{\partial x^i},\frac{\partial}{\partial x^j}]=0$ and $h_*[X,Y]=[h_*X,h_*Y]$, one obtains the necessary condition $[X_i,X_j]=0$ for all $i,j\in\{1,\ldots,k\}$. It turns out that this condition is also locally sufficient.

26. Theorem (Straightening Out) *Let W be an open subset of $\mathbb{R}^n$ and suppose that $X_1,\ldots,X_k$, $k\leq n$, are smooth vector fields on W that are linearly independent at every point of W. If further, $[X_i,X_j]=0$ for all $i,j=1,\ldots,k$, then for every $a\in W$, there are open sets, U around $\mathbf{0}\in\mathbb{R}^n$ and V around a, and a smooth diffeomorphism $h{:}U\to V$ so that $h(\mathbf{0})=a$, and $h_*(\frac{\partial}{\partial x^i})=X_i$ for $i = 1,\ldots,k$.*

Note that for $k=1$, the bracket condition is automatically satisfied, therefore the theorem implies that if $X_1(a)\neq\mathbf{0}$, then a local diffeomorphism h from a neighborhood of a to a neighborhood of $\mathbf{0}\in\mathbb{R}^n$ exists with $h_*(\frac{\partial}{\partial x^1})=X_1$.

Proof. By a translation in $\mathbb{R}^n$, we may assume that $a=\mathbf{0}\in\mathbb{R}^n$. Further, using an invertible linear map of $\mathbb{R}^n$ (a smooth diffeomorphism), we may assume that $X_i(\mathbf{0}) = \frac{\partial}{\partial x^i}(\mathbf{0})$ for $i=1,\dots,k$. We define a mapping Ψ with domain an open neighborhood of $\mathbf{0}\in\mathbb{R}^n$ as follows:

$$\Psi(x^1,\dots,x^n) = (\Phi^1_{x^1}\circ\cdots\circ\Phi^k_{x^k})(0,\dots,0,x^{k+1},\dots,x^n) \tag{2.45}$$

where Φ^i is the flow for X_i, $i=1,\dots,k$. For $x^1,\dots,x^k$ sufficiently small, $\Phi^1_{x^1}\circ\cdots\circ\Phi^k_{x^k}$ is defined, and if, in addition, $x^{k+1},\dots,x^n$ are small enough, the right-hand side of (2.45) can be defined. Thus we can take an open neighborhood of $\mathbf{0}$ for the domain of Ψ. The map Ψ is smooth since it is the composition of a subspace projection followed by flows of smooth vector fields. We check that $D\Psi(0)=\mathbb{1}_{\mathbb{R}^n}$ by examining the matrix in the standard basis. If $j\in\{1,\dots,k\}$, then column j is the partial derivative of Ψ with respect to x^j keeping the other variables fixed. So this is just $\frac{\partial}{\partial x^j}(\mathbf{0})$ in view of the initial assumption $X_j(\mathbf{0})=\frac{\partial}{\partial x^j}(\mathbf{0})$. On the other hand, letting $x^1=\cdots=x^k=0$, $\Phi^1_0\circ\cdots\circ\Phi^k_0$ will be the identity map, so the claim is established. The inverse function theorem (Appendix II) guarantees the existence of open sets U and V around $\mathbf{0}\in\mathbb{R}^n$ so that Ψ maps U onto V, and the inverse $h\colon V\to U$ of $\Psi|_U$ is also smooth. We wish to show $h_*X_i=\frac{\partial}{\partial x^i}$ for $i=1,\dots,k$. For this purpose, it suffices to show that the flows of $\frac{\partial}{\partial x^i}$ and X_i correspond under Ψ. Denoting the flow of $\frac{\partial}{\partial x^i}$ by Θ^i, we have

$$\begin{aligned}(\Psi\circ\Theta^i_t)(x^1,\dots,x^n) &= \Psi(x^1,\dots,x^{i-1},x^i+t,x^{i+1},\dots,x^n)\\ &= \Phi^1_{x^1}\circ\cdots\circ\Phi^i_{x^i}\circ\Phi^i_t\circ\cdots\circ\Phi^k_{x^k}(0,\dots,0,x^{k+1},\dots,x^n)\end{aligned}$$

Now using the assumption $[X_i,X_j]=0$, $i,j=1,\dots,k$, the flows Φ^i and Φ^j commute, so we can move Φ^i_t all the way to the left to obtain

$$\begin{aligned}(\Psi\circ\Theta^i_t)(x^1,\dots,x^n) &= \Phi^i_t\circ\Phi^1_{x^1}\circ\cdots\circ\Phi^k_{x^k}(0,\dots,0,x^{k+1},\dots,x^n)\\ &= \Phi^i_t\circ\Psi(x^1,\dots,x^n)\end{aligned}$$

proving the correspondence of flows. □

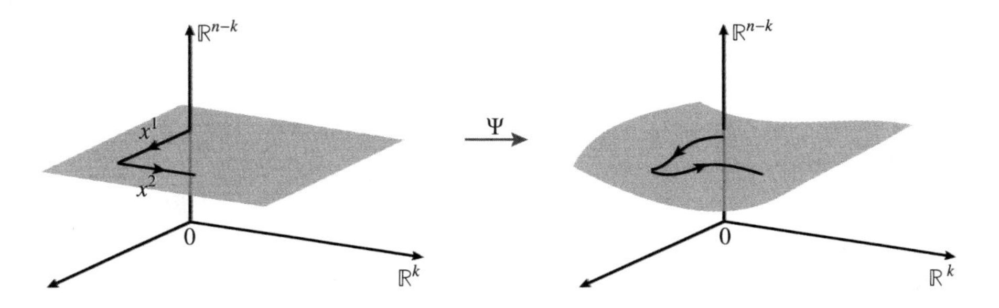

Figure 9. Straightening out

EXERCISES

2.1 Show that the function α defined in Lemma 12 is C^∞ and its derivatives of all orders at 0 are zero.

2.2 Consider the time-dependent differential equation $\frac{dx}{dt}=t$ on $\mathbb{R}$. Show that the group property $\Phi_{s+t}=\Phi_s\circ\Phi_t$ does not hold here (see the note at the end of Subsection 11). Look at the geometry of solution curves in (t,x) plane.

2.3 Let X bc a C^1 vector field on $\mathbb{R}^n$ and suppose that $f:\mathbb{R}^n\to\mathbb{R}$ is a C^1 function. If f is everywhere positive (respectively negative), then fX is called an ***orientation-preserving*** (respectively, ***orientation-reversing***) ***reparametrization*** of X.
(a) If fX is a reparametrization of X, show that the solution curves of fX are reparametrizations of solution curves of X.
(b) A function $\phi:\mathbb{R}^n\to\mathbb{R}$ is called ***proper*** if for every compact subset K of $\mathbb{R}$, the inverse image $\phi^{-1}(K)$ is compact. Suppose that ϕ is proper and $M>0$ exists so that for every integral curve $\gamma:I\to\mathbb{R}$ of X, one has $|(\phi\circ\gamma)'(t)|\leq M$ for all $t\in\mathbb{R}$, then X is complete.
(c) Let X be a C^1 vector field on $\mathbb{R}^n$. Show that an everywhere positive C^1 function $f:\mathbb{R}^n\to\mathbb{R}$ exists so that fX is complete.

2.4 Let (r,θ) be polar coordinates in $\mathbb{R}^2-\{\mathbf{0}\}$. What is the relation of $\frac{\partial}{\partial r}$ and $\frac{\partial}{\partial\theta}$ to the vector fields in Example (b) of Subsection 25? Let (ρ,φ,θ) be spherical coordinates in $\mathbb{R}^3-\{z\text{-axis}\}$, i.e., $x=\rho\sin\varphi\cos\theta$, $y=\rho\sin\varphi\sin\theta$, and $z=\rho\cos\varphi$. Define $\{\frac{\partial}{\partial\rho},\frac{\partial}{\partial\varphi},\frac{\partial}{\partial\theta}\}$ as dual to $\{d\rho,d\varphi,d\theta\}$ and compute them in terms of $\{\frac{\partial}{\partial x},\frac{\partial}{\partial y},\frac{\partial}{\partial z}\}$.

2.5 Let U be an open subset of $\mathbb{R}^n$, and C^1 functions $f_1,\ldots,f_n:U\to\mathbb{R}$ be so that $\{df_1(x),\ldots,df_n(x)\}$ is linearly independent at every $x\in U$. Define $\frac{\partial}{\partial f_i}$ and show that $(f_1,\ldots,f_n)$ represents a local coordinate change.

2.6 Let $f,g:\mathbb{R}\to\mathbb{R}$ be smooth functions, and consider vector fields $X=f(y)\frac{\partial}{\partial x}$ and $Y=g(x)\frac{\partial}{\partial y}$ on $\mathbb{R}^2$. Find necessary and sufficient conditions on f and g so that there exists a smooth diffeomorphism $h:\mathbb{R}^2\to\mathbb{R}^2$ with $h_*(\frac{\partial}{\partial x})=X$ and $h_*(\frac{\partial}{\partial y})=Y$.

2.7 Let $D:C^\infty(\mathbb{R})\to C^\infty(\mathbb{R})$ be a derivation. Show that there exists $\varphi\in C^\infty(\mathbb{R})$ so that $D(f)=\varphi\cdot f'$, where f' denotes the derivative of f.

2.8 Let $\mathbf{M}$ and $\mathbf{N}$ be $n\times n$ real matrices and consider the linear vector fields $X_{\mathbf{M}}$ and $X_{\mathbf{N}}$ on $\mathbb{R}^n$ associated respectively to $\mathbf{M}$ and $\mathbf{N}$ (see Example (a) of Subsection 4). Show that $[X_{\mathbf{M}},X_{\mathbf{N}}]$ is a linear vector field and compute its associated matrix.

2.9 Let C be a constant vector field on $\mathbb{R}^2$ and $Z=-y\frac{\partial}{\partial x}+x\frac{\partial}{\partial y}$. Show that $[C,Z]$ is a constant vector field on $\mathbb{R}^2$ obtained by rotating the vector field C by $\frac{\pi}{2}$ in the positive direction.

2.10 Let X and Y be smooth vector fields in $\mathbb{R}^3$ and $X\times Y$ be the cross-product of X and Y. Prove the following formula in which **curl** and **div** are the usual concepts of "curl" and "divergence" in $\mathbb{R}^3$:

$$[X, Y] = (\mathbf{div}Y)X - (\mathbf{div}X)Y - \mathbf{curl}(X\times Y)$$

2.11 Let X be a smooth vector field on $\mathbb{R}^n$.
(a) If $[X, \frac{\partial}{\partial x^i}]=\mathbf{0}$ for $i=1,\ldots,n$, show that X is a constant vector field.
(b) If $[X, \frac{\partial}{\partial x^i}]=\mathbf{0}$ for $i=1,\ldots,n-1$, show that X has no non-zero periodic solutions.

2.12 Let X be a smooth vector field on an open subset U of $\mathbb{R}^n$ and $f{:}U\to\mathbb{R}$ be a smooth function.
(a) If for $a\in U$, $(X\cdot f)(a)>0$, show that f is *locally increasing* at a, in the sense that if $\Phi^a{:}I\to U$ is the trajectory of X with $\Phi^a(0)=a$, then $f\circ\Phi^a$ is an increasing function on some interval $]-\varepsilon,+\varepsilon[\subset U$. A similar result holds for $(X\cdot f)(a)<0$.
(b) Suppose that $X\cdot f>0$ throughout U. Show that X has no periodic solutions in U.
(c) Consider the *gradient vector field* ∇f defined as $\nabla f(x)=\sum_{i=1}^n \frac{\partial f}{\partial x^i}\frac{\partial}{\partial x^i}$ on U. Show that ∇f has no non-zero periodic solutions.

2.13 We denote the points of $\mathbb{R}^{2n}$ by $(q^1,\ldots,q^n;p^1,\ldots,p^n)$, and consider a smooth function $H{:}\mathbb{R}^{2n}\to\mathbb{R}$. The *Hamiltonian vector field* associated to H is defined as

$$X_H = \sum_{i=1}^n \frac{\partial H}{\partial p^i}\frac{\partial}{\partial q^i} - \sum_{i=1}^n \frac{\partial H}{\partial q^i}\frac{\partial}{\partial p^i}$$

(a) Show that H is constant along the trajectories of X_H ("conservation of energy").
(b) For smooth functions H and K, show that H is constant along the trajectories of X_K if and only if K is constant along the trajectories of X_H.
(c) Show that if K is constant along the trajectories of X_H, then $[X_H, X_K]=0$.

2.14 Let $\alpha{:}\mathbb{R}\to\mathbb{R}$ be the auxiliary function constructed in Lemma 12, and define f by $f(t) = (\sin t)\alpha(t)$. Consider the following vector field on $\mathbb{R}^2$.

$$X(x,y) = (xf(r))\frac{\partial}{\partial x} + (yf(r))\frac{\partial}{\partial y}$$

where $r = \sqrt{x^2+y^2}$. Describe the trajectories of X and show that the vector field is complete.

2.15 Consider the following vector field in $\mathbb{R}^2$:

$$X = \exp(xy^2)\frac{\partial}{\partial x}$$

Describe the maximal integral curves γ with initial condition $\gamma(0)=(0,c)$. Show that there is a single such maximal integral curve with domain the entire real line $\mathbb{R}$.

2.16 Show that the following vector field on $\mathbb{R}^2$ is complete.

$$((y^2+1)\sin x)\frac{\partial}{\partial x} + ((x^2+1)\sin y)\frac{\partial}{\partial y}$$

2.17 Suppose $f\in C^\infty(\mathbb{R}^2)$ is bounded, and $X(x,y)=\frac{\partial}{\partial x}+f(x,y)\frac{\partial}{\partial y}$. Show that there is a C^∞ diffeomorphism of $\mathbb{R}^2$ onto itself so that $h_*X=\frac{\partial}{\partial x}$.

2.18 (Bendixson's Criterion) For a C^1 vector field $X=P\frac{\partial}{\partial x}+Q\frac{\partial}{\partial y}$ in the plane, we define the *divergence* of X by $\mathrm{div}(X)=\frac{\partial P}{\partial x}+\frac{\partial Q}{\partial y}$. Suppose that $\mathrm{div}(X)\geq 0$ throughout the plane but is not identically zero on any open subset of the plane. Show that X has no non-zero closed orbits.

2.19 Consider the vector field in the plane given as

$$f(x)\frac{\partial}{\partial x}+g(x,y)\frac{\partial}{\partial y}$$

where f and g are smooth functions. Show that this vector field has no non-zero closed orbits.

2.20 A C^2 function $f:\mathbb{R}^2\to\mathbb{R}$ is called *harmonic* if $\mathrm{div}(\nabla f)=0$. Let X be a smooth vector field in the plane with a non-zero closed orbit γ. If f is harmonic on $\mathbb{R}^2$, show that ∇f is tangent to the image of γ in at least two points.

2.21 Let X and Y be smooth vector fields on an open subset U of $\mathbb{R}^n$ such that $[X,Y]=0$. If Φ, Ψ and Θ are the flows, respectively, of X, Y and $X+Y$, show that $\Theta_t=\Psi_t\circ\Phi_t$.

2.22 Let X and Y be smooth vector fields on an open subset U of $\mathbb{R}^2$, which are linearly independent throughout U. Show that for every $a\in U$, there are smooth nowhere zero real-valued functions f and g defined in a neighborhood $V\subset U$ of a so that $[fX,gY]=0$.

2.23 Let U be an open subset of $\mathbb{R}^m$ and

$$\frac{dx}{dt}=A(t)\cdot x$$

a time-dependent linear system of differential equations on U, i.e., $A(t)$ is an $m\times m$ matrix. If $A(t)$ is defined and continuous for t in an interval $[a,b]$, show that the solutions of the system are also defined on $[a,b]$. Deduce that if $A(t)$ is defined and continuous for all $t\in\mathbb{R}$, then the solutions are defined for all $t\in\mathbb{R}$.

2.24 (Continuation of Exercise 2.23) Let $U=\mathbb{R}^m$ and suppose that $A(t)$ is defined and continuous for t in some interval I.
(a) Show that the set of solutions $I\to U$ of the system form a real vector space of dimension m.
(b) For $t_1,t_2\in I$, transition maps $\Psi^{t_1}_{t_2}:\mathbb{R}^m\to\mathbb{R}^m$ are defined as follows. Let $x\in\mathbb{R}^m$ and φ be the solution that $\varphi(t_1)=x$, then

$$\Psi^{t_1}_{t_2}(x)=\varphi(t_2)$$

Show that $\Psi^{t_1}_{t_2}$ is linear.

2.25 Let X be a C^2 vector field on an open subset U of $\mathbb{R}^n$ with associated flow Φ and a zero at $a \in U$.
(a) Given $r>0$, show there is an open subset V of a on which Φ_t is defined for all $t \in]-r, +r[$.
(b) Since $\Phi_t(a)=a$ for all t, the tangent maps $\eta(t)=T_a\Phi_t$ define a C^1 curve in the space of linear maps of T_aU to itself. We denote the derivative $\frac{d\eta}{dt}(0)$ by $X'(a)$ and call it the *Hessian* of X at a. Suppose $X=\sum_{i=1}^n u^i \frac{\partial}{\partial x^i}$ with $u^i(a)=0$ for $i=1,\ldots,n$. Show that $X'(a)$ is given by the matrix $[\frac{\partial u^i}{\partial x^j}]$.
(c) If X is a gradient vector field (see Exercise 2.12) with $X(a)=0$, show that $X'(a)$ is symmetric.
(d) If X is a Hamiltonian vector field (see Exercise 2.13) with $X(a)=0$, show that the trace of $X'(a)$ is zero. (Hint: Show that if λ is an eigenvalue of multiplicity m, then so is $-\lambda$.)

CHAPTER 3

Tensor Fields: Local Theory

In Chapter 2, we studied vector fields on an open subset U of $\mathbb{R}^n$. We proceeded by first installing a vector space T_xU at each point x of U, which we called the tangent space to U at x, and then defined a vector field on U to be a function that assigns to each x in U an element of T_xU. This can be viewed as defining a *field* of $(0, 1)$-tensors on U. By constructing the various tensor spaces of Chapter 1 on each T_xU, we proceed in the present chapter to define general tensor fields of type (p, q) on U. Special attention will be focused on anti-symmetric covariant tensor fields, known as *differential forms*, which will play an important role in much of the book.

A. Basic Constructions

Given a vector space V over $\mathbb{R}$, one can construct the vector space L^p_qV and its variants over $\mathbb{R}$ as in Chapter 1. Performing these constructions for each $V{=}T_aU$, where U is an open subset of $\mathbb{R}^n$ and $a{\in}U$, one obtains vector spaces such as $(T_aU)^*$, $L^p_q(T_aU)$, and $\Lambda^p(T_aU)$. Using the generic notation $F(V)$ to denote any of these spaces constructed from the given vector space V, we note that the natural isomorphism of $T_aU{=}\{a\}{\times}\mathbb{R}^n{\cong}\mathbb{R}^n$ given by $(a, v){\mapsto}v$ allows us to make the following natural identification throughout the discussion:

$$F(T_aU) \cong \{a\} \times F(\mathbb{R}^n) \tag{3.1}$$

where all the vector space operations are carried out in the second component. We will use the following notation:

$$T^*_aU \;\text{ for }\; (T_aU)^* \tag{3.2}$$

$$(L^p_q)_aU \;\text{ for }\; L^p_q(T_aU) \tag{3.3}$$

$$(\Lambda^p)_aU \;\text{ for }\; \Lambda^p(T_aU) \tag{3.4}$$

Adopting the Chapter 2 notation of

$$\frac{\partial}{\partial x_j}(a) = (a, e_j)$$

$$dx^i(a) = (a, e^i)$$

we obtain the following bases:

$$\text{For } (L^p_q)_aU:\; dx^{j_1}(a) \otimes \cdots \otimes dx^{j_p}(a) \otimes \frac{\partial}{\partial x^{i_1}}(a) \otimes \cdots \otimes \frac{\partial}{\partial x^{i_q}}(a), \quad 1{\le}i_\mu, j_\nu{\le}n \tag{3.5}$$

$$\text{For } (\Lambda^p)_aU:\; dx^{j_1}(a) \wedge \cdots \wedge dx^{j_p}(a), \quad 1{\le}j_1{<}\cdots{<}j_p{\le}n \tag{3.6}$$

Just as in Chapter 2, we form the disjoint union of the constructed vector spaces at the points of U to form a "bundle" of same type constructs:

$$T^*U = \bigcup_{x\in U} T_x^*U \cong U \times (\mathbb{R}^n)^* \quad (\textbf{\textit{Cotangent Bundle}} \text{ of U}) \tag{3.7}$$

$$L_q^pU = \bigcup_{x\in U} (L_q^p)_x(U) \cong U \times L_q^p(\mathbb{R}^n) \quad (\textbf{\textit{(p,q)-Tensor Bundle}} \text{ of } U) \tag{3.8}$$

$$\Lambda^pU = \bigcup_{x\in U} (\Lambda^p)_xU \cong U \times \Lambda^p(\mathbb{R}^n) \quad (\textbf{\textit{pth Exterior Bundle of U}}) \tag{3.9}$$

Each of these spaces is endowed with the product topology or, equivalently, the topology of an open subset of $\mathbb{R}^n \times F(\mathbb{R}^n)$, which is a finite-dimensional real vector space.

A ***(p,q)-tensor field*** on U is a map $\alpha : U \to L_q^pU$ such that $\alpha(x) \in (L_q^p)_x(U)$ for each $x \in U$. A field is also known as a ***cross-section*** of the relevant bundle. We can then write

$$\alpha(x) = (x, \sum_{i_\mu, j_\nu} \alpha_{i_1\cdots i_p}^{j_1\cdots j_q}(x) e^{i_1} \otimes \cdots \otimes e^{i_p} \otimes e_{j_1} \otimes \cdots \otimes e_{j_q}) \tag{3.10}$$

or, equivalently,

$$\alpha = \sum_{i_\mu, j_\nu} \alpha_{i_1\cdots i_p}^{j_1\cdots j_q} dx^{i_1} \otimes \cdots \otimes dx^{i_p} \otimes \frac{\partial}{\partial x^{j_1}} \otimes \cdots \otimes \frac{\partial}{\partial x^{j_q}} \tag{3.11}$$

where the coefficients $\alpha_{i_1\cdots i_p}^{j_1\cdots j_q}$ are functions from U to $\mathbb{R}$. Since the first component of the right-hand side of (3.10) is smooth, being the value of the identity function, properties such as "smooth", "C^r" or "continuous" are completely determined by the coefficient functions $\alpha_{i_1\cdots i_p}^{j_1\cdots j_q}$. A cross-section of Λ^pU is called a ***differential p-form*** , or simply a ***p-form***. A differential p-form α can be represented as

$$\alpha(x) = (x, \sum_{i_1<\cdots<i_p} \alpha_{i_1\cdots i_p}(x) e^{i_1} \wedge \cdots \wedge e^{i_p}) \tag{3.12}$$

or, equivalently,

$$\alpha = \sum_{i_1<\cdots<i_p} \alpha_{i_1\cdots i_p} dx^{i_1} \wedge \cdots \wedge dx^{i_p} \tag{3.13}$$

where the $\alpha_{i_1\cdots i_p}$ are functions from U to $\mathbb{R}$. Just as above, a differential p-form α is smooth, C^r or continuous, if and only if all the coefficients $\alpha_{i_1\cdots i_p}$ are. As the simplest examples, we describe differential 0- and 1-forms. A 0-form, being a cross-section of $\Lambda^0U = U \times \mathbb{R}$, is in the form $x \mapsto (x, f(x))$, and can therefore be identified with a real-valued function f on U. A one-form has the representation

$$\alpha = \sum_{i=1}^{n} \alpha_i dx^i$$

where the α_i are real-valued functions. An example is the differential of a real-valued function f, introduced in Chapter 2, namely

$$df = \sum_{i=1}^{n} \frac{\partial f}{\partial x^i} dx^i$$

We noted in Chapter 2 that the set of smooth vector fields over U was a module over the algebra $C^\infty(U)$ of smooth real-valued functions on U. The same can be said of the set of smooth cross-sections of any of the bundles above. The module structure over $C^\infty(U)$ results from the C^∞ nature of coefficient function and the $\mathbb{R}$-vector space structure at each point of U. We are especially interested in the space of smooth differential p-forms, which we shall henceforth denote by $\Omega^p U$. The module of smooth (p, q)-tensor fields over U will be denoted by $\mathcal{T}^p_q U$. From (3.5) and (3.6), we see that bases for these modules over $C^\infty(U)$ are provided by

(3.14) For $\mathcal{T}^p_q U$: $dx^{j_1} \otimes \cdots \otimes dx^{j_p} \otimes \dfrac{\partial}{\partial x^{i_1}} \otimes \cdots \otimes \dfrac{\partial}{\partial x^{i_q}}, \quad 1 \le i_\mu, j_\nu \le n$

(3.15) For $\Omega^p U$: $dx^{j_1} \wedge \cdots \wedge dx^{j_p}, \quad 1 \le j_1 < \cdots < j_p \le n, \ p \ge 1$

There is an alternative to the definition of modules such as $\mathcal{T}^p_q U$ and $\Omega^p U$, which is equivalent to the one in terms of cross-sections. We shall describe this alternative in the case of $\Omega^p U$, the other cases being completely similar. For each $a \in U$, and each natural number p, we define $(T_a U)^p$ to be the p-fold direct product $T_a U \times \cdots \times T_a U$, which is a vector space over $\mathbb{R}$ of dimension pn. For $p=0$, $(T_a U)^0$ is defined to be $\{a\} \times \{\mathbf{0}\}$. As usual, all vector space operations take place in the second component, and natural isomorphisms are obtained as

$$(T_a U)^p \cong \{a\} \times (\mathbb{R}^n)^p \cong \mathbb{R}^{np}$$

We proceed as before to take the "bundle," or the disjoint union

$$\bigcup_{a \in U} (T_a U)^p \cong U \times \mathbb{R}^{np}$$

This will be denoted by $(TU)^p$ and will be given the product topology or, equivalently, the topology of an open subset of $\mathbb{R}^N$, where $N=n+np$. Thus it makes sense to speak of smooth functions $(TU)^p \to \mathbb{R}$. Now let $\hat{\Omega}^p U$ be the set of smooth functions $(TU)^p \to \mathbb{R}$, the restriction of which to each $(T_a U)^p$ is anti-symmetric and covariant, i.e., each such restriction is an anti-symmetric covariant p-tensor on $T_a U$. For $p=0$, $\hat{\Omega}^0 U$ is naturally identified with $C^\infty(U)$. For $\alpha, \beta \in \hat{\Omega}^p U$ and $f \in C^\infty(U)$, defining $\alpha+\beta$ and $f\alpha$ by their restrictions for fixed a as

$$(\alpha + \beta)|_{(T_a U)^p} = \alpha|_{(T_a U)^p} + \beta|_{(T_a U)^p}$$
$$(f\alpha)|_{(T_a U)^p} = f(a)(\alpha|_{(T_a U)^p})$$

and noting that the sum and product of smooth functions is smooth, we see that $\hat{\Omega}^p U$ becomes a module over $C^\infty(U)$. A natural isomorphism of $C^\infty(U)$ modules $\Omega^p U$ and

$\hat{\Omega}^p U$ will now be described. For $\alpha \in \Omega^p U$ as in (3.12) or (3.13), we define $\hat{\alpha}$ in the following way. If $w_1, \ldots, w_p \in T_x U$, then

$$\hat{\alpha}(w_1, \ldots, w_p) = \sum_{i_1 < \cdots < i_p} \alpha_{i_1 \cdots i_p}(x)(dx^{i_1} \wedge \cdots \wedge dx^{i_p}(w_1, \ldots, w_p))$$

Definitions ensure that $\hat{\alpha} \in \hat{\Omega}^p U$. Conversely, suppose $\beta \in \hat{\Omega}^p U$ is given, we wish to define $\check{\beta} \in \Omega^p U$, which provides an inverse to $\alpha \mapsto \hat{\alpha}$. Let $x \in U$, we define $\check{\beta}(x)$ by giving its value on a p-tuple $(w_1, \ldots, w_p)$, where each w_i is an element of $T_x U$:

$$(\check{\beta}(x))(w_1, \ldots, w_p) = \beta(w_1, \ldots, w_p)$$

Again it follows from the definitions that $\hat{\beta} \in \Omega^p(U)$, and that the two operations $\alpha \mapsto \hat{\alpha}$ and $\beta \mapsto \check{\beta}$ are inverses. We will use either definition of differential form, as convenient, dispensing with the symbols $\hat{}$ and $\check{}$ altogether. In both cases, the functional coefficients $\alpha_{i_1 \cdots i_p}$ determine the form.

B. Pointwise Operations

In Chapter 1, we introduced several operations such as the tensor product, the wedge product and the contraction. Given an open subset U of $\mathbb{R}^n$ and a point $x \in U$, these operations automatically extend to fields of the same category by performing the operations in vector spaces installed at each x in U. For smooth fields, $\mathbb{R}$-linearity at individual points will translate into $C^\infty(U)$-linearity for the induced field operations. We will deal exclusively with smooth differential forms in this section, but the methods employed have appropriate extensions to other contexts.

If α is a cross-section of $\Lambda^p U$, $p \geq 1$, and $X_1, \ldots, X_p$ are vector fields on U, then the real-valued function $\alpha(X_1, \ldots, X_p)$ on U is defined by

$$(\alpha(X_1, \ldots, X_p))(x) = \alpha(x)(X_1(x), \ldots, X_p(x)), \quad x \in U \tag{3.16}$$

The following preliminary lemma will be of great utility throughout.

1. Lemma *Let α be a cross-section of $\Lambda^p U$, $p \geq 1$. Then α is smooth if and only if for all smooth vector fields $X_1, \ldots, X_p$, the function $\alpha(X_1, \ldots, X_p)$ is smooth.*

Proof. Writing α in the form (3.13), α is smooth if and only if all functional coefficients $\alpha_{i_1 \cdots i_p}$ are smooth. First suppose α is smooth, and vector fields $X_1, \ldots, X_p$ are also smooth. Writing X_i as $X_i = \sum_{j=1}^n X_i^j \frac{\partial}{\partial x^j}$, we know from Chapter 2 that the functions X_i^j are smooth. Now

$$\begin{aligned}\alpha(X_1,\ldots,X_p) &= \sum_{i_1<\cdots<i_p} \alpha_{i_1\cdots i_p} dx^{i_1}\wedge\cdots\wedge dx^{i_p}(X_1,\ldots,X_p)\\ &= \sum_{i_1<\cdots<i_p} \alpha_{i_1\cdots i_p} \det\left[dx^{i_\mu}(X_\nu)\right]_{\mu,\nu}\\ &= \sum_{i_1<\cdots<i_p} \alpha_{i_1\cdots i_p} \det\left[X_\nu^{i_\mu}\right]_{\mu,\nu}\end{aligned}$$

The right-hand side is a combination under addition and multiplication of smooth functions, and is hence smooth.

Conversely, suppose that for any p smooth vector fields $X_1,\ldots,X_p$, the function $\alpha(X_1,\ldots,X_p)$ is smooth. In particular, letting $X_k=\frac{\partial}{\partial x^{j_k}}$, for $k = 1,\ldots,p$, where $j_1<\cdots<j_p$, we obtain

$$\begin{aligned}\alpha\left(\frac{\partial}{\partial x^{j_1}},\ldots,\frac{\partial}{\partial x^{j_p}}\right) &= \sum_{i_1<\cdots<i_p} \alpha_{i_1\cdots i_p} dx^{i_1}\wedge\cdots\wedge dx^{i_p}\left(\frac{\partial}{\partial x^{j_1}},\ldots,\frac{\partial}{\partial x^{j_p}}\right)\\ &= \alpha_{j_1\cdots j_p}\end{aligned}$$

which shows that the coefficients of α are smooth. □

We now extend the operations of wedge product and interior product, which were discussed in Section E of Chapter 1, to differential forms. As usual, we work on an open subset U of $\mathbb{R}^n$. Let $\alpha\in\Omega^p U$, $\beta\in\Omega^q U$ and $X\in\mathfrak{X}(U)$. We define $\alpha\wedge\beta$ and $i_X\alpha$ as follows.

$$(\alpha\wedge\beta)(x) = \alpha(x)\wedge\beta(x) \tag{3.17}$$

$$(i_X\alpha)(x) = i_{X(x)}\alpha(x) \tag{3.18}$$

Thus $\alpha\wedge\beta$ and $i_X\alpha$ are, respectively, cross-sections of the bundles $\Lambda^{p+q}U$ and $\Lambda^{p-1}U$. We must show they are smooth, i.e., $\alpha\wedge\beta\in\Omega^{p+q}(U)$ and $i_X\alpha\in\Omega^{p-1}U$. For $\alpha\wedge\beta$, let $X_1,\ldots,X_{p+q}$ be smooth vector fields on U. Then using (3.16), (3.17) and the definition of the wedge product, we have

$$\begin{aligned}((\alpha\wedge\beta)(X_1,\ldots,X_{p+q}))(x) &= (\alpha(x)\wedge\beta(x))(X_1(x),\ldots,X_{p+q}(x))\\ &= \frac{1}{p!q!}\sum_{\sigma\in S_{p+q}} \varepsilon(\sigma)\alpha(x)(X_{\sigma(1)}(x),\ldots,X_{\sigma(p)}(x))\beta(x)(X_{\sigma(p+1)}(x),\ldots,X_{\sigma(p+q)}(x))\\ &= \frac{1}{p!q!}\sum_{\sigma\in S_{p+q}} \varepsilon(\sigma)(\alpha(X_{\sigma(1)},\ldots,X_{\sigma(p)}))(x)(\beta(X_{\sigma(p+1)},\ldots,X_{\sigma(p+q)}))(x)\end{aligned}$$

It follows from the smoothness of α, β, the X_i and the lemma that each summand is smooth, hence the sum is smooth, and the smoothness of $\alpha\wedge\beta$ follows from the lemma.

Likewise for $i_X\alpha$, assuming first that $p\geq 2$, let $X_2,\ldots,X_p$ be smooth vector fields on U, then

$$\begin{aligned}(i_X\alpha(X_2,\ldots,X_p))(x) &= ((i_{X(x)}\alpha(x))(X_2(x),\ldots,X_p(x))\\ &= \alpha(x)(X(x),X_2(x),\ldots,X_p(x))\\ &= (\alpha(X,X_2,\ldots,X_p))(x)\end{aligned}$$

The smoothness of $\alpha, X, X_2, \ldots, X_p$ together with the lemma prove that $i_X\alpha$ is smooth. If p=1, one has $i_X\alpha$=$\alpha(X)$, and again the lemma proves the smoothness of the real-valued function $\alpha(X)$. For p=0, $i_X\alpha$=0 by convention.
The following is now an immediate consequence of corresponding properties in Section E of Chapter 1.

2. Elementary Properties of Wedge Product and Interior Product
(a) *Wedge product* $\Omega^p\times\Omega^q\to\Omega^{p+q}$ *is* $C^\infty(U)$*-bilinear.*
(b) *For* $\alpha\in\Omega^p U$ and $\beta\in\Omega^q$, $\alpha\wedge\beta=(-1)^{pq}\beta\wedge\alpha$.
(c) $(\alpha\wedge\beta)\wedge\gamma=\alpha\wedge(\beta\wedge\gamma)$.
(d) $i_X\alpha$ *is* $C^\infty(U)$*-linear with respect to each of* X *and* α.
(e) $i_X\circ i_Y=-i_Y\circ i_X$.
(f) *For* $\alpha\in\Omega^p U$ *and* $\beta\in\Omega^q(U)$*, one has* $i_X(\alpha\wedge\beta)=i_X\alpha\wedge\beta+(-1)^p\alpha\wedge i_X\beta$.

For future reference, we also record the useful computational rule corresponding to formula (1.39) of Chapter 1.

3. Basic Example Using the standard coordinates $(x^1,\ldots,x^n)$ on the open subset U of $\mathbb{R}^n$:

$$i_{\frac{\partial}{\partial x^k}}(dx^{i_1}\wedge\cdots\wedge dx^{i_p}) = \begin{cases}0 & \text{if } k\notin\{i_1,\ldots,i_p\}\\ (-1)^{\nu-1}dx^{i_1}\wedge\cdots\wedge\widehat{dx^{i_\nu}}\wedge\cdots dx^{i_p} & \text{if } k=i_\nu\end{cases} \tag{3.19}$$

where the symbol $\widehat{*}$ indicates the deletion of $*$.

As in the case of spaces $\Lambda^p V$, where we formed the graded algebra $\Lambda^* V$ over $\mathbb{R}$ (see 14j in Chapter 1), it is convenient now to form the graded exterior algebra of smooth differential forms on U over $C^\infty(U)$, i.e.,

$$\Omega^* U = \bigoplus_p \Omega^p U \tag{3.20}$$

Of course, $\Omega^p(U)$={**0**} for p>n and for p<0, and $\Omega^0 U$=$C^\infty(U)$. The wedge product

$$\wedge : \Omega^* U\times\Omega^* U\to\Omega^* U$$

satisfies the distributive laws

$$\begin{aligned}\alpha\wedge(\beta+\gamma) &= (\alpha\wedge\beta)+(\alpha\wedge\gamma)\\ (\alpha+\beta)\wedge\gamma &= (\alpha\wedge\gamma)+(\beta\wedge\gamma)\end{aligned}$$

Finally, we recall that a linear map $f{:}E{\to}E'$ of vector spaces induces a linear map $f^*{:}\Lambda^*E'{\to}\Lambda^*E$, or more precisely, a sequence of linear maps $\Lambda^pE'{\to}\Lambda^pE$. Explicitly, given $\lambda{\in}\Lambda^pE'$, $f^*\lambda{\in}\Lambda^pE$ is defined by

$$(f^*\lambda)(u_1,\ldots,u_p) = \lambda(f(u_1),\ldots,f(u_p))$$

where $u_1,\ldots,u_p$ are elements of E. Now given open subsets $U{\subset}\mathbb{R}^n$, $V{\subset}\mathbb{R}^m$ and a smooth map $\varphi{:}U{\to}V$, φ itself is not linear, but induces a "field of linear maps" namely the tangent maps $T_x\varphi{:}T_xU{\to}T_{\varphi(x)}V$, one for each $x{\in}U$. Each $T_x\varphi$ then induces a linear map

$$(T_x\varphi)^* : \Lambda^*(T_{\varphi(x)}V) \to \Lambda^*(T_xU)$$

which substitutes for f^* above. We now put these $(T_x\varphi)^*$s together to produce a map $\Omega^*V{\to}\Omega^*U$, which we temporarily denote by $\Omega^*\varphi$. For $\alpha{\in}\Omega^pV$, we must have $\Omega^*\varphi(\alpha){\in}\Omega^pU$. Let us define

$$(\Omega^*\varphi(\alpha))(x) = (T_x\varphi)^*\alpha(\varphi(x)) \tag{3.21}$$

This produces a cross-section, namely $\Omega^*\varphi(\alpha)$, for which smoothness must be established. We use the criterion of Lemma 1. Let $X_1,\ldots,X_p$ be smooth vector fields on U. Then using (3.16) and (3.21), we obtain

$$\begin{aligned}(\Omega^*\varphi(\alpha)(X_1,\ldots,X_p))(x) &= \Big((\Omega^*\varphi(\alpha))(x)\Big)(X_1(x),\ldots,X_p(x))\\ &= (T_x\varphi)^*\Big(\alpha(\varphi(x))\Big)(X_1(x),\ldots,X_p(x))\\ &= \Big(\alpha(\varphi(x))\Big)\Big(T_x\varphi(X_1(x)),\ldots,T_x\varphi(X_p(x))\Big)\\ &= \Big(\alpha(\varphi(x))\Big)\Big((T\varphi\circ X_1)(x),\ldots,(T\varphi\circ X_p)(x)\Big)\end{aligned}$$

Now all of α, φ, $T\varphi$ and the X_i are smooth, and for each x, they are combined by an operation that is $\mathbb{R}$-linear in each component, so the smoothness of $\Omega^*\varphi(\alpha)$ is established. We should also remark here that in the case $p{=}0$, $(T_x\varphi)^0{=}\mathbb{1}_\mathbb{R}$ by convention of Chapter 1, Section C, so if $\alpha{=}f{\in}\Omega^0V{=}C^\infty V$, then by (3.21) we obtain

$$\Omega^*\varphi(f) = f\circ\varphi \tag{3.22}$$

Further, if $\alpha{\in}\Omega^pV$ and $\beta{\in}\Omega^qV$, then by 14i of Chapter 1, we have

$$(T_x\varphi)^*(\alpha(x)\wedge\beta(x)) = (T_x\varphi)^*(\alpha(x))\wedge(T_x\varphi)^*(\beta(x))$$

Therefore, we can write

$$\Omega^*\varphi(\alpha\wedge\beta) = \Omega^*\varphi(\alpha)\wedge\Omega^*\varphi(\beta)$$

From now on we shall use the more common notation φ^* for $\Omega^*\varphi$. The context will make the intention clear. Even later, a third setting for the use of this notation will appear! All uses of the notation φ^* are referred to as the ***pullback***. Using the new notation, we can now summarize the above discussion in the following.

4. Properties of the Pullback
Let U and V be open subsets, respectively, of $\mathbb{R}^n$ and $\mathbb{R}^m$ and $\varphi: U \to V$ be a smooth mapping. Then the map

$$\varphi^* : \Omega^* V \to \Omega^* U$$

defined by

$$(\varphi^* \alpha(x))(w_1, \ldots, w_p) = \alpha(\varphi(x))(T_x\varphi(w_1), \ldots, T_x\varphi(w_p))$$

for $w_1, \ldots, w_p \in T_x U$, satisfies the following properties for $f \in C^\infty(V)$ and $\alpha, \beta \in \Omega^* V$:
(a) $\varphi^*(f) = f \circ \varphi$
(b) $\varphi^*(f\alpha) = (f \circ \varphi)\, \varphi^* \alpha$
(c) $\varphi^*(\alpha + \beta) = \varphi^*(\alpha) + \varphi^*(\beta)$
(d) $\varphi^*(\alpha \wedge \beta) = \varphi^*(\alpha) \wedge \varphi^*(\beta)$

Proof. These properties were all explained above. Note that (a) is a restatement of (3.22), and it can also be considered a special case of (d) since $C^\infty(V) = \Omega^0 V$, and wedge product becomes the ordinary product in this case. □

C. Exterior Derivative

In this and the following section, we will introduce operations on $\Omega^* U$ that have no analogue for single vector spaces. These are derivative-like local operations, they depend on the value of the differential form in an arbitrarily small neighborhood of the point under consideration. As we shall see, the "exterior derivative," to be discussed in this section, embodies standard operators "gradient", "curl" and "divergence" of elementary vector analysis in Euclidean space $\mathbb{R}^3$. The following lemma incorporates the definition.

5. Lemma *Let U be an open subset of $\mathbb{R}^n$. There is a unique sequence of maps $d_p: \Omega^p U \to \Omega^{p+1} U$, $p = 0, 1, \ldots$, that satisfies the following four conditions:*
(a) $d_0 = d$.
(b) Each d_p is $\mathbb{R}$-linear.
(c) $d_{p+1} \circ d_p = 0$ for all p.
(d) If $\alpha \in \Omega^p U$ and $\beta \in \Omega^q U$, then $d_{p+q}(\alpha \wedge \beta) = (d_p \alpha) \wedge \beta + (-1)^p \alpha \wedge (d_q \beta)$.

Proof. We begin by presenting a sequence d_p that satisfies the four conditions. We know that any smooth p-form is uniquely represented as a sum of terms of the form

$$f dx^{i_1} \wedge \cdots \wedge dx^{i_p}, \quad i_1 < \cdots < i_p \tag{3.23}$$

To define a map $\Omega^p U \to \Omega^{p+1} U$, it then suffices to specify the effect on single terms of the form (3.23), and to postulate that the map preserves sums. We propose to define

$$d_p(f dx^{i_1} \wedge \cdots \wedge dx^{i_p}) = df \wedge dx^{i_1} \wedge \cdots \wedge dx^{i_p} \tag{3.24}$$

and stipulate that d_p is additive. Note that d_p is well-defined since the ordering $i_1 < \cdots < i_p$, and the fact that the terms $dx^{i_1} \wedge \cdots \wedge dx^{i_p}$ form a basis for smooth p-forms over $C^\infty(U)$, allow no other representation for a term of the form (3.23). Of course, we have to check that the right-hand side of (3.24) is actually smooth, but writing $df = \sum_{i=1}^{n} \frac{\partial f}{\partial x^i} dx^i$, expanding according to distributive law, and rearranging into standard form, we see that the coefficients of basis elements $dx^{j_1} \wedge \cdots \wedge dx^{j_{p+1}}$ are either zero or $\pm \frac{\partial f}{\partial x^i}$, which are smooth. We note here, for future use, that although (3.24) was defined for $i_1 < \cdots < i_p$, the definition remains valid for any permutation σ of indices since the effect of permutation will be multiplication by $\varepsilon(\sigma)$, and the wedge product is $\mathbb{R}$-linear in each component, pushing $\varepsilon(\sigma)$ out of the wedge product.

We now check that the stated conditions are satisfied by the sequence (d_p). Conditions (a) and (b) are immediate. Condition (c) turns out to be a restatement of the symmetry of the second derivative, or the equality of mixed partial deivatives, for C^2 functions. Since the d_p are additive, it suffices to check (c) for single terms $f dx^{i_1} \wedge \cdots \wedge dx^{i_p}$.

$$\begin{aligned}
d_{p+1}(d_p(f dx^{i_1} \wedge \cdots \wedge dx^{i_p})) &= d_{p+1}(df \wedge dx^{i_1} \wedge \cdots \wedge dx^{i_p}) \\
&= d_{p+1}\Big(\sum_{\mu=1}^{n} \frac{\partial f}{\partial x^\mu}(dx^\mu \wedge dx^{i_1} \wedge \cdots \wedge dx^{i_p})\Big) \\
&= \sum_{\nu=1}^{n} \sum_{\mu=1}^{n} \frac{\partial^2 f}{\partial x^\nu \partial x^\mu} dx^\nu \wedge dx^\mu \wedge dx^{i_1} \wedge \cdots \wedge dx^{i_p}
\end{aligned}$$

Now for indices $\mu = \nu$, the corresponding term vanishes since $dx^\nu \wedge dx^\nu = 0$, and for $\mu \neq \nu$, we have $dx^\nu \wedge dx^\mu = -dx^\mu \wedge dx^\nu$, while $\frac{\partial^2 f}{\partial x^\nu \partial x^\mu} = \frac{\partial^2 f}{\partial x^\mu \partial x^\nu}$, so the terms vanish pairwise, and (c) is proved.

To prove (d), it suffices by (b) to prove the formula for $\alpha = f dx^{i_1} \wedge \cdots \wedge dx^{i_p}$ and $\beta = g dx^{j_1} \wedge \cdots \wedge dx^{j_q}$, where f and g are smooth functions.

$$\begin{aligned}
d_{p+q}(\alpha \wedge \beta) &= d_{p+q}(fg dx^{i_1} \wedge \cdots \wedge dx^{i_p} \wedge dx^{j_1} \wedge \cdots \wedge dx^{j_q}) \\
&= (g df + f dg) \wedge dx^{i_1} \wedge \cdots \wedge dx^{i_p} \wedge dx^{j_1} \wedge \cdots \wedge dx^{j_q} \\
&= g df \wedge dx^{i_1} \wedge \cdots \wedge dx^{i_p} \wedge dx^{j_1} \wedge \cdots \wedge dx^{j_q} \\
&\quad + f dg \wedge dx^{i_1} \wedge \cdots \wedge dx^{i_p} \wedge dx^{j_1} \wedge \cdots \wedge dx^{j_q} \\
&= (df \wedge dx^{i_1} \wedge \cdots \wedge dx^{i_p}) \wedge (g dx^{j_1} \wedge \cdots \wedge dx^{j_q}) \\
&\quad + (-1)^p (f dx^{i_1} \wedge \cdots \wedge dx^{i_p}) \wedge (dg \wedge dx^{j_1} \wedge \cdots \wedge dx^{j_q}) \\
&= (d_p \alpha) \wedge \beta + (-1)^p \alpha \wedge (d_q \beta)
\end{aligned}$$

Finally, we have to prove uniqueness. Suppose (D_p) is another sequence with the same four properties. Then $D_0 f = df$ for $f \in C^\infty(U)$ by (a). By (b), it suffices to show the equality of the effect of d_p and D_p on single terms $f dx^{i_1} \wedge \cdots \wedge dx^{i_p}$. First, a simple induction using property (c) shows that $D_p(dx^{i_1} \wedge \cdots \wedge dx^{i_p}) = 0$, then using property (d), we have

$$\begin{aligned} D_p(f dx^{i_1} \wedge \cdots \wedge dx^{i_p}) &= D_0 f \wedge dx^{i_1} \wedge \cdots \wedge dx^{i_p} + f D_p(dx^{i_1} \wedge \cdots \wedge dx^{i_p}) \\ &= df \wedge dx^{i_1} \wedge \cdots \wedge dx^{i_p} \\ &= d_p(f dx^{i_1} \wedge \cdots \wedge dx^{i_p}) \end{aligned}$$

as required. □

From now on, we will drop the subscript p from d_p and simply use d as an operator from $\Omega^* U$ to itself, one that increases the degree of each constituent p-form by one.

One seemingly weak aspect of the definition of d is that (3.24) makes apparent use of standard coordinates in $\mathbb{R}^n$. The following proposition will show that (3.24) holds quite generally. In this connection, see also the subsequent coordinate-free definition later in this section.

6. Lemma *Let U be an open subset of $\mathbb{R}^n$, and $f, g_1, \ldots . g_p \in C^\infty(U)$. Then*

$$d(f dg_1 \wedge \cdots \wedge dg_p) = df \wedge dg_1 \wedge \cdots \wedge dg_p$$

Proof. As we remarked in the ending paragraph of the previous proof, a simple induction using properties (c) and (d) shows that

$$d(dg_1 \wedge \cdots \wedge dg_p) = 0$$

The assertion then follows from property (d). □

7. Coordinate-free Approach to d

A coordinate-free definition of the exterior derivative, expressing its value on vector fields, is available. If $\alpha \in \Omega^p U$ and $X_1, \ldots, X_{p+1}$ are smooth vector fields, then

$$\begin{aligned} (3.25) \quad d\alpha(X_1, \ldots, X_{p+1}) = &\sum_{i=1}^{p+1} (-1)^{i-1} X_i \cdot \alpha(X_1, \ldots, \widehat{X_i}, \ldots, X_{p+1}) \\ &+ \sum_{i<j} (-1)^{i+j} \alpha([X_i, X_j], X_1, \ldots, \widehat{X_i}, \ldots, \widehat{X_j}, \ldots, X_{p+1}) \end{aligned}$$

The proof of this general formula has been relegated to Exercise 3.8 at the end of this chapter. We will only use the case p=1 of the formula in this book, for which we now provide a direct proof. Let $\alpha \in \Omega^1 U$, and suppose X and Y are smooth vector fields. Then

$$(3.26) \qquad d\alpha(X, Y) = X \cdot \alpha(Y) - Y \cdot \alpha(X) - \alpha[X, Y]$$

Proof. Both sides are additive with respect to α, so it suffices to verify the formula for $\alpha = f dg$, where f and g are smooth real-valued functions. We have

$$\begin{aligned}(d(f dg))(X, Y) &= (df \wedge dg)(X, Y)\\ &= df(X)dg(Y) - df(Y)dg(X)\\ &= (X \cdot f)(Y \cdot g) - (Y \cdot f)(X \cdot g)\end{aligned}$$

On the other hand,

$$\begin{aligned}X \cdot ((f dg)(Y)) - Y \cdot ((f dg)(X)) - (f dg)[X, Y]&\\ = X \cdot (f(Y \cdot g)) - Y \cdot (f(X \cdot g)) - f(X \cdot (Y \cdot g) - Y \cdot (X \cdot g))&\\ = (X \cdot f)(Y \cdot g) + fX \cdot (Y \cdot g) - (Y \cdot f)(X \cdot g) - fY \cdot (X \cdot g)&\\ - fX \cdot (Y \cdot g) + fY \cdot (X \cdot g)&\\ = (X \cdot f)(Y \cdot g) - (Y \cdot f)(X \cdot g)&\end{aligned}$$

and the verification is complete. □

The following commutativity relation between the exterior derivative and the pullback generalizes the coordinate-independence of the exterior derivative and will play a key role throughout the subject. It should be reminiscent of the change of variable in single integrals where the change of variable need not be one-to-one.

8. Theorem *Let U and V be open subsets, respectively, of $\mathbb{R}^n$ and $\mathbb{R}^m$ and $\varphi : U \to V$ a smooth map with induced pullback $\varphi^* : \Omega^* V \to \Omega^* U$. Then*

$$d \circ \varphi^* = \varphi^* \circ d$$

Proof. d and φ^* are both additive, so it suffices to prove the relation for single terms $f dx^{i_1} \wedge \cdots \wedge dx^{i_p}$. Note first that for $p=0$ and $f \in \Omega^0 V = C^\infty(V)$, we have

$$\begin{aligned}d(\varphi^*(f)) &= d(f \circ \varphi)\\ &= df \circ T\varphi\\ &= \varphi^*(df)\end{aligned}$$

Therefore,

$$\begin{aligned}(d \circ \varphi^*)(f dx^{i_1} \wedge \cdots \wedge dx^{i_p}) &= d(\varphi^*(f)\, \varphi^*(dx^{i_1}) \wedge \cdots \wedge \varphi^*(dx^{i_p}))\\ &= d(\varphi^*(f)\, d(\varphi^* x^{i_1}) \wedge \cdots \wedge d(\varphi^* x^{i_p}))\\ &= d(\varphi^* f) \wedge d(\varphi^* x^{i_1}) \wedge \cdots \wedge d(\varphi^* x^{i_p})\\ &= \varphi^*(df) \wedge \varphi^*(dx^{i_1}) \wedge \cdots \wedge \varphi^*(dx^{i_p})\\ &= \varphi^*(df \wedge dx^{i_1} \wedge \cdots \wedge dx^{i_p})\\ &= (\varphi^* \circ d)(f dx^{i_1} \wedge \cdots \wedge dx^{i_p})\end{aligned}$$

proving the theorem. □

9. Analogy with Vector Analysis in $\mathbb{R}^3$

In classical vector analysis of $\mathbb{R}^3$ with coordinates (x, y, z), the vector fields $\frac{\partial}{\partial x}$, $\frac{\partial}{\partial y}$ and $\frac{\partial}{\partial z}$ are generally denoted by **i**, **j** and **k**. We are going to use the ordered bases (dx, dy, dz) for $\Omega^1\mathbb{R}^3$, $(dy\wedge dz, dz\wedge dx, dx\wedge dy)$ for $\Omega^2\mathbb{R}^3$ and $\{dx\wedge dy\wedge dz\}$ for $\Omega^3\mathbb{R}^3$. The components of the three operators **grad**, **curl** and **div** turn out to be the components of the exterior derivatives of 0-, 1- and 2-forms in $\mathbb{R}^3$. Let us look at the analogy.

(a) For a 0-form (= real-valued function) f:

$$\mathbf{grad} f = \frac{\partial f}{\partial x}\mathbf{i} + \frac{\partial f}{\partial y}\mathbf{j} + \frac{\partial f}{\partial z}\mathbf{k}$$

$$df = \frac{\partial f}{\partial x}dx + \frac{\partial f}{\partial y}dy + \frac{\partial f}{\partial z}dz$$

(b) For a 1-form $\alpha = Pdx + Qdy + Rdz$, or analogously, for the vector field with like components $\mathbf{F} = P\mathbf{i} + Q\mathbf{j} + R\mathbf{k}$:

$$\mathbf{curl}\ \mathbf{F} = (\frac{\partial R}{\partial y} - \frac{\partial Q}{\partial z})\mathbf{i} + (\frac{\partial P}{\partial z} - \frac{\partial R}{\partial x})\mathbf{j} + (\frac{\partial Q}{\partial x} - \frac{\partial P}{\partial y})\mathbf{k}$$

$$d\alpha = (\frac{\partial R}{\partial y} - \frac{\partial Q}{\partial z})(dy \wedge dz) + (\frac{\partial P}{\partial z} - \frac{\partial R}{\partial x})(dz \wedge dx) + (\frac{\partial Q}{\partial x} - \frac{\partial P}{\partial y})(dx \wedge dy)$$

(c) For a 2-form $\beta = Ady\wedge dz + Bdz\wedge dx + Cdx\wedge dy$, or analogously, for the vector field with like components $\mathbf{G} = A\mathbf{i} + B\mathbf{j} + C\mathbf{k}$:

$$\mathbf{div}\ \mathbf{G} = \frac{\partial A}{\partial x} + \frac{\partial B}{\partial y} + \frac{\partial C}{\partial z}$$

$$d\beta = (\frac{\partial A}{\partial x} + \frac{\partial B}{\partial y} + \frac{\partial C}{\partial z})dx \wedge dy \wedge dz$$

We are leaving the computation of exterior derivatives above to the reader as a straightforward exercise (see Exercise 3.1 at the end of the chapter). It is important to note also that the classical formulas **curl(grad**f**)=0**, and **div(curl F)**=0 can be interpreted as instances of $d\circ d=0$. In both interpretations, the proof comes down to the equality of mixed partial derivatives.

In view of the analogy above, differential forms and the exterior derivative may be seen as an extension of vector analysis to arbitrary dimensions. We will also see in the chapter on integration that these tools enable us to formulate a general change of variable formula for integrals in higher dimensions, analogous to the one-variable case. The efficacy of differential forms in this context may be attributed to the fact that as anti-symmetric tensors, they are sensitive to orientation at each point. The exterior derivative, introduced by Elie Cartan, in a sense "factors out" the redundant information in higher order derivatives manifested in their symmetry. For an approach based on this idea, which defines the exterior derivative directly as the anti-symmetrization of ordinary derivative, see Exercise 3.7 or consult Henri Cartan's book [**6**].

D. Lie Derivative

In Chapter 2 we encountered twice the idea of "rate of change" along the trajectories of a vector field, where a derivative-like limit of quotients came into play. The first instance was that of $X \cdot f$, Theorem 17, and the second was that of the Lie bracket, Theorem 23. This can be generalized to arbitrary tensor fields of both covariant and contravariant type, and is known as the "Lie derivative." We will confine ourselves below to the case we need in the sequel, namely the Lie derivative of differential forms.

Let U be an open subset of $\mathbb{R}^n$, X a smooth vector field on U with flow Φ, $\alpha \in \Omega^p U$ and $x \in U$. We will show that the following limit always exists in $\Lambda^p(T_x U)$ and is smooth with respect to x.

$$\lim_{t \to 0} \frac{(\Phi_t{}^* \alpha)(x) - \alpha(x)}{t} \tag{3.27}$$

Thus an element of $\Omega^p U$ will be defined, which we denote by $L_X \alpha$ and will call the ***Lie derivative*** of α with respect to X. In short-hand notation, we will often write

$$L_X \alpha = \lim_{t \to 0} \frac{\Phi_t{}^* \alpha - \alpha}{t}$$

The following two lemmas will provide the necessary tools for the proof.

10. Lemma *(a) If for $\alpha, \beta \in \Omega^p U$, $L_X \alpha$ and $L_X \beta$ exist, then $L_X(\alpha + \beta)$ exists and*

$$L_X(\alpha + \beta) = L_X \alpha + L_X \beta$$

(b) If for $\alpha \in \Omega^p U$ and $\beta \in \Omega^q U$, $L_X \alpha$ and $L_X \beta$ exist, then $L_X(\alpha \wedge \beta)$ exists and

$$L_X(\alpha \wedge \beta) = L_X \alpha \wedge \beta + \alpha \wedge L_X \beta$$

Proof. The proofs are replicas of the elementary calculus treatments of the sum and product of differentiable functions. We will only present the proof of (b), the proof of (a) being more straightforward. We write

$$\begin{aligned} \frac{\Phi_t{}^*(\alpha \wedge \beta) - (\alpha \wedge \beta)}{t} &= \frac{\Phi_t{}^* \alpha \wedge \Phi_t{}^* \beta - \alpha \wedge \beta}{t} \\ &= \frac{\Phi_t{}^* \alpha \wedge (\Phi_t{}^* \beta - \beta)}{t} + \frac{(\Phi_t{}^* \alpha - \alpha) \wedge \beta}{t} \\ &= \Phi_t{}^* \alpha \wedge \frac{\Phi_t{}^* \beta - \beta}{t} + \frac{\Phi_t{}^* \alpha - \alpha}{t} \wedge \beta \end{aligned}$$

Now inserting x as the argument of each differential form and letting $t \to 0$, we obtain the desired result. □

11. Lemma *For $f \in \Omega^0 U$, $L_X f$ and $L_X(df)$ exist and*
(a) $L_X f = X \cdot f$
(b) $L_X(df) = d(X \cdot f)$

Proof. The first formula is a restatement of Theorem 17, Chapter 2. For the second, writing $\Phi_t{}^* df(x)$ as $d(\Phi_t{}^*(f))$ by virtue of Theorem 8, the formula amounts to changing the order of differentiation of $\Phi(t, x)$ from $\frac{\partial}{\partial t} \circ \frac{\partial}{\partial x}$ to $\frac{\partial}{\partial x} \circ \frac{\partial}{\partial t}$, which is permissible since Φ is C^∞. □

With these lemmas on hand, the existence of the limit in (3.27) now follows from the fact that any smooth p-form is a sum of terms of the form $f dx^{i_1} \wedge \cdots \wedge dx^{i_p}$.

12. Cartan's Formula

$$L_X = d \circ i_X + i_X \circ d$$

Proof. We denote the right-hand side of the formula temporarily by L'_X. To show the identity, it suffices to prove that Lemmas above hold for L'_X as well. Part (a) of Lemma 10 holds by the $\mathbb{R}$-linearity of both d and i_X. We verify (b) by direct computation.

$$\begin{aligned} L'_X(\alpha \wedge \beta) &= d(i_X\alpha \wedge \beta + (-1)^p \alpha \wedge i_X\beta) + i_X(d\alpha \wedge \beta + (-1)^p \alpha \wedge d\beta) \\ &= (d \circ i_X)\alpha \wedge \beta + (-1)^{p-1} i_X\alpha \wedge d\beta + (-1)^p d\alpha \wedge i_X\beta + \alpha \wedge (d \circ i_X)\beta \\ &+ (i_X \circ d)\alpha \wedge \beta + (-1)^{p+1} d\alpha \wedge i_X\beta + (-1)^p i_X\alpha \wedge d\beta + \alpha \wedge (i_X \circ d)\beta \\ &= (d \circ i_X + i_X \circ d)\alpha \wedge \beta + \alpha \wedge (d \circ i_X + i_X \circ d)\beta \\ &= L'_X(\alpha) \wedge \beta + \alpha \wedge L'_X(\beta) \end{aligned}$$

For Lemma 11(a)

$$\begin{aligned} L'_X(f) &= 0 + i_X(df) \\ &= df(X) \\ &= X \cdot f \end{aligned}$$

Likewise, for Lemma 11(b)

$$\begin{aligned} L'_X(df) &= (d \circ i_X \circ d)(f) \\ &= d(X \,.\, f) \end{aligned}$$

and the proof is complete. □

13. Elementary Properties of the Lie Derivative
(a) *$L_X\alpha$ is $\mathbb{R}$-linear in each of X and α.*
(b) *For $f \in C^\infty(U)$, and $\alpha \in \Omega^p U$,*

$$L_{fX}\alpha = f L_X\alpha + df \wedge i_X\alpha$$

(c) *For $f \in C^\infty(U)$, and $\alpha \in \Omega^p U$,*

$$L_X(f\alpha) = f L_X\alpha + (X \cdot f)\alpha$$

(d) $d \circ L_X = L_X \circ d$.

(e) *If* $h: U \to V$ *is a smooth diffeomorphism, then*

$$h^* \circ L_{h_* X} = L_X \circ h^*$$

(f) *Denoting* $L_X \circ L_Y - L_Y \circ L_X$ *by* $[L_X, L_Y]$, *we have*

$$L_{[X,Y]} = [L_X, L_Y]$$

Proof. (a) follows from the $\mathbb{R}$-linearity of d and i_X. We use Cartan's formula to prove (b).

$$\begin{aligned} L_{fX}\alpha &= (d \circ i_{fX})\alpha + (i_{fX} \circ d)\alpha \\ &= d(f(i_X\alpha)) + f(i_X(d\alpha)) \\ &= df \wedge i_X\alpha + f(d \circ i_X)\alpha + f(i_X \circ d)\alpha \\ &= df \wedge i_X\alpha + fL_X\alpha \end{aligned}$$

(c) follows from Lemma 10 and Lemma 11. For (d), the application of Cartan's formula shows that both sides are equal to $d \circ i_x \circ d$. To prove (e), we recall from (2.26) of Chapter 2 that if Φ_t is the flow of X, then the flow of $h_* X$ is $h \circ \Phi_t \circ h^{-1}$. So

$$\begin{aligned} L_{h_* X}\alpha &= \lim_{t \to 0} \frac{(h \circ \Phi_t \circ h^{-1})^*\alpha - \alpha}{t} \\ &= \lim_{t \to 0} \frac{(h^{-1})^*(\Phi_t^* \circ h^*)\alpha - (h^{-1})^*(h^*\alpha)}{t} \\ &= (h^{-1})^* \lim_{t \to 0} \frac{\Phi_t^*(h^*\alpha) - h^*\alpha}{t} \\ &= (h^{-1})^* L_X(h^*\alpha) \end{aligned}$$

which is equivalent to the desired result. Finally, for (f), we note that the operator $[L_X, L_Y]$ satisfies the assertions of Lemma 10; it therefore suffices to compare its values with those of $L_{[X,Y]}$ on f and df:

$$\begin{aligned} (L_X \circ L_Y - L_Y \circ L_X)(f) &= L_X(Y \cdot f) - L_Y(X \cdot f) \\ &= X \cdot (Y \cdot f) - Y \cdot (X \cdot f) \\ &= L_{[X,Y]}(f) \end{aligned}$$

The fact that the Lie derivative and d commute by (d), together with the above, show that the effects of $L_{[X,Y]}$ and $[L_X, L_Y]$ on df are the same. □

14. Computation of Lie Derivative In view of 13, the computation of $L_X\alpha$ can be carried once one knows how to compute

$$L_{\frac{\partial}{\partial x^i}}(f dx^{i_1} \wedge \cdots \wedge dx^{i_p})$$

By (c):

$$L_{\frac{\partial}{\partial x^i}}(f dx^{i_1} \wedge \cdots \wedge dx^{i_p}) = (L_{\frac{\partial}{\partial x^i}} f) dx^{i_1} \wedge \cdots \wedge dx^{i_p} + f L_{\frac{\partial}{\partial x^i}}(dx^{i_1} \wedge \cdots \wedge dx^{i_p})$$

Now

$$L_{\frac{\partial}{\partial x^i}}(dx^{i_1}\wedge\cdots\wedge dx^{i_p}) = d(i_{\frac{\partial}{\partial x^i}}(dx^{i_1}\wedge\cdots\wedge dx^{i_p})) + i_{\frac{\partial}{\partial x^i}}(d(dx^{i_1}\wedge\cdots\wedge dx^{i_p}))$$

But $d(dx^{i_1}\wedge\cdots\wedge dx^{i_p}) = 0$, and by (3.19), $i_{\frac{\partial}{\partial x^i}}(dx^{i_1}\wedge\cdots\wedge dx^{i_p})$ is a constant $(p-1)$-form, so the first term of the above also vanishes. Finally, since by Lemma 11(a), $L_{\frac{\partial}{\partial x^i}} f = \frac{\partial f}{\partial x^i}$, we end up with

$$L_{\frac{\partial}{\partial x^i}}(f dx^{i_1}\wedge\cdots\wedge dx^{i_p}) = \frac{\partial f}{\partial x^i}\, dx^{i_1}\wedge\cdots\wedge dx^{i_p} \tag{3.28}$$

The following is the analogue of the identical vanishing of the bracket of vector fields (Chapter 2, Corollary 24).

15. Vanishing of Lie Derivative

Let U be an open subset of $\mathbb{R}^n$, X a smooth vector field on U with flow Φ and $\alpha\in\Omega^p U$. Then $L_X\alpha=0$ if and only if $\Phi_t^*\alpha=\alpha$ wherever Φ_t is defined.

Proof. For $x\in U$, Φ_t is defined for $|t|$ sufficiently small in a neighborhood of x. If $\Phi_t^*\alpha=\alpha$, then the definition of Lie derivative implies that $L_X\alpha(x)=0$. On the other hand, suppose that $L_X\alpha$ is everywhere zero. Fixing $x\in U$, we look at the curve $\gamma(t)=\Phi_t^*\alpha(x)$ in $\Lambda^p(T_xU)$. We have:

$$\begin{aligned}\gamma'(t) &= \lim_{h\to 0}\frac{\Phi_{t+h}^*\alpha(x)-\Phi_t^*\alpha(x)}{h}\\ &= \lim_{h\to 0}\frac{\Phi_h^*(\Phi_t^*\alpha(x))-\Phi_t^*\alpha(x)}{h}\\ &= L_X(\Phi_t^*\alpha)(x)\\ &= \Phi_t^*(L_{(\Phi_t)_*X}\alpha)(x) \quad \text{by 13(e)}\end{aligned}$$

But Φ is the flow of X, so $(\Phi_t)_*X=X$, so the right-hand side vanishes by the hypothesis $L_X\alpha=0$. Thus $\gamma'(t)=0$, $\gamma(t)$ is a constant, and by setting $t=0$, $\Phi_t^*\alpha(x)=\Phi_0^*\alpha(x)=\alpha(x)$. □

16. Divergence of a Vector Field

Recall from Chapter 1 that for a vector space V of dimension n, a non-zero element of $\Lambda^n V$ is called a volume element. Given an open subset U of $\mathbb{R}^n$, one may choose a volume element $\omega(x)$ for each vector space T_xU. This cross-section of $\Lambda^n U$ is considered a "volume element for U" if it is at least continuously varying with respect to x. Since we are dealing mainly with smooth data, we actually define a ***volume element*** for U to be a nowhere vanishing element of $\Omega^n U$, i.e., we assume smoothness. The ***standard volume element*** (also called the ***Euclidean volume element***) for U is $\omega_0=dx^1\wedge\cdots\wedge dx^n$. Since each $\Lambda^n(T_xU)$ is one-dimensional, any volume element for U can be represented as $V(x)(dx^1\wedge\cdots\wedge dx^n)$, where $V:U\to\mathbb{R}$ is a smooth non-vanishing function. Let $X=\sum_{i=1}^n X^i\frac{\partial}{\partial x^i}$ be a smooth vector field on

U. In classical vector analysis, the real-valued function $\mathbf{div}X = \sum_{i=1}^{n} \frac{\partial X^i}{\partial x^i}$ is usually called the "divergence of X". The value of this function at each point x is supposed to indicate whether regions of small volume around that point tend to expand (if $\mathbf{div}X(x)>0$) or contract (if $\mathbf{div}X(x)<0$). In other words, $\mathbf{div}X(x)$ is the rate of change of volume at the point x under the flow of the vector field X. We can now describe this notion for general volume elements, using the Lie derivative, in a way that the above interpretation becomes immediately clear. Let $\omega \in \Omega^n U$ be a volume element for U and X be a smooth vector field on U. The definition (3.27) expresses the Lie derivative $L_X\omega$ as the rate of change of ω under the flow of X. Further, since $\Omega^n U$ is one-dimensional over $C^\infty(U)$, we have $L_X\omega = f\omega$, where $f \in C^\infty(U)$. We call this function f, the ***divergence of X relative to the volume element*** ω, and denote it by $\mathbf{div}_\omega X$. In other words,

$$L_X\omega = (\mathbf{div}_\omega X)\omega \tag{3.29}$$

We can now derive a simple formula for divergence. Using Cartan's formula and the fact that $d\omega=0$, we have $L_X\omega = (d \circ i_X)\omega$. Letting $\omega = V dx^1 \wedge \cdots \wedge dx^n$ and $X = \sum_{k=1}^{n} X^k \frac{\partial}{\partial x^k}$, we obtain

$$\begin{aligned} i_X(V dx^1 \wedge \cdots \wedge dx^n) &= \sum_{k=1}^{n} X^k \, i_{\frac{\partial}{\partial x^k}}(V dx^1 \wedge \cdots \wedge dx^n) \\ &= \sum_{k=1}^{n} (-1)^{k-1} X^k V (dx^1 \wedge \cdots \widehat{dx^k} \cdots \wedge dx^n) \end{aligned}$$

Therefore,

$$\begin{aligned} (d \circ i_X)\omega &= \sum_{l=1}^{n} \sum_{k=1}^{n} (-1)^{k-1} \frac{\partial (X^k V)}{\partial x^l} (dx^l \wedge dx^1 \wedge \cdots \widehat{dx^k} \cdots \wedge dx^n) \\ &= \sum_{j=1}^{n} \frac{\partial (X^j V)}{\partial x^j} dx^1 \wedge \cdots \wedge dx^n \end{aligned}$$

So the following formula for divergence results

$$\mathbf{div}_\omega X = \frac{1}{V} \sum_{j=1}^{n} \frac{\partial (X^j V)}{\partial x^j} \tag{3.30}$$

which includes the Euclidean case for $V=1$.

E. Riemannian Metrics

In this last section of Chapter 3, we move away from anti-symmetric tensor fields and discuss an important family of symmetric tensor fields known as "Riemannian metrics". As always, we work in an open subset U of $\mathbb{R}^m$. A covariant 2-tensor field ρ is called a ***Riemannian metric*** if for each $x \in U$, $\rho(x)$ is an inner product on T_xU. As explained in Section A, there are two equivalent interpretations for ρ. One can regard ρ as a cross-section of L^2U, one such that each bilinear $\rho(x): T_xU \times T_xU \to \mathbb{R}$

is symmetric and positive-definite. The other interpretation is to regard ρ as a map $(TU)^2 \to \mathbb{R}$, the restriction of which to each $T_xU \times T_xU$, denoted by $\rho(x)$, is bilinear, symmetric and positive-definite. In either approach, the adjectives continuous, C^r, smooth and the like make sense for the mapping ρ. Using the representation (3.11), we write ρ as

$$\rho = \sum_{i,j=1}^{m} g_{ij} dx^i \otimes dx^j \tag{3.31}$$

where g_{ij} are real-valued functions on U, and for each $x \in U$, the matrix $[g_{ij}(x)]$ is symmetric and its eigenvalues are all positive. ρ is smooth (C^r, continuous) if and only if the functions g_{ij} satisfy that same property.

The appearance of the word "metric" in this context requires an explanation. Briefly for the moment, the existence of an inner product in each tangent space allows one to define the length of a tangent vector $w \in T_xU$ as $|w|_\rho = (\rho(x)(w, w))^{\frac{1}{2}}$. Now if $\gamma : [a, b] \to U$ is a path with $\gamma(a) = A$ and $\gamma(b) = B$, the length of γ is defined as in calculus by

$$\int_a^b |\gamma'(t)|_\rho \, dt$$

Taking the infimum of the lengths over paths from A to B will provide a metric d_ρ in the sense of metric space for the set U. This will be the setting for "Riemannian geometry", which will be encountered in later chapters. We also point out that an inner product on tangent space T_xU enables one to measure angles between non-zero tangent vectors emanating from x through

$$|w_1|_\rho |w_2|_\rho \cos \angle(w_1, w_2) = \rho(w_1, w_2)$$

Thus it makes sense to speak of a ***unit cube*** in T_xU as a set

$$K_x = \Big\{ \sum_{i=1}^{m} t_i w_i : 0 \le t_i \le 1, i = 1, \dots, m \Big\}$$

where $w_1, \dots, w_m$ are n mutually perpendicular vectors of length one in T_xU. It turns out that there is a unique volume element for U, known as the ***Riemannian volume element*** (associated to the Riemannian metric ρ), which assigns value 1 to each such unit cube provided $(w_1, \dots, w_m)$ has the same orientation as the standard basis $(\frac{\partial}{\partial x^1}(x), \dots, \frac{\partial}{\partial x^m}(x))$. We proceed to produce this volume element, but first we revisit the familiar Gram-Schmidt orthogonalization method in a slightly revised form.

17. Lemma (Parametrized Gram-Schmidt Procedure) *Let U be an open subset of $\mathbb{R}^m$, and suppose that a smooth Riemannian metric ρ on U is given. We consider smooth vector fields Y_i on U, $i = 1, \dots, m-1$, so that at each $x \in U$, the set $\{Y_1(x), \dots, Y_{m-1}(x)\}$ is linearly independent. Then there is a set of m smooth vector fields $X_1, \dots, X_m$ on U with the following properties:*
(i) For each $x \in U$, the ordered m-tuple $(X_1(x), \dots, X_m(x))$ is a positively oriented orthonormal basis for T_xU, where orthonormality is with respect to the inner product

$\rho(x)$ on T_xU, and positive orientation is with respect to some given orientation of $\mathbb{R}^m$, e.g., the standard orientation.
(ii) For each $x \in U$ and each $i = 1, \ldots, m-1$, the i-tuples $(X_1(x), \ldots, X_i(x))$ and $(Y_1(x), \ldots, Y_i(x))$ span the same linear subspace of T_xU and have the same orientation.

Proof. We proceed in exactly the same fashion as the standard Gram-Schmidt. First let $Z_1 = Y_1$. Z_2 is then obtained from Y_2 by subtracting its orthogonal projection on Z_1

$$Z_2 = Y_2 - \frac{\rho(Y_2, Z_1)}{\rho(Z_1, Z_1)} Z_1$$

Proceeding inductively, and assuming $Z_1, \ldots, Z_{i-1}$, $i < m-1$, have been constructed, we construct Z_i by subtracting from Y_i its orthogonal projections on $Z_1, \ldots, Z_{i-1}$

$$Z_i = Y_i - \sum_{j<i} \frac{\rho(Y_i, Z_j)}{\rho(Z_j, Z_j)} Z_j \tag{3.32}$$

Note that the coefficients are smooth functions. One checks that $Z_1, \ldots, Z_i$ are mutually orthogonal and have the same linear span as $Y_1, \ldots, Y_i$. Further, we observe from (3.32) that the matrix of the $Z_1, \ldots, Z_i$ with respect to $Y_1, \ldots, Y_i$ is lower-triangular and all diagonal entries are equal to 1. Therefore, the assertions of (ii) hold if we substitute Z_j for X_j. Thus the smooth vector fields $Z_1, \ldots, Z_{m-1}$ are constructed. The construction of Z_m is similar to defining cross-product of two vectors in $\mathbb{R}^3$. Suppose that the standard orientation of $\mathbb{R}^m$ is stipulated in (i). For each $1 \leq j \leq m-1$, we write

$$Z_j(x) = \sum_{i=1}^{m} a^i_j(x) \frac{\partial}{\partial x^i}(x)$$

We tabulate the coefficients of Z_j as the entries of the jth column of an $m \times (m-1)$ matrix, and define

$$Y_m = \sum_{i=1}^{m} a^i_m(x) \frac{\partial}{\partial x^i}(x)$$

where $a^i_m(x)$ is $(-1)^{m+i}$ times the determinant of the $(m-1) \times (m-1)$ matrix obtained by deleting the ith row of the above matrix. Computing the determinant of the $m \times m$ matrix $[a^i_j(x)]$ by expanding relative to the mth column, we obtain a positive real number, showing that the ordered m-tuple $(Z_1, \ldots, Z_{m-1}, Y_m)$ is positively oriented. Subtracting the orthogonal projections of Y_m on $Z_1, \ldots, Z_{m-1}$ from Y_m, we obtain Z_m. The ordered m-tuple $(Z_1, \ldots, Z_m)$ is orthogonal and positively oriented. Finally, dividing each Z_i by its ρ-length, we obtain the X_i as sought. □

Consider the vector fields $\frac{\partial}{\partial x^1}, \ldots, \frac{\partial}{\partial x^m}$ on U. At every point $x \in U$, the vectors $\frac{\partial}{\partial x^i}(x)$, $i = 1, \ldots, m$, form a basis for T_xU, however this set need not be orthonormal relative to the inner product $\rho(x)$. Applying the above lemma, we obtain an everywhere orthonormal set of basis vector fields $(X_1, \ldots, X_m)$ with the same orientation as the standard basis $(\frac{\partial}{\partial x^1}, \ldots, \frac{\partial}{\partial x^m})$. At each $x \in U$, let $(\xi^1(x), \ldots, \xi^m(x))$

be the dual basis for $(X_1(x), \ldots, X_m(x))$. The 1-forms ξ^i, thus defined, are smooth because their values on smooth basis vector fields X_j are constant, hence smooth, functions. We define

$$\omega_\rho = \xi^1 \wedge \cdots \wedge \xi^m \tag{3.33}$$

This m-form is indeed a volume element, as the $\xi^1(x), \ldots, \xi^m(x)$ form a basis for T_x^*U for all x. Further, $\omega_\rho(x)$ assigns unit volume to the unit cube determined by $(X_1(x), \ldots, X_m(x))$. We show that $\omega_\rho(x)$ does not depend on the specific choice of similarly oriented orthonormal bases $(X_1(x), \ldots, X_m(x))$ for T_xU. Let $(Y_1(x), \ldots, Y_m(x))$ be an orthonormal basis for T_xU with the same orientation as $(X_1(x), \ldots, X_m(x))$, and consider its dual basis $(\eta^1(x), \ldots, \eta^m(x))$. Suppose f is the linear map from T_xU to itself with $f(Y_i(x))=X_i(x)$, hence $f^*(\xi^i(x))=\eta^i(x)$. Then f is an orientation-preserving orthogonal transformation of T_xU, which implies that $\det f=1$. Therefore,

$$\begin{aligned}
\eta^1(x) \wedge \cdots \wedge \eta^m(x) &= f^*(\xi^1(x)) \wedge \cdots \wedge f^*(\xi^m(x)) \\
&= f^*(\xi^1(x) \wedge \cdots \wedge \xi^m(x)) \\
&= (\det f)(\xi^1(x) \wedge \cdots \wedge \xi^m(x)) \\
&= \xi^1(x) \wedge \cdots \wedge \xi^m(x)
\end{aligned}$$

Thus ω_ρ is well-defined and uniquely determined by the specified property. We now express ω_ρ as a functional multiple of the standard Euclidean volume element $\omega_0 = dx^1 \wedge \cdots \wedge dx^m$. Let us write

$$\omega_\rho = V\omega_0 \tag{3.34}$$

The real-valued function V must be everywhere positive as $(\xi^1(x), \ldots, \xi^m(x))$ has the same orientation as the standard basis. We will compute V in terms of components g_{ij} of the Riemannian metric ρ by expressing the standard basis in terms of the orthonormal basis $(X_1, \ldots, X_m)$

$$\frac{\partial}{\partial x^i} = \sum_{k=1}^m c_i^k X_k$$

Evaluating the two sides of (3.34) on $(\frac{\partial}{\partial x^1}, \ldots, \frac{\partial}{\partial x^m})$, we obtain

$$\begin{aligned}
V &= (\xi^1 \wedge \cdots \wedge \xi^m)(\sum_k c_1^k X_k, \ldots, \sum_k c_m^k X_k) \\
&= \det\left[\xi^i(\sum_k c_j^k X_k)\right] \\
&= \det[c_j^i]
\end{aligned}$$

On the other hand,

$$
\begin{aligned}
g_{ij} &= \rho\,(\frac{\partial}{\partial x^i}, \frac{\partial}{\partial x^j}) \\
&= \rho\,(\sum_k c_i^k X_k, \sum_k c_j^k X_k) \\
&= \sum_{k,l} c_i^k c_j^l\, \rho(X_k, X_l) \\
&= \sum_k c_i^k c_j^k
\end{aligned}
$$

Writing this in matrix notation

$$[g_{ij}] = [c_j^i] \cdot [c_j^i]^T$$

But $\det\,[c_j^i]^T = \det\,[c_j^i]$, so

$$\det\,[c_j^i] = \sqrt{\det[g_{ij}]}$$

Therefore, the following expression for ω_ρ is obtained:

$$\omega_\rho = \sqrt{\det[g_{ij}]}\; dx^1 \wedge \cdots \wedge dx^m \tag{3.35}$$

Note that because the symmetric matrix $[g_{ij}]$ is positive-definite, $\det[g_{ij}] > 0$. One sometimes writes g for $\det[g_{ij}]$. Then, using (3.30), the formula for divergence of a vector field $X = \sum_{i=1}^m X^i \frac{\partial}{\partial x^i}$ relative to the Riemannian metric ρ takes the form

$$\mathbf{div}_{\omega_\rho} X = \frac{1}{\sqrt{g}} \sum_{i=1}^m \frac{\partial (X^i \sqrt{g})}{\partial x^i} \tag{3.36}$$

In Exercise 3.14 below, as well as in later chapters of the book, we treat the concept of "gradient" relative to a Riemannian metric. The notion of "curl" in $\mathbb{R}^3$ is also affected by a change of Riemannian metric (see Exercise 3.16).

EXERCISES

3.1 Perform the computation of $d\alpha$ and $d\beta$ in Subsection 9 dealing with vector analysis in $\mathbb{R}^3$.

3.2 Work out the expressions of the standard volume element and the Euclidean metric in $\mathbb{R}^3$ in terms of cylindrical and spherical coordinates.

3.3 Let U be an open subset of $\mathbb{R}^m$, ω a volume element, X a smooth vector field and $f \in C^\infty(U)$. Show that

$$\mathbf{div}_\omega(fX) = f\mathbf{div}_\omega X + X \cdot f$$

3.4 Let U and V be open subsets of $\mathbb{R}^m$ and $h:U \to V$ be a smooth diffeomorphism. If X is a smooth vector field on U, show that

$$i_X \circ h^* = h^* \circ i_{h_* X}$$

Use this formula to give an alternate proof of 13e.

3.5 Let U be an open subset of $\mathbb{R}^m$ with X and Y smooth vector fields on U. Show that

$$i_{[X,Y]} = [L_X, i_Y]$$

Use this formula to give an alternate proof of 13f.

3.6 Let U be an open subset $\mathbb{R}^m$. A subring I of $\Omega^*(U)$ is a ***differential ideal*** if the following conditions are satisfied: (i) $\alpha \in I$ and $\beta \in \Omega^*(U)$ imply $\beta \wedge \alpha \in I$, (ii) $dI \subset I$ and (iii) if $\alpha = \alpha_0 + \cdots + \alpha_m$ is in I, where $\alpha_i \in \Omega^i(U)$, then $\alpha_i \in I$ for all i.
(a) For an ideal I, show that if $\alpha \in I$ and $\beta \in \Omega^*(U)$, then $\alpha \wedge \beta \in I$.
(b) A smooth vector field X on U is called a ***Cauchy characteristic*** for the ideal I if $i_X I \subset I$. Show that the Cauchy characteristic vector fields of I form a Lie algebra.

3.7 Represent $\alpha \in \Omega^P U$ as $\alpha(x) = (x, \bar{\alpha}(x))$ and $v \in T_x U$ as $(x, \bar{v})$, where $\bar{\alpha}(x) \in \Lambda^p \mathbb{R}^m$ and $\bar{v} \in \mathbb{R}^m$. Prove that

$$d\alpha(v_1, \ldots, v_{p+1}) = \sum_{i=1}^{p+1} (-1)^{i-1} \Big((D\bar{\alpha}(x))(\bar{v}_i) \Big) (\bar{v}_1, \ldots, \hat{\bar{v}}_i, \ldots, \bar{v}_{p+1})$$

where D denotes the usual derivative as in Appendix II. (Hint: Denote the right-hand side by $d'\alpha(v_1, \ldots, v_{p+1})$ and show that d' satisfies the four conditions of Lemma 5.)

3.8 Verify the general coordinate-free expression for the exterior derivative, formula (3.25) of the text. (Hint: Note that it suffices to look at the case $\alpha = f dx^{i_1} \wedge \cdots \wedge dx^{i_p}$ and $X_k = g_k \frac{\partial}{\partial x^{j_k}}$. There are two cases to consider depending on whether $\{i_1, \ldots, i_p\}$ is a subset of $\{j_1, \ldots, j_{p+1}\}$ or not.)

3.9 Let U be an open subset of $\mathbb{R}^m$, $\alpha \in \Omega^p U$ and $X, X_1, \ldots, X_p$ be smooth vector fields on U. Prove that

$$(L_X \alpha)(X_1, \ldots, X_p) = X \cdot \alpha(X_1, \ldots, X_p) + \sum_{i=1}^{p} \alpha(X_1, \ldots, [X_i, X], \ldots, X_p)$$

3.10 In $\mathbb{R}^3$ with the standard Euclidean metric, consider smooth vector fields X, Y and Z with the property that X is perpendicular to $\mathbf{curl}X$, Y and Z. Show that X is perpendicular to $[Y, Z]$.

3.11 Let $Z = (-y)\frac{\partial}{\partial x} + (x)\frac{\partial}{\partial y}$ and $\omega = V dx \wedge dy$ be a smooth volume element on $\mathbb{R}^2$. Show that $L_Z \omega = 0$ if and only if $V(x, y)$ depends only on $x^2 + y^2$.

3.12 Suppose $\alpha \in \Omega^p \mathbb{R}^n$ satisfies $d\alpha = 0$ and $i_{\frac{\partial}{\partial x^n}} \alpha = 0$. Prove that α can be written as

$$\sum_{i_1 < \cdots < i_p} a_{i_1 \cdots i_p} dx^{i_1} \wedge \cdots \wedge dx^{i_p}$$

where $i_p < n$ and $\frac{\partial a_{i_1 \cdots i_p}}{\partial x^n} = 0$.

3.13 (Assuming familiarity with Exercise 1.10) Let U be an open subset of $\mathbb{R}^n$ and suppose β is a smooth covariant 2-tensor field such that $\beta(x)$ has rank n for each $x \in U$. Therefore for each x, an isomorphism $\beta(x)^\flat : T_x U \to T_x^* U$ is defined. We denote this family of isomorphisms by $\beta^\flat : TU \to T^* U$ and represent its inverse by $\beta_\sharp$. Thus $\beta^\flat|_{T_x U} = \beta(x)^\flat$, and $\beta_\sharp|_{T_x^* U} = (\beta(x)^\flat)^{-1}$. Show that $\beta^\flat$ induces an isomorphism of $C^\infty(U)$ modules $\mathfrak{X}(U)$ and $\Omega^1 U$; in particular, smooth vector fields are mapped to smooth 1-forms and vice versa. We shall denote the image of $X \in \mathfrak{X}(U)$ in $\Omega^1 U$ by $X^\flat$, and the image of $\alpha \in \Omega^1 U$ in $\mathfrak{X}(U)$ by $\alpha_\sharp$. Denote $\beta(\frac{\partial}{\partial x^i}, \frac{\partial}{\partial x^j}) = (\beta^\flat(\frac{\partial}{\partial x^j}))(\frac{\partial}{\partial x^i})$ by β_{ij}, and write the inverse matrix of $[\beta_{ij}]$ as $[\beta^{ij}]$. If $X = \sum_{j=1}^n X^j \frac{\partial}{\partial x^j}$, show that

$$X^\flat = \sum_{i=1}^{n} \left(\sum_{j=1}^{n} \beta_{ij} X^j \right) dx^i$$

Likewise, if $\alpha = \sum_{i=1}^n \alpha_i dx^i$, then

$$\alpha_\sharp = \sum_{j=1}^{n} \left(\sum_{i=1}^{n} \beta^{ij} \alpha_i \right) \frac{\partial}{\partial x^j}$$

3.14 (Based on Exercise 3.13) Let ρ be a smooth Riemannian metric on U. We know from Exercise 1.10b that each $\rho(x)$ has rank n, hence the situation of Exercise 3.13 is obtained. Suppose $f : U \to \mathbb{R}$ is a smooth function. Then $(df)_\sharp$ is called the ***gradient*** of f relative to the Riemannian metric ρ, and is denoted by $\nabla_\rho f$.
(a) Show that for any vector field X,

$$df(X) = \rho(X, \nabla_\rho f)$$

(b) Express $\nabla_\rho f$ in terms of the metric and the partial derivatives of f. Show that a gradient vector field has no periodic solutions.

(c) Let f be a C^2 real-valued function on U. The ***Laplacian*** of f relative to the Riemannian metric ρ, denoted by $\Delta_\rho f$, is defined as $\mathbf{div}_{\omega_\rho}(\nabla_\rho f)$. Compute the formula for the Laplacian in terms of the g_{ij} and the partial derivatives (up to order two) of f.

3.15 (Based on Exercise 3.13) Consider the 2-form $\sigma = \sum_{i=1}^n dq^i \wedge dp^i$ on $\mathbb{R}^{2n}$ with coordinates $(q^1, \ldots, q^n; p^1, \ldots, p^n)$.
(a) Show that the rank of each $\sigma(x)$ is $2n$, hence the isomorphism of Exercise 3.13 is obtained.
(b) Let $H:\mathbb{R}^{2n} \to \mathbb{R}$ be smooth. Show that

$$(dH)_\sharp = \sum_{i=1}^n \frac{\partial H}{\partial p^i}\frac{\partial}{\partial q^i} - \sum_{i=1}^n \frac{\partial H}{\partial q^i}\frac{\partial}{\partial p^i}$$

The vector field $(dH)_\sharp$ is also denoted by X_H and called the ***Hamiltonian vector field*** associated to H (Cf. Exercise 2.13).
(c) Show that $L_{X_H}\sigma = 0$.
(d) Show that n-fold wedge product $\sigma^n = \sigma \wedge \cdots \wedge \sigma$ is a volume element for $\mathbb{R}^{2n}$ and that $\mathbf{div}_{\sigma^n} X_H = 0$.

3.16 (Based on Exercises 3.13 and 1.15) Let U be an open subset of $\mathbb{R}^n$ and ρ be a smooth Riemannian metric for U. For each $x \in U$, the Hodge star isomorphism $*:\Lambda^p(T_xU) \to \Lambda^{n-p}(T_xU)$ is defined as in Exercise 1.15.
(a) Show that $*$ induces an isomorphism of $C^\infty(U)$ modules $\Omega^p U \to \Omega^{n-p} U$; in particular, α is smooth if and only if $*\alpha$ is smooth.
(b) Let X be a smooth vector field on an open subset U of $\mathbb{R}^3$. We define $\mathbf{curl}_\rho X$ by

$$\mathbf{curl}_\rho X = (*\, d(X^\flat))_\sharp$$

Express $\mathbf{curl}_\rho X$ in terms of the metric and the partial derivatives of the components of X.

3.17 Referring to Exercises 3.14 and 3.16 above, are the following identities valid for an arbitrary smooth Riemannian metric ρ?
(a) $\mathbf{curl}_\rho(\nabla_\rho f) = \mathbf{0}$
(b) $\mathbf{div}_{\omega_\rho}(\mathbf{curl}_\rho X) = 0$

3.18 Let $\mathcal{P}$ be the subset

$$\{(x^1, \ldots, x^n) : x^i > 0, i = 1, \ldots, n\}$$

of $\mathbb{R}^n$. We define the Riemannian metric ρ on $\mathcal{P}$ by

$$\sum_{i=1}^n \frac{\sum_j x^j}{x^i}\, dx^i \otimes dx^i$$

Show that the straight line $\frac{x^1}{a_1} = \cdots = \frac{x^n}{a_n}$ is perpendicular (with respect to ρ) to the hyperplanes $x^1 + \cdots + x^n = constant$.

3.19 Consider the upper half plane $\mathbb{U}=\{(x, y):y > 0\}$ with the ***Poincaré metric*** $(y)^{-2}(dx\otimes dx+dy\otimes dy)$. Show that each of the following transformations is an ***isometry*** of this metric, i.e., the inner product of tangent vectors at a point does not change under the tangent map of these transformations. We use $z=x + iy$ for points of $\mathbb{U}$.
$z \mapsto z + t$, where $t\in\mathbb{R}$,
$z \mapsto kz$, where $k>0$, and
$z \mapsto -(z)^{-1}$

3.20 Consider the following vector field on $\mathbb{U}$ (see Exercise 3.19):

$$X = (1 + x^2 - y^2)\frac{\partial}{\partial x} + (2xy)\frac{\partial}{\partial y}$$

(a) Show that $\mathbf{div}_\omega X=0$, where ω is the volume element associated with the Poincaré metric.
(b) Try to identify the integral curves of X; they are all closed orbits except for one singular point. (Hint: Notice that the two components are the real and imaginary parts of $1+z^2$, where $z=x+iy$, or use an integrating factor to solve the corresponding system of differential equations.)
(c) Show that the flow maps Φ_t of this vector field are isometries relative to the Poincaré metric.

Part III

Global

CHAPTER 4

Manifolds, Tangent Bundle

The idea that countless number of geometries can be legitimate subjects of mathematical discourse, and each geometry has a manifold as its habitat, was formulated in Bernhard Riemann's famous Habilitation lecture of 1854. The word "manifold" (English for the German *Mannigfaltigkeit*) was actually a short for "many-fold-extended-quantity," in today's jargon a set that can be locally parametrized by a certain number of numerical values, i.e., by a subset of $\mathbb{R}^m$. This substrate served only as a medium for implementing the tools of calculus and did not impose any preferred geometry on the manifold. Geometry proper started when notions such as distance and angle came into play. Riemann suggested that these were most conveniently introduced by stipulating an inner product for tangent vectors originating at each point of the manifold, a Riemannian metric in the language of the previous chapter. Through familiarity with the work of Gauss and others on curved surfaces and his own research in complex analysis of multi-valued functions, Riemann was well aware that one set of parameters was often not adequate for covering the entire extent of a manifold, but the state of development of mathematics at the time did not provide an adequate language for a lucid global description. Although the development of Riemann's approach to geometry continued through the adoption of ad hoc global tools as needed, a completely adequate definition of manifold had to await the development of point-set topology. Hermann Weyl's definition of a Riemann surface in 1913[1] set the tone for the global treatment of manifolds, a task that was accomplished in the 1930s through the work of Hassler Whitney and others. In Part II of the book, Chapters 2 and 3, we developed much of the calculus-based local tools that are to be subsequently used. These were, in turn, partially grounded in linear and multilinear algebra that was treated in Chaper 1. Beginning with this chapter, and throughout Part III, we will introduce and develop the global language that serves not only geometry (to be briefly encountered in Part IV), but also other fields of mathematics such as differential topology, dynamical systems, Lie groups and differential equations.

A. Topological Manifolds

Our intuitive picture of an m-dimensional manifold is a smooth object that can be broken into (overlapping) parts, each of which can be parametrized by m independent

[1]See his 1913 book *Die Idee der Riemannschen Fläche*, English translation *The Concept of a Riemann Surface*, Addison-Wesley, Reading, MA 1964.

parameters. Curves in a plane or in space are one-dimensional manifolds, while surfaces in space are examples are two-dimensional manifolds (see Figure 1). To

FIGURE 1. Examples of manifolds

begin the formal approach, we first introduce the notion of a topological manifold. We wish to consider topological spaces M with the following property: Every $a \in M$ has an open neighborhood homeomorphic to an open set of $\mathbb{R}^m$. This condition seems to make M sufficiently $\mathbb{R}^m$-like to allow the use of m independent parameters for representing a piece of the topological space. But certain pathologies and inconveniencies may creep in. The definition below requires that the topological space be Hausdorff and second-countable. In the Appendix to this chapter, we discuss the reasons for adding these conditions.

1. Definition Let m be a non-negative integer. A ***topological m-manifold*** is a Hausdorff and second-countable topological space M with the following property: For each $a \in M$, there is an open neighborhood of a homeomorphic to an open subset of $\mathbb{R}^m$.

This definition calls for some explanation. Is the integer m in the definition uniquely determined? Suppose a point $a \in M$ has two open neighborhoods, U and

V, with homeomorphisms $\xi{:}U{\to}U'$ and $\eta{:}V{\to}V'$, where U' and V' are, respectively, open subsets of $\mathbb{R}^m$ and $\mathbb{R}^n$. Then letting $W{=}U{\cap}V$, the map $\eta{\circ}\xi^{-1}$ defines a homeomorphism from $\xi(W)$ to $\eta(W)$. It is an easy consequence of Brouwer's so-called *Invariance of Domain Theorem* (see, e.g., [**12**]) that if an open subset of $\mathbb{R}^m$ is homeomorphic to an open subset of $\mathbb{R}^n$, then necessarily $m{=}n$ (see Exercise 2). This shows that the integer m, called the ***dimension of** M **at*** a, is uniquely determined at each point a of M. In fact, if the dimension is m at a point a, and $\xi{:}U{\to}U'{\subset}\mathbb{R}^m$ a homeomorphism, then the dimension of M at $b{\in}U$ is also m. Thus dimension is a locally constant function. This shows that each connected component of M admits a unique dimension. When we speak of a toplogical m-manifold, we mean that all connected components have dimension m.

A pair (U,ξ), where U is an open subset of M and $\xi{:}U{\to}U'$ a homeomorphism of U onto an open subset U' of $\mathbb{R}^m$, is called a **chart** for M. The inverse $\xi^{-1}{:}U'{\to}U$ is called a ***coordinate system*** on U. For $a{\in}U$, $(x^1,\dots,x^m){=}\xi(a)$ are called the ***local coordinates*** of a with respect to the chart (U,ξ).

A geographic map ("chart") operates the same way (see Figure 2); it maps a piece of the earth's surface onto a flat sheet of paper (subset of $\mathbb{R}^2$). Local coordinates in

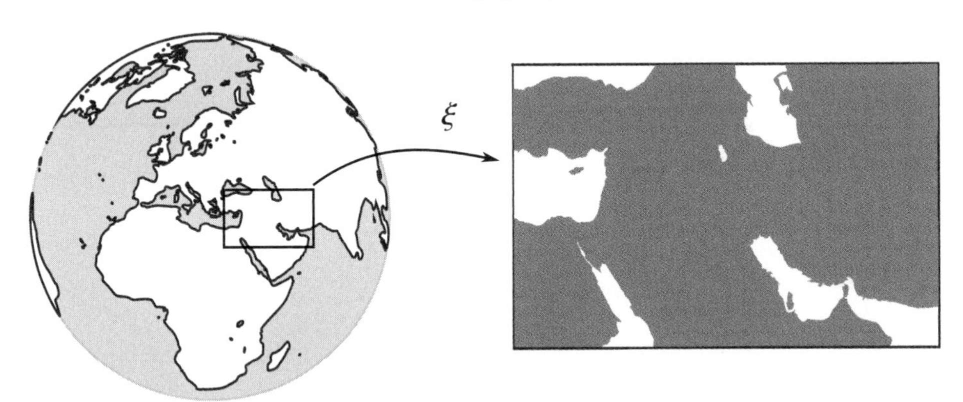

FIGURE 2. A chart

this case are the geographic longitude and latitude. The geographic analogy is carried even further. A collection of charts $(U_\alpha,\xi_\alpha)_\alpha$ is called an ***atlas*** for the topological manifold M if the union of the U_α covers M.

2. Examples The following important examples will be encountered throughout the book.

(a) Let U be an open subset of $\mathbb{R}^m$. The atlas consisting of the single chart $(U,\mathbb{1}_U)$ establishes U as a topological m-manifold. The linear space $M(p,q;\mathbb{R})$ of real $p{\times}q$ matrices is homeomorphic to $\mathbb{R}^{pq}$ and is hence a topological manifold of

dimension pq. The set GL$(n,\mathbb{R})$ of invertible $n{\times}n$ matrices, being an open subset of $M_n(\mathbb{R})$=M$(n,n;\mathbb{R})$, is a topological manifold of dimension n^2. GL$(n,\mathbb{R})$ is open because it is the inverse image of the open subset $\mathbb{R}^{\times}$ of non-zero real numbers under the continuous map det $: M(n,\mathbb{R})\to\mathbb{R}$.
(b) Countable sets with discrete topology are the only examples of zero-dimensional topological manifolds. Zero-dimensional manifolds are modeled on $\mathbb{R}^0=\{0\}$, so every chart must have a single point in its domain. The set of rational numbers, $\mathbb{Q}$, with induced topology from $\mathbb{R}$, is thus not a zero-dimensional manifold.
(c) Let $f: W\to\mathbb{R}^n$ be a continuous function, where W is an open subset of $\mathbb{R}^m$, and consider the graph

$$\Gamma f = \{(x, f(x)) : x \in W\}$$

We endow Γf with the subspace topology from $W{\times}\mathbb{R}^n$. With this topology, Γf is a topological m-manifold. Hausdorff-ness and second-countability are inherited from $W{\times}\mathbb{R}^n$ as subspace. The singleton consisting of the chart $(\Gamma f, \xi)$, where ξ is the restriction to Γf of the projection $W \times \mathbb{R}^n \to W$ on the first coordinate, constitutes an atlas. Note that ξ is one-to-one and onto, and the projection map is continuous with continuous inverse $x\mapsto(x, f(x))$ relative to the product topology.
(d) Consider the ***unit*** $\boldsymbol{m}$***-sphere***

$$S_m = \{(x^1, \ldots, x^{m+1})\in\mathbb{R}^{m+1} : \sum_{i=1}^{m+1}(x^i)^2 = 1\}$$

We show S_m is a topological m-manifold. In fact, we will construct two different atlases; it will be helpful to have both of these around.
(i) For each i=1, ..., $m + 1$, consider the two open hemispheres

$$U_i^{\pm} = \{(x^1, \ldots, x^{m+1}) \in S_m : x_i \gtrless 0\}$$

and mappings $\xi_i^{\pm}: U_i^{\pm}\to\mathbb{R}^m$ defined by

$$\xi_i^{\pm}(x^1, \ldots, x^{m+1}) = (x^1, \ldots \widehat{x^i} \ldots x^{m+1})$$

Since each open hemisphere is the graph of a continuous function from the open unit disk in $\mathbb{R}^m$ to $\mathbb{R}$, each $\xi_i^{\pm}$ is a homeomorphism (see Example (c) above), and an atlas consisting of $2(m + 1)$ charts is obtained for S_m.
(ii) The alternative atlas consists of two charts. Define the ***north*** and ***south poles*** of S_m, respectively, by N=(0, ..., 0, 1) and S=(0, ..., 0, −1). The ***stereographic projections*** σ_N and σ_S are, respectively, projections of S_m from N and S onto $\mathbb{R}^m$ considered as the set of points of $\mathbb{R}^{m+1}$ with zero last coordinate (see Figure 3). Analytically, using similar triangles, $\sigma_N: S_m-\{N\}\to\mathbb{R}^m$, and $\sigma_S: S_m-\{S\}\to\mathbb{R}^m$ are given by

$$\sigma_N(x^1, \ldots, x^{m+1}) = (1 - x^{m+1})^{-1}(x^1, \ldots, x^m) \tag{4.1}$$

$$\sigma_S(x^1, \ldots, x^{m+1}) = (1 + x^{m+1})^{-1}(x^1, \ldots, x^m) \tag{4.2}$$

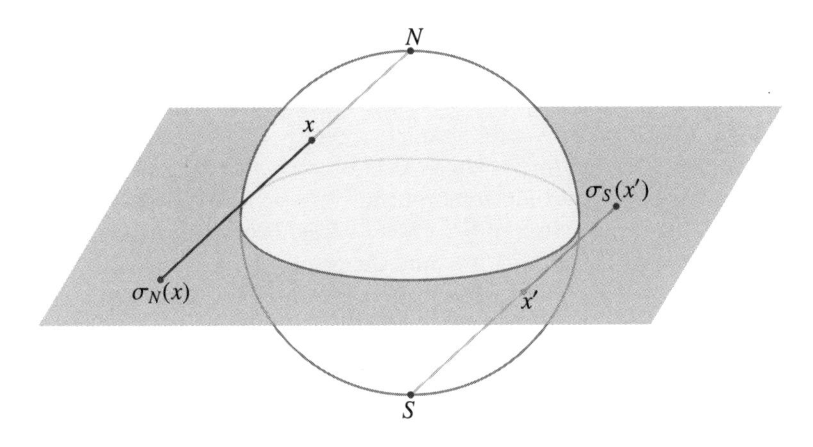

FIGURE 3. Stereographic projections

Both σ_N and σ_S are continuous, one-to-one and onto. Inverses are also easily calculated and seen to be continuous. In fact, elementary computation yields

$$\sigma_N{}^{-1}(y^1, \ldots, y^m) = (q+1)^{-1}(2y^1, \ldots, 2y^m, s-1) \tag{4.3}$$

$$\sigma_S{}^{-1}(y^1, \ldots, y^m) = (q+1)^{-1}(2y^1, \ldots, 2y^m, 1-s) \tag{4.4}$$

where $q = \sum_{i=1}^m (y^i)^2$.

(e) Unlike the graph of a continuous function examined in Example (c), the image set of a function need not be a manifold, even if the function is smooth and one-to-one. Consider the function $f:]-1, \infty[\to \mathbb{R}^2$ defined by $f(t) = (t^2-1, t^3-t)$. f is smooth and one-to-one on its domain. We show that the image set

$$S = \{(t^2-1, t^3-t) : -1 < t < +\infty\}$$

with topology as a subset of $\mathbb{R}^2$ is not a topological manifold (see Figure 4). In fact,

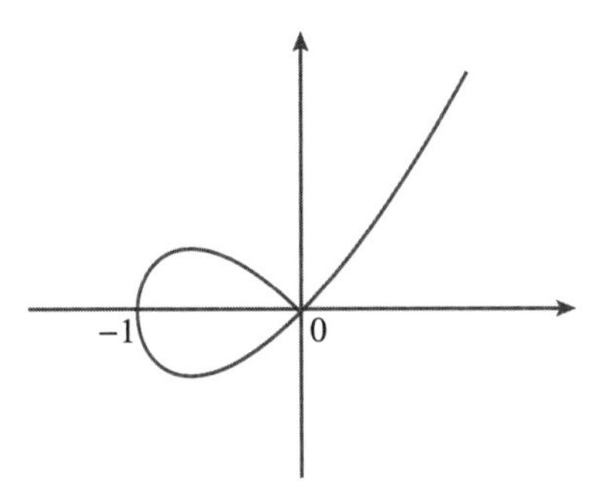

FIGURE 4. Image set not a manifold

no open neighborhood of $(0,0) \in S$, however small, is homeomorphic to an open set in $\mathbb{R}$ (or of any $\mathbb{R}^m$ for that matter). If we delete $(0,0)$ from its small open neighborhood

in S, we are left with three disjoint sets each homeomorphic to an open interval, while removing a point from an open interval yields two disjoint open intervals.
(f) Let M and N be, respectively, m-dimensional and n-dimensional topological manifolds. We consider $M\times N$ with product topology and show that it is an $(m+n)$-dimensional manifold. Let $(a,b)\in M\times N$. Choosing charts (U,ξ) and (V,η) around a and b, respectively, we have homeomorphisms $\xi:U\rightarrow U'$, U' an open subset of $\mathbb{R}^m$, and $\eta:V\rightarrow V'$, V' an open subset of $\mathbb{R}^n$. Then $\xi\times\eta$ defined by $(\xi\times\eta)(x,y)=(\xi(x),\eta(y))$ defines a homeomorphism from $U\times V$ onto the open subset $U'\times V'$ of $\mathbb{R}^m\times\mathbb{R}^n\cong\mathbb{R}^{m+n}$. Further, Hausdorff and second-countable properties hold for the product of two such spaces. $M\times N$ is known as a ***product manifold***. This construction can be extended to the topological product of any finite number of manifolds.

An especially useful product is the m-fold product of the unit circle S_1 with itself. The product manifold $S_1\times\cdots\times S_1$ is known as the ***m-torus***, and will be denoted by $\mathbb{T}^m$. Thus $\mathbb{T}^1$ will be another notation for S_1. We will meet various representations (homeomorphic copies) of $\mathbb{T}^m$ in this book. As examples, we consider two variants below.
(i) Starting with $\mathbb{T}^1=S_1$, note that this is homeomorphic to the quotient space of the closed interval $[0,1]$ under the identification $0\sim 1$[2]. Generalizing this, consider the unit m-cube $[0,1]\times\cdots\times[0,1]$ and pass on to the quotient topological space under the equivalence relation

$$(s^1,\ldots,s^m)\sim(t^1,\ldots,t^m)\iff|s^i-t^i|=0\text{ or }1,\ \forall i=1,\ldots,m$$

This quotient space is homeomorphic to $\mathbb{T}^m$. A familiar case is the identification of

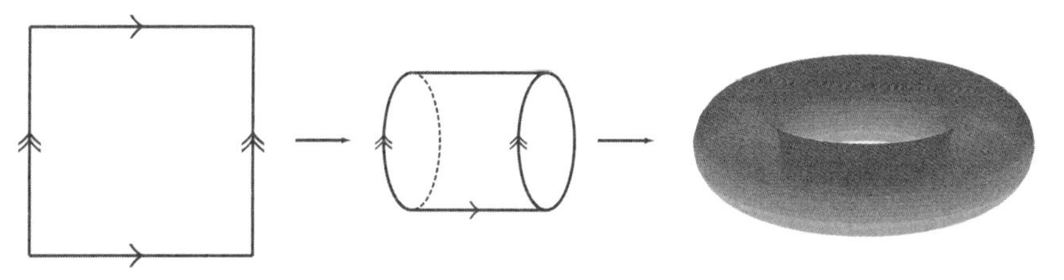

FIGURE 5. Two-torus as quotient

pairs of opposite sides of a rectangle (topologically equivalent to unit square) for the construction of the two-torus (Figure 5).
(ii) Following on the previous example, one can represent the two-torus in $\mathbb{R}^3$ in many shapes and sizes. Let $a>b>0$. By rotating a circle of radius b in $\mathbb{R}^3$ around an axis in the plane of the circle, which is at a distance a from the center of the circle, a two-torus is generated. Using cylindrical coordinates (r,θ,z) in $\mathbb{R}^3$, the equation

$$(r-a)^2+z^2=b^2$$

describes the case where a circle of radius b in a vertical plane with center on xy-plane is rotated around the z-axis. Each point of this surface can be uniquely represented

[2] See Section C of Chapter 5 for a review of relevant facts about quotient spaces.

by two angles θ and φ (see Figure 6). This indicates the correspondence with $S_1 \times S_1$.

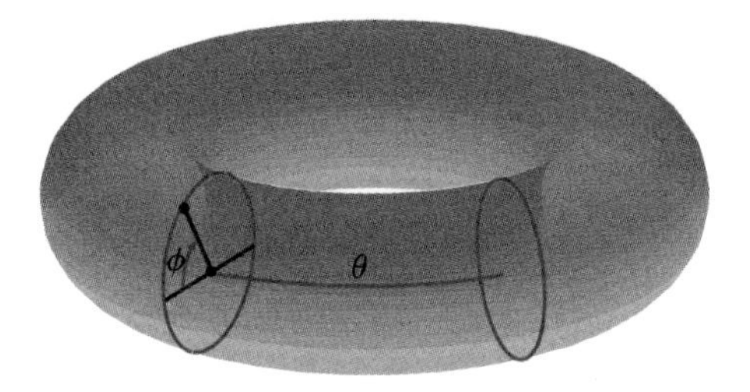

FIGURE 6. Two-torus as $S_1 \times S_1$

g) Figure 7 displays a number of closed knots in $\mathbb{R}^3$. These are all homeomorphic to the unit circle and are hence topological 1-manifolds. An especially well-studied

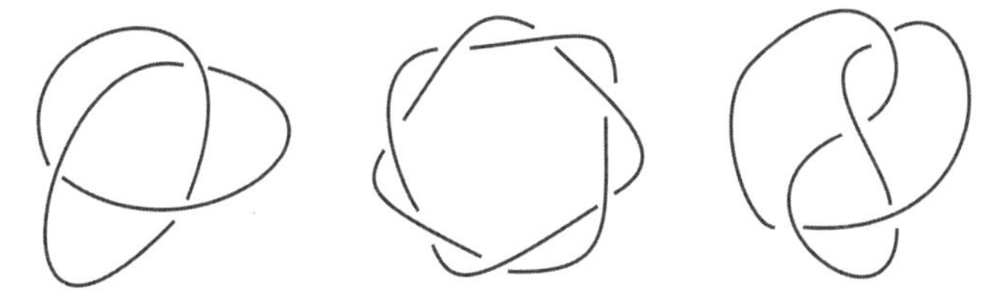

FIGURE 7. Examples of knots

family of knots, known as ***torus-knots***, will now be described in the context of the representation (i) of the 2-torus in example above. Taking the unit square $0 \le x \le 1, 0 \le y \le 1$, with opposite pairs of sides identified as described in the preceding example, we consider the trace of the straight line $y = \frac{p}{q}\, x$, where p and q are relatively prime (non-zero) integers (see Figure 8 for the case $p=2, q=3$). The image of this straight line in the quotient space (=2-torus) will be a topological circle, which will be "knotted" if $|p|>1$ and $|q|>1$. In fact, stretching the unit square and pasting the opposite sides into the torus of Example f,(ii), this closed curve will travel p times around the central circle $r=a$ and q times around the z-axis before closing (or the other way around, depending on the direction of stretching).

On the other hand, replacing the straight line $y = \frac{p}{q} x$ above with $y = \mu x$, where μ is an irrational number, the image of the straight line will not close, but will spiral densely on the torus (see Exercise 4.10). With the subspace topology inherited from the torus, the spiral will not be a one-dimensional topological manifold since a small neighborhood of each point of the spiral will contain an infinite number of disjoint copies of topological open intervals.

h) (Real Projective Spaces) We define a relation $\sim$ on $\mathbb{R}^{m+1} - \{\mathbf{0}\}$ by

$$(x^1, \ldots, x^{m+1}) \sim (y^1, \ldots, y^{m+1}) \iff y^i = \lambda x^i \text{ for all } i = 1, \ldots, m+1$$

for some non-zero λ in $\mathbb{R}$. $\sim$ is an equivalence relation, the equivalence class of $(x^1, \ldots, x^{m+1})$ is denoted by $[x^1 : \cdots : x^{m+1}]$, and the set of equivalence classes is called

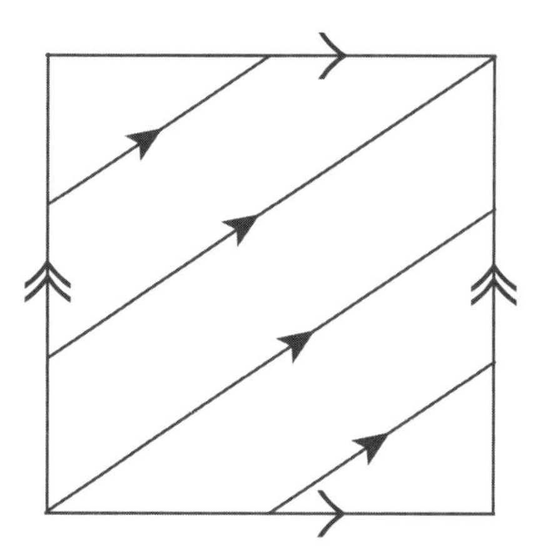

FIGURE 8. A torus knot

the ***real projective*** ***m-space*** and is denoted by $\mathbb{RP}(m)$. We furnish $\mathbb{RP}(m)$ with the quotient topology. General criteria from quotient topology show that this topology is both second-countable and Hausdorff[3]. An atlas consisting of $m+1$ charts (U_i, ξ_i) is provided as follows.

$$U_i = \{[x^1 : \cdots : x^{m+1}] : x_i \neq 0\}$$
$$\xi_i[x^1 : \cdots : x^{m+1}] = (\frac{x_1}{x_i}, \ldots, \frac{x_{i-1}}{x_i}, \frac{x_{i+1}}{x_i}, \ldots, \frac{x_{m+1}}{x_i})$$

The mappings ξ_i are well-defined, one-to-one and onto $\mathbb{R}^m$. The fact they are homeomorphisms follows from general properties of maps from quotient spaces[4]. Thus $\mathbb{RP}(m)$ is a topological m-manifold.

We collect below, for future reference, some consequences of the topological hypotheses in the definition of topological manifold.

3. Elementary Properties of Topological Manifolds *Let M be a topological manifold. Then the following hold.*
(a) *M is locally-connected and locally-path-connected.*
(b) *Connected components and path components of M coincide, and they are both open and closed.*
(c) *M is locally compact.*
(d) *Every cover of M by open sets has a countable subcover.*

[3] See Section C of Chapter 5 for a review of relevant facts about quotient spaces.
[4] *ibid*

(e) *M can be written as a countable union, $M=\bigcup_n K_n$, where each K_n is compact and $K_n \subset int(K_{n+1})$ for all n.*

PROOF. We recall that M is called locally-connected (respectively, locally path-connected) if for every $x \in M$ and every open set U containing x, an open connected (respectively, open path-connected) set V exists with $x \in V \subset U$. Let W be an open neighborhood of x that admits a homeomorphism $h: W \rightarrow W'$ with W' an open subset of $\mathbb{R}^m$. Then $W \cap U$ is an open neighborhood of x, and $h(W \cap U)$ is an open neighborhood of $h(x)$ in $\mathbb{R}^m$. By the local connectedness (respectively, local path-connectedness) of $\mathbb{R}^m$, an open neighborhood V' of $h(x)$ exists in $\mathbb{R}^m$ with $V' \subset h(W \cap U)$, which is connected (respectively, path-connected). Then $V = h^{-1}(V') \subset W \cap U \subset U$ establishes the property at x. Regarding (b), the following are general facts from point-set topology:

- Connected components are always closed.
- Connected components of locally connected spaces are also open.
- Path-connectedness (respectively, local path-connectedness) implies connectedness (respectively, local connectedness).
- If a space is connected and locally path-connected, then it is path-connected.

From these facts, (b) follows. For (c), recall that a space is locally compact if every point has a compact neighborhood. Every open subset of $\mathbb{R}^m$ is locally compact. Since each point of M has an open neighborhood homeomorphic to an open subset of $\mathbb{R}^m$, M is locally compact. (d) is a general fact of point-set topology for second-countable spaces.

We prove (e), which is a valid proposition for any topological space that is second-countable and locally compact. For every $x \in M$, there is, by local compactness, an open set W_x and a compact set L_x with $x \in W_x \subset L_x$. By (d), there is a countable subcover by $W_i = W_{x_i}$, $i = 1, 2, \ldots$. We construct the sequence (K_i) as follows. To begin, let $K_1 = L_1 = L_{x_1}$. By compactness, a finite number of W_i's cover K_1. Let K_2 be the (finite) union of corresponding L_i's together with L_{n_1}, where n_1 is the smallest integer j so that W_j is not a member of the finite cover above for K_1. This is compact, being the union of a finite number of compact sets. Further, the interior of this union contains the corresponding open sets, so $K_1 \subset int(K_2)$. Now supposing that $K_1, \ldots, K_p$ have been constructed, we show how to construct K_{p+1}. By compactness of K_p, we can take a finite number of W_i's that cover K_p. We let K_{p+1} be the union of the corresponding L_i's together with L_{n_p}, where n_p is the smallest j so that W_j has not been used in constructions up to K_p. By the same argument as before, K_{p+1} is compact and K_P is contained in the interior of K_{p+1}. Note that $n_p \geq p$; this fact guarantees that the union of K_n's will exhaust M.

□

4. Remark One can also show that every topological manifold is metrizable, i.e., it admits a metric whose topology is the same as the given topology of the manifold. This may be seen as follows. We first note that a topological m-manifold M is *locally metrizable*, i.e., every point of M has a neighborhood that is metrizable. This follows from the existence of local homeomorphisms with $\mathbb{R}^m$. Next by a theorem of point set topology (see [**22**]), a Hausdorff and second-countable space that is locally metrizable is in fact metrizable.

B. Smooth Manifolds

As we explained in the opening paragraph to this chapter, manifolds are to serve as the turf on which the tools of analysis will be deployed. Let M be a topological m-manifold. An elementary requirement from differential calculus is to give meaning to the differentiability of a function $f{:}M\to\mathbb{R}$ at a point $a\in M$. The only tool we have available to transfer the notion of differentiability from $\mathbb{R}^m$ to manifolds is the use of charts.

5. Preliminary Definitions

Suppose $\xi{:}U\to U'$ is a chart with $a\in U$. It is natural to consider f differentiable at a in case $f\circ\xi^{-1}$ is differentiable at $\xi(a)$. But suppose $\eta{:}V\to V'$ is another chart with $a\in V$, and $W=U\cap V$. For the proposed definition of differentiability to be independent of the particular chart, differentiability of $f\circ\xi^{-1}$ at $\xi(a)$ must be equivalent to the differentiability of $f\circ\eta^{-1}$ at $\eta(a)$. Now

$$f\circ\xi^{-1} = (f\circ\eta^{-1})\circ(\eta\circ\xi^{-1})$$

where the composition $\eta\circ\xi^{-1}$ is a homeomorphism from $\xi(W)$ to $\eta(W)$. There is in general no reason to expect a homeomorphism from one open set to another, in this case from $\xi(W)$ to $\eta(W)$, to be differentiable. We must therefore work with a more limited atlas for which the maps $\eta\circ\xi^{-1}$, known as ***transition maps*** or ***coordinate changes***, are diffeomorphisms. For $r\geq 0$, we define an atlas $\mathcal{A}$ for M to be a C^r ***atlas*** if for any pair of charts $(U_1,\xi_1),(U_2,\xi_2)\in\mathcal{A}$ with non-empty $U_1\cap U_2$, the coordinate change $\xi_2\circ\xi_1{}^{-1}$ is a C^r diffeomorphism. Thus, of course, in this case, $\xi_1\circ\xi_2{}^{-1}=(\xi_2\circ\xi_1{}^{-1})^{-1}$ will also be a C^r diffeomorphism. The pair $(M,\mathcal{A})$ will be called a C^r ***manifold***. A C^0 manifold is just a topological manifold, C^∞ manifolds are also known as ***smooth manifolds***, and by a C^ω ***manifold***, we mean a real-analytic manifold.

Now if $(M,\mathcal{A})$ is a C^r manifold, and $s\leq r$, it makes sense to speak of a function $f{:}M\to\mathbb{R}$ to be C^s (relative to the atlas $\mathcal{A}$) at a point $a\in M$. For if (U_1,ξ_1) and (U_2,ξ_2) are elements of $\mathcal{A}$ with $a\in U_1\cap U_2$, then $f\circ\xi_1{}^{-1}$ is C^s at $\xi_1(a)$ if and only if $f\circ\xi_2{}^{-1}$ is C^s at $\xi_2(a)$. In the same vein, let $(M,\mathcal{A})$ and $(N,\mathcal{B})$ be C^r manifolds, $f{:}M\to N$ continuous at $a\in M$ (as a map of topological spaces M and N), and $s\leq r$. f is said to be C^s at the point a (relative to the atlases $\mathcal{A}$ and $\mathcal{B}$), if there are charts $(U,\xi)\in\mathcal{A}$ with $a\in U$, and $(V,\eta)\in\mathcal{B}$ with $f(a)\in V$, so that $\eta\circ f\circ\xi^{-1}$ is C^s at $\xi(a)$. f is called C^s on a subset S of M if f is C^s at every point of S.

6. Remarks and Conventions

(a) Let $\mathcal{A}$ and $\mathcal{B}$ be C^r atlases for M and N, respectively. For the phrase "$f:M\to N$ is C^s relative to the atlases $\mathcal{A}$ and $\mathcal{B}$," we sometimes write "$f:(M,\mathcal{A})\to(N,\mathcal{B})$ is C^s." When there is no ambiguity about choice of atlases, we simply write "$f:M\to N$ is C^s."
(b) In the definition of $f:(M,\mathcal{A})\to(N,\mathcal{B})$ being C^s at a, we assumed beforehand that f is continuous at a. This is necessitated by the following possibility. Suppose $(U,\xi)\in\mathcal{A}$ with $a\in U$, and $(V,\eta)\in\mathcal{B}$ with $f(a)\in V$. Unless we assume that f is continuous at a, there is no guarantee that the domain of $\eta\circ f\circ\xi^{-1}$ will include an open neighborhood of the point $\xi^{-1}(a)$, so that C^s-ness may even be discussed.
(c) It follows from the definition of C^r atlas that our definition of C^s mapping, $s\leq r$, depends only on the choice of atlases, not on particular charts.
(d) If $(M,\mathcal{A})$ is a C^r manifold, then the identity map $\mathbb{1}_M:(M,\mathcal{A})\to(M,\mathcal{A})$ is C^r.
(e) Let $(M,\mathcal{A})$, $(N,\mathcal{B})$ and $(P,\mathcal{C})$ be C^r manifolds with $f:(M,\mathcal{A})\to(N,\mathcal{B})$ and $g:(N,\mathcal{B})\to(P,\mathcal{C})$, C^s mappings, $s\leq r$, then $g\circ f:(M,\mathcal{A})\to(P,\mathcal{C})$ is C^s.

Both (d) and (e) are immediate consequences of the definition.

7. Examples

(a) Let W be an open subset of $\mathbb{R}^m$, and consider the atlas consisting of the single chart $(W,\mathbb{1}_W)$. The only change of variable here is the identity mapping of W, hence we have a C^r manifold for any r. More generally, an open subset W of any C^r manifold $(M,\mathcal{A})$ admits a C^r atlas by taking intersection of W with domains of charts of $\mathcal{A}$ (Exercise 4.5). An important remark here is that if $(M,\mathcal{A})$ is a C^r manifold and $(U,\xi)\in\mathcal{A}$, then the homeomorphism $\xi:U\to\xi(U)$ is actually a C^r diffeomorphism of U onto $\xi(U)$, where U and $\xi(U)$ are considered C^r manifolds as above. The reason is that $\xi\circ\xi^{-1}$ is the identity map.
(b) Let $(M,\mathcal{A})$ be a C^r manifold and $h:M\to N$ a homeomorphism of topological spaces. For every chart $(U,\xi)\in\mathcal{A}$, we define a chart $(\bar{U},\bar{\xi})$ for N by

$$\bar{U}=h(U),\quad \bar{\xi}=\xi\circ h^{-1}$$

Now if (V,η) is another chart in $\mathcal{A}$ with $U\cap V$ non-empty, and $\bar{V}=h(V)$, $\bar{\eta}=\eta\circ h^{-1}$, we obtain $\bar{\eta}\circ\bar{\xi}^{-1}=\eta\circ\xi^{-1}$. It follows that N also admits a C^r atlas. In particular, looking at Example 2c in Section A, we note that the graph of any *continuous* function $\mathbb{R}^m\to\mathbb{R}^n$ can be made into a C^∞ manifold, a fact that may sound unintuitive at first sight. But note that there is no topological distinction between $\mathbb{R}^m$ and the graph of a continuous function $\mathbb{R}^m\to\mathbb{R}^n$. The possible dents and corners on the graph of a non-differentiable continuous function will be manifested when we discuss "submanifolds" in the next chapter.
(c) It is a routine matter to check that the product manifold $M\times N$ (see Example 2f of Section A) of two C^r manifolds can be made into a C^r manifold (Exercise 4.17, also Subsection 18).

(d) We show that each of the atlases constructed in 2d of Section A for the unit m-sphere is a smooth (in fact, C^ω) atlas.
(i) Using the notation of the aforementioned example, if $i{\neq}j$, then $U_i^{\pm}{\cap}U_j^{\pm}$ is non-empty. We work out one example of coordinate change. For $i{<}j$,

$$\begin{aligned}\xi_j^+\circ(\xi_i^-)^{-1}(t^1,\dots,t^m) &= \xi_j^+\left(t^1,\dots,t^{i-1},-\sqrt{1-\sum(t^k)^2},t^i,\dots,t^m\right)\\ &= \left(t^1,\dots,t^{i-1},-\sqrt{1-\sum(t^k)^2},t^i,\dots,\widehat{t^{j-1}},\dots,t^m\right)\end{aligned}$$

Now since $(t^1,\dots,t^m)$ belongs to the *open* unit disk in $\mathbb{R}^m$, the expression under the radical sign does not vanish, and the map is C^ω. Other instances of change of variable are of exactly the same nature and the atlas is C^ω.
(ii) For the second atlas (stereographic projection), there are two changes of variable to check. Using the notation of Example 2d(ii) with $q{=}\sum(y^i)^2$, we obtain

$$\sigma_N\circ{\sigma_S}^{-1}(y^1,\dots,y^m)=\frac{1}{q}(y^1,\dots,y^m) \tag{4.5}$$

This is known as ***inversion*** (with respect to the unit sphere). Since the change of variable is taking place outside $\{N,S\}$, $(y^1,\dots,y^m){\neq}(0,\dots,0)$, $q{\neq}0$, and the map is C^ω. In fact, we have $\sigma_N{\circ}{\sigma_S}^{-1}{=}\sigma_S{\circ}{\sigma_N}^{-1}$, and the C^ω nature of atlas is established.
(e) We check that $\mathbb{RP}(m)$ with the atlas described in Example 2h is a C^ω manifold. Using the notation of that example, we work out the change of variable on $U_i{\cap}U_j$, $i{<}j$. This is the set of equivalence classes $[x^1:\cdots:x^{m+1}]$ with $x^i{\neq}0$ and $x^j{\neq}0$. We have

$$\begin{aligned}\xi_j\circ\xi_i^{-1}(t^1,\dots,t^m) &= \xi_j[t^1:\cdots:t^{i-1}:1:t^i:\cdots:t^m]\\ &= \frac{1}{t^j}(t^1,\dots,t^{i-1},1,t^i,\dots,\widehat{t^{j-1}},\dots,t^m)\end{aligned}$$

which expresses a C^ω map since $t^j{\neq}0$.
(f) We give an example of a C^1 atlas that is not C^2. Let $M{=}\mathbb{R}$ and consider an atlas of two charts $\mathcal{A}{=}\{(\mathbb{R},\mathbb{1}_{\mathbb{R}}),(\mathbb{R},\xi)\}$, where $\xi{:}\mathbb{R}{\to}\mathbb{R}$ is defined as follows

$$\xi(t)=\begin{cases}t^2+t & \text{if } t\geq 0\\ t & \text{if } t\leq 0\end{cases}$$

ξ is a C^1 diffeomorphism of $\mathbb{R}$ onto itself that is not C^2, since the derivative of ξ is not differentiable at $t{=}0$. It follows that $\mathcal{A}$ is C^1, but not C^2.

C. Smooth Structures

In the previous section, we establishcompatibleed that once a C^r atlas for a topological manifold M is fixed, with $r{\geq}1$, then the notion of differentiability of a function $M{\to}\mathbb{R}$ can be unambiguously defined relative to that atlas. A natural question is, under what circumstances do different atlases provide the same notion of differentiability? As a concrete example, in the case of the unit m-sphere, we provided two quite distinct smooth atlases, one with $2(m+1)$ charts, and another with just two. Is the concept of differentiability of a real-valued function on the unit sphere dependent on the choice of the atlas? Recalling the criterion that led to the definition of a C^r atlas for $r \geq 1$, one can make headway into streamlining this discussion. Consider two charts (U, ξ) and (V, η) for a topological manifold M. We say these two charts are ***C^r-compatible*** in case either $U \cap V$ is empty, or the coordinate-change mapping $\eta \circ \xi^{-1} : \xi(U \cap V) \to \eta(U \cap V)$ is a C^r diffeomorphism. Now if $\mathcal{A}$ and $\mathcal{B}$ are two C^r atlases for M so that any $(U, \xi) \in \mathcal{A}$ is C^r-compatible with any $(V, \eta) \in \mathcal{B}$, then the same reasoning that led us to the definition of C^r-atlas shows that the notion of a function on M being C^s, for $s \leq r$, is the same whether $\mathcal{A}$ or $\mathcal{B}$ is used. For this reason, it is natural to extend a given C^r atlas as much as possible to include all C^r-compatible charts. Thus if $\mathcal{A}$ is a C^r atlas for M, we let $\widetilde{\mathcal{A}}$ be the maximal C^r atlas that contains $\mathcal{A}$. We show this maximal atlas is unique. Suppose we annex to $\mathcal{A}$ any chart that is C^r-compatible with charts of $\mathcal{A}$. We have to show that the new expanded collection is still a C^r atlas. The only thing to verify is that any two additions (U, ξ) and (V, η) are also C^r-compatible. If $U \cap V$ is empty, there is nothing to prove, otherwise, let a be a point in $U \cap V$. There is then a chart (W, ζ) in $\mathcal{A}$ with $a \in W$. Now since both (U, ξ) and (V, η) are C^r-compatible with (W, ζ), the map $\eta \circ \xi^{-1} = (\eta \circ \zeta^{-1}) \circ (\zeta \circ \xi^{-1})$ is C^r at a. Therefore, $\widetilde{\mathcal{A}}$ is a C^r atlas. Note that this atlas can no longer be expanded because any new chart would have to be C^r-compatible with charts of $\mathcal{A}$, and hence already a member of $\widetilde{\mathcal{A}}$.

A maximal C^r-atlas for M will be called a C^r ***structure*** for M. A topological manifold has a unique C^0 structure, namely the collection of all possible charts for M. From now on, when we speak of a C^r manifold $(M, \mathcal{A})$, we always assume that either $\mathcal{A}$ is maximal or its maximal extension may be used whenever necessary.

8. Examples Consider the atlas of $\mathbb{R}^m$ consisting of the single chart $(\mathbb{R}^m, \mathbb{1}_{\mathbb{R}^m})$. The extension of this atlas to the maximal atlas is called the ***standard smooth structure*** on $\mathbb{R}^m$. The same terminology is used for open subsets W of $\mathbb{R}^m$, when the singleton atlas $\{(W, \mathbb{1}_W)\}$ is maximally extended.

Several questions come to mind:

Question 1. Does every topological manifold admit a C^1 structure?

Question 2. Given a C^r structure $\mathcal{A}$ for M, $r \geq 1$, and $s > r$, is there a subset $\mathcal{B}$ of $\mathcal{A}$ that is a C^s atlas for M?

Question 3. Can there be more than one C^1 structure for a topological manifold?

The first examples of topological manifolds that are not "smoothable," i.e., do not admit a C^1 structure, were discovered around 1960 and were of dimension ≥ 8. More recently, 4-dimensional examples have been found. The subject of smoothing is studied in differential and algebraic topology.

The answer to Question 2 is positive. In fact, any C^1 structure contains a smooth (C^∞) structure. For this reason and due to the convenience of working with smooth data, we will be mainly working with smooth manifolds in this book. A deeper result is that a C^∞ structure always contains a real-analytic (C^ω) structure. This result of Whitney and the previously mentioned result were among the early theorems of differential topology. For a modern treatment of Question 2, see [**13**].

As posed, there is a trivial positive answer to Question 3. Let $\mathcal{A}$ be a maximal C^1 atlas for M and suppose $h{:}M\to M$ is a homeomorphism, which is not differentiable as a map $(M,\mathcal{A})\to(M,\mathcal{A})$. A new atlas $\mathcal{A}'$ is defined for M as follows. For each $(U,\xi)\in\mathcal{A}$, we define $(U',\xi')\in\mathcal{A}'$ by

$$U' = h(U),\ \xi' = \xi \circ h^{-1}$$

Now $\mathcal{A}'$ is a C^1 atlas for M, for if (U',ξ') and (V',η') are constructed as above with $U'\cap V'$ non-empty, then $\eta'\circ\xi'^{-1}=\eta\circ\xi^{-1}$ is C^1. We show that $\mathcal{A}'$ is *maximal.* Let (V,η) be a C^1-related addition to $\mathcal{A}'$. For any $(U',\xi')\in\mathcal{A}'$ with $U'\cap V$ non-empty, $\eta\circ(\xi')^{-1}=\eta\circ h\circ\xi^{-1}$ is a C^1 diffeomorphism from $\xi'(U'\cap V)$ to $\eta(U'\cap V)$. It follows that $(h^{-1}(V),\eta\circ h)$ is C^1-related to any $(U,\xi)\in\mathcal{A}$ with $U\cap h^{-1}V$ non-empty. Since $\mathcal{A}$ is maximal, $(h^{-1}(V),\eta\circ h)\in\mathcal{A}$, implying that $(V,\eta)\in\mathcal{A}'$; this proves that $\mathcal{A}'$ is maximal. Now, in fact, $\mathcal{A}$ and $\mathcal{A}'$ are *distinct* maximal atlases since a chart of the form (U',ξ') belongs to $\mathcal{A}'$ but not to $\mathcal{A}$. Note, however, that $h{:}(M,\mathcal{A})\to(M,\mathcal{A}')$ is a C^1 diffeomorphism. To overcome this artificial distinction, we define two maximal C^1 atlases $\mathcal{A}$ and $\mathcal{A}'$ for M to be ***C^1-equivalent*** in case there is a C^1 diffeomorphism $\varphi{:}(M,\mathcal{A})\to(M,\mathcal{A}')$. We can then fine-tune Question 3 in the following manner:

Question 3′**.** Can a topological manifold admit two non-equivalent C^1 structures?

This turns out to be a highly non-trivial question. In 1956, J. Milnor showed that S_7 possesses 28 non-equivalent C^1 structures! The first examples of non-smoothable manifolds, mentioned earlier, are related to Milnor's discovery. In 1982, M. Freedman showed that $\mathbb{R}^4$ possesses an uncountable number of non-equivalent C^1 structures. It had been known earlier that the standard C^1 structure of $\mathbb{R}^n$ is the only possible one for $n\neq 4$.

On a much lighter note, we are asking the reader in Exercise 4.8 to show that the two atlases described for the unit sphere in Example 2d extend to the same maximal atlas.

9. Remark Classification of one-dimensional and two-dimensional manifolds is quite classical. Any smooth one-dimensional manifold is diffeomorphic to either S_1 or $\mathbb{R}$ (see [**18**] or [**20**]). Complete classification of two-dimensional manifolds (surfaces) is also well-known ([**13**] and [**22**]). The last 50 years have witnessed crucial breakthroughs in the theory of 3- and 4-dimensional manifolds, subjects of intensive current research. Important advances in the classification of manifolds of dimension 5 and higher began in the 1960s.

Before ending this section, we will describe a very useful elementary tool, called the *partition of unity*, for piecing together local data defined on charts to create a global object on the manifold. Examples of the concrete use of this tool will appear in future chapters. As the name "partition of unity" suggests, this is a method of averaging with variable and smoothly varying weights over the manifold. This device works in the C^∞ case but is not valid for real-analytic data, as essential use will be made of non-analytic C^∞ auxiliary bump functions that were described in Lemma 12 of Chapter 2.

We begin with two general definitions. An open cover of a topological space is ***locally finite*** if every point of the space has a neighborhood that intersects only a finite number of the members of the cover. A cover $\mathcal{B}$ is a ***refinement*** of a cover $\mathcal{A}$ if every member of $\mathcal{B}$ is a subset of some member of $\mathcal{A}$. We call an atlas $\mathcal{A}$ for a manifold M of dimension m a ***handy atlas*** if the following hold:
(i) $\mathcal{A}$ is countable and locally finite.
(ii) For each $(U,\xi)\in\mathcal{A}$, $\xi(U)$ contains the closed ball $\bar{B}_2(\mathbf{0})$ of radius 2 around $\mathbf{0}$ in $\mathbb{R}^m$.
(iii) The inverse images of the unit open ball of radius 1, $\xi^{-1}(B_1(\mathbf{0}))$, for all $(U,\xi)\in\mathcal{A}$, form a cover of M.

10. Lemma *Let M be a C^r manifold with maximal atlas $\widetilde{\mathcal{A}}$, $r\geq 0$, and suppose that $(U_\alpha)_\alpha$ is an open cover of M. Then $\widetilde{\mathcal{A}}$ contains a handy atlas that is a refinement of $(U_\alpha)_\alpha$.*

Proof. According to Subsection 3(e), there is a sequence (K_i), i=1, 2, ... of compact subsets of M with properties $K_i\subset int(K_{i+1})$ and $M=\bigcup_i K_i$. Letting $K_i=\emptyset$ for $i\leq 0$, we denote the closure of $K_{i+1}-K_i$ by C_i, for i=0, 1, ... The sets C_i are also compact and cover M. For each $x\in C_i$, we pick an index $\alpha(x)$ so that $x\in U_{\alpha(x)}$. By substituting a smaller open neighborhood of x for $U_{\alpha(x)}$, if necessary, we may assume that $U_{\alpha(x)}$ is the domain of a chart $(U_{\alpha(x)},\xi_{\alpha(x)})$. We claim there exists a chart $(V_{x,i},\xi_{x,i})$ in $\widetilde{\mathcal{A}}$ so that

$$V_{x,i}\subset U_{\alpha(x)}\cap(int(K_{i+2})-K_{i-1})$$
$$\xi_{x,i}(V_{x,i})\supset\bar{B}_2(\mathbf{0}),\ \xi_{x,i}(x)=\mathbf{0}$$

The first requirement follows from the fact that we can substitute for $U_{\alpha(x)}$, its intersection with the open set $int(K_{i+2})-K_{i-1}$, and restrict $\xi_{\alpha(x)}$ to this possibly smaller set. For the second requirement, we may follow $\xi_{\alpha(x)}$ with a translation and a linear expansion in $\mathbb{R}^m$ to ensure the condition. Both translations and invertible linear maps are analytic diffeomorphisms and do not reduce differentiability. Now consider open sets $W_{x,i}=\xi_{x,i}{}^{-1}(B_1(\mathbf{0}))$ for all $x\in C_i$. These sets cover the compact set C_i, so a finite number can be selected that will cover C_i. Overall, a countable number of $W_{x,i}$'s will cover M, which we renumber as W_j, j=1, 2, ... The corresponding $(V_{x,i}, \xi_{x,i})$'s will be renumbered as (V_j, ξ_j), j=1, 2, ... The collection (V_j) is locally finite since V_j=$V_{x,i}$ can only intersect those $V_{y,k}$ with $i-2 \leq k \leq i+2$, of which only a finite number have been selected. The family (V_j, ξ_j) is the desired handy atlas. □

11. Theorem (Partition of Unity) *Let $(M, \widetilde{\mathcal{A}})$ be a C^r manifold, $0 \leq r \leq \infty$, and suppose that $(U_\alpha)_\alpha$ is an open cover of M. Then there exists a countable locally finite open cover $(V_i)_i$ of M and a countable family of C^r functions θ_i:$M \to \mathbb{R}$ with the following properties:*
i) $(V_i)_i$ is a refinement of $(U_\alpha)_\alpha$.
ii) Each V_i is the domain of a chart in $\widetilde{\mathcal{A}}$.
iii) The **support** *of each θ_i, defined as the closure of the set $\{x \in M : \theta_i(x) \neq 0\}$ is contained in V_i.*
iv) For each $x \in M$, $\sum_i \theta_i(x) = 1$.

PROOF. We construct sets V_i as in the lemma so that $(V_i)_i$ is a locally finite refinement of $(U_\alpha)_\alpha$, each V_i is the domain of a chart (V_i, ξ_i) in $\widetilde{\mathcal{A}}$ and the sets $W_i = \xi_i^{-1}(B_1(\mathbf{0}))$ form an open cover of M. We know from Lemma 12 of Chapter 2 that a C^∞ function $\delta : \mathbb{R}^m \to [0, 1]$ exists that takes value 1 on $B_1(\mathbf{0})$ and has value 0 outside $B_2(\mathbf{0})$. We define δ_i:$M \to \mathbb{R}$ by

$$\delta_i(x) = \begin{cases} \delta(\xi_i(x)) & \text{if } x \in V_i \\ 0 & \text{if } x \notin V_i \end{cases}$$

δ_i is C^r on V_i, being the composition of a C^r mapping and a C^∞ function. Moreover, since $\xi_i(V_i) \supset \bar{B}_2(\mathbf{0})$, and δ vanishes outside $B_2(\mathbf{0})$, extending δ_i by value 0 outside V_i yields a C^r function. Now the sets V_i form a locally finite family, so for a given $x \in M$, the value of $\delta_i(x)$ is zero except for a finite number of indices i. On the other hand, since the sets W_i cover M, $\delta_i(x)$=1 for at least one index i, so $\sum_i \delta_i(x) > 0$. We then define

$$\theta_i(x) = \frac{\delta_i(x)}{\sum_i \delta_i(x)}$$

These functions satisfy the required properties. □

We refer to $(\theta_i)_i$ as a C^r ***partition of unity*** subordinate to $(U_\alpha)_\alpha$. In practice, we often use the following corollary.

12. Corollary *Let $(M, \widetilde{\mathcal{A}})$ be a C^r manifold, $0 \leq r \leq \infty$, and suppose that $(U_i)_i$ is a (countable) locally finite open cover of M. Then there exists a family of C^r functions θ_i:$M \to \mathbb{R}$ with the following properties:*
i) The support of θ_i is contained in U_i.
ii) For each $x \in M$, $\sum_i \theta_i(x) = 1$.

PROOF. We note initially that any locally finite open cover of a second-countable topological space (e.g., a manifold) is necessarily countable. Suppose that $(U_\alpha)_{\alpha \in I}$ is a locally finite open cover with I uncountable. For each α, we pick a point $x_\alpha \in U_\alpha$. Some x_α will then be a limit point for this set of points (see [**16**], p.48) and any neighborhood of this point will have non-empty intersection with U_α for infinitely many α, contradicting local-finiteness. Going back to the main proof and using Theorem 11, we consider for each U_i those V_i^j that refine U_i with corresponding θ_i^j as in Theorem 11. Letting $\phi_i = \sum_j \theta_i^j$, note that the support of ϕ_i is contained in U_i. Now the functions

$$\theta_i(x) = \frac{\phi_i(x)}{\sum_i \phi_i(x)}$$

satisfy the requirements. □

D. The Tangent Bundle

In the previous two sections, we set up the machinery necessary to discuss the differentiability of a map from a manifold to $\mathbb{R}$ or from one manifold to another. What we did not discuss was what the actual "derivative" of such a map would be. The goal of this section is to complete the picture by providing an appropriate derivative-like concept.

13. Exploratory Discussion

Let $(M, \mathcal{A})$ be a C^r m-manifold, $r \geq 1$, and suppose that $f : M \to \mathbb{R}$ is differentiable at a point $a \in M$. How should one define the derivative of f at a? We know how to do this if the domain of f were an open subset of $\mathbb{R}^m$. So a natural attempt is to use charts of the given atlas to transfer the problem to $\mathbb{R}^m$. Suppose $(U, \xi) \in \mathcal{A}$ with $a \in U$. Then $f \circ \xi^{-1}$ is the local coordinate representation of f, which we know is differentiable at $\xi(a)$. So a candidate for the derivative of f at a is the derivative of $f \circ \xi^{-1}$ at $\xi(a)$, which is a linear map from $\mathbb{R}^m$ to $\mathbb{R}$:

$$D(f \circ \xi^{-1})(\xi(a)) : \mathbb{R}^m \to \mathbb{R}$$

Now taking another representation via a chart $(V, \eta) \in \mathcal{A}$, $a \in V$, we obtain a similar candidate:

$$D(f \circ \eta^{-1})(\eta(a)) : \mathbb{R}^m \to \mathbb{R}$$

If this approach were to be succesful, these two linear maps should coincide. However, the chain rule gives

$$D(f \circ \xi^{-1})(\xi(a)) = D(f \circ \eta^{-1})(\eta(a)) \circ D(\eta \circ \xi^{-1})(\xi(a)) \tag{4.6}$$

For this to hold, the derivative of change-of-coordinate function $\eta \circ \xi^{-1}$ at the point $\xi(a)$ would have to be the identity map, something not to be expected for an arbitrary pair of charts in the maximal atlas. A step in resolving this apparently disheartening state of affairs is to recall the local case, Section A of Chapter 2, where we supplanted the derivative by the "tangent map" that was defined, not on $\mathbb{R}^m$, but on the tangent space at the given point. We now aim to define at each point a of M, a tangent space, i.e., a vector space of the same dimension as M, which will serve as the domain of an appropriately defined tangent map of f at a.

How is one to construct a "tangent space" to M at a point of M, given that M is an abstract object? If M were a smooth surface sitting in $\mathbb{R}^3$, we would have the familiar picture of the tangent plane $\mathcal{P}$ as the flat approximation to the surface at the point a (see Figure 9). Now suppose we flatten out a piece of the surface around a as

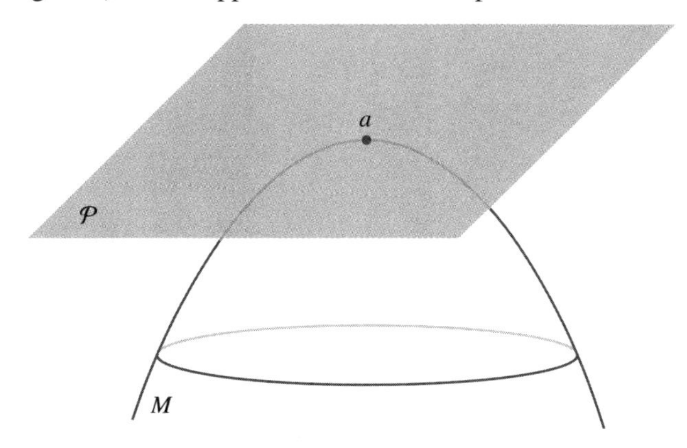

FIGURE 9. Tangent plane

in a geographic map. Mathematically, this translates as taking a chart for M around the point a. The tangent plane to the surface would then be expected to follow along and lie flat on the straightened-out surface. The vectors in the tangent plane would be mapped to vectors in the plane with the origin at the image of the point a. Let us try to justify this more convincingly. Any tangent vector to the surface, i.e., any vector $\mathbf{z}$ in $\mathcal{P}$ emanating from a, can be realized as the velocity vector of a curve α whose image lies on the surface M. Our chart maps the image of the curve α to a curve γ on the flat $\mathbb{R}^2$ plane carrying the point a to the point p. Then the velocity vector $\mathbf{z}$ of α at a would correspond to the velocity vector $\mathbf{v}$ of γ at the point p. But since the image of the curve γ lies in flat $\mathbb{R}^2$, its velocity vector at p would lie flat on this surface as a vector originating from p. The tangent plane $\mathcal{P}$ will then correspond to the tangent plane to the $\mathbb{R}^2$ at point p (as defined in Chapter 2). Now suppose we take a different chart around a (see Figure 10). In this chart, the curve α is mapped to the curve λ, the point a is carried to q and the velocity vector $\mathbf{z}$ to $\mathbf{w}$. What is the relationship between $\mathbf{v}$ and $\mathbf{w}$? The passage from the image of one chart to the image of the other is just what we have been calling change-of-coordinates, or transition map. Let us denote it with h. Then

$$\lambda(t) = h(\gamma(t)) \tag{4.7}$$

with, say, $\gamma(t_0) = p$ and $\lambda(t_0) = h(\gamma(t_0)) = q$. Now $\mathbf{v} = \gamma'(t_0)$ and $\mathbf{w} = \lambda'(t_0)$, so using the chain rule to differentiate (4.7), we obtain

$$\mathbf{w} = (Dh(p))(\mathbf{v}) \tag{4.8}$$

In other words, one representation of the tangent vector to M at a is carried to another representation by the derivative of the coordinate change. This key idea leads us to

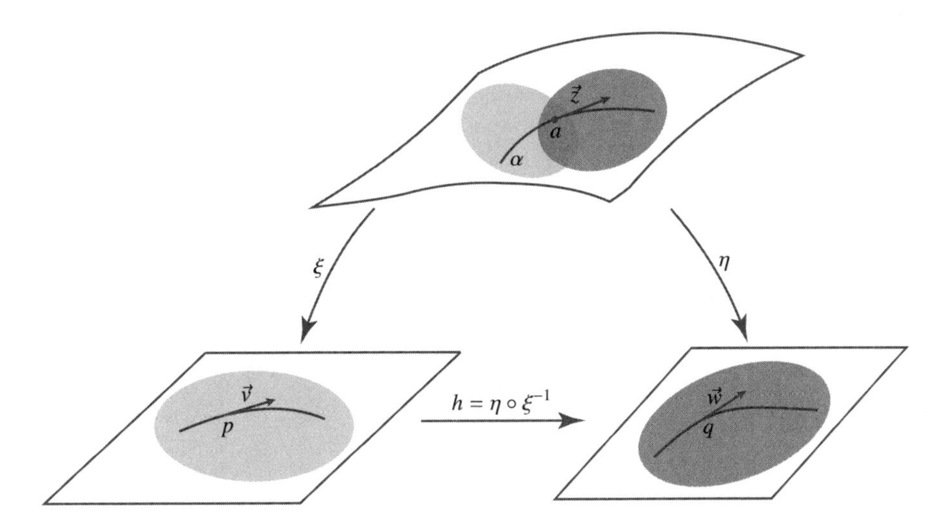

FIGURE 10. Two representations of a tangent vector

the formal definition of tangent space in the abstract setting. An arbitrary m-manifold M is not given as sitting in any known or familiar space; it is a topological space with a given atlas of C^r-compatible charts. We have no direct visualization of the "tangent space" to M at a point $a \in M$, but we have as many would-be representations of this elusive object as the charts containing the point a. For every such chart (U, ξ), the tangent space to $\mathbb{R}^m$ at $\xi(a)$ is one such potential representation. We can integrate all these representations into one by identifying tangent vectors in different representations that correspond under the derivative map of the coordinate change from one to the other. We will then call the integrated abstract object the "tangent space to M at a". With this in mind, we first lay out a clean-cut general setting, and then proceed with the precise definition of the tangent space, as well as the tangent map of a differentiable map.

14. Identification Lemma
Let $(E_\alpha)_{\alpha \in I}$ be a family of real vector spaces of the same dimension and suppose for each pair $(\alpha, \beta) \in I \times I$, a linear isomorphism $\lambda_\alpha^\beta : E_\alpha \to E_\beta$ is given so that for all $\alpha, \beta, \gamma \in I$,

$$\lambda_\beta^\gamma \circ \lambda_\alpha^\beta = \lambda_\alpha^\gamma \tag{4.9}$$

Then there exists a real vector space E, and an isomorphism $i_\alpha : E_\alpha \to E$ for each $\alpha \in I$, such that

$$i_\alpha = i_\beta \circ \lambda_\alpha^\beta, \ \forall \alpha, \beta \in I \tag{4.10}$$

Further, if for each $\alpha\in I$, a linear map $L_\alpha:E_\alpha\to\mathbb{R}$ is given that satisfies the compatibility condition

$$L_\alpha = L_\beta \circ \lambda_\alpha^\beta,\ \forall\alpha,\beta\in I \tag{4.11}$$

then a linear map $L:E\to\mathbb{R}$ exists with the property

$$L_\alpha = L\circ i_\alpha,\ \forall\alpha\in I \tag{4.12}$$

Before embarking on the proof, we note the following two immediate consequences of (4.9):

$$\lambda_\alpha^\alpha = \mathbb{1}_{E_\alpha},\ \forall\alpha\in I \tag{4.13}$$

$$\lambda_\beta^\alpha = (\lambda_\alpha^\beta)^{-1},\ \forall\alpha,\beta\in I \tag{4.14}$$

The first identity results from setting two of α,β,γ equal in (4.9), and then the second by letting $\gamma=\alpha$.

Proof. We define a relation $\sim$ on the disjoint union $\widetilde{E}$ of E_α's in the following manner. If $e_\alpha\in E_\alpha$ and $e_\beta\in E_\beta$, then $e_\alpha\sim e_\beta$ if and only if $\lambda_\alpha^\beta(e_\alpha)=(e_\beta)$. It follows from (4.13), (4.14) and (4.9) that $\sim$ is an equivalence relation with each equivalence class $[e_\alpha]$ containing exactly one element from each E_α. We denote the set of equivalence classes by E. For each α, we define $i_\alpha:E_\alpha\to E$ by $i_\alpha(e_\alpha)=[e_\alpha]$. (4.10) holds by virtue of the definition of $\sim$. Since i_α is a one-to-one correspondence, we can use it to transfer the linear structure of E_α onto E. This structure is well-defined, i.e., it is independent of any particular α, on account of (4.10) and the fact that each λ_α^β is an isomorphism. Finally, $L:E\to\mathbb{R}$ is defined by $L[e_\alpha]=L_\alpha(e_\alpha)$. That this is well-defined follows from (4.11) and the definition of $\sim$. The identity (4.12) is virtually the definition of L. □

We now go back to the task of defining the tangent space. Let $(U_\alpha,\xi_\alpha)_{\alpha\in I}$ be the family of all charts in the maximal atlas of the C^r manifold M such that $a\in U_\alpha$. For each $\alpha\in I$, let E_α be the tangent space to $\mathbb{R}^m$ at $\xi_\alpha(a)$. Explicitly, we recall from Chapter 2 that E_α is the set of ordered pairs $(\xi_\alpha(a),e)$, where e runs through the elements of $\mathbb{R}^m$. E_α is in one-to-one correspondence with $\mathbb{R}^m$ via the map $k_\alpha:(\xi_\alpha(a),e)\mapsto e$. Using this correspondence, E_α is a real linear space of dimension m (isomorphic to $\mathbb{R}^m$) by operations that ignore the first component:

$$(\xi_\alpha(a),e_1)+(\xi_\alpha(a),e_2) = (\xi_\alpha(a),e_1+e_2),\ \forall e_1,e_2\in\mathbb{R}^m$$
$$r(\xi_\alpha(a),e) = (\xi_\alpha(a),re),\ \forall r\in\mathbb{R}$$

Thus k_α establishes a linear isomorphism between E_α and $\mathbb{R}^m$. For $\alpha,\beta\in I$, let us denote the coordinate change, i.e., the restriction of $\xi_\beta\circ\xi_\alpha{}^{-1}$ to $\xi_\alpha(U_\alpha\cap U_\beta)$, by ξ_α^β. Then we define $\lambda_\alpha^\beta:E_\alpha\to E_\beta$ by

$$\lambda_\alpha^\beta(\xi_\alpha(a),e) = \left(\xi_\beta(a), D\xi_\alpha^\beta(\xi_\alpha(a))(e)\right) \tag{4.15}$$

The reader should note that (4.15) is just an abstract rendition of (4.8). Now for $\alpha,\beta,\gamma\in I$, we have $\xi_\alpha^\gamma=\xi_\beta^\gamma\circ\xi_\alpha^\beta$ on $\xi_\alpha(U_\alpha\cap U_\beta\cap U_\gamma)$, and by the chain rule, the relation (4.9) holds. The existence of m-dimensional vector space E and isomorphisms i_α satisfying (4.10) follows from the lemma. In fact, explicitly, E is the set of equivalence classes of the relation $\sim_a$ defined by

$$(4.16) \qquad (\xi_\alpha(a), e) \sim_a \left(\xi_\beta(a), D\xi_\alpha^\beta(\xi_\alpha(a))(e)\right)$$

In the language of Chapter 2, Section A, we are using the isomorphism from $T_{\xi_\alpha(a)}\mathbb{R}^m$ to $T_{\xi_\beta(a)}\mathbb{R}^m$, induced by the tangent map of the coordinate change ξ_α^β at the point $\xi_\alpha(a)$, to identify the two tangent spaces. E can then be regarded as the result of these identifications, i.e., the set of equivalence classes $[(\xi_\alpha(a), e)]_a$, where $[.]_a$ denotes equivalence class under $\sim_a$. The map i_α sends each element of $T_{\xi_\alpha(a)}\mathbb{R}^m$ to its equivalence class. Finally, we rename E the ***tangent space of** M **at** a*, and denote it by T_aM.

Now we can place the various candidates for the derivative of $f{:}M\to\mathbb{R}$ in proper perspective. Let

$$L_\alpha = T_{\xi_\alpha(a)}(f\circ\xi_\alpha{}^{-1}) : T_{\xi_\alpha}\mathbb{R}^m = E_\alpha \to \mathbb{R}$$

With this definition, relation (4.6) is transformed into (4.11), and we are in position to use the second part of the lemma. Thus a linear map $L{:}T_aM\to\mathbb{R}$ is obtained that satisfies (4.12). In the present context, this means that L is the result of the identification of all candidates for the derivative of f at a as a single linear map $T_aM\to\mathbb{R}$. We rename this map , and call it the ***differential of** f **at** a*. If f is differentiable at every point of M, then df (the ***differential of f***) assigns to each $a\in M$, a linear map $T_aM\to\mathbb{R}$, i.e., an element of the dual of T_aM.

The above discussion accomplishes our quest for finding a substitute for the notion of derivative of a real-valued function. To carry this project further, we will need to look more carefully at the totality of all tangent spaces assigned to various points on the manifold.

15. Manifold Structure of the Tangent Bundle

Having defined a tangent space T_xM at every point x of a C^r manifold M, $r\geq 1$, we proceed, as in Chapter 2, to form their disjoint union

$$TM = \bigsqcup_{x\in M} T_xM$$

which we call, as before, the ***tangent bundle of** M*. More generally, for an open subset N of M, which is a C^r manifold in its own right, see Subsection 7(a), we have

$$TN = \bigsqcup_{x\in N} T_xM$$

In the local case, TU, where U was an open subset of $\mathbb{R}^m$, every point w of TU was an ordered pair $w=(x, v)$, with $x\in U$ and $v\in\mathbb{R}^m$. In other words, there were two well-defined projections on first and second components, $\pi_1=\tau_U{:}w\mapsto x$, and $\pi_2{:}w\mapsto v$. In the case of TM, there is a natural projection on the first component, called the ***tangent bundle projection***, $\tau_M{:}TM{\to}M$, which maps every $w{\in}T_xM$ to x, because TM is the *disjoint* union of T_xM's, and each $w{\in}TM$ belongs to a uniquely defined T_xM. However, defining projection on the second component turns out to be problematic. Even though any m-dimensional real vector space is isomorphic to $\mathbb{R}^m$, there is no canonical or basis-independent recipe for an isomorphism. It will turn out that, in general, no homeomorphism $TM{\to}M{\times}\mathbb{R}^m$ exists that maps each T_xM to $\{x\}{\times}\mathbb{R}^m$. This will be a major point of departure from the local theory developed earlier, and will be at the core of the introduction of "vector bundles" in Chapter 6. For this reason, it is essential that the reader pay careful attention to what is to follow.

We proceed to define a topology for TM that will be consistent with our local treatment in Chapter 2. For an open subset U of $\mathbb{R}^m$, it was natural there to assign the product topology to TU on account of the set equality $TU{=}U{\times}\mathbb{R}^m$. Now if (U,ξ) is a chart in the maximal C^r atlas of M, (4.16) allows us to define a one-to-one correspondence between $T(\xi(U))$ and TU by

$$(\xi(x), v) \to [(\xi(x), v)]_x, \; x{\in}U, \; v{\in}\mathbb{R}^m \tag{4.17}$$

where $[.]_x$ is the equivalence class under $\sim_x$. Note that for a fixed x, this mapping is a vector space isomorphism between the tangent space to $\mathbb{R}^m$ at $\xi(x)$ (as defined in Chapter 2), and T_xM. In particular, $(\xi(x), \mathbf{0})$ corresponds to the zero element $\mathbf{0}_x$ of T_xM. Using this one-to-one correspondence, we transfer the product topology on $T(\xi(U)){=}\xi(U){\times}\mathbb{R}^m$ to TU. Since ξ is a homeomorphism of U onto $\xi(U)$, it follows that TU is also homeomorphic to $U{\times}\mathbb{R}^m$. To recapitulate, we have now arrived at the following situation. For each chart (U,ξ) in the maximal C^r atlas of M, a homeomorphism $TU{\to}\xi(U){\times}\mathbb{R}^m$ exists, the restriction of which to each T_xM is a linear isomorphism from T_xM to $T_{\xi(x)}U$. To define a compatible topology for the entire TM, we invoke a general topological concept.

16. Lemma (Sum Topology) *Let X be a set, and $X{=}\bigcup_{\alpha\in I} X_\alpha$. Suppose that on each X_α, a topology $\mathfrak{T}_\alpha$ is given. We define a family $\mathfrak{T}$ of subsets of X by*

$$U \in \mathfrak{T} \iff U \cap X_\alpha \in \mathfrak{T}_\alpha, \; \forall\alpha{\in}I$$

Then $\mathfrak{T}$ is a topology on X (called the **sum topology***). Further, suppose that the following conditions hold:*
(i) For all $\alpha,\beta{\in}I$, the set $X_\alpha{\cap}X_\beta$ belongs to both $\mathfrak{T}_\alpha$ and $\mathfrak{T}_\beta$, and
(ii) the induced topologies on $X_\alpha{\cap}X_\beta$ from $\mathfrak{T}_\alpha$ and $\mathfrak{T}_\beta$ coincide.
Then $\mathfrak{T}_\alpha{\subset}\mathfrak{T}$ for all $\alpha{\in}I$.

Proof. That $\emptyset, X{\in}\mathfrak{T}$ follows from the fact that $\emptyset, X_\alpha{\in}\mathfrak{T}_\alpha$ for each α. Suppose $\{U_\mu\}_\mu$ is a collection of elements of $\mathfrak{T}$, then $U_\mu{\cap}X_\alpha{\in}\mathfrak{T}_\alpha$ for all $\alpha{\in}I$, so

$$(\bigcup_\mu U_\mu) \cap X_\alpha = \bigcup_\mu (U_\mu \cap X_\alpha) \in \mathfrak{T}_\alpha$$

since each $\mathcal{T}_\alpha$ is a topology. So $\mathcal{T}$ is closed under arbitrary unions. Likewise, if $U_1, \ldots, U_n$ are in $\mathcal{T}$, then

$$(U_1 \cap \cdots \cap U_n) \cap X_\alpha = (U_1 \cap X_\alpha) \cap \cdots \cap (U_n \cap X_\alpha) \in \mathcal{T}_\alpha$$

and $\mathcal{T}$ is closed under finite intersections. Therefore, $\mathcal{T}$ is a topology on X. For the second part, we first note that each X_α is in $\mathcal{T}$. This follows from (i) and the definition of $\mathcal{T}$. Now if $U{\in}\mathcal{T}_\alpha$, we must show $U{\cap}X_\beta{\in}\mathcal{T}_\beta$ for all β. But

$$U \cap X_\beta = U \cap (X_\alpha \cap X_\beta) \subset X_\alpha \cap X_\beta$$

and $U{\cap}X_\beta$ is open in $X_\alpha{\cap}X_\beta$ relative to the induced topology from $\mathcal{T}_\alpha$. It follows from (ii) that $U{\cap}X_\beta$ is open in $X_\alpha{\cap}X_\beta$ relative to the topology induced from X_β. But $X_\alpha{\cap}X_\beta$ is open in $\mathcal{T}_\beta$ by (i), so $U{\cap}X_\beta$ is open in X_β, i.e., it is a member of $\mathcal{T}_\beta$. □

Going back to the tangent bundle, if $\{U_\alpha\}_{\alpha\in I}$ is the set of charts of the maximal C^r atlas of M, then $TM{=}\bigcup_{\alpha\in I} TU_\alpha$, and we have already defined a topology for each TU_α. We then assign the sum topology to TM as in the first part of the lemma and check that conditions (i) and (ii) also satisfied here. For (i), first note that $TU_\alpha{\cap}TU_\beta{=}T(U_\alpha{\cap}U_\beta)$. If $U_\alpha{\cap}U_\beta{=}\emptyset$, then (i) holds, otherwise $\xi_\alpha(U_\alpha{\cap}U_\beta)$ is an open subset of $\xi_\alpha(U_\alpha)$, so $\xi_\alpha(U_\alpha{\cap}U_\beta){\times}\mathbb{R}^m$ is an open subset of $\xi_\alpha(U_\alpha){\times}\mathbb{R}^m$, and the same reasoning holds replacing α by β. Since TU is homeomorphic to $\xi(U){\times}\mathbb{R}^m$ for any chart (U,ξ), (i) follows. (ii) is immediate since the topology induced from both TU_α and TU_β on $T(U_\alpha{\cap}U_\beta)$ is the product topology $(U_\alpha{\cap}U_\beta){\times}\mathbb{R}^m$. Therefore, by the lemma, open subsets of each TU_α are also open in TM.

We next show that if M is a C^r manifold, $r{\geq}1$, of dimension m, then TM can be made into a C^{r-1} manifold of dimension $2m$ in a natural way. Consider a chart (U,ξ) in the maximal C^r atlas of M. In (4.17) we constructed a one-to-one correspondence between TU and $T(\xi(U)){=}\xi(U){\times}\mathbb{R}^m$, through which we assigned a topology to TU. Let us denote the inverse of (4.17) by $T\xi$. Thus $T\xi$ is a homeomorphism from TU onto $\xi(U){\times}\mathbb{R}^m$.

17. Theorem *Let M be a C^r manifold of dimension m, $r{\geq}1$. Then TM has the structure of a C^{r-1} manifold of dimension $2m$. In fact, if $\mathcal{A}$ is a C^r atlas for M and $(U,\xi){\in}\mathcal{A}$, then the set of $(TU,T\xi)$ is a C^{r-1} atlas for TM.*

Proof. Let us first check the topological prerequisites for TM to be a manifold. Note that for each chart (U,ξ), TU is both Hausdorff and second-countable, having the product topology of two Hausdorff and second-countable spaces. Now take two points $w,w'{\in}TM$. If there is a chart (U,ξ) in the maximal atlas so that $w,w'{\in}TU$, we can separate them by open sets of TU, and these sets are also open in TM. If w and w' do not belong to the same TU, then, in particular, $\tau_M(w){\neq}\tau_M(w')$, and we can find disjoint open subsets U and U' of M so that $\tau_M(w){\in}U$ and $\tau_M(w'){\in}U'$. It follows that $w{\in}TU$ and $w'{\in}TU'$, with $TU{\cap}TU'{=}\emptyset$. Thus TM is Hausdorff.

For second-countability of M, first note that by Lemma 10, the maximal C^r atlas of M contains a countable sub-atlas $(U_i)_i$, $i=1,2,\ldots$. Now each TU_i, being second-countable, possesses a countable basis of open sets that we denote by $\{W_{i,1}, W_{i,2}, \ldots\}$. We claim that the sets $W_{i,j}$ form a basis for the topology of TM. If W is an open subset of TM, we write $W=\bigcup_i(W\cap TU_i)$. But each $W\cap TU_i$ is open in TU_i, so it can be written as a union of $W_{i,j}$'s, and the assertion follows.

Finally, given a C^r atlas $\mathcal{A}$ for M, for charts $(U,\xi)\in\mathcal{A}$, we have constructed the charts $(TU, T\xi)$ for the topological manifold TM. We have to show that these charts are C^{r-1}-compatible. For $(U,\xi),(V,\eta)\in\mathcal{A}$, if $U\cap V=\emptyset$, then $TU\cap TV=\emptyset$, and there is nothing to prove. Suppose that $U\cap V\neq\emptyset$, we investigate the change of variable. Let x be an arbitrary point in $U\cap V$. Then

$$\begin{aligned}
T\eta\circ T\xi^{-1}(\xi(x),v) &= T\eta[(\xi(x),v)]_x \\
&= T\eta[(\eta(x), D(\eta\circ\xi^{-1})(\xi(x))(v))]_x \\
&= (\eta(x), D(\eta\circ\xi^{-1})(\xi(x))(v)) \\
&= ((\eta\circ\xi^{-1})(\xi(x)), D(\eta\circ\xi^{-1})(\xi(x))(v))
\end{aligned}$$

Now $\eta\circ\xi^{-1}$ is C^r, being a change of variable in the C^r atlas $\mathcal{A}$, so its derivative, $D(\eta\circ\xi^{-1})$, is C^{r-1}. It follows that the atlas constructed for TM is a C^{r-1} atlas, and the proof is complete. □

18. Example In Exercise 4.17, we are asking the reader to prove that if M and N are C^r manifolds, then $M\times N$ can be endowed with a C^r manifold structure. Moreover, if $r\geq 1$, there is a natural C^{r-1} homeomorphism

$$C : T(M\times N) \to TM\times TN$$

In fact, if (U,ξ) and (V,η) are charts for the m-manifold M and the n-manifold N, respectively, then a chart $(U\times V, \xi\times\eta)$ is defined for $M\times N$ by $(\xi\times\eta)(x,y)=(\xi(x),\eta(y))\in\mathbb{R}^{m+n}$, and

$$C : [(\xi\times\eta)(x,y),(u,v)]_{(x,y)} \mapsto ([\xi(x),u]_x, [\eta(y),v]_y) \tag{4.18}$$

where $u\in\mathbb{R}^m$ and $v\in\mathbb{R}^n$. The main point here is that C is well-defined. For if (U',ξ') and (V',η') are charts, respectively, around x and y, and

$$[(\xi\times\eta)(x,y),(u,v)]_{(x,y)} = [(\xi'\times\eta')(x,y),(u',v')]_{(x,y)}$$

then $u'=D(\xi'\circ\xi^{-1})(\xi(x))(u)$ and $v'=D(\eta'\circ\eta^{-1})(\xi(y))(v)$, and therefore the right-hand side of (4.18) is well-defined.

For a C^r manifold M, if $r\geq 2$, one can go on to construct a manifold $T(TM)=T^2M$, which will be a C^{r-2} manifold, and the process can go on r times until one loses differentiability. Working with C^∞ manifolds, one need not keep track of the degree of smoothness.

Now to complete the picture we have constructed, we discuss the tangent map of a differentiable function from one manifold to another. Consider a C^r m-manifold

M and a C^r n-manifold N, with maximal atlases $\mathcal{A}$ and $\mathcal{B}$, respectively, and $f{:}M{\to}N$ a continuous map. In Subsection 5, we discussed what it means for f to be C^s, when $s{\leq}r$. Assuming $r{\geq}1$, we now wish to define the tangent map $Tf{:}TM{\to}TN$ just as in the local case of Chapter 2. Let $a{\in}M$; we first construct the ***tangent linear map*** $T_af{:}T_aM{\to}T_{f(a)}N$ at a. Let $(U,\xi){\in}\mathcal{A}$ and $(V,\eta){\in}\mathcal{B}$ be charts with $a{\in}U$ and $f(a){\in}V$. Any element w of T_aM can be represented as $[(\xi(a),v)]_a$ for appropriate $v{\in}\mathbb{R}^m$. We define T_af by

$$T_af[(\xi(a),v)]_a = [T_{\xi(a)}(\eta\circ f\circ\xi^{-1})(\xi(a),v)]_{f(a)} \tag{4.19}$$

More explicitly,

$$T_af[(\xi(a),v)]_a = [(\eta(f(a)), D(\eta\circ f\circ\xi^{-1})(\xi(a))(v)]_{f(a)} \tag{4.20}$$

The abundance of parentheses should not confuse or dishearten the reader, as the idea is actually quite simple. A look at Figure 11 should help. A tangent vector w to

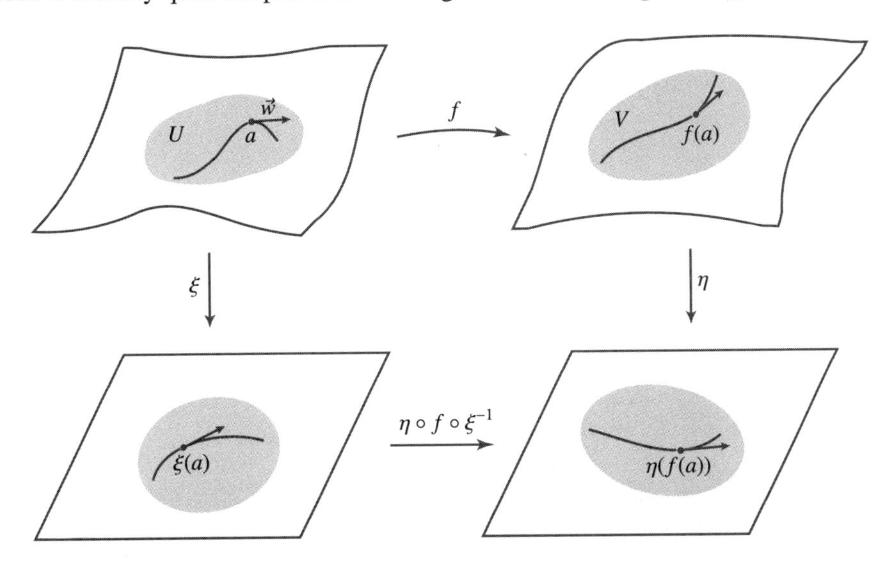

FIGURE 11. Representing the tangent map

M at a is actually an equivalence class. A representative of this equivalence class is shown in Figure 11 as a vector $(\xi(a),v)$ originating at the point $\xi(a)$ in $\mathbb{R}^m$. The map f is represented in local charts by $\eta{\circ}f{\circ}\xi^{-1}$. This map takes the point $\xi(a)$ to $\eta(f(a))$, and its derivative at $\xi(a)$ takes v to $D(\eta{\circ}f{\circ}\xi^{-1})(\xi(a))(v)$, which we have pictured as originating from $\eta(f(a))$. This vector is a representative of what is to be $T_af(w)$.

Of course, we have to check that T_af is well-defined, but the machinery that has been set up makes this immediate. Note that if $(\bar{U},\bar{\xi})$ and $(\bar{V},\bar{\eta})$ are alternative charts to (U,ξ) and (V,η), respectively, then

$$\bar{\eta}\circ f\circ\bar{\xi}^{-1} = (\bar{\eta}\circ\eta^{-1})\circ(\eta\circ f\circ\xi^{-1})\circ(\xi\circ\bar{\xi}^{-1})$$

and coordinate change at either a or $f(a)$ does not change the equivalence class. Note also that T_af is linear since it is locally represented by the linear map $T_{\xi(a)}(\eta\circ f\circ\xi^{-1})$. Finally, if $f:M\to N$ is differentiable at every point of M, then the ***tangent map*** $Tf:TM\to TN$ is defined by restriction

$$Tf|_{T_xM} = T_xf$$

This definition makes sense as TM is the *disjoint* union of T_xM's.

19. Elementary Properties of the Tangent Map

All manifolds are C^r, with $r\geq 1$.

(i) *If (U,ξ) is in the maximal C^r atlas of M, then the map ξ is C^r.*
(ii) *If $f: M\to N$ is C^s, $1\leq s\leq r$, then Tf is C^{s-1}.*
(iii) *The identity mapping, $\mathbb{1}_M$, is C^r.*
(iv) *(Chain rule) If $f:M\to N$ and $g:N\to P$ are C^s (respectively, differentiable), then $g\circ f$ is C^s (respectively, differentiable), and $T(g\circ f)=Tg\circ Tf$.*
(v) *For a differentiable function $f:M\to\mathbb{R}$, $df=\pi_2\circ Tf$, where $\pi_2:T\mathbb{R}=\mathbb{R}\times\mathbb{R}\to\mathbb{R}$ is projection on the second component.*
(vi) *For differentiable functions $\varphi:M\to N$ and $f:N\to\mathbb{R}$, $d(f\circ\varphi)=df\circ T\varphi$.*

Note that if M is C^r, and N is C^s, it makes sense to speak of a map $f:M\to N$ be C^p only if $p\leq min\{r,s\}$. The reason is that the degree of differentiability must be invariant under change of coordinates in both the domain and the target spaces.

Proof. For (i), we can regard the target space ($\mathbb{R}^n$ or $\xi(U)$) as a manifold with the standard C^∞ structure. Now if (V,η) is a chart in the maximal C^r atlas of M with $U\cap V\neq\emptyset$, $\xi\circ\eta^{-1}$ is C^r on $\eta(U\cap V)$, and thus (i) is proved. Now by virtue of the local representation (4.20), we note that being C^s is a local property, and therefore assertions (ii)-(iv) follow from their local counterparts (see Chapter 2, Section A). (v) is an immediate consequence of definitions of df and Tf and the fact that $T\mathbb{R}=\mathbb{R}\times\mathbb{R}$. Finally, (vi) is a corollary of (iv) and (v). □

Looking back, recall that the definition of tangent space at a point a on a manifold M was motivated by the idea that the tangent space should be the space of all velocity vectors at the point a of curves on the manifold passing through that point. If coordinate t is used to designate points of $\mathbb{R}$, recall also that $\frac{d}{dt}$ denotes the unit tangent vector field on $\mathbb{R}$ in the positive direction. For a differentiable curve $\gamma:I\to M$, we define the velocity vector $\gamma'(t_0)$ by

$$\gamma'(t_0) = T\gamma(\frac{d}{dt}(t_0)) \tag{4.21}$$

If M is an open subset of $\mathbb{R}^m$, this is consistent with the definition (2.14) of Chapter 2. The following gives *a posteriori* justification to the aforementioned intuition about tangent vectors, or at least confirms that our definition has achieved its aim.

20. *Let M be a C^1 manifold and $a \in M$. Then T_aM consists precisely of the set of velocity vectors at a to differentiable curves on M that pass through a.*

PROOF. First let $\gamma : I \to M$ be a differentiable curve with $\gamma(t_0) = a$. Since the open interval I is a manifold and $\frac{d}{dt}(t_0) \in T_{t_0}I$, it follows from the definition of $\gamma'(t_0)$ that $\gamma'(t_0) \in T_aM$. Conversely, let $w \in T_aM$, and suppose (U, ξ) is a chart around a. Then $T\xi(w) = (\xi(a), v)$, where $v \in \mathbb{R}^m$, $m = \dim M$. Consider the straight-line curve λ defined from an open interval J around 0 to $\xi(U)$ by $\lambda(t) = \xi(a) + tv$. Thus $\lambda(0) = \xi(a)$ and $\lambda'(0) = (\xi(a), v)$. Letting $\gamma = \xi^{-1} \circ \lambda$, we have $\gamma(0) = a$ and $\gamma'(0) = w$, so the converse is proved. □

We mentioned repeatedly in our discussion that, in general, there is no homeomorphism $TM \to M \times \mathbb{R}^m$ that takes each T_xM linearly onto $\{x\} \times \mathbb{R}^m$. Manifolds for which such a homeomorphism exists are called ***parallelizable***. The simplest non-parallelizable manifold is the two-dimensional sphere, S_2. In later chapters, there will be more occasion to discuss this problem, but a full discussion requires tools of algebraic topology. For the anticipating reader, Exercises 4.18-4.21 are devoted to some elementary facts and examples about the subject.

A remark about two canonically defined maps is in order. Recall that $\tau_M : TM \to M$, the tangent bundle projection sends every $w \in T_xM$ to x. In the opposite direction, the ***zero-section*** of M, $\zeta_M : M \to TM$ is defined as the map that sends each $x \in M$ to the zero, $\mathbf{0}_x$, of the vector space T_xM.

21. *If M is a C^r manifold, then τ_M and ζ_M are C^{r-1} maps.*

PROOF. Since TM is a C^{r-1} manifold, the highest degree of smoothness that can be expected is C^{r-1}. On the other hand, in local representation, we have C^∞ maps $\xi \circ \tau_M \circ (T\xi)^{-1} : (x, v) \mapsto x$, and $T\xi \circ \zeta_M \circ (\xi)^{-1} : x \mapsto (x, \mathbf{0})$, both of which are C^∞. □

Note Our discussion in this chapter was confined to *real* manifolds. One can replace the model space $\mathbb{R}^m$ by $\mathbb{C}^m$ and smooth change of coordinates by complex-analytic ones to arrive at *complex analytic manifolds*. A complex analytic manifold of complex dimension one is also known as a *Riemann surface*. Complex manifolds of complex dimension m can also be regarded as real manifolds of dimension $2m$. Here different concepts of tangent bundle come into play, a subject that will not be covered in the present book.

Appendix

Here we discuss the rationale for requiring the conditions "Hausdorff" and "second-countable" in the definition of a manifold. It may seem that being Hausdorff should follow from being locally like $\mathbb{R}^m$, but Hausdorff-ness is actually a global condition. The following example or its variants are often given to justify this requirement.

Example 1. Adjoin a new point $0'$ to $\mathbb{R}$ to obtain a set $X=\mathbb{R}\cup\{0'\}$. We define a topology on X via the following basis: Open intervals of $\mathbb{R}$, and sets of the form $]a,0[\cup\{0'\}\cup]0,b[$, where $a<0$ and $b>0$. The latter type of set is obtained through replacing 0 by $0'$ in open intervals that contain 0. It is straightforward to check that this is a basis for a topology. This topology is non-Hausdorff since open sets containing 0 and $0'$ necessarily intersect. However, any point of X has an open neighborhood homeomorphic to an open set of $\mathbb{R}$ (see Figure 12).

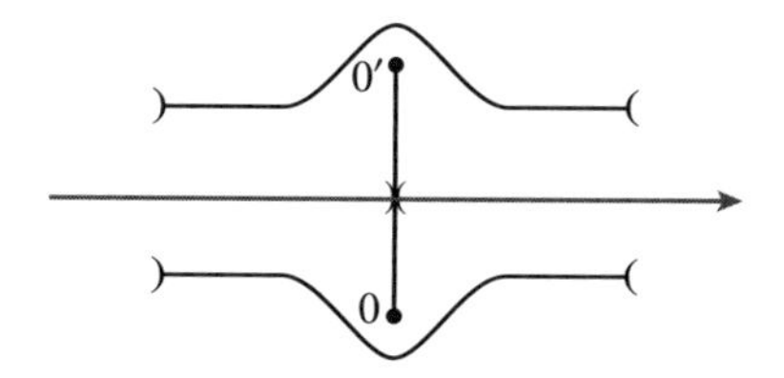

FIGURE 12. Non-Hausdorff pathology

Regarding the condition of second-countability, all we really need is that each connected component of the manifold be second-countable; separate connected components of a manifold do not interact in any way. As we saw in Section C, it was necessary to use second-countability (in each connected component) to construct a Partition of Unity, a tool that will be of great utility throughout the book. The following oft-cited example of a connected Hausdorff topological space that is locally homeomorphic to $\mathbb{R}$, yet is not second-countable, is known as the *long line*.

Example 2. We will outline the "construction" of the example; the reader is asked in Exercise 4.1 at the end of the chapter to supply details. Let A be a well-ordered *uncountable* set with the property that for each $a\in A$, the set of predecessors of a is countable. By well-ordering, this set can be listed uncountably as $A=\{1,2,3,\ldots\}$. The positive integers constitute an initial segment of this well-ordered set, but the set continues beyond them by reason of uncountability of A. Again by well-ordering, each $a\in A$ possesses a unique immediate successor (the least element among the set of members of A strictly greater than a). We will denote the successor of a by a^+. For each $a\in A$, we consider a copy of the open unit interval, which we denote by I_a, and imagine placing it so as to fill the gap between a and a^+; thus $I_a=]a,a^+[$. We adopt the usual ordering within this unit interval and further stipulate that $a<x<a^+$ for each $x\in I_a$. Finally, we add as the initial open unit segment, another copy of $]0,1[$, call it I_0, taken with the usual ordering and the added assumption $x<1$ for all $x\in I_0$. The union of A and the added intervals with the ordering described above and extended transitively will be denoted by L. The ***long line***, or, more accurately, the ***open long half-line***, is L with the order topology. L is then Hausdorff, but it is not second-countable. Further, every point of L has an open neighborhood homeomorphic to a subinterval of $\mathbb{R}$, and L is connected.

EXERCISES

4.1 Complete the description of the long line by showing the existence of an uncountable well-ordered set in which the set of predecessors of every member is countable. (Hint: If S is an uncountable well-ordered set with element s that has an uncountable number of predecessors, consider the set of predecessors of s that possess only a countable number of predecessors.)

4.2 Brouwer's Invariance of Domain Theorem states the following: If W is an open subset of $\mathbb{R}^p$ and $h:W \to \mathbb{R}^p$ is a one-to-one and continuous map, then h is a homeomorphism onto its image; in particular, $h(W)$ is open in $\mathbb{R}^p$. Deduce the claim that if an open subset of $\mathbb{R}^m$ is homeomorphic to an open subset of $\mathbb{R}^n$, then necessarily $m=n$.

4.3 Show that for a connected C^1 manifold, the well-definedness of dimension can be inferred without the use of the Brouwer Invariance Theorem. (Hint: Chain rule for the change of coordinate map.)

4.4 Let $\mathcal{A}$ and $\mathcal{A}'$ be C^1 atlases for M. Show that $\mathbb{1}_M:(M,\mathcal{A}) \to (M,\mathcal{A}')$ is a C^1 diffeomorphism if and only if $\mathcal{A}$ and $\mathcal{A}'$ have the same maximal C^1 extension.

4.5 Let M be a C^r manifold with maximal atlas $\mathcal{A}$, and suppose W is an open subset of M. Consider the collection $\mathcal{B}$ of those $(U,\xi) \in \mathcal{A}$ where $U \subset W$. Show that $\mathcal{B}$ is a maximal C^r atlas for W. Further, $\mathcal{B}$ is the unique maximal C^r atlas for W with the property that the inclusion map $(W,\mathcal{B}) \hookrightarrow (M,\mathcal{A})$ is C^r.

4.6 Let M and N be topological manifolds, and $f:M \to N$ a continuous map with graph Γ as a subspace of $M \times N$.
(a) Suppose M is a C^r manifold with atlas $\mathcal{A}$. For every chart $(U,\xi) \in \mathcal{A}$, consider (W,ζ), where $W=\{(x,f(x)) : x \in U\}$, and $\zeta = \xi \circ \pi_1$, where π_1 is the projection on the first component. Show that the set of such (W,ζ) is a C^r atlas for Γ.
(b) If $f : M \to M$ is a homeomorphism of M, with M a C^r manifold, then the graph of f is also the graph of f^{-1} and can be made into a C^r manifold in two ways as in (a). Find a necessary and sufficient condition for the two C^r atlases of Γ so constructed to have the same maximal extension.

4.7 For any given $r \geq 1$, give an example of a C^r manifold that is not C^{r+1}.

4.8 Show that the two atlases defined for S_m in Example 2d have the same C^∞ maximal extension.

4.9 The ***antipodal map*** of S_m is the map that sends each x to $-x$. Show carefully that the antipodal map is C^∞.

4.10 Let θ be an irrational number, $0<\theta<1$, and define θ_n to be $n\theta - \lfloor n\theta \rfloor$, where $\lfloor x \rfloor$ denotes the integer part of x. Prove *Jacobi's theorem*: $\{\theta_n\}_{n\in\mathbb{Z}}$ is dense in $[0, 1]$. Use this theorem to show that in Example 2g, the spiral is dense in the two-torus.

4.11 Let $f:\mathbb{R}\to\mathbb{R}$ be a continuous and everywhere positive function. Write $y=f(x)$ and consider the graph of this function in the xy-plane. By revolving this graph in the xyz-space around the x-axis, we obtain M, a ***surface of revolution***. Consider the following four charts $(U^{\pm},\xi_{\pm})$, $(V^{\pm},\eta_{\pm})$ for M.

$$U^{\pm} = M \cap \{y \gtrless 0\}$$
$$V^{\pm} = M \cap \{z \gtrless 0\}$$

with $\xi_{\pm}$ perpendicular projection on xz-plane, and $\eta_{\pm}$ perpendicular projection on xy-plane. Show that if f is a C^r function, then the charts above define a C^r atlas for M and that any two such surfaces of revolution are C^r diffeomorphic.

4.12 (Grassmann manifolds) This exercise deals with a generalization of $\mathbb{RP}(m)$, the projective space. We will turn the set $\mathrm{G}(k,m)$ of k-dimensional linear subspaces of $\mathbb{R}^m$, $k\leq m$, into a smooth manifold of dimension $k(m-k)$.
(a) Let $M(p,q)$ be the set of real $p\times q$ matrices. Show that the subset $\mathcal{L}_k$ of real $k\times m$ matrices of rank k is an open subset of $M(k,m)$, hence a smooth manifold of dimension km.
(b) For $\mathrm{A},\mathrm{B}\in\mathcal{L}_k$, define $\mathrm{A}\sim\mathrm{B}$ if and only if there is an invertible $k\times k$ matrix G so that GA=B. $\sim$ is an equivalence relation. Show that the set of equivalence classes can be identified with the set $\mathrm{G}(k,m)$ of k-dimensional linear subspaces of $\mathbb{R}^m$. (Fix a basis for $\mathbb{R}^m$, represent each element of $\mathbb{R}^m$ as a row vector and each set of k linearly independent vectors as an element of $\mathcal{L}_k$. Now if $\mathrm{A}\in\mathcal{L}_k$ represents a basis for an element of $\mathrm{G}(k,m)$, then GA is a basis for the same element.)
(c) We construct an atlas of $\binom{m}{k}$ charts (U_i,ξ_i) for $\mathrm{G}(k,m)$ as follows. For each multi-index $i=(i_1,\ldots,i_k)$, $1\leq i_1<\cdots<i_k\leq m$, let U_i be the set of equivalence classes [A], where A is an element of $\mathcal{L}_k$ for which the the columns indexed by i form an invertible $k\times k$ matrix G. Define $\xi_i:U_i\to M(k,m-k)$ by deleting the k columns of A indexed by i. Show that ξ is a well-defined homeomorphism onto $M(k,m-k)$, and $\mathrm{G}(k,m)$ is a topological manifold of dimension $k(m-k)$.
(d) Show that the above atlas is a C^∞ atlas.

4.13 Let M be a C^r manifold, $r\geq 1$, and $f : M\to\mathbb{R}$ a differentiable function. Show that if $a\in M$ is a point of local maximum or local minimum for f, then $df(a)=0$.

4.14 Let M be a C^r manifold and S a subset of M. A function $f:S\to\mathbb{R}$ is called C^s, $0\leq s\leq r$, if for every $a\in S$ there is an open neighborhood U of a in M and a C^s function $g:U\to\mathbb{R}$ so that the restrictions of g and f agree on $S\cap U$. If A is a closed subset of M and $f:A\to\mathbb{R}$ is C^s, show that f has a C^s extension from M to $\mathbb{R}$.

4.15 Let M be a C^r manifold, $r\geq 0$.
(a) If K is a closed subset and L an open subset of M with $K\subset L$, show that there is a C^r function $\theta:M\to[0,1]$ which takes value 1 in K and value 0 outside L.
(b) Show that there is a C^r everywhere positive function δ from M to $\mathbb{R}$ with the following property: For every $\epsilon>0$, there exists a compact subset K of M so that $\delta(x)<\epsilon$ for all $x\notin K$.

4.16 Let M be a connected smooth manifold with p and q points of M. Show that there is a smooth curve $\gamma{:}I{\to}M$ that passes through p and q.

4.17 Complete the discussion of Example 7c and Sub-section 18 by showing that if M and N are C^r manifolds, then $M{\times}N$ can be made into a C^r manifold in a natural way, and if $r{\geq}1$, then $T(M{\times}N)$ is homeomorphic to $TM{\times}TN$.

4.18 Let M be a C^1 manifold. By a continuous vector field on M, we mean a mapping $X{:}M{\to}TM$ so that $\tau_M{\circ}X{=}\mathbb{1}_M$. Suppose the dimension of M is m. Show that M is parallelizable if and only if there exist m continuous vector fields on M that are everywhere linearly independent. According to a theorem of Brouwer, every continuous vector field on S_2 vanishes at some point. Deduce that S_2 is not parallelizable.

4.19 (Assuming the result of 4.18) Show that the m-torus is parallelizable.

4.20 Let M be a C^r manifold, $r{\geq}2$. Show that if M is parallelizable, then so is TM.

4.21 (Based on 4.18) Show that S_3 is parallelizable. (Here take the “tangent space” to S_3 at a point a to be the three-dimensional affine subspace of $\mathbb{R}^4$ perpendicular to the radius vector. This is the common intuitive notion of tangent space and will be justified in Theorem 10, relation (5.7), of Chapter 5.)

CHAPTER 5

Mappings, Submanifolds and Quotients

In this chapter, we develop methods for constructing new smooth manifolds out of given ones. In particular, we will investigate conditions under which level sets and images of smooth mappings admit manifold structure. Our main tools here will be the Inverse Function Theorem and the Rank Theorem from analysis, for which the reader may consult Appendix II. In Section A, we will actually review the Rank Theorem and discuss its utility in constructing submanifolds.

A. Submanifolds

Consider integers m and n with $0 \le n \le m$. Throughout we will regard $\mathbb{R}^n$ as the subset of $\mathbb{R}^m$ consisting of the ordered m-tuples with the last $m - n$ components set to zero. This particular placement of $\mathbb{R}^n$ in $\mathbb{R}^m$ will be our primal local model for n-dimensional submanifolds of m-dimensional manifolds.

1. Definition Let M be a C^r manifold of dimension m, $r \ge 0$, N a subset of M, and $0 \le s \le r$. N is called an n-dimensional C^s ***submanifold*** of M, provided that for every $a \in N$, there is an open subset U of M with $a \in U$, and a C^s diffeomorphism ξ of U onto an open subset of $\mathbb{R}^m$, so that

(5.1) $$\xi(U \cap N) = \xi(U) \cap \mathbb{R}^n$$

This can be equivalently expressed as

(5.2) $$U \cap N = \xi^{-1}(\mathbb{R}^n)$$

In other words, the piece of N that lies in U can be flattened out by a C^s diffeomorphism to look as a piece of $\mathbb{R}^n$ sitting in $\mathbb{R}^m$ (see Figure 1). In the case

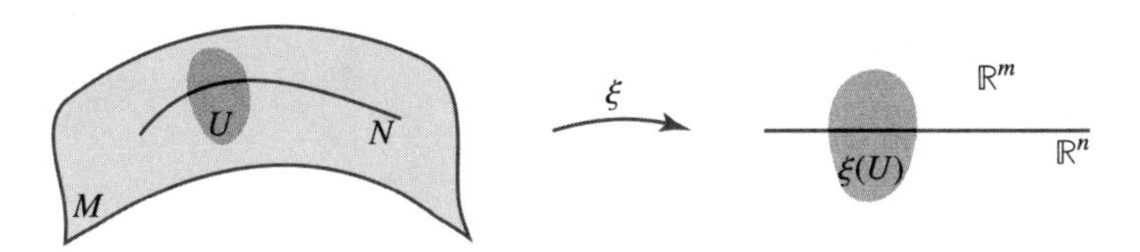

FIGURE 1. A submanifold

$s=r$, (U, ξ) is just a chart in the maximal C^r atlas of M.

2. Examples

(a) Let N be an open subset of M and $a \in N$. For any chart (V, η) of any maximal atlas of M with $a \in V$, we let $U = V \cap N$ and $\xi = \eta|_U$. Then (U, ξ) is in the same maximal atlas, and (5.1) is satisfied with $n = m$. So any open subset of a C^r m-manifold is a C^r submanifold of dimension m. A remarkable example occurs in the case of real square matrices, or the space of linear maps from a real vector space to itself. The set of all $m \times m$ real matrices, $M_m(\mathbb{R})$, may be identified with $\mathbb{R}^{m^2}$, regarding the m^2 entries as coordinates. An open subset of $M_m(\mathbb{R})$ is $\mathrm{GL}(m, \mathbb{R})$, the group of invertible real $m \times m$ matrices. This set is open because it is the inverse image of the open subset $\mathbb{R} - \{0\}$ of $\mathbb{R}$ under the continuous function $\det : M_m(\mathbb{R}) \to \mathbb{R}$. The same considerations apply to any real vector space V of finite dimension m. Any two norms on V are equivalent and give rise to a homeomorphism between V and $\mathbb{R}^m$. The set $\mathrm{GL}(V)$ of invertible linear maps from V to V is an open subset of the m^2-dimensional vector space $L(V, V)$ of all linear maps $V \to V$, and is hence a smooth submanifold of dimension m^2.

(b) For $m \geq 0$, we show that S_m is a smooth submanifold of $\mathbb{R}^{m+1}$ of dimension m. For $1 \leq k \leq m+1$, let

$$U_k^+ = \{(x^1, \ldots, x^{m+1}) \in \mathbb{R}^{m+1} : 0 < x^k < 2, \; \textstyle\sum_{i \neq k} (x^i)^2 < 1\}$$

$$\psi_k^+(x^1, \ldots, x^{m+1}) = (x^1, \ldots, x^{k-1}, x^k - \sqrt{1 - \sum_{i \neq k} (x^i)^2}, x^{k+1}, \ldots, x^{m+1})$$

ψ_k^+ is smooth since $x^k \neq 0$, and therefore the quantity under the square root sign is non-zero. Defining $\sigma_k : \mathbb{R}^{m+1} \to \mathbb{R}^{m+1}$ by

$$\sigma_k(x^1, \ldots, x^{m+1}) = (x^1, \ldots, \widehat{x^k}, \ldots, x^{m+1}, x^k)$$

and letting $\xi_k^+ = \sigma_k \circ \psi_k^+$, we look at the effect of ξ_k^+ on $U_k^+ \cap S_m$ (see Figure 2). For

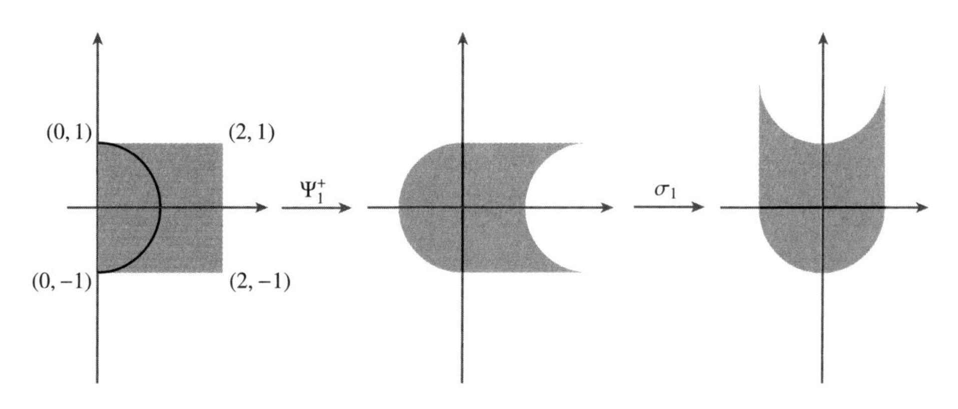

FIGURE 2. S_1 as a submanifold of $\mathbb{R}^2$

$(x^1,\ldots,x^{m+1})\in U_k^+\cap S_m$, we have

$$\begin{aligned}\xi_k^+(x^1,\ldots,x^{m+1}) &= \sigma_k(x^1,\ldots,x^{k-1},0,x^{k+1},\ldots,x^{m+1}) \\ &= (x^1,\ldots,x^{k-1},x^{k+1},\ldots,x^{m+1},0)\end{aligned}$$

Considering U_k^- for $-2< x^k < 0$ and defining ψ_k^- by changing the sign in front of the square root in the kth component of the formula for ψ_k^+, we obtain $m + 1$ additional charts that together with the above establish S_m as a smooth submanifold of $\mathbb{R}^{m+1}$.

(c) Let $f{:}\mathbb{R}^m\to\mathbb{R}^n$ be C^r. We show that the graph Γ of f is a C^r submanifold of $\mathbb{R}^m\times\mathbb{R}^n\cong\mathbb{R}^{m+n}$ of dimension m. Consider the single chart $(\mathbb{R}^{m+n},\xi)$ defined by

$$\xi(x,y) = (x, y - f(x))$$

This is indeed a C^r diffeomorphism with C^r inverse $\eta(x,y){=}(x,y + f(x))$. Now if $z{=}(x,y)\in\Gamma$, then $y{=}f(x)$, so $\xi(z){=}(x,\mathbf{0})\in\mathbb{R}^{m+n}$ and (5.1) is satisfied.

(d) Let E be an affine n-dimensional subspace of $\mathbb{R}^m$. By an invertible affine transformation $A{:}\mathbb{R}^m\to\mathbb{R}^m$, i.e., a composition of translation and invertible linear map, we can map E onto the linear subspace of $\mathbb{R}^m$ consisting of m-tuples $(x^1,\ldots,x^n,0,\ldots,0)$. Since A is a C^∞ diffeomorphism, the single chart $(\mathbb{R}^m,A)$ establishes E as a smooth submanifold of $\mathbb{R}^m$.

(e) Generalizing example (d), let $h{:}M\to M'$ be a C^r diffeomorphism of C^r manifolds. If N is a C^s submanifold of M, $s\leq r$, then $N'{=}h(N)$ is a C^s submanifold of M'. For if (U,ξ) satisfies (5.1) for N, then $(h(U),\xi\circ h^{-1})$ can be used for N' in M'.

(f) We look at some examples that are not submanifolds. $\mathbb{Q}$ is not even a C^0 zero-dimensional submanifold of $\mathbb{R}$. If $q\in\mathbb{Q}$, there is no open neighborhood U of q in $\mathbb{R}$ that can be mapped homeomorphically to a neighborhood of $\mathbb{R}^0{=}\{0\}$ in $\mathbb{R}$ so that $U\cap\mathbb{Q}$ is mapped to $\{0\}$. For a similar reason, the toral spiral in Example 2g of Chapter 4 is not a one-dimensional C^0 submanifold of the two-torus. For a different kind of example, consider the graph Γ of the absolute value function $x\mapsto|x|$ as a subspace of $\mathbb{R}^2$. We know that Γ can be made into a C^∞ manifold by virtue of being homeomorphic to $\mathbb{R}$ (see Example 7b of Chapter 4). However, Γ is not a C^1 submanifold of $\mathbb{R}^2$. Suppose there is an open neighborhood U of $(0,0)$ in $\mathbb{R}^2$ and a C^1 diffeomorphism ξ of U onto an open neighborhood V of $\mathbb{R}^2$, which sends $U\cap\Gamma$ onto $V\cap(\mathbb{R}\times\{0\})$. We may take U to be an open ball, so that $U\cap\Gamma$ is connected. Then $\xi(U\cap\Gamma)$ will be an open interval in $\mathbb{R}$, say $]t_1,t_2[$, with $\xi(0,0){=}t_0$ in $]t_1,t_2[$. This interval can be regarded at the image of a C^1 curve $\gamma(t){=}t$, $t_1{<}t{<}t_2$. Note that the velocity vector of this curve is non-zero, in fact $\gamma'(t){=}\frac{d}{dt}$. Since ξ is assumed to be a diffeomorphism, it follows that $(\xi^{-1}\circ\gamma)'(t_0)\neq\mathbf{0}$. But it is impossible for Γ to have a non-zero tangent at $(0,0)$ in view of the continuity of the tangent. This shows that Γ is not a C^1 submanifold of $\mathbb{R}^2$. The argument above raises other possibilities that are treated in Exercise 5.2. In the same vein, we are asking the reader in Exercise 5.1 to give an example of a C^1 submanifold of $\mathbb{R}^2$ that is not a C^2 submanifold.

Going back to the definition of a C^s submanifold and the notation used in the definition, the pairs $(U\cap N, \xi|_{U\cap N})$ form a C^s atlas for N, which may be maximally extended to produce a C^s structure for N. This will be what we have in mind when we speak of the C^s submanifold structure of N. A submanifold of dimension n of a manifold of dimension m is also referred to as a submanifold of ***codimension*** $m-n$. The following discussion will show that often codimension appears more directly in actual examples.

Some of the best-known and the most amenable submanifolds to study are level sets of functions. The so-called Rank Theorem describes sufficient conditions under which such submanifolds appear. We will transcribe below a version of this theorem suited to our needs. For a C^1 map of manifolds $f:P\to Q$, the ***rank of f at the point*** $p\in P$ will mean the rank of the linear map T_pf. We will denote the zero element of $\mathbb{R}^m$ by $\mathbf{0}_m$.

3. Rank Theorem *Let P and Q be C^r manifolds of dimension p and q, respectively, and let $f:P\to Q$ be a C^r map, $r\geq 1$, which has the same rank at every point of P. Then for each $a\in P$, there are charts (U,ϕ) around a, and (V,ψ) around $b=f(a)$, respectively, with $\phi(a)=\mathbf{0}_p$ and $\psi(b)=\mathbf{0}_q$, so that $\psi\circ f\circ\phi^{-1}$ is linear on its domain, i.e., it is the restriction of a linear map $\mathbb{R}^p\to\mathbb{R}^q$.*

Proof. This is an almost immediate consequence of the Rank Theorem of analysis (Appendix II). Let $\mathcal{A}$ and $\mathcal{B}$ be the maximal C^r atlases of P and Q, respectively. Take any pair of charts $(U_1,\phi_1)\in\mathcal{A}$, $(V_1,\psi_1)\in\mathcal{B}$ with $a\in U_1$, $b\in V_1$. Now $\psi_1\circ f\circ\phi_1{}^{-1}$ is defined on $\phi_1(U_1\cap f^{-1}(V_1))$ and has constant rank there. So, applying the Rank Theorem, we obtain open neighborhoods $U_2\subset U_1$ of $\phi_1(a)$, U_3 of $\mathbf{0}_p$, $V_2\subset V_1$ of $\psi_1(b)$, V_3 of $\mathbf{0}_q$, and C^r diffeomorphisms $\phi_2:U_2\to U_3$, $\psi_2:V_2\to V_3$, so that $\psi_2\circ(\psi_1\circ f\circ\phi_1{}^{-1})\circ\phi_2{}^{-1}$ is linear. Letting $\phi=\phi_2\circ\phi_1$, $\psi=\psi_2\circ\psi_1$, and reducing the size of neighborhoods, if necessary, to allow for all compositions to be defined, we obtain the desired conclusion. □

The main point here is linearization. Level sets of linear maps fit into a very simple universal configuration that is the key to understanding the local configuration of level sets of nonlinear maps of constant rank. Recall from linear algebra that if $L:\mathbb{R}^p\to\mathbb{R}^q$ is a linear map of rank k, then the non-empty level sets of L are parallel affine subspaces of codimension k (or dimension $p-k$) that fill up $\mathbb{R}^p$. One of these, the one that passes through the origin, is the kernel of L, and the rest are obtained by parallel-translating the kernel (see Figure 3). There is a further simplification of linear maps that comes in handy. Given any linear map of rank k from $\mathbb{R}^p$ to $\mathbb{R}^q$, we can make linear changes of basis in the domain and the target spaces to bring the linear map into the form of a projection $(x^1,\ldots,x^p)\mapsto(x^1,\ldots,x^k,0,\ldots,0)$. Thus we will henceforth assume that the linear representation $\psi\circ f\circ\phi^{-1}$ of f in the theorem is in the form

$$(\psi\circ f\circ\phi^{-1})(x^1,\ldots,x^p) = (x^1,\ldots,x^k,0,\ldots,0) \tag{5.3}$$

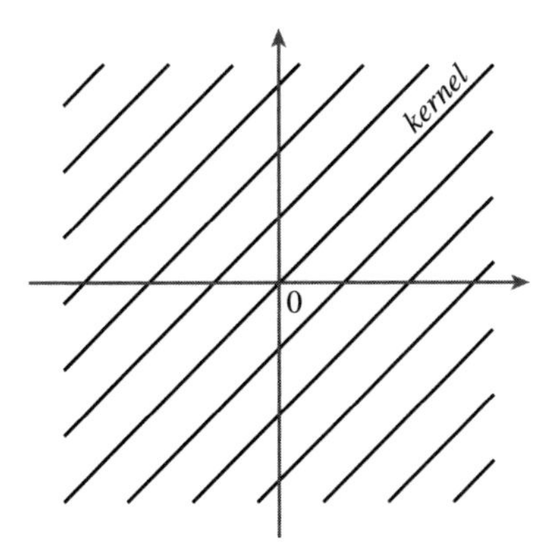

FIGURE 3. Level sets of a linear map

The general structure of level sets of linear maps applies to $\psi \circ f \circ \phi^{-1}$, except that the domain is limited to an open subset $U' = \phi(U)$ of $\mathbb{R}^p$ (see Figure 4). Keeping

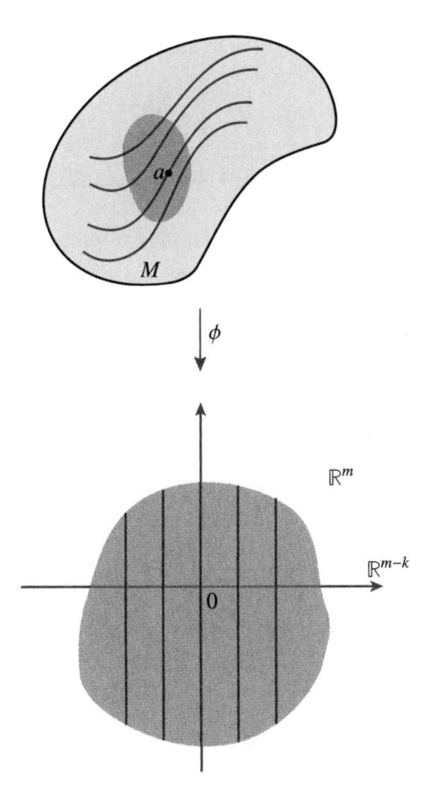

FIGURE 4. Level sets of a nonlinear constant rank map

this picture in mind, we derive some important consequences of the Rank Theorem below.

4. Theorem *Let P and Q be C^r manifolds of dimension p and q, respectively, and let $f:P\to Q$ be a C^r map, $r\geq 1$, of constant rank k throughout P. Then:*

(a) Every non-empty level set $f^{-1}(b)$ is a C^r submanifold of P of codimension k.

(b) Each point $a\in P$ has an open neighborhood W so that $W'=f(W)$ is a C^r submanifold of Q, C^r diffeomorphic to an open disk of dimension k. In fact, W can be chosen so that there is a C^r diffeomorphism $h:W'\times D\to W$, D an open disk in $\mathbb{R}^{p-k}$, and for each $c\in W'$,

$$h(\{c\}\times D) = f^{-1}(c)\cap W$$

Proof. The whole argument consists of transplanting the local linear picture into the manifolds P and Q via the diffeomorphisms ϕ^{-1} and ψ^{-1} (see Figure 5). We take an arbitrary point $a\in f^{-1}(b)\subset P$, and apply Theorem 3. Around $\mathbf{0}=\phi(a)$ in the open neighborhood $\phi(U\cap f^{-1}(V))$, we can take a neighborhood of the form $W_1\times D$, where W_1 and D are open disks in $\mathbb{R}^k$ and $\mathbb{R}^{p-k}$, respectively. If we identify $W_1\times\{\mathbf{0}_{p-k}\}$ with $W_1\times\{\mathbf{0}_{q-k}\}$, the linearization $\psi\circ f\circ\phi^{-1}$ will be given by projection $W_1\times D\to W_1$ on the first component. Letting $h=\phi^{-1}|_{W_1\times D}$, $W=h(W_1\times D)$, and noting that $W'=f(W)=\psi^{-1}(W_1\times\{\mathbf{0}_{q-k}\})$, assertion (b) of the theorem is seen to hold.

For part (a), consider the (smooth) linear isomorphism $\sigma:\mathbb{R}^p\to\mathbb{R}^p$ given by $(x^1,\dots,x^p)\mapsto(x^{k+1},\dots,x^p,x^1,\dots,x^k)$, and let $\xi=\sigma\circ\phi$. The chart (W,ξ) around a maps $f^{-1}(b)\cap W$ onto $\xi(W)\cap\mathbb{R}^{p-k}$ as required for a submanifold chart. Since a was an arbitrary point of $f^{-1}(b)$, the same argument works at every point of $f^{-1}(b)$, and the proof is complete. □

We revisit two earlier examples as applications of Theorem 4.

5. Examples (a) Let $P=\mathbb{R}^{m+1}-\{\mathbf{0}\}$, and define $f:P\to\mathbb{R}$ by $f(x^1,\dots,x^{m+1}) = \sum_i(x^i)^2$. f has constant rank 1 throughout P. For every positive c, the level set is non-empty and is the m-sphere of radius $\sqrt{c}$.

(b) We give an alternative proof of the fact that the graph Γ of a C^r function $f:\mathbb{R}^m\to\mathbb{R}^n$ is a C^r submanifold of $\mathbb{R}^{m+n}$. Consider $F:\mathbb{R}^{m+n}\to\mathbb{R}^n$ defined by $F(x,y)=y-f(x)$, where $x\in\mathbb{R}^m$ and $y\in\mathbb{R}^n$. This map has constant rank n since the matrix of the derivative of F with respect to the standard bases is of the form

$$\begin{bmatrix} * & \mathrm{I}_n \end{bmatrix}$$

where I_n is the $n\times n$ identity matrix. It follows from Theorem 4 that $\Gamma=F^{-1}(\mathbf{0})$ is a C^r submanifold.

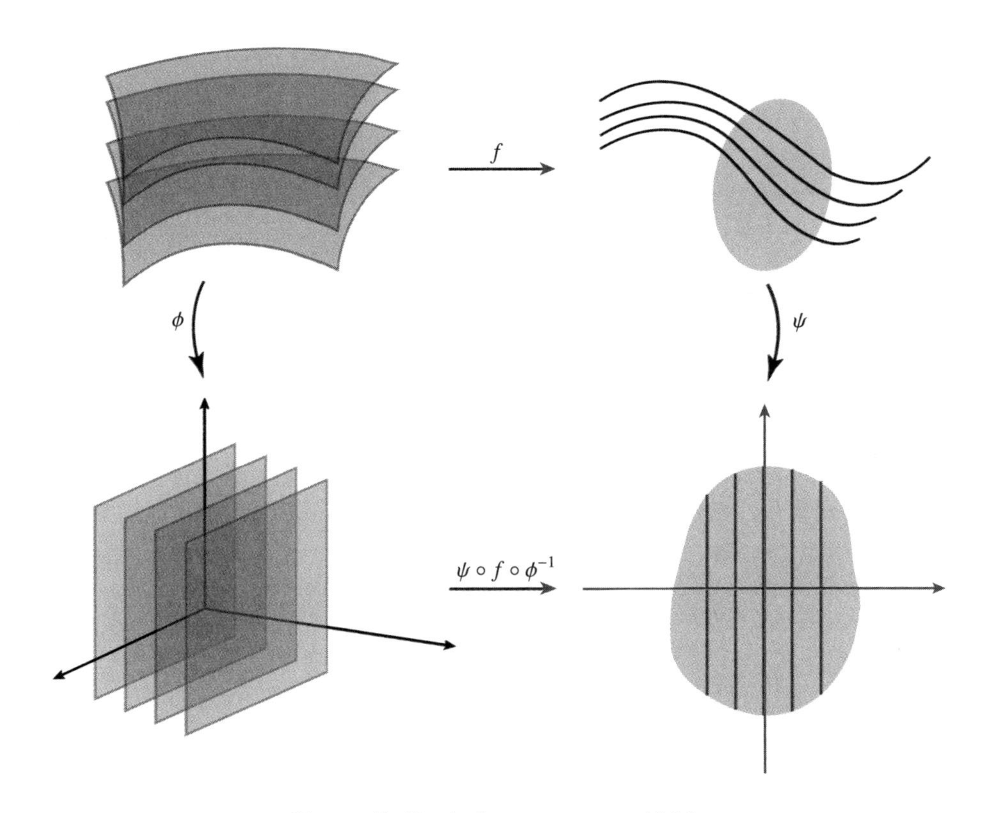

FIGURE 5. Rank theorem on manifolds

We now discuss a corollary of Theorem 4 often encountered in literature. Suppose $f{:}P\to Q$ is a C^1 map. A point a of P is called a ***regular point*** for f if T_af is onto. Otherwise, a is called a ***critical point***. A point $b\in Q$ is a ***regular value*** if there are no critical points in $f^{-1}(b)$, in particular, if $f^{-1}(b)$ is empty. Otherwise, b will be called a ***critical value***. A level set of f is a ***regular level set*** if it contains no critical points.

6. Lemma *Let P and Q be C^r manifolds of dimension p and q, respectively, and let $f{:}P\to Q$ be a C^1 map. Setting $k=\min\{p,q\}$, the set of points x of P, where the rank of T_xf is k, is an open subset of P, possibly empty.*

PROOF. Note that k is the maximum possible rank for T_xf. Suppose that for $a\in P$, the rank of T_af is k. Let (U,ξ) be a chart around a and (V,η) a chart around $f(a)$. Then the derivative of $\eta\circ f\circ\xi^{-1}$ at $\xi(a)$ has rank k. The $q\times p$ matrix of this linear map relative to the standard basis has a $k\times k$ sub-matrix with non-zero determinant. Since f is assumed C^1, the entries of the above matrix are continuous with respect to a. It follows that the determinant of the $k\times k$ sub-matrix in the same position will stay non-zero for x in a neighborhood of a. Thus the derivative of $T_x f$ will have rank k for x in a neighborhood of a. □

7. Theorem (Regular Value Theorem) *Let P and Q be C^r manifolds of dimension p and q, respectively, and suppose that $f:P \to Q$ is a C^r map. Any regular level set of f is a C^r submanifold of P of codimension q.*

PROOF. Let S be the set of regular points for f in P. By Lemma 6, S is an open subset, hence also a C^r submanifold of P. Any regular level set $f^{-1}(b)$ is a subset of S, so $f^{-1}(b) = g^{-1}(b)$, where $g = f|_S$. But g is a C^r map of constant rank q, so $f^{-1}(b) = g^{-1}(b)$ is a C^r submanifold of S of codimension q by Theorem 4. Since S is open in P, $f^{-1}(b)$ is also a C^r submanifold of P. □

B. Immersions, Submersions and Embeddings

In this section, we discuss various types of mappings between manifolds. Thoughout we assume that P and Q are C^r manifolds of dimension p and q, respectively, with $r \geq 1$, and $f:P \to Q$ is a C^r map. f is called an ***immersion at*** $a \in P$ (respectively, a ***submersion at*** $a \in P$) if $T_a f$ is one-to-one (respectively, onto). For f to be an immersion (respectively, a submersion) at a, it is necesssary that $p \leq q$ (respectively, $p \geq q$). In the terminology of the last section, if f is a submersion at a, then a is a regular point, and vice versa. We say that f is an ***immersion*** (respectively, ***submersion***), if f is an immersion (respectively, submersion) at each point of its domain. Immersions and submersions are maps of constant rank (p and q, respectively). By Theorem 3 of the last section, if $a \in P$ and $b = f(a) \in Q$, then there exist charts (U, ϕ) around a and (V, ψ) around b so that the local representation of f, i.e., $\psi \circ f \circ \phi^{-1}$, has the following form:

$$(5.4) \qquad (x^1, \ldots, x^p) \mapsto (x^1, \ldots, x^p, 0, \ldots, 0) \qquad \text{(immersion)}$$

$$(5.5) \qquad (x^1, \ldots, x^p) \mapsto (x^1, \ldots, x^q) \qquad \text{(submersion)}$$

8. Examples

(a) *Regular curves.* By a ***regular curve*** on a manifold M, we mean a C^1 immersion $\gamma:I \to M$, where I is an interval in $\mathbb{R}$. The ***velocity vector*** of γ for $t \in I$ is $\gamma'(t) = T\gamma(\frac{d}{dt}(t))$. The requirement that the tangent map have rank 1 is equivalent to the velocity vector being non-zero. If γ is a regular curve, it follows from part (b) of Theorem 4 that for each interior point t_0 of I, there is an open interval J, $t_0 \in J \subset I$, such that $\gamma(J)$ is a one-dimensional submanifold of M, with $\gamma|_J:J \to \gamma(J)$ a C^1 diffeomorphism.

(b) There is no requirement that an immersion itself be one-to-one. A simple example is the ***alpha curve*** $\alpha:\mathbb{R} \to \mathbb{R}^2$ defined by

$$\alpha(t) = (t^2 - 1, t^3 - t)$$

We have $\alpha'(t) \neq \mathbf{0}$, but $\alpha(-1) = \alpha(1) = (0, 0)$ (see Figure 6). Note that while the self-

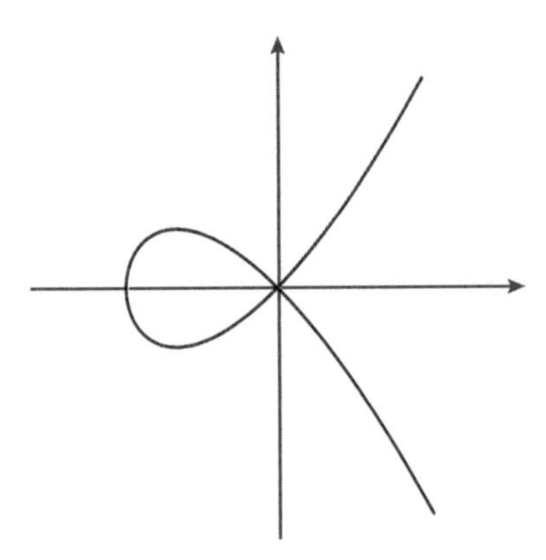

FIGURE 6. The alpha curve

intersection causes the image of γ not to be a manifold (see also Example 2e of Chapter 4), there is no contradiction with part (b) of Theorem 4. Taking for example $t_0=-1$, the image of restriction of α to $]-\infty, 0[$ is a C^1 submanifold of $\mathbb{R}^2$.

(c) Even if an immersion is one-to-one, the *global* image need not be a submanifold. Example 2e, and the dense spiral in Example 2g of Chapter 4, display different examples of how this can fail.

(d) Let M be a C^r manifold, $r\geq 2$. Then the zero-section, ζ_M, and the tangent-bundle projection, τ_M (see 12e of Chapter 4), are examples of C^{r-1} immersion and submersion, respectively. Note that the local representation of these maps fits (5.4) and (5.5), respectively.

The following generalization of Regular Value Theorem is a source of many examples of manifolds.

9. Theorem *Let P and Q be C^r manifolds and $f{:}P\to Q$ a C^r submersion onto Q. If N is a C^r submanifold of Q of codimension k, then $M=f^{-1}(N)$ is a C^r submanifold of P of codimension k.*

PROOF. Let a be an arbitrary point of M. To prove that M is a C^r submanifold of P, it suffices to show there is an open neighborhood U of a in P so that $M\cap U$ is a C^r submanifold of U, for then a submanifold chart around a in U will also be a submanifold chart for M in P. Let $b=f(a)$. Since N is a C^r codimension k submanifold of Q, there is a chart (V,η) around b so that $\eta(V\cap N)=\eta(V)\cap\mathbb{R}^{q-k}$, where $q=\dim Q$. Here we have identified $\mathbb{R}^{q-k}$ with the linear subspace of $\mathbb{R}^q$ consisting of q-tuples with the last k components set to zero. Now consider $\mathbb{R}^k$ as the complementary subspace, i.e., the set q-tuples with the first $q-k$ components equal to zero, and let π be the projection $(x^1,\dots,x^q)\mapsto(x^{q-k+1},\dots,x^q)$. The composition

$\pi\circ\eta\circ f$, defined on the open neighborhood $U=f^{-1}(V)$ of a, is a submersion, and $M\cap U=(\pi\circ\eta\circ f)^{-1}(\mathbf{0}_k)$. It follows from the Regular Value Theorem that $M\cap U$ is a C^r submanifold of U of codimension k. □

Addendum on the Tangent Space

We will now discuss the tangent space at a point of a submanifold. For $n\leq m$, consider $\mathbb{R}^n$ as a C^∞ submanifold of $\mathbb{R}^m$ consisting of real m-tuples with the last $m-n$ components set to zero, and let $j:\mathbb{R}^n\hookrightarrow\mathbb{R}^m$ be the inclusion map, which is linear. Then $Tj:T\mathbb{R}^n=\mathbb{R}^n\times\mathbb{R}^n\to T\mathbb{R}^m=\mathbb{R}^m\times\mathbb{R}^m$ is given by the pair of inclusions (j, j). Now if N is a C^1 submanifold of M, it follows that the inclusion map $N\hookrightarrow M$ induces a natural injective map $TN\to TM$ which identifies each T_xN, $x\in N$, with a subspace of T_xM. We shall henceforth regard T_xN as a linear subspace of T_xM via this natural identification. The following complements Theorem 9 and the Regular Value Theorem.

10. Theorem *In the situation of Theorem 9, let $a\in M$ with $b=f(a)\in N$. Then*

$$T_aM = (T_af)^{-1}(T_bN) \tag{5.6}$$

In the situation of Theorem 7, if $a\in M=f^{-1}(b)$, then

$$T_aM = \ker(T_af) \tag{5.7}$$

Proof. We first consider the case of the Regular Value Theorem. Since M is mapped to the zero-dimensional manifold {b}, its tangent space T_aM is mapped to $\mathbf{0}_b$, i.e., $T_aM\subset\ker(T_af)$. But T_aM has codimension q, and so does $\ker(T_af)$ because T_af is onto, therefore $T_aM=\ker(T_af)$.

Now for (5.6), using the notation in the proof of Theorem 9, it follows from above that $T_aM=\ker(\pi\circ T_b\eta\circ T_af)$. Noting that $\ker(\pi\circ T_b\eta)$ is exactly T_bN, the result follows. □

We will pursue the topic of submersions further in the next section in connection with the notion of a quotient manifold. Some notable examples will be presented there.

Recall from the examples in 2f that the image of an immersion need not be a submanifold even if the immersion is one-to-one. A C^s immersion $f:P\to Q$ is called a C^s ***embedding*** if f is a homeomorphism of P onto $f(P)$, where $f(P)$ is taken with the subspace topology from Q. In general, a map $f:P\to Q$ of topological spaces is called an ***embedding*** if f is a homeomorphism from P onto $f(P)$, where $f(P)$ is given the subspace topology from Q.

11. Theorem *Let Q be a C^r manifold, $r\geq 1$, and suppose that S is a subspace of Q. Then S is a C^s submanifold of Q, $0\leq s\leq r$, if and only if S is the image of a C^s embedding into Q.*

PROOF. If S is a C^s submanifold of Q of dimension p, then the inclusion mapping $S \hookrightarrow Q$ is the desired C^s embedding. This is because given $a \in S$ and a C^s submanifold chart (U, ξ) around a, the local representation of the inclusion map will be $(x^1, \ldots, x^p) \mapsto (x^1, \ldots, x^p, 0, \ldots, 0)$. Conversely, suppose $S = f(P)$, where f is a C^s embedding of a p-dimensional manifold P into Q. Take $b = f(a)$ in S, we wish to construct a submanifold chart for S around b. By the Rank Theorem, there are charts (U, ϕ) for P around a, and (V, ψ) for Q around b, so that

$$\phi(a) = \mathbf{0}_p, \quad \psi(b) = \mathbf{0}_q$$

and

$$(\psi \circ f \circ \phi^{-1})(z^1, \ldots, z^p) = (z^1, \ldots, z^p, 0, \ldots, 0)$$

Since f is an embedding, $f(U)$ is an open subset of S, hence there exists an open subset W of Q so that $f(U) = W \cap S$. The pair $(W \cap V, \psi|_{W \cap V})$ is the desired chart. □

12. Examples

(a) Suppose $f: P \to Q$ is a one-to-one C^r immersion. If P is compact, then f is an embedding. This follows from the following general fact of point-set topology: Any one-to-one continuous mapping of a compact space P into a Hausdorff space Q is an embedding. To prove this assertion, it suffices to show that the image of every open set in P is open in $f(P)$ with subspace topology from Q. But f is one-to-one, so this assertion is equivalent to showing that the image of every closed set in P is closed in $f(P)$ with subspace topology from Q. Since P is compact, every closed subset C of P is compact, but f is continuous, so $f(C)$ is compact in Q. Now Q is Hausdorff, so every compact subset of Q is closed, proving the assertion.

(b) Let M and N be C^r manifolds of dimension m and n, respectively, and $f: M \to N$ a C^r map. Then we show that the graph Γ of f is a C^r, m-dimensional submanifold of the product manifold $M \times N$. This generalizes Example 2c. By Theorem 11, it suffices to show that the map $F: M \to M \times N$ by $F(x) = (x, f(x))$ is a C^r embedding. This map is C^r as each component is, and it is one-to-one. To show it is an immersion, we take a chart (U, ξ) for M around $a \in M$, and a chart (V, η) around $b = f(a)$ and look at the matrix of the local representative of $T_a F$ relative to the charts (U, ξ) and $(U \times V, \xi \times \eta)$ around a and (a, b) (see 2e of Chapter 4 for notation). This has the form

$$\begin{bmatrix} \mathrm{I}_m \\ \mathrm{A} \end{bmatrix}$$

where I_m is the $m \times m$ identity matrix, which has rank m, and A represents the derivative of $\eta \circ f \circ \xi^{-1}$ at $\xi(a)$. This matrix has rank m, therefore F is an immersion. It then suffices to show that the image of every open set U of M in Γ is open. But $U \times N$ is open in product topology and $F(U) = (U \times N) \cap \Gamma$, so the assertion is proved. An important special case is the graph of the identity map $\mathbb{1}_M$, known as the ***diagonal***, and denoted by Δ_M. The diagonal consists of pairs (x, x) in $X \times X$.

(c) Let M be a C^r manifold, $r \geq 1$. Then the zero-section ζ_M is a C^{r-1} embedding of M into TM. In fact, we already observed in Example 8d that ζ_M is a C^{r-1} immersion. If

U is open subset of M, then $\zeta_M(U)=\tau_M{}^{-1}(U)\cap\zeta_M(M)$, and $\tau_M{}^{-1}(U)$ is an open subset of TM, so $\zeta_M(U)$ is an open subset of $\zeta_M(M)$ with the induced topology. In fact, quite often, we wish to treat M as a submanifold of TM; we then identify the image of the zero-section as a copy of M.

A famous theorem of Whitney asserts that every C^r m-manifold, $m\geq 1$, admits a C^r embedding in $\mathbb{R}^{2m}$. Below we prove a much weaker proposition; we show how a *compact* m-manifold can be embedded in $\mathbb{R}^N$, with N sufficiently large.

13. Theorem *Let M be a compact C^r manifold, $1\leq r\leq\infty$. For N sufficiently large, there is a C^r embedding of M in $\mathbb{R}^N$.*

PROOF. Let $\dim M=m$. Since M is assumed compact, Example 12a above shows that it is only necessary to look for one-to-one immersions. By Lemma 10 of Chapter 4, M admits a handy atlas, which can be taken to be finite on account of the compactness of M. Thus we have an atlas

$$\mathcal{A} = \{(U_1, \xi_1), \ldots, (U_k, \xi_k)\}$$

such that $\bar{B}_2(\mathbf{0})\subset\xi_i(U_i)$ for all $i=1,\ldots,k$, and the sets $\xi_i{}^{-1}(B_1(\mathbf{0}))$, $i=1,\ldots,k$, form an open cover for M. We consider a C^∞ function δ on $\mathbb{R}^m$ that has value 1 on $B_1(\mathbf{0})$ and vanishes outside $B_2(\mathbf{0})$. Then for $i=1,\ldots,k$, the functions $\delta_i:M\to\mathbb{R}$ and $f_i:M\to\mathbb{R}^m$ are defined as follows:

$$\delta_i(x) = \begin{cases} \delta(\xi_i(x)) & \text{for } x \in U_i \\ 0 & \text{for } x \notin U_i \end{cases}$$

and

$$f_i(x) = \delta_i(x)\xi_i(x)$$

Since δ is C^∞ and vanishes outside $B_2(\mathbf{0})$, the functions δ_i and f_i are as smooth as the functions ξ_i, i.e., they are C^r. Now we define $f:M\to\mathbb{R}^{k+km}$ by

$$f(x) = (\delta_1(x), \ldots, \delta_k(x), f_1(x), \ldots, f_k(x))$$

This is a C^r function as all the components are C^r; we show it is an immersion. Each $x\in M$ belongs to some $\xi_i{}^{-1}(B_1(\mathbf{0}))$, wherein $f_i(x)=\xi_i(x)$. Since ξ_i is a diffeomorphism, it has rank m, so f must have rank at least m, and in fact equal to m, since $\dim M=m$. Finally, to show f is one-to-one, suppose $f_i(x)=f_i(y)$. There is an index $i\in\{1,\ldots,k\}$, such that $x\in\xi_i{}^{-1}(B_1(\mathbf{0}))$. Now if $y\in\xi_i{}^{-1}(B_1(\mathbf{0}))$, then $\xi_i(y)=\xi_i(x)$, and since ξ_i is a diffeomorphism, $y=x$. But if y is not a member of $\xi_i{}^{-1}(B_1(\mathbf{0}))$, then $\delta_i(y)\neq 1$, while $\delta_i(x)=1$. So f is one-to-one. □

The above proof is not valid for C^ω manifolds in view of the use of non-analytic bump functions. Nevertheless, Whitney Embedding Theorem still holds on the basis of deeper analytic tools. For a thorough discussion of embeddings into $\mathbb{R}^n$, see **[13]**.

C. Quotient Manifolds

We start out by recalling in detail facts of quotient topology that will be crucial for our discussion. Let X be a topological space and $\sim$ an equivalence relation on X. We denote the set of equivalence classes of elements of X under $\sim$ by $X/\sim$, and define the ***quotient projection***, $q:X\to X/\sim$ as the map that associates to each $x\in X$ its equivalence class $[x]$. The ***quotient topology*** on $X/\sim$ is defined as follows: A set W in $X/\sim$ is open if and only if $q^{-1}(W)$ is open in X. It is straightforward to check that this is indeed a topology on $X/\sim$.

14. Elementary Properties of Quotient Topology *Let X and Y be topological spaces, $\sim$ an equivalence relation on X, $q:X\to X/\sim$ the quotient projection, and suppose that $X/\sim$ is endowed with the quotient topology. Then:*
(a) *q is continuous, and in fact the quotient topology is the largest topology on $X/\sim$ that makes q continuous.*

(b) *Let $f:X/\sim\to Y$ be a map. If $f\circ q$ is continuous, then so is f.*

(c) *Let $f:X/\sim\to Y$ be a map. If $f\circ q$ is open, then so is f.*

(d) *If $g:X\to Y$ is continuous and is constant on each equivalence class, then a unique continuous map $f:X/\sim\to Y$ exists so that $g=f\circ q$.*

(e) *Suppose that q is open. Then $X/\sim$ is Hausdorff if and only if $\sim$ is a closed subspace of $X\times X$ (with the product topology).*

PROOF. Since every subset of $X/\sim$ with open inverse image in X under q is defined as open in quotient topology, this is the largest possible topology that makes q continuous. For (b), let V be open in Y, then $(f\circ q)^{-1}(V)$ is open in X. But $(f\circ q)^{-1}(V)=q^{-1}(f^{-1}(V))$, so $f^{-1}(V)$ is open in $X/\sim$, proving the continuity of f. For (c), let W be open in $X/\sim$, then $q^{-1}(W)$ is open in X. But $f(W)=(f\circ q)(q^{-1}(W))$, so $f(W)$ is open in Y by the hypothesis. For (d), the condition $g=f\circ q$ and the constancy of g on equivalence classes force the unique well-defined definition $f[x]=g(x)$. f is then continuous by (b). Assume q is open. Consider points $[x]$ and $[y]$ in $X/\sim$. Then $[x]\neq[y]$ if and only if $(x,y)\notin\sim$. Now $\sim$ is closed if and only if $X\times X-\sim$ is open. By the definition of product topology, the latter is open if and only if there exist open neighborhoods U of x and V of y so that $U\times V\subset X\times X-\sim$. This last condition is equivalent to $q(U)\cap q(V)=\emptyset$, from which (e) follows □

15. Important Case: Group Action
Let G be a group. A ***(left) action*** by G on a set X is a map $\Phi:G\times X\to X$ that satisfies the following:
(i) $\Phi(e,x)=x$, for all $x\in X$, where e is the neutral element of the group G.
(ii) $\Phi(g,\Phi(h,x))=\Phi(gh,x)$, for all $x\in X$ and all $g,h\in G$.

We also denote $\Phi(g,x)$ by $\Phi_g(x)$, $\Phi^x(g)$ or simply gx. Φ_e is the identity map of X by (i). Letting $h{=}g^{-1}$ in (ii), we note that Φ_g is a bijective map of X with inverse $\Phi_{g^{-1}}$. For a fixed $x{\in}X$, the set $\{\Phi^x(g){:}g{\in}G\}$ is called the ***orbit*** of x, and is denoted by Gx.

A group action $\Phi{:}G{\times}X{\to}X$ gives rise to an equivalence relation $\sim_\Phi$ on X as follows:

$$x{\sim_\Phi}y \iff \text{there is } g \in G \text{ with } gx{=}y$$

Reflexivity follows from $\phi_e{=}\mathbb{1}_X$, symmetry from the fact that $\Phi_g{}^{-1}{=}\Phi_{g^{-1}}$, and transitivity results from (ii). In fact, the equivalence classes are just the orbits of the elements of X under the action of G. We denote the set of orbits, i.e., the quotient space, by X/Φ.

16. Lemma *Let X be a topological space, G a group and $\Phi{:}G{\times}X{\to}X$ a group action. If each Φ_g is continuous, then the quotient projection $q{:}X{\to}X/\Phi$ is an open map.*

Proof. Note that if each Φ_g is continuous, then Φ_g is a homeomorphism with inverse $\Phi_{g^{-1}}$. Let U be an open subset of X, we must show $q(U)$ is open in X/Φ. By the definition of quotient topology, we must show that $q^{-1}(q(U))$ is open in X. But

$$q^{-1}(q(U)) = \bigcup_{x\in U} Gx = \bigcup_{g\in G} \Phi_g(U)$$

and each Φ_g is a homeomorphism, so $\Phi_g(U)$ is open, and so is the union over $g{\in}G$, proving the assertion. □

It was noted above that if each Φ_g is continuous, then in fact each Φ_g is a homeomorphism. If this condition holds for an action of the group G on a topological space X, we say that ***G operates on X by homeomorphisms***. This lemma together with 14e above will be useful later in checking Hausdorff and second-countable properties for "quotient manifolds," a subject we are now ready to take up.

17. Theorem *Let M be a C^r manifold, $\sim$ an equivalence relation on M, and suppose that $M/{\sim}$ is given the quotient topology. Then there is at most one C^r manifold structure on $M/{\sim}$ relative to which the quotient projection $q{:}M{\to}M/{\sim}$ is a C^r submersion.*

The above C^r structure, if it exists, will be called the ***quotient manifold structure*** for $M/{\sim}$. The proof of the theorem will be based on two lemmas.

(a) **Lemma** *Let M and N be C^r manifolds and $f{:}M{\to}N$ a C^r surjective submersion. Then for every $y{\in}N$, there is an open neighborhood W of y in N and a C^r map $g{:}W{\to}M$ so that $f{\circ}g{=}\mathbb{1}_W$.*

Proof. This is a general corollary of the Rank Theorem. Let $\dim M = m$, $\dim N = n$, with $m \geq n$. Take $x \in M$ so that $f(x) = y$. There are charts (U, ξ) and (V, η) around x and y, respectively, with $\xi(x) = \mathbf{0} \in \mathbb{R}^m$ and $\eta(y) = \mathbf{0} \in \mathbb{R}^n$ so that $\eta \circ f \circ \xi^{-1}$ maps $(z^1, \ldots, z^m)$ to $(z^1, \ldots, z^n)$. Letting s be the map $(z^1, \ldots, z^n) \mapsto (z^1, \ldots, z^n, 0, \ldots, 0)$, then $g = \xi^{-1} \circ s \circ \eta$, suitably restricted, is the desired local C^r right-inverse for f. □

(b) **Lemma** *Let M be a C^r manifold, $\sim$ an equivalence relation on M, $M/\!\sim$ endowed with the quotient topology, and suppose that $M/\!\sim$ admits a C^r manifold structure relative to which the quotient projection $q : M \to M/\!\sim$ is a C^r submersion. If N is a C^r manifold and $f : M/\!\sim \to N$ a continuous map, then f is C^r if and only if $f \circ q$ is.*

Proof. Of course, if f is C^r, then the composition $f \circ q$ also is; the main point is the converse. So suppose $f \circ q$ is C^r and take an arbitrary point $y \in M/\!\sim$. Since being C^r is a local property, it suffices to show that f is C^r in an open neighborhood of y. By the previous lemma, there is an open neighborhood W of y in $M/\!\sim$ and a C^r map $g : M/\!\sim \to M$ so that $q \circ g = \mathbb{1}_W$. Then the composition $(f \circ q) \circ g = f|_W$ is C^r. □

We can now provide a proof of uniqueness in the theorem. Let $\mathcal{A}$ be the C^r atlas for M, and suppose that $M/\!\sim$, with the quotient topology, admits two C^r atlases $\mathcal{A}'$ and $\mathcal{A}''$ so that both $q : (M, \mathcal{A}) \to (M/\!\sim, \mathcal{A}')$ and $q : (M, \mathcal{A}) \to (M/\!\sim, \mathcal{A}'')$ are C^r submersions. Since the two versions of $M/\!\sim$ have the same topology, the identity map $\mathbb{1}_{M/\sim}$ is continuous. It follows from the above lemma that the identity map

$$\mathbb{1}_{M/\sim} : (M/\!\sim, \mathcal{A}') \to (M/\!\sim, \mathcal{A}'')$$

is a C^r diffeomorphism. This means that any chart in $\mathcal{A}'$ is C^r-compatible with any chart in $\mathcal{A}''$, therefore the two atlases define the same C^r structure for $M/\!\sim$, proving the theorem.

18. Examples We will revisit the example of projective spaces from the previous chapter and discuss some variants. Another notable example, namely the m-torus, will be treated in the next section, in the context of "covering spaces."

(a) **(Real Projective Spaces)** We recall the construction of $\mathbb{RP}(m)$ Example 2h of Chapter 4, and recast in detail that construction in the present framework. Let $\mathbb{R}^\times$ be the multiplicative group of non-zero real numbers. There is an action of $\mathbb{R}^\times$ on $\mathbb{R}^{m+1} - \{\mathbf{0}\}$ by

$$(\lambda, (x^1, \ldots, x^{m+1})) \mapsto (\lambda x^1, \ldots, \lambda x^{m+1})$$

The orbit of $(x^1, \ldots, x^{m+1})$ is denoted by $[x^1 : \cdots : x^{m+1}]$, and the quotient space, called the real projective m-space, is denoted by $\mathbb{RP}(m)$. Geometrically, each equivalence class consists of points on a straight line in $\mathbb{R}^{m+1}$ that passes through the origin, with the origin removed, and $\mathbb{RP}(m)$ is the set of all such lines. Since for each $\lambda \in \mathbb{R}^\times$, the map $(x^1, \ldots, x^{m+1}) \mapsto (\lambda x^1, \ldots, \lambda x^{m+1})$ is continuous, the quotient projection is an open map by Lemma 16. It follows that $\mathbb{RP}(m)$ is second-countable. We show that the associated equivalence relation, $\sim$, on $(\mathbb{R}^{m+1}-\{\mathbf{0}\})\times(\mathbb{R}^{m+1}-\{\mathbf{0}\})$ is a closed subspace. Consider the function

$$F : (\mathbb{R}^{m+1}-\{\mathbf{0}\}) \times (\mathbb{R}^{m+1}-\{\mathbf{0}\}) \to \mathbb{R}^m$$

defined by

$$F((x^1,\dots,x^{m+1}),(y^1,\dots,y^{m+1})) = (x^1y^2 - x^2y^1, x^2y^3 - x^3y^2 \dots, x^my^{m+1} - x^{m+1}y^m)$$

F is continuous and $\sim$ is precisely $F^{-1}(0,\dots,0)$, so $\sim$ is closed. It follows from Lemma 16 and 14e that $\mathbb{RP}(m)$ is Hausdorff. For each $i=1,\dots,m+1$, we define $U_i\subset\mathbb{RP}(m)$ by

$$U_i = \{[x^1:\cdots:x^{m+1}] : x^i \neq 0\}$$

U_i is open in $\mathbb{RP}(m)$ since its inverse image under the quotient projection is open in $\mathbb{R}^{m+1}-\{\mathbf{0}\}$. Now $\xi_i:U_i\to\mathbb{R}^m$ is defined by

$$\xi_i[x^1:\cdots:x^{m+1}] = (x^1/x^i,\dots,\widehat{x^i/x^i},\dots,x^{m+1}/x^i)$$

ξ_i is one-to-one and onto $\mathbb{R}^m$. That it is continuous and open follows from 14b and c, because the composition with the quotient projection is the map

$$(x^1,\dots,x^{m+1}) \mapsto (x^1/x^i,\dots,\widehat{x^i/x^i},\dots,x^{m+1}/x^i) \tag{5.8}$$

defined on the subset of $(\mathbb{R}^{m+1}-\{\mathbf{0}\})$ with $x^i\neq0$, to $\mathbb{R}^m$, and this map is both continuous and open. Hence ξ_i is a homeomorphism. In Example 7e of Chapter 4, we verified that the atlas given by (U_i,ξ_i), $i=1,\dots,m+1$, is actually C^ω. To show that $\mathbb{RP}(m)$ is a quotient manifold of $\mathbb{R}^{m+1}-\{\mathbf{0}\}$, it suffices to show that the quotient projection is a submersion. In local coordinates over U_i, the quotient projection is given by (5.8). This is a submersion because for each fixed non-zero x^i, it is a surjective linear map onto $\mathbb{R}^m$.

(b) **(Complex Projective Spaces)** Identifying the m-tuple $(z^1,\dots,z^m)$ of complex numbers as the $2m$-tuple of real numbers $(x^1,y^1,\dots,x^m,y^m)$, where $z^k=x^k+iy^k$, we look at $\mathbb{C}^m$ as another name for the real smooth manifold $\mathbb{R}^{2m}$, the advantage being that the use of complex numbers simplifies the description of some operations and mappings. Let $\mathbb{C}^\times$ be the multiplicative group of non-zero complex numbers. We imitate the construction of $\mathbb{RP}(m)$ by defining a group action $\Phi:\mathbb{C}^\times\times(\mathbb{C}^{m+1}-\{\mathbf{0}\})\to\mathbb{C}^{m+1}-\{\mathbf{0}\}$ by $(\lambda,(z^1,\dots,z^{m+1}))\mapsto(\lambda z^1,\dots,\lambda z^{m+1})$. The equivalence class of $(z^1,\dots,z^{m+1})$ will be denoted by $[z^1:\cdots:z^{m+1}]$. The set of equivalence classes will be called the ***complex projective*** m***-space***, and will be denoted by $\mathbb{CP}(m)$. Geometrically, each equivalence class consists of points on a "complex line" going through the origin with the origin itself removed (a punctured plane), and $\mathbb{CP}(m)$ is the set of all such "lines." We endow this set with the quotient topology received from $\mathbb{C}^{m+1}-\mathbf{0}$, which, as before, makes this a Hausdorff and second-countable topological space. $\mathbb{CP}(m)$ can be turned into a smooth $2m$-dimensional manifold by constructing an atlas of $m+1$ charts (U_i,ξ_i) exactly as in the real case, and verifying that the coordinate changes, which are formally the same as in the real case, are in fact real-analytic. Further, the quotient projection is a smooth submersion making $\mathbb{CP}(m)$ a quotient manifold. Our discussion below in (c) and (d) will show, among other things, that $\mathbb{CP}(1)$ is diffeomorphic to S_2, and that projective spaces, both real and complex, are compact and connected.

(c) We show that the standard smooth structure of S_m can be realized as a quotient structure descending from $\mathbb{R}^{m+1}-\{\mathbf{0}\}$. The quotient mapping will be the natural projection $\pi:\mathbb{R}^{m+1}-\{\mathbf{0}\}\to S_m$ by $x\mapsto x/|x|$. This is indeed surjective and smooth; we have to check it is a submersion. But the restriction of π to the submanifold S_m of $\mathbb{R}^{m+1}-\{\mathbf{0}\}$ is the identity map, which implies that the tangent map will be surjective, and the claim is proved. Recall that we showed earlier in Example 2b, as well as in Example 5a, that the standard smooth structure of S_m is the same as its submanifold structure in $\mathbb{R}^{m+1}$.

(d) We describe alternative constructions for $\mathbb{RP}(m)$ and $\mathbb{CP}(m)$, using the unit spheres in $\mathbb{R}^{m+1}$ and in $\mathbb{R}^{2(m+1)}\cong\mathbb{C}^{m+1}$, respectively, in place of $\mathbb{R}^{m+1}-\{\mathbf{0}\}$ and $\mathbb{C}^{m+1}-\{\mathbf{0}\}$. Beginning with $\mathbb{R}^{m+1}-\{\mathbf{0}\}$ and the group action of $\mathbb{R}^{\times}$ on this space, each equivalence class contains precisely one pair of antipodal points $\{x,-x\}$ from S_m. This restriction can be viewed as the action of two-element multiplicative group $\{1,-1\}$ on S_m, which gives rise to the quotient topological space $\mathbb{RP}(m)$. Denoting the quotient maps $\mathbb{R}^{m+1}-\{\mathbf{0}\}\to\mathbb{RP}(m)$ and $S_m\to\mathbb{RP}(m)$, respectively, by q and q', we have $q=q'\circ\pi$, where π is the quotient map $\mathbb{R}^{m+1}-\{\mathbf{0}\}\to S_m$ described in c above. It follows from Lemma b that q' is smooth. Finally, by applying the chain rule to $q=q'\circ\pi$, we infer from the surjectivity of the tangent map of q that the tangent map of q' is also surjective. Thus $\mathbb{RP}(m)$ can be regarded as the quotient manifold of S_m under the identification of antipodal points. In completely analogous manner, we consider the unit sphere S_{2m+1} in $\mathbb{R}^{2(m+1)}-\{\mathbf{0}\}\cong\mathbb{C}^{m+1}-\{\mathbf{0}\}$. In this case, the restriction of the equivalence relation to the unit sphere may be viewed as the result of the action of the circle group, i.e., the multiplicative group $S_1=\{e^{it}:t\in\mathbb{R}\}$ in the complex plane, on the unit sphere $\{(z^1,\ldots,z^{m+1}) : |z^1|^2+\cdots+|z^{m+1}|^2=1\}$. The rest of the argument is the repetition of above and we obtain a quotient manifold submersion $S_{2m+1}\to\mathbb{CP}(m)$. This map is sometimes referred to as the ***Hopf fibration***. Now since spheres are compact and connected, then both $\mathbb{RP}(m)$ and $\mathbb{CP}(m)$, as continuous images of spheres, are compact and connected.

(e) We show that $\mathbb{CP}(1)$ is smoothly diffeomorphic to S_2, therefore S_2 may be viewed as the quotient of S_3 under an action of the circle group. This is the original Hopf fibration and is of fundamental importance in topology. Let us retrace the construction of $\mathbb{CP}(1)$, known also as the ***Riemann sphere***. There are two charts (U_1,ξ_1) and (U_2,ξ_2) defined by

$$U_1 = \{[z^1{:}z^2] : z^1\neq 0\},\ \ \xi_1[z^1{:}z^2]=z^2/z^1 \in \mathbb{C}$$
$$U_2 = \{[z^1{:}z^2] : z^2\neq 0\},\ \ \xi_2[z^1{:}z^2]=z^1/z^2 \in \mathbb{C}$$

Taking $z^1=1$, we see that $\xi_1(U_1)=\mathbb{C}$. Likewise, ξ_2 maps U_2 onto $\mathbb{C}$. Writing $z=z^2/z^1$, we emphasize that we are working with real manifolds by rewriting the chart maps as

$$\xi_1[z^1:z^2] = (Re(z), Im(z)) \in \mathbb{R}^2$$
$$\xi_2[z^1:z^2] = (Re(z^{-1}), Im(z^{-1})) \in \mathbb{R}^2$$

Note that for $[z^1:0]=[w^1:w^2]$, it is necessary and sufficient that $w^2=0$, $z^1 \neq 0$ and $w^1 \neq 0$, so there is a single element of the form $[z^1:0]$ in $\mathbb{CP}(1)$. Likewise, there is a single element of the form $[0:z^2]$ in $\mathbb{CP}(1)$. The latter is known as ***the point at infinity*** and denoted by ∞. Therefore, we can write

$$\mathbb{CP}(1) = U_1 \cup \{\infty\} = U_2 \cup \{[1:0]\}$$

In this representation, U_1 is identified with $\mathbb{C}$ and $[1:0]$ corresponds to the zero of $\mathbb{C}$.[1] Recalling the stereographic maps σ_N and σ_S for S_2 from Example 2d and Example 7d of Chapter 4, we can now define the desired diffeomorphism $h:\mathbb{CP}(1) \to S_2$. Let

$$h[z^1:z^2] = \begin{cases} {\sigma_N}^{-1}(z) & \text{if } z^1 \neq 0 \\ {\sigma_S}^{-1}(\bar{z}^{-1}) & \text{if } z^2 \neq 0 \end{cases}$$

We have to check that the two expressions agree on the intersection $U_1 \cap U_2$. Using (4.5) from Chapter 4 for $\sigma_N \circ {\sigma_S}^{-1}$, this is immediately verified. Further, the map is smooth on each open set U_i and the inverse is given by stereographic projection (composed with conjugation in $\mathbb{C}$ in one case), which is smooth. This concludes the description of Hopf fibration $S_3 \to S_2$. For some interesting properties of this map, see Exercise 5.20 at the end of this chapter.

D. Covering Spaces

Some important examples of quotient manifolds arise in connection with "covering spaces." The proper framework for the study of covering spaces is algebraic topology; in fact, not all the spaces involved in this subject are manifolds. Nevertheless, on account of interaction with the topic of quotient manifolds, there is some merit in an independent discussion here.

Let M be a topological space. By a ***covering space over*** M, we mean a topological space E and a continuous map $p:E \to M$ that satisfy the following condition: For every $x \in M$, there is an open neighborhood U of x so that $p^{-1}(U)$ is a disjoint union of open sets $(U_\alpha)_\alpha$, and the restriction of p to each U_α is a homeomorphism onto U.

A neighborhood U of x as above will be called an ***evenly covered*** neighborhood of x. Here is some standard terminology:

[1]For readers familiar with complex analysis, the Riemann sphere is an example of a *Riemann surface*, or a complex manifold of dimension 1. The charts (U_1, ξ_1) and (U_2, ξ_2) provide complex coordinates from $\mathbb{C}$, and change of variables on $U_1 \cap U_2$ is provided by the map $[z_1:z_2] \mapsto [z_2:z_1]$. This map corresponds to $z \mapsto 1/z$ in $\mathbb{C}^\times = \mathbb{C} - \{0\}$.

- E: the ***total space*** of the covering
- M: the ***base space*** of the covering
- p: the ***covering projection*** and
- for each $x \in M$, $p^{-1}(x)$ is the ***fiber over*** x

The following are elementary consequences of the definition.

(i) *p is a surjective local homeomorphism; in particular, it is an open map.*

(ii) *The relation of "belonging to the same fiber" is an equivalence relation on E, and M is the quotient space of E under this equivalence relation, with p the quotient projection.*

(iii) *If M is connected, then all fibers have the same cardinality.*

To see these, first note that the existence of an evenly covered neighborhood implies both the surjectivity of p and that it is a local homeomorphism. Further, every local homeomorphism is an open map. Now suppose $p:E \to M$ is any surjective, open and continuous map of topological spaces. The relation of belonging to the same subset of a set, in particular the relation of belonging to the same $p^{-1}(x)$, is an equivalence relation. Suppose that for a subset U of M, the set $p^{-1}(U)$ is open in E. Then since $U=p(p^{-1}(U))$, and p is an open map, it follows that M has the quotient topology and p is the quotient projection. For (iii), we note that the evenly-covered property implies that the cardinality of $p^{-1}(x)$ is a locally constant function of x, and a locally constant function on a connected domain is constant.

It should be noted that not every surjective local homeomorphism is a covering map. For example, let $E = S_1 - \{1\}$ (unit circle in $\mathbb{C}$ with the point $\{1\}$ removed), $M=S_1$, and $p:E \to M$ by $z \mapsto z^2$. This is a surjective local homeomorphism, but the point $1 \in M$ has no evenly covered neighborhood. For note that if $\varepsilon > 0$ is small, then the inverse image of the arc $I =]e^{-i\varepsilon}, e^{i\varepsilon}[$ on S_1 is the disjoint union of three disjoint arcs, two of which do not project *onto* I. Example 19e below will show that under suitable smoothness assumption and compactness of the domain, such examples cannot occur.

19. Examples

(a) Let $E=M \times D$ be the product of a space M with a discrete topological space D, and suppose p is projection on the first component. With product topology on E, we obtain the ***product covering***. If $p:E \to M$ is a product covering and $p':E' \to M$ another covering with $H:E \to E'$ a homeomorphism that satisfies $p=p' \circ H$, then $p':E' \to M$ is called a ***trivial covering (space)***.

(b) Looking back at Example 18d of the previous section in the case of *real* projective spaces, the map q' (in the notation of that example) that sends $\pm x \in S_m$ to $[x]=\{x, -x\} \in \mathbb{RP}(m)$ defines a covering space. For $a \in \mathbb{RP}(m)$, we obtain an evenly covered neighborhood as follows. Letting $a=[a^1:\cdots:a^{m+1}]$, there is index i so that $a^i \neq 0$. Let, as before, U_i be the set of $[x^1:\cdots:x^{m+1}]$ with $x^i \neq 0$. Then $p^{-1}(U_i)=U_i^+ \cup U_i^-$, where U_i^+ (respectively, U_i^-) is the set of $(x^1, \ldots, x^{m+1}) \in S_m$ with $x^i>0$ (respectively, $x^i<0$). The two sets U_i^+ and U_i^- are disjoint and the restriction of p to each is a homeomorphism onto U_i.

(c) For any integer $n \geq 1$, consider the map $p_n: S_1 \to S_1$ by $p_n(z)=z^n$, where elements of S_1 are regarded as complex numbers of modulus one. This maps wraps the unit circle around itself n times. Let I be an arc of the unit circle defined by $\alpha<\theta<\beta$, where $0<\beta-\alpha<2\pi$. Then I is an evenly covered open set with inverse image consisting of n disjoint arcs $(1/n)(\alpha+2k\pi)<\theta<(1/n)(\beta+2k\pi)$, $k=0, 1, \ldots, n-1$. For a variation of this example, see Exercise 5.25(b).

(d) Let $p: \mathbb{R}^m \to \mathbb{T}^m$ be the map $(x^1, \ldots, x^m) \mapsto (e^{2\pi i x^1}, \ldots, e^{2\pi i x^m})$. If $t=(t_1, \ldots, t_m) \in \mathbb{T}^m$ with $t_k=e^{2\pi i x^k}$, let $U=U_1 \times \cdots \times U_m$, where U_k is the open arc of $S_1=\mathbb{T}^1$ between $e^{2\pi i(x^k-1/2)}$ and $e^{2\pi i(x^k+1/2)}$. Then $p^{-1}(U)$ is the disjoint union of open cubes in $\mathbb{R}^m$ of the form $V_{n_1} \times \cdots \times V_{n_m}$, where $V_{n_k}=]x^k+n_k-1/2, x^k+n_k+1/2[$, $n_k \in \mathbb{Z}$. Each of these open cubes is mapped homeomorphically by p onto U.

(e) Let P and Q be C^r manifolds of the same dimension, $r \geq 1$, with P compact, and suppose that $f: P \to Q$ is a C^r map. We know from Lemma 6 that the set of regular points is an open subset of P, therefore the set C of critical points is closed. But P is assumed compact, so C and $f(C)$ are compact, and it follows that the set R of regular values of f, i.e., $Q-f(C)$, is open in this case. Of course, $f^{-1}(R)$ may be empty. But letting $E=f^{-1}(R)$, $M=f(E)$, and $p=f|_E$, we claim that $p: E \to M$ defines a covering space provided E is not empty. Since R is open in Q, E is an open subset of P; in fact, $E \subset P-C \subset P$. It follows from the Inverse Function Theorem that $M=f(E)$ is an open subset of Q; $M=f(P) \cap R \subset R \subset Q$. Thus E and M are C^r manifolds of the same dimension. By the Inverse Function Theorem, $p=f|_E$ is a local diffeomorphism. Let $a \in M$. Since $f^{-1}(a) \subset E$ is closed in P, it is compact, and it has no limit points since p is a local diffeomorphism. Therefore, $p^{-1}(a)$ is a finite set $\{a_1, \ldots, a_k\}$. For each i, there is an open neighborhood U_i of a_i so that the restriction of p to U_i is a diffeomorphism onto an open neighborhood of a. We may assume that the U_i's are disjoint. Let

$$V = \bigcap_{i=1}^{k} f(U_i) - f(P - \bigcup_{i=1}^{k} U_i)$$

It follows from the construction that V is an evenly covered neighborhod of a, and the claim is proved.

(f) A corollary of the above is the following: Suppose P and Q are C^r manifolds of the same dimension with $r \geq 1$, and $f: P \to Q$ is a C^r immersion. If P is compact and Q is connected, then f defines a covering, $Q=f(P)$, and therefore Q is compact. Note that with the notation of the previous example, we have $E=P$, and $f=p$ is a covering map from P to $f(P)$. But since P is compact, $f(P)$ is closed in Q. $f(P)$ is also open by the Inverse Function Theorem, so by the connectedness of Q, we obtain $Q=f(P)$. Examples (b) and (c) above are special instances of this proposition.

Our main result in this section concerns sufficient conditions under which the quotient manifold construction defines a covering space. We will begin by providing names for two types of group actions. A group action $\Phi:G\times X\to X$ is said to be ***free*** if any relation $\Phi(g,x)=x$ implies that g is the identity element of the group. Thus if Φ is a free action and $g\neq e$, then Φ_g has no fixed points. Now suppose X is a topological space and Φ is an action of a group G on X by homeomorphisms. Φ is said to be ***properly discontinuous*** if for every pair of compact subsets K and L of X, the relation $g(K)\cap L\neq\emptyset$ can hold for only a finite number of $g\in G$.

20. Lemma *Let X be a Hausdorff and locally compact topological space with $\Phi:G\times X\to X$ a properly discontinuous action of the group G on X by homeomorphisms. Then X/Φ is Hausdorff and locally compact.*

Proof. We know from Lemma 16 that the quotient projection $X\to X/\Phi$ is an open map for group actions by homeomorphisms. The quotient projection is also always continuous. The image of a locally compact space under a continuous and open map is always locally compact. To show X/Φ is Hausdorff, we take points x and y in X so that $[x]\neq[y]$, we must find disjoint open neighborhoods of $[x]$ and $[y]$ in X/Φ. It suffices to find neighborhoods (not necessarily open) U and V of x and y, respectively, so that $h_1(U)\cap h_2(V)=\emptyset$ for all $h_1,h_2\in G$. This, in turn, is equivalent to finding U and V so that $g(U)\cap V=\emptyset$ for all $g\in G$. Let K and L be arbitrary compact neighborhoods of x and y, respectively. We know that $g(K)\cap L=\emptyset$ unless $g\in S$, where $S=\{g_1,\dots,g_k\}$, a finite subset of G. Since x and y are not in the same orbit and X is Hausdorff, there are open neighborhoods A_i and B_i of x and y, respectively, so that $g_i(A_i)\cap B_i=\emptyset$, for all $i=1,\dots,k$. We now let $U=K\cap\bigcap_{i=1}^k A_i$ and $V=L\cap\bigcap_{i=1}^k B_i$, and claim that these are desired neighborhoods. For if $g\notin S$, then $g(U)\cap V=\emptyset$ since $g(K)\cap L=\emptyset$; and if $g=g_i$ for some i, then $g(U)\cap V=\emptyset$ since $g_i(A_i)\cap B_i=\emptyset$. □

Turning now to our main result, sufficient conditions will be established under which a group action leads to the construction of a covering space and a quotient manifold structure for the base of the covering. Exercise 5.23 provides a construction in the opposite direction.

21. Theorem *Let X be a Hausdorff and locally compact topological space, G a group and Φ a free and properly discontinuous action of G on X by homeomorphisms. Then the quotient projection $q{:}X{\to}X/\Phi$ defines a covering space. Further, if X is a C^r manifold, $r{\geq}1$, and if all Φ_g are C^r diffeomorphisms, then X/Φ admits a (unique) C^r structure as a quotient manifold of X, i.e., a C^r structure that makes q a submersion.*

Proof. For the first part, the crucial point is to prove that given any $x{\in}X$, there is an open neighborhood V of x so that the sets $g(V)$, for $g{\in}G$, are pairwise disjoint. Once this is proved, then $U{=}q(V)$ is an evenly covered neighborhood of $[x]$ with inverse image under q consisting of the sets $g(V)$, $g{\in}G$. To prove the claim, we start out by taking a compact neighborhood K for x. Since the action is properly discontinuous, we have $g(K){\cap}K{=}\emptyset$ except for $g{\in}\{g_1,\dots,g_k\}{\subset}G$. But the action is also free, so $g_ix{\neq}x$ for $i{=}1,\dots,k$. It follows that for each i, there are open neighborhoods V_i of x and W_i of g_ix so that $V_i{\cap}W_i{=}\emptyset$. Further, we may assume that the W_i are pairwise disjoint. We now let

$$V = int(K) \cap \bigcap_{i=1}^{k} (V_i \cap g_i^{-1} W_i)$$

The open set V has the desired property.

Now assume, in addition, that X is a C^r manifold, and G acts by diffeomorphisms on X. It follows from the above lemma that X/Φ possesses the topological prerequisites for being a manifold. We construct a C^r atlas for X/Φ as follows. For an evenly covered open set U in X/Φ, suppose that the disjoint open sets U_α are the inverse images of U under q. For any pair of indices α and α', it follows from the construction of evenly covered neighborhood above that, there is a (necessarily unique because of the free action) $g^\alpha_{\alpha'}{\in}G$ so that $g^\alpha_{\alpha'}(U_\alpha){=}U_{\alpha'}$. If for some α, the set U_α is the domain of a chart (U_α,ξ_α) in the maximal C^r atlas of X, then $(U_{\alpha'},\xi_\alpha{\circ}\Phi_{g^{\alpha'}_\alpha})$ will be a chart with domain $U_{\alpha'}$ since $\Phi_{g^{\alpha'}_\alpha}$ is assumed to be a diffeomorphism. We may actually assume then, by reducing the size of U if necessary, that all U_α are domains of charts (U_α,ξ_α) in the maximal C^r atlas of X. Denote the inverse of the homeomorphism $q|_{U_\alpha}$ by s_α. Then letting $\bar{\xi}_\alpha{=}\xi_\alpha{\circ}s_\alpha$, each $(U,\bar{\xi}_\alpha)$ is a chart for X/Φ. We must verify that these charts are C^r-compatible. Suppose that $(U,\bar{\xi}_\alpha)$ and $(V,\bar{\eta}_\beta)$ are two charts as above with $W{=}U{\cap}V{\neq}\emptyset$. Taking $y{\in}W$, there is $x{\in}U_\alpha$ so that $y{=}[x]$. Now restricting to W and its pre-images, $s_\beta{\circ}s_\alpha^{-1}(x){=}\Phi_{g^\alpha_\beta}(x){\in}V_\beta$ (see Figure 7). We look at the coordinate change on the intersection W:

$$\begin{aligned}\bar{\eta}_\beta \circ (\bar{\xi}_\alpha)^{-1} &= \eta_\beta \circ s_\beta \circ s_\alpha^{-1} \circ \xi_\alpha^{-1}\\ &= \eta_\beta \circ \Phi_{g^\alpha_\beta} \circ \xi_\alpha^{-1}\end{aligned}$$

The above is a composition of C^r maps relative to the atlas of X, hence the C^r-compatibility is proved. Finally, q is a submersion because it is a local diffeomorphism, and the theorem is proved. □

We point out that the theorem is also valid for $r{=}0$ (topological manifolds), where only part of the above reasoning is necessary.

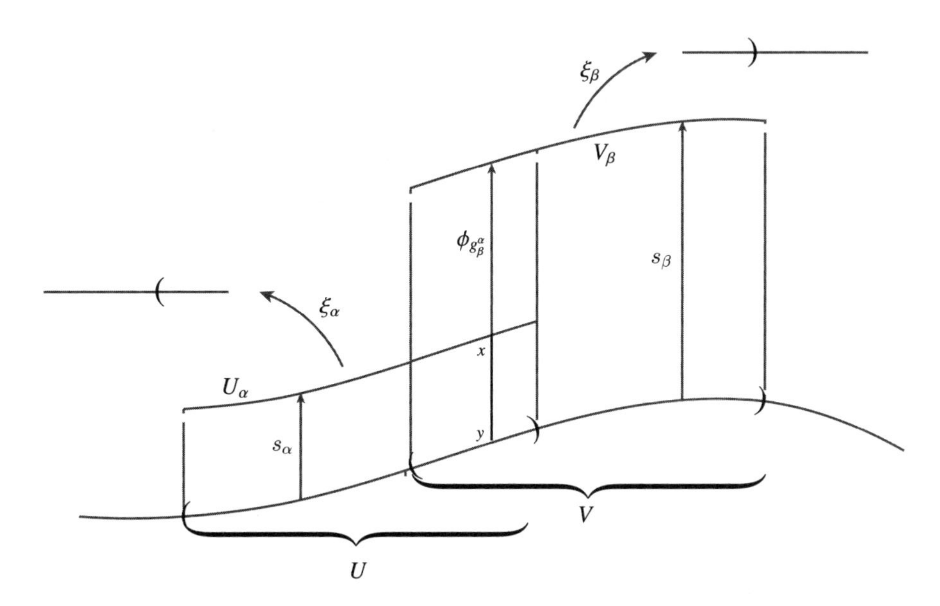

FIGURE 7. Manifold structure on the base

22. Examples

(a) The action of two-element multiplicative group {1,-1} on S_m satisfies the requirements of the theorem and yields a covering $S_m \to \mathbb{RP}(m)$ with the real projective m-space appearing as the quotient manifold of the m-sphere.

(b) The additive group $\mathbb{Z}^m$ acts freely and properly discontinuously by diffeomorphisms on $\mathbb{R}^m$ by

$$((n^1, \ldots, n^m), x^1, \ldots, x^m)) \mapsto (x^1 + n^1, \ldots, x^m + n^m)$$

to yield $\mathbb{T}^m$ as a smooth quotient manifold of $\mathbb{R}^m$ (this is an alternative look at Example 19d above).

(c) Example 19c can be realized as the action of cyclic group of order n on the unit circle. Representing the cyclic group of order n by the nth roots of unity under multiplication as complex numbers, the quotient projection can be viewed as the mapping p_n.

(d) Consider the horizontal strip $E = \mathbb{R} \times]-1, 1[$. We define an action of $(\mathbb{Z}, +)$ on E by

$$\Phi(n, (x, y)) = (x + n, (-1)^n y)$$

This action satisfies the requirements of the theorem (see Figure 8). The resulting quotient manifold is the (open) ***Möbius band***.

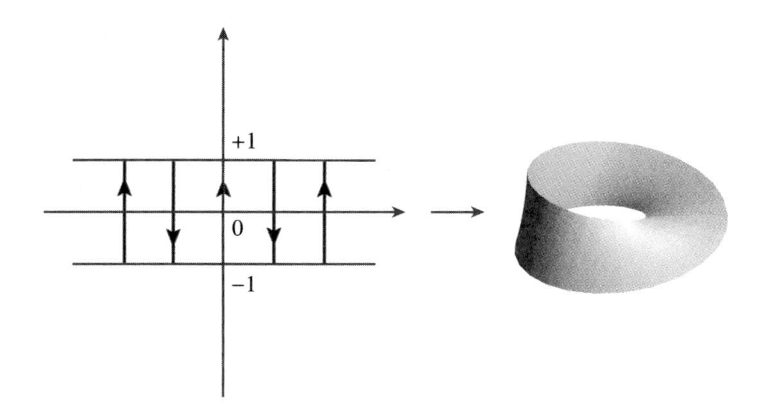

FIGURE 8. Möbius band

(e) We start with the infinite cylinder $E = \mathbb{R} \times S_1$ and consider the action of $(\mathbb{Z}, +)$ on E by

$$\Phi\big(n, (x, (\cos t, \sin t))\big) = (x + n, (\cos t, (-1)^n \sin t))$$

It is again straightforward to verify the conditions of the theorem. The quotient manifold is known as the ***Klein Bottle***. This two-dimensional manifold cannot be embedded in $\mathbb{R}^3$, but an immersion appears in Figure 9 (see also Exercise 5.17). Topologically, the Klein Bottle may also be viewed as the quotient of the compact

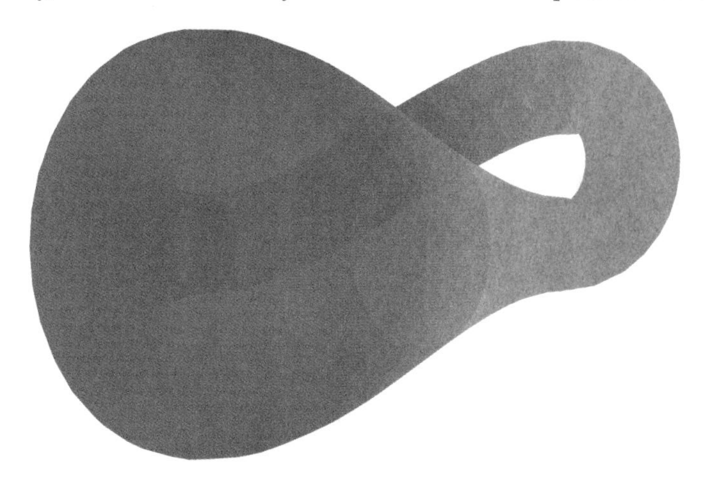

FIGURE 9. Klein Bottle

cylinder $[0, 1] \times S_1$ with the two ends identified by $(0, (\cos t, \sin t)) \sim (1, (\cos t, -\sin t))$, hence it is a compact manifold.

(f) The so-called "spherical manifolds," obtained as quotients of the action of finite groups of rotations in $\mathbb{R}^4$ on the 3-sphere S_3, are of importance in topology. A class of such examples is provided by the ***lens spaces*** $L(p,q)$, which we shall now describe. Let p and q be relatively prime positive integers. The cyclic group of order p acts on S_3 as follows. Represent the cyclic group as the set of pth roots of unity $\{e^{\frac{2\pi i}{p}k}:k=1,\ldots,p\}$ under multiplication as complex numbers. If ω is an element of this group, and $(z_1,z_2)\in S_3\subset\mathbb{C}^2\cong\mathbb{R}^4$, then the action Φ is defined by

$$\Phi(\omega,(z_1,z_2)) = (\omega z_1, \omega^q z_2)$$

The base space of the corresponding covering space, i.e., the quotient manifold, is denoted by $L(p,q)$. Various generalizations of this construction are studied both for S_{2n+1} with $n>1$, and for other finite-group actions on S_3.

EXERCISES

5.1 Give an example of a C^1 submanifold of $\mathbb{R}^2$ that is not a C^2 submanifold.

5.2 Give an example of a C^1 homeomorphism of $\mathbb{R}^2$ onto itself that maps $\mathbb{R}\times\{0\}$ onto the graph of $x\to|x|$. Is there a diffeomorphism (not requiring the continuity of the derivative) that accomplishes this?

5.3 Suppose N is a C^r submanifold of a C^r manifold M, and $f:N\to\mathbb{R}$ is a C^r function. Show that f can be extended to a C^r function from an open neighborhood of N in M to $\mathbb{R}$. If further, N is a closed submanifold of M, show that a C^r extension to all of M exists. (Cf. Exercise 4.14.)

5.4 Let N be a C^s submanifold of a C^r manifold M, $s\leq r$. Prove or disprove: Every chart of the maximal C^s atlas of N is the restriction of a chart of the maximal atlas of M to N.

5.5 Determine whether the composition of two immersions, two submersions and two embeddings are of the same type.

5.6 State and prove a generalization of Theorem 9 to the case where the mapping is of constant rank, but not necessarily a submersion.

5.7 A mapping $f:X\to Y$ of topological spaces is called *proper* if for every compact subset K of Y, the set $f^{-1}(K)$ is compact in X. Show that every proper one-to-one immersion is an embedding. Give an example of an embedding that is not proper.

5.8 Let N be a smooth codimension 2 submanifold of a smooth manifold M. Suppose that X is a smooth vector field defined on an open neighborhood of N in M that is nowhere tangent to N, and denote the flow of X by Φ. Show that there is a positive continuous function $\delta:N\to\mathbb{R}$ so that the set $\{\Phi(t,x) : x\in N, |t|<\delta(x)\}$ is a smooth codimension 1 submanifold of M. (Hint: Use 3e of Chapter 4 to write N as an expanding union of compact subsets K_n, and for each n show that there is a positive number δ_n so that the set $\{\Phi(t,x) : x\in int(K_n), |t|<\delta_n\}$ is a codimension 1 submanifold of M.)

5.9 Let U be an open subset of $\mathbb{R}^n$, K a compact set contained in U, and $f:U\to\mathbb{R}^m$ a smooth map that is an immersion at every point of K. Show that if f is one-to-one on K, then it is one-to-one on an open neighborhood of K.

5.10 Let M and N be two compact, connected and smooth manifolds of the same dimension. Show that every one-to-one immersion $M\to N$ is a diffeomorphism of M and N.

5.11 Provide a smooth embedding of $S_m\times S_n$ into $\mathbb{R}^{m+n+1}$.

5.12 For $n<m$, show how $\mathbb{RP}(n)$ can be embedded into $\mathbb{RP}(m)$. Prove that $\mathbb{RP}(m)$ can be written as the disjoint union of two submanifolds, one diffeomorphic to $\mathbb{R}^m$ and

another diffeomorphic to $\mathbb{RP}(m-1)$. In this context, $\mathbb{RP}(m)$ is regarded as the compactification of $\mathbb{R}^m$ by the addition of a "projective hyperplane at infinity."

5.13 Imitate the discussion of Hopf fibration to show that $\mathbb{CP}(m)$ is a quotient manifold of S_{2m+1}. As in the previous exercise, show that $\mathbb{CP}(m)$ can be written as the disjoint union of two submanifolds, one diffeomorphic to $\mathbb{C}^m$ and another diffeomorphic to $\mathbb{CP}(m-1)$.

5.14 Show that the real projective plane is the disjoint union of a closed two-dimensional disk and a manifold diffeomorphic to the open Möbius band (Example 22d). (Hint: Before passing S_2 down to the quotient, mark off an open "equatorial zone").

5.15 Referring to Exercise 5.12, where $\mathbb{RP}(m)$ is regarded as the compactification of $\mathbb{R}^m$ with the addition of a copy of $\mathbb{RP}(m-1)$ at infinity, show that there is no smooth map $f{:}\mathbb{RP}(m)\to\mathbb{R}$ for which the projective hyperplane at infinity is a regular level set.

5.16 Cosider the map $\mathbb{R}^3-\{\mathbf{0}\}\to\mathbb{R}^5$ by

$$f(x, y, z) = (xy, yz, zx, x^2 - y^2, x^2 + y^2 + z^2 - 1)$$

Show that f is an immersion. Using f, construct an embedding of $\mathbb{RP}(2)$ into $\mathbb{R}^4$. (Hint: Look at the restriction of f to the unit sphere.)

5.17 Let $a > b > 0$, and consider the map $\mathbb{R}^2\to\mathbb{C}^2\cong\mathbb{R}^4$ by

$$f(x, y) = (g^2(x, y), g(x, y)h(y))$$

where $g(x, y)=(a + b\cos y)e^{ix}$, and $h(y) = \sin y+i\sin 2y$. Use f to construct an embedding of the Klein Bottle (Example 22e) into $\mathbb{R}^4$ (example due to W. Rudin).

5.18 Let $\mathfrak{M}_k$ be the subspace of linear maps of rank k in the space of linear maps $\mathbb{R}^n\to\mathbb{R}^m$. Show that $\mathfrak{M}_k$ is a smooth submanifold of codimension $(m-k)(n-k)$. (Hint: Given an element f of $\mathfrak{M}_k$, choose bases for $\mathbb{R}^n$ and $\mathbb{R}^m$ so that f has a matrix representation with $k\times k$ identity matrix in the upper left corner and zeros elsewhere. Keeping these bases, a neighborhood of f is represented by matrices of the form

$$\begin{bmatrix} A & B \\ C & D \end{bmatrix}$$

where A is an invertible $k\times k$ matrix and $CA^{-1}B=D$.)

5.19 Let M be a C^1 manifold, and suppose that a C^1 map $f{:}M\to M$ satisfies $f\circ f=f$. Show that $f(M)$ is a closed C^1 submanifold of M. In this situation, $f(M)$ is called a C^1 ***retract*** of M. (Hint: Note that the restriction of f to $f(M)$ is the identity map. At every point x of $f(M)$, consider the splitting of T_xM into the kernel and the image of T_xf.)

5.20 For Hopf fibration $S_3 \to S_2$, show that the inverse image of any point p in S_2 is a geometric circle in S_3, and that if p and q are distinct points, then the two circles are "linked," in the sense that each circle pierces the disk bounded by the other circle. (Hint: You may wish to consider convenient points p and q to check the claim, and then show that this picture is general.)

5.21 Let Φ be an action of a group G on the topological space X by homeomorphisms. Sometimes the action is defined to be ***properly discontinuous*** if for every compact subset K of X, the relation $g(K) \cap K \neq \emptyset$ can hold for only a finite number of $g \in G$. Show that this definition is equivalent to the one given in the text.

5.22 Plücker embedding (Based on Exercise 4.12, Chapter 4, and notation therein)
(a) Revisit Exercise 4.12 and exhibit $\mathrm{G}(k, m)$ as the quotient manifold of $\mathcal{L}_k$ under the action of the group $\mathrm{GL}(k, \mathbb{R})$.
(b) For $\mathrm{A} \in \mathcal{L}_k$ and multi-index $i=(i_1, \ldots, i_k)$, let $p_i = \det(A_i)$, where A_i is the $k \times k$ matrix formed by the columns of A corresponding to i. The real numbers p_i are known as the ***Plücker coordinates*** of A. If $\mathrm{G} \in \mathrm{GL}(k, \mathbb{R})$, show that the Plücker coordinates (q_i) of GA are related to (p_i) by $q_i = (\det G) p_i$ for all i. Deduce that a map $\pi : \mathrm{G}(k, m) \to \mathbb{RP}(\binom{m}{k} - 1)$ is obtained.
(c) Show that π above is a smooth embedding.

5.23 Suppose that E and M are topological manifolds and a continuous map $p : E \to M$ defines a covering space. Show that if M has the structure of a C^r manifold, $r \geq 1$, then one can construct a C^r structure for E that makes p a submersion.

5.24 Suppose that E and M are topological manifolds and the continuous map $p : E \to M$ is surjective, proper and a local homeomorphism. Show that $p : E \to M$ defines a covering space and that for every $x \in M$, $p^{-1}(x)$ is finite. (For the definition of *proper* see Exercise 5.7 above.)

5.25 Show that the following are covering spaces.
(a) $\mathbb{C} \to \mathbb{C}^{\times} = \mathbb{C} - \{0\}$ by $z \mapsto e^z$.
(b) For n, a natural number, the map $\mathbb{C}^{\times} \to \mathbb{C}^{\times}$ by $z \mapsto z^n$.

5.26 Consider the maps of the Riemann sphere into itself given by $f(z) = \frac{az+b}{cz+d}$ (so-called ***Möbius maps*** or ***linear fractional transformations***), where a, b, c, d are given complex numbers with $ad - bc \neq 0$. Here we define $f(\infty) = a/c$ and $f(-d/c) = \infty$ if $c \neq 0$. If $c = 0$, then necessarily $d \neq 0$, and we define $f(\infty) = \infty$. Show that f is a smooth diffeomorphism of S_2 onto itself. (The point ∞ was defined in Example 18e.)

5.27 ***Rational functions*** on the Riemann sphere are generalizations of Möbius maps, where we allow the numerator and the denominator to be arbitrary polynomials in z subject to the condition that the two polynomials have no common factor. The point ∞ is treated similarly as above. Show that each rational function f is a smooth map, the set of critical values of f is finite, and the restriction of f to the set of regular points defines a covering space over the set of regular values. What is the cardinality of $f^{-1}(p)$ for a regular value p?

CHAPTER 6

Vector Bundles and Fields

The major aim of this chapter is to extend the concepts of vector and tensor fields discussed over open subsets of $\mathbb{R}^m$ in Chapters 2 and 3 to globally defined notions on an entire manifold. Such *fields* are most conveniently introduced as cross-sections of *vector bundles*. An important example of a vector bundle is the tangent bundle, which was already treated in Chapter 4. The construction of the tangent bundle there will guide us through the discussion here. Globalization of fields will bring out some new features that will be discussed in Sections C and D. In Section E, we will study both the local and the global aspects of the so-called *plane fields*, also known as *distributions*; these may be considered higher-dimensional analogues of vector fields.

A. Basic Constructions

We confine ourselves to the study of vector bundles in the category of manifolds. Vector bundles over more general spaces are studied in algebraic topology. Recall from Example 2a of Chapter 5 that if F is a finite-dimensional real vector space, then the group $\mathrm{GL}(F)$ of invertible linear maps $F\to F$ is a smooth manifold.

1. Definition Let F be a finite-dimensional real vector space. A (real) C^r ***vector bundle of fiber type*** F, $r\geq 0$, consists of the following data:
(i) C^r manifolds E and M, and a C^r surjective map $\pi:E\to M$,
(ii) a collection of indexed pairs (U_α, Φ_α), such that $(U_\alpha)_\alpha$ is an open cover for M, and $\Phi_\alpha:\pi^{-1}(U_\alpha)\to U_\alpha\times F$ is a C^r diffeomorphism so that $\pi_1\circ\Phi_\alpha=\pi$, where π_1 is the projection of $U_\alpha\times F$ on the first component, and
(iii) for each pair of indices α and β with $U_\alpha\cap U_\beta\neq\emptyset$, C^r maps $\phi_\alpha^\beta:U_\alpha\cap U_\beta\to\mathrm{GL}(F)$ so that:

$$\Phi_\beta\circ\Phi_\alpha{}^{-1}(x,u)=\left(x,(\phi_\alpha^\beta(x))(u)\right) \tag{6.1}$$

By condition (ii), the diagram below is commutative.

$$\begin{array}{ccccc} (U_\alpha\cap U_\beta)\times F & \xleftarrow{\Phi_\alpha} & \pi^{-1}(U_\alpha\cap U_\beta) & \xrightarrow{\Phi_\beta} & (U_\alpha\cap U_\beta)\times F \\ & \searrow^{\pi_1} & \downarrow{\scriptstyle\pi} & {}^{\pi_1}\swarrow & \\ & & U_\alpha\cap U_\beta & & \end{array} \tag{6.2}$$

This implies that an element (x, u) of $(U_\alpha \cap U_\beta) \times F$ is mapped to an element of the form (x, u') on the right. Condition (iii) requires that u and u' be related by an invertible linear map $\phi_\alpha^\beta(x)$. Further, since condition (ii) shows that the diffeomorphism Φ_α maps $\pi^{-1}(x)$ bijectively onto $\{x\} \times F$, a vector space structure, isomorphic to F, is induced on $\pi^{-1}(x)$. This linear structure is independent of the particular (U_α, Φ_α) because $\phi_\alpha^\beta(x)$ is a linear isomorphism. Thus in a vector bundle, each $\pi^{-1}(x)$ possesses a well-defined linear structure isomorphic to F.

The following terminology is used for the elements of the above definition:
- E: ***total space***, M: ***base space***, π: ***(vector bundle) projection***,
- $E_x = \pi^{-1}(x)$: ***fiber over*** x,
- property (ii) defines the ***local product structure***,
- a pair (U_α, Φ_α) that satisfies (ii) is called a ***vector bundle chart*** (in short, a ***VB-chart***), and a collection of charts (U_α, Φ_α) so that the sets U_α cover M is called a ***vector bundle atlas*** (in short, a ***VB-atlas***),
- the maps ϕ_α^β are ***transition maps***, and
- if dim $F = k$, the vector bundle is said to be of ***rank*** k.

We will speak alternatively of a vector bundle (E, π, M), a vector bundle E over M, or even simply, a vector bundle E, if the rest of the data are clearly understood. If E and E' are two C^r vector bundles over M of fiber type F, then a C^s ***vector bundle isomorphism*** (in short, a C^s ***VB-isomorphism***), $0 \leq s \leq r$, is a C^s diffeomorphism, $E \to E'$ the restriction of which to each E_x is a linear isomorphism onto E'_x. A ***cross-section*** of the vector bundle (E, π, M) is a map $\sigma: M \to E$ with the property $\pi \circ \sigma = \mathbb{1}_M$. In other words, σ assigns to every $x \in M$ an element of E_x. Being a map of manifolds, adjectives such as continuous, C^r and smooth may be used for cross-sections as appropriate.

2. Examples

(a) Let M be a C^r manifold and set $E = M \times F$, with π projection on the first component. This is a ***product vector bundle***. Any vector bundle over M that is C^s isomorphic to the product bundle is called a C^s ***trivial vector bundle***. If $(b_1, \ldots, b_k)$ is a basis for the vector space F, then the cross-sections $\tilde{b}_i$, $i = 1, \ldots, k$, of the product vector bundle, defined by $\tilde{b}_i(x) = (x, b_i)$, form a vector space basis for $\{x\} \times F$ at every $x \in M$. If E is a trivial vector bundle over M with $\Phi: E \to M \times F$ an isomorphism, then the cross-sections σ_i, $i = 1, \ldots, k$, defined by $\sigma_i(x) = \Phi^{-1}(x, b_i)$ form a vector space basis for E_x at every $x \in M$.

(b) If M is a C^{r+1} manifold of dimension m, we have seen in Chapter 4 that the tangent bundle TM of M is a C^r vector bundle of rank m over M. A cross-section of TM is a ***(tangent) vector field***. In the terminology of the above example, parallelizable manifolds (see the comments after Subsection 20 in Chapter 4) are those manifolds for which the tangent bundle is C^0 trivial. Thus an m-manifold is parallelizable if and only if it admits m continuous tangent vector fields that are linearly independent everywhere (see also Exercise 4.18 of Chapter 4).

(c) Let M and N be C^r manifolds, $r \geq 1$. Example 18 and Exercise 4.17 of Chapter 4 can now be expressed by statement that there is a canonical VB-isomorphism C from $T(M \times N)$ to $TM \times TN$ as vector bundles over $M \times N$.

3. On the Nature of a Vector Bundle

Before continuing with the construction that will yield our main examples, it is worthwhile to explore further the definition of a vector bundle. We note the following fundamental property of transition maps:

$$\phi^\gamma_\beta(x) \circ \phi^\beta_\alpha(x) = \phi^\gamma_\alpha(x) \text{ for all } x \in U_\alpha \cap U_\beta \cap U_\gamma \neq \emptyset \tag{6.3}$$

This identity follows from the fact that if $U_\alpha \cap U_\beta \cap U_\gamma \neq \emptyset$, then the equality $\Phi_\gamma \circ \Phi_\alpha^{-1} = (\Phi_\gamma \circ \Phi_\beta^{-1}) \circ (\Phi_\beta \circ \Phi_\alpha^{-1})$ holds on $(U_\alpha \cap U_\beta \cap U_\gamma) \times F$. We also point out a couple of consequences of (6.3), namely,

$$\phi^\alpha_\alpha(x) = \mathbb{1}_F \text{ for all } x \in U_\alpha \tag{6.4}$$

$$\phi^\alpha_\beta(x) = (\phi^\beta_\alpha(x))^{-1} \text{ for all } x \in U_\alpha \cap U_\beta \neq \emptyset \tag{6.5}$$

The first identity is obtained by setting $\beta = \alpha$ in (6.3), and the second by letting $\gamma = \alpha$. (6.3) should be reminiscent of the derivative of coordinate change (the chain rule) for manifolds and of the Identification Lemma in Chapter 4, which led to the construction of the tangent space and the tangent bundle.

The analogy with the definition of a manifold has been noted through the use of the words "chart" and "atlas." We saw in the case of manifolds that distinct C^r atlases could define the "same" manifold, where "sameness" was conceived as the identity of the notion of differentiable functions on the manifold. This identity was formulated by introducing the notion of "C^r structure" or "maximal C^r atlas," an atlas that included all C^r-compatible charts with the given atlas. Here, too, we have to fine-tune our understanding of the identity of a vector bundle so as to avoid artificial distinctions and to be able to optimize the choice of VB-atlas without changing the vector bundle. Looking back at the definition of a vector bundle, we call the C^r atlas $\mathcal{A} = \{(U_\alpha, \Phi_\alpha)_\alpha\}$ ***maximal*** if no chart (U, Φ) may be added to $\mathcal{A}$ so that condition (iii) is maintained. Just as in the case of manifolds, we show that any given C^r atlas $\mathcal{A}$ can be extended to a unique maximal C^r atlas $\widetilde{\mathcal{A}}$. For any C^r chart (U_μ, Φ_μ), we let $(U_\mu, \Phi_\mu) \in \widetilde{\mathcal{A}}$ if and only if $\mathcal{A} \cup \{(U_\mu, \Phi_\mu)\}$ satisfies (iii). To show that $\widetilde{\mathcal{A}}$ satisfies (iii), one need only check (iii) for two new additions (U_μ, Φ_μ) and (U_ν, Φ_ν) such that $U_\mu \cap U_\nu \neq \emptyset$, i.e., we must define C^r map $\phi^\nu_\mu : U_\mu \cap U_\nu \to \mathrm{GL}(F)$ so that

$$\Phi_\nu \circ \Phi_\mu{}^{-1}(x, u) = \big(x, (\phi^\nu_\mu(x))(u)\big) \tag{6.6}$$

Let $x \in U_\mu \cap U_\nu$. There is a chart $(U_\alpha, \Phi_\alpha) \in \mathcal{A}$ so that $x \in U_\alpha$. By the criterion of construction of $\widetilde{\mathcal{A}}$, C^r functions ϕ^ν_α and ϕ^α_μ exist that satisfy (iii). On $U_\mu \cap U_\nu \cap U_\alpha$, we define $\phi_\mu^\nu(x) = \phi_\alpha^\nu(x) \circ \phi_\mu^\alpha(x)$. Being the bilinear pairing of C^r functions, this is C^r. Further, the identity (6.6) is satisfied since

$$\Phi_\nu \circ \Phi_\mu^{-1} = (\Phi_\nu \circ \Phi_\alpha^{-1}) \circ (\Phi_\alpha \circ \Phi_\mu^{-1})$$

in a neighborhood of x. Finally, this definition is independent of the particular chart (U_α, Φ_α), for if $(U_\beta, \Phi_\beta) \in \mathcal{A}$ with $x \in U_\beta$, then

$$(\Phi_\nu \circ \Phi_\alpha^{-1}) \circ (\Phi_\alpha \circ \Phi_\mu^{-1}) = (\Phi_\nu \circ \Phi_\beta^{-1}) \circ (\Phi_\beta \circ \Phi_\mu^{-1})$$

in a neighborhood of x. It thus follows that ϕ_μ^ν is well-defined and C^r on $U_\mu \cap U_\nu$, and (iii) is satisfied. To sum up, we have shown that any C^r atlas has a uniquely defined maximal extension. This will also be referred to as the C^r ***VB-structure*** of the vector bundle.

4. Remark One often encounters a slicker definition of a vector bundle in mathematical literature as follows. A C^r vector bundle $\pi: E \to M$ (of fiber type F) consists of the following data:
(i) C^r manifolds E and M, and a C^r surjective map $\pi: E \to M$, and
(ii) a vector space structure on each *fiber* $E_x = \pi^{-1}(x)$, $x \in M$, so that
(iii) for each $x \in M$, there exists an open neighborhood U of x and a C^r diffeomorphism $\Phi: \pi^{-1}(U) \to U \times F$ with Φ mapping each vector space E_x isomorphically onto $\{x\} \times F$.

The fact that a vector space structure for each E_x is pre-ordained here immediately establishes the existential nature of the vector bundle in theory. In actual practice and examples, there is usually no way to define vector space structure on fibers except by having local product structures on hand and transferring the vector space structure from F. Nevertheless, this definition is often convenient for conceptual discussions and proving theorems. Let us explain why the two definitions are equivalent. Using our original definition, we showed that each E_x receives a well-defined vector space structure, and the other requirements of the new definition follow immediately. Conversely, given the latter definition, each point of M belongs to at least one VB-chart. Compatibility of charts follows from the fact that each local product diffeomorphism provides an isomorphism of fibers with F. Extension to maximal atlas takes place with reference to the given vector space structure on the fibers.

Regardless of which definition one is using, a C^r vector bundle of fiber type F may be visualized as a collection of disjoint copies of F, parametrized by points x of the base manifold M so that the vector space operations vary from fiber to fiber in a C^r manner. This is how we originally encountered the tangent bundle of an m-manifold, as the disjoint union of tangent spaces installed at every point of the manifold, each isomorphic to $\mathbb{R}^m$. We are going to show that the considerations that lie at the basis of the construction of the tangent bundle in Chapter 4 are quite universal, and can serve as a recipe for the construction of all vector bundles.

In general, suppose M is a C^r manifold, (U_α) is an open cover for M, and for each pair of indices α and β such that $U_\alpha \cap U_\beta \neq \emptyset$, a C^s map $\phi_\alpha^\beta : U_\alpha \cap U_\beta \to \mathrm{GL}(F)$ is given, $0 \leq s \leq r$, so that the ϕ_α^β satisfy (6.3). Then the family (ϕ_α^β) is called a C^s ***1-cocycle on M with value in*** $\mathbf{GL}(F)$, subordinate to the cover (U_α). By imitating the construction in Chapter 4, we will show that every C^s 1-cocycle determines an essentially unique C^s vector bundle for which the 1-cocycle constitutes the collection of transition maps.

5. Theorem *Let M be a C^r manifold, $r \geq 0$, (U_α) an open cover for M, $0 \leq s \leq r$, and (ϕ_α^β) a C^s 1-cocycle with value in $GL(F)$. Then there is a C^s vector bundle of fiber type F over M, unique up to C^s VB-isomorphism, for which the cover (U_α) is a VB-atlas and the (ϕ_α^β) are the transition maps.*

Proof. For each fixed $x \in M$, consider those indices α such that $x \in U_\alpha$, and let $E_\alpha = \{(x, \alpha, u) : u \in F\}$. E_α is turned into a real vector space isomorphic to F by ignoring the first two components and performing the vector space operations on the third component. Letting $\widetilde{E}_x$ be the disjoint union of these vector spaces E_α, we define a relation $\sim_x$ on $\widetilde{E}_x$ by

$$(x, \alpha, u) \sim_x (x, \beta, v) \iff v = (\phi_\alpha^\beta(x))(u) \tag{6.7}$$

This is an equivalence relation by virtue of (6.4), (6.5) and (6.3). Defining an isomorphism $\lambda_\alpha^\beta : E_\alpha \to E_\beta$ by

$$\lambda_\alpha^\beta(x, \alpha, u) = \big(x, \beta, (\phi_\alpha^\beta(x))(u)\big)$$

we are precisely in the situation of the Identification Lemma in Chapter 4. We thus obtain a k-dimensional real vector space E_x and isomorphisms $i_\alpha : E_\alpha \to E_x$ that satisfy $i_\alpha = i_\beta \circ \lambda_\alpha^\beta$. E_x will be the *fiber over x* for the vector bundle under construction, the disjoint union of the E_x for $x \in M$ will be the *total space* E, and $\pi : E \to M$ is defined by $\pi[x, \alpha, u] = x$. Letting E_α be the subset of E comprised of E_x with $x \in U_\alpha$, we have

$$E_\alpha = U_\alpha \times \{\alpha\} \times F \cong U_\alpha \times F$$

We can thus equip E_α with the product topology, and topologize E with the ensued *sum topology* as in the case of the tangent bundle. Checking the topological requirements for E to serve as a manifold follows verbatim the first two paragraphs of the proof of Theorem 17 in Chapter 4, and will not be repeated here.

The manifold structure of E is also obtained in a similar manner as in the case of the tangent bundle as we shall now demonstrate. Fix a linear isomorphism $i : F \to \mathbb{R}^k$, where $k = \dim F$. Let $[x, \alpha, u]$ be a point of E. Either U_α itself is the domain of a C^r manifold chart (U_α, ξ_α), or by taking its intersection with the domain of such a chart, we assume that (U_α, ξ_α) is a C^r chart for M containing x. A manifold chart for E will be (E_α, η_α), where $E_\alpha = \pi^{-1}(U_\alpha)$, and $\eta_\alpha[x, \alpha, u] = (\xi_\alpha(x), i(u))$. Suppose (U_α, ξ_α) and (U_β, ξ_β) are two C^r charts for M with $x \in U_\alpha \cap U_\beta$. Then the change of variable for the induced charts (E_α, η_α) and (E_β, η_β) will be in the form

$$\eta_\beta \circ \eta_\alpha^{-1}(\xi_\alpha(x), i(u)) = \Big(\xi_\beta(x), (i \circ \phi_\alpha^\beta(x))(u)\Big)$$
$$= \Big((\xi_\beta \circ \xi_\alpha^{-1})(\xi_\alpha(x)), (i \circ \phi_\alpha^\beta(x) \circ i^{-1})(i(u))\Big)$$

This being C^s, the C^s manifold structure of E is established.

To prove uniqueness up to isomorphism, suppose E and E' are both C^s vector bundles over M of fiber type F with VB-atlases (U_α, Φ_α) and (U_α, Ψ_α), respectively, which possess the same set of transition functions, i.e.,

$$\Phi_\beta \circ \Phi_\alpha^{-1}(x, u) = \Big(x, (\phi_\alpha^\beta(x))(u)\Big) = \Psi_\beta \circ \Psi_\alpha^{-1}(x, u)$$

Then

$$\Psi_\beta^{-1} \circ \Phi_\beta = \Psi_\alpha^{-1} \circ \Phi_\alpha \tag{6.8}$$

Now for each index α, $\Psi_\alpha^{-1} \circ \Phi_\alpha$ is a C^s VB-isomorphism from E_α to E'_α. To define a C^s VB-isomorphism from E to E', it suffices to show that on non-empty overlaps $E_\alpha \cap E_\beta$, the isomorphisms indexed by α and β coincide, but this is precisely the content of (6.8), and the claim is proved. □

We will introduce below some of the important examples of vector bundles that we will be dealing with throughout the rest of the book. Analogues of these for product bundles have been introduced earlier in Section A of Chapter 3. The 1-cocycles we present in each case describe how the representation of elements of the total space of the bundle change when we pass from one local product representation to an overlapping one.

6. Examples

(a) The tangent bundle of a C^{r+1} manifold was our original model of a C^r vector bundle. If $\mathcal{A}=(U_\alpha, \xi_\alpha)_\alpha$ is a C^{r+1} atlas for an m-manifold M, then the corresponding 1-cocycle is

$$\phi_\alpha^\beta(x) : U_\alpha \cap U_\beta \to GL(m, \mathbb{R})$$
$$(\phi_\alpha^\beta(x))(u) = \Big(D(\xi_\beta \circ \xi_\alpha^{-1})(\xi_\alpha(x))\Big)(u)$$

(b) For $i=1, \ldots, p$, let (E_i, π_i, M) be a vector bundle of fiber type F_i, with 1-cocycle $((\phi_i)_\alpha^\beta)$ subordinate to the open cover (U_α), same cover for all i. The ***direct sum*** (or the ***Whitney sum***) $(E_1 \oplus \cdots \oplus E_k, \pi_1 \oplus \cdots \oplus \pi_p, M)$ is a vector bundle of fiber type $F_1 \times \cdots \times F_p$ defined by ψ_α^β as follows:

$$\psi_\alpha^\beta(x) = (\phi_1)_\alpha^\beta(x) \oplus \cdots \oplus (\phi_p)_\alpha^\beta(x)$$

If each $(\phi_i)_\alpha^\beta : U \to \mathrm{GL}(F_i)$ is C^r, then so is $\psi_\alpha^\beta : U \to \mathrm{GL}(F_1 \times \cdots \times F_p)$, and it is a straightforward matter to check that (6.3) is satisfied. If the 1-cocycles are given as subordinate relative to different open covers, we take a common refinement and proceed as above.

In the particular case $E_i{=}TM$, we denote the total space by $(TM)^p$ and the projection by τ_M^p. Here, a typical element of the total space is an ordered p-tuple of tangent vectors at the *same* point x. We denote the fiber of $(TM)^p$ over x by $(T_xM)^p$.

(c) Let (E, π, M) be a C^r vector bundle of fiber type F with 1-cocycle (ϕ_α^β) subordinate to the cover (U_α). The ***dual bundle*** (E^*, π^*, M) is defined by a cocycle (ψ_α^β), subordinate to the same open cover, as follows:

$$\begin{aligned} \psi_\alpha^\beta(x) &: U_\alpha \cap U_\beta \to GL(F^*) \\ \psi_\alpha^\beta(x) &= (\phi_\beta^\alpha(x))^* \\ &= \left((\phi_\alpha^\beta(x))^{-1}\right)^* \end{aligned}$$

If $\phi_\alpha^\beta{:}U{\to}\mathrm{GL}(F)$ is C^r, then so is $\psi_\alpha^\beta{:}U{\to}\mathrm{GL}(F^*)$ since the matrices with respect to dual bases are related by inversion and transposition. The identity (6.3) then holds for ψ_α^β, and this is consistent with the change of coordinates for the dual vector space. In particular, for the tangent bundle (TM, τ_M, M) of M, the dual bundle is called the ***cotangent bundle of M***, and is denoted by (T^*M, τ_M^*, M). The fiber $(T^*M)_x$ is usually denoted by T_x^*M and can be regarded as the dual space to T_xM. The elements of T_x^*M are the ***cotangent vectors*** to M at x.

(d) Combining the previous two examples, we have the important bundle with total space $(TM)^p \oplus (T^*M)^q$ over the manifold M. An element of the total space is a $(p{+}q)$-tuple of p tangent and q cotangent vectors at the same point of M.

(e) We can generalize the construction in (c) above. Suppose we are given a C^r vector bundle (E, π, M) of fiber type F with 1-cocycle (ϕ_α^β) subordinate to the cover (U_α). Recall from Chapter 1 the definitions of the spaces L^pF and Λ^pF, and the convention that $L^0F{=}\Lambda^0F{=}\mathbb{R}$. Recall also that any linear map $f{:}F{\to}F$ induces linear maps $L^pf{:}L^pF{\to}L^pF$ and $\Lambda^pf{:}\Lambda^pF{\to}\Lambda^pF$, both of which are contravariant, i.e.,

$$\begin{aligned} L^p(f \circ g) &= L^p(g) \circ L^p(f) \\ \Lambda^p(f \circ g) &= \Lambda^p(g) \circ \Lambda^p(f) \end{aligned}$$

By convention $L^0f{=}\Lambda^0f{=}\mathbb{1}_{\mathbb{R}}$. It follows in exactly the same manner as in Example c that the families of maps defined as

$$(L^p\gamma)_\alpha^\beta(x) = L^p(\phi_\beta^\alpha(x))$$

and

$$(\Lambda^p\gamma)_\alpha^\beta(x) = \Lambda^p(\phi_\beta^\alpha(x))$$

define C^r 1-cocycles. The corresponding vector bundles will be called, respectively, the ***p-covariant tensor bundle*** and ***exterior p-bundle*** of (E, π, M). The total spaces will be denoted, respectively, by L^pE and Λ^pE. The fiber $(L^pE)_x$ can be viewed as $L^p(E_x)$, and the fiber $(\Lambda^pE)_x$ as $\Lambda^p(E_x)$. Recalling that for a vector space F of dimension k, $\Lambda^pF{=}\{\mathbf{0}\}$ for $p{>}k$, we may construct the direct sum of exterior p-bundles of (E, π, M), simply called the ***exterior bundle*** of (E, π, M), and denoted by

(Λ^*E, π^*, M).

We have $L^1E{=}\Lambda^1E{=}E^*$. Further, $L^0E{=}\Lambda^0E$ is a trivial bundle, which can be canonically identified with $M{\times}\mathbb{R}$. The latter claim follows from the fact that for $p{=}0$, $L_0E{=}\Lambda^0E{=}\mathbb{R}$, $(L^0)^\beta_\alpha(x){=}(\Lambda^0)^\beta_\alpha(x){=}\mathbb{1}_\mathbb{R}$, hence by the construction method of Theorem 5, an identification with $M{\times}\mathbb{R}$ is obtained.

(f) In similar manner to the construction of L^pE, recalling the covariance of L_q, we define the ***q-contravariant tensor bundle*** of (E, π, M) by the 1-cocycle

$$(L_q)^\beta_\alpha(x) = L_q(\phi^\beta_\alpha(x))$$

Finally, the (mixed) ***(p,q) tensor bundle*** L^p_qE of (E, π, M) is defined by the direct sum $L^p_qE{=}L^pE{\oplus}L_qE$.

In the special case of the tangent bundle $E{=}TM$, we simply write Λ^pM, Λ^*M and L^p_qM instead of $\Lambda^p(TM)$, $\Lambda^*(TM)$ and $L^p_q(TM)$, and call them, respectively, the ***exterior p-bundle of*** M, the ***exterior bundle of*** M and the (p, q) ***tensor bundle of*** M.

In addition to the above constructions, one can consider sub-structures, quotients and homomorphisms of vector bundles. We will consider here vector sub-bundles; homomorphisms are treated in Exercise 6.2 at the end of this chapter. Let $\pi{:}E{\to}M$ be a C^r vector bundle of fiber type F, with $\dim F{=}k$. Suppose E' is a subset of E, and the following conditions hold:
(i) For each $x{\in}M$, $E_x{\cap}E'{=}E'_x$ is a linear subspace of E_x of dimension l.
(ii) For every $a{\in}M$, there is an open neighborhood V of a, and C^r cross-sections $\sigma_1, \ldots, \sigma_l$ of E defined on V, so that for each $x{\in}V$, $\sigma_1(x), \ldots, \sigma_l(x)$ span E'_x.
We then say that E' is a C^r ***vector sub-bundle*** of E of rank l. The following gives justification to the "sub-bundle" terminology we have used.

7. Lemma *Let E' be a C^r vector sub-bundle defined as above. Then:*
(a) E' is a closed C^r submanifold of E.
(b) There is a linear subspace F' of F, $\dim F'{=}l$, so that $\pi|_{E'}{:}E'{\to}M$ is a C^r vector bundle of fiber type F'.

Proof. Let $a{\in}M$ and consider an open neighborhood V of a as in the definition. Since $\pi{:}E{\to}M$ is a C^r vector bundle of fiber type F, there is an open neighborhood U of a and a C^r diffeomorphism $\Phi{:}\pi^{-1}(U){\to}U{\times}F$ providing the local-product structure. Let $V'{=}U{\cap}V$ and denote the restriction of Φ to $\pi^{-1}(V')$ by the same notation Φ. Thus $\Phi{:}\pi^{-1}(V'){\to}V'{\times}F$ so that for each $x{\in}V'$, Φ maps the vector space E_x isomorphically onto $\{x\}{\times}F$. Fix an ordered basis $(b_1, \ldots, b_k)$ for F. For $x{\in}V'$ and $i{=}1, \ldots, l$, let $b'_i(x)$ be defined by

$$\Phi(\sigma_i(x)) = (x, b'_i(x))$$

Let $\mathcal{L}(F, F)$ be the space of linear maps from F to F. We define a map $f{:}V'{\to}\mathcal{L}(F, F)$ by specifying the value of $f(x)$ on each basis element b_i. To begin, for $i{=}1, \ldots, l$, we let

$$(f(x))(b_i) = b'_i(x)$$

For $x{=}a$, we pick $b'_{l+1}, \dots, b'_k$ in F so that the set $\{b'_1(a), \dots, b'_l(a), b'_{l+1}, \dots, b'_k\}$ is linearly independent. We then complete the definition of f by letting

$$(f(x))(b_i) = b'_i, \quad \text{for } l{+}1{\le}i{\le}k$$

By the choice of $b'_{l+1}, \dots, b'_k$, $f(a){\in}\mathrm{GL}(F)$. Since $\mathrm{GL}(F)$ is an open subset of $\mathcal{L}(F,F)$, it follows from the continuity of the σ_i and Φ that there is an open neighborhood W of a, $W{\subset}V'$, so that for all $x{\in}W$, one has $f(x){\in}\mathrm{GL}(F)$. Therefore, confining Φ further to $\pi^{-1}(W)$, this diffeomorphism gives rise to a C^r map $f{:}W{\to}\mathrm{GL}(F)$. Now define $\Psi{:}W{\times}F{\to}W{\times}F$ by $\Psi(x,u){=}\big(x,(f(x))(u)\big)$. Ψ is bijective since its restriction to each $\{x\}{\times}F$ acts as an element of $\mathrm{GL}(F)$. The derivative of this function at (x,u) is given by the $(m{+}k){\times}(m{+}k)$ matrix

$$\begin{bmatrix} I_m & \mathbf{0} \\ * & f(x) \end{bmatrix}$$

Since this matrix is invertible, Ψ is a C^r diffeomorphism by the Inverse Function Theorem. The diffeomorphism $\Phi'{=}\Psi^{-1}{\circ}\Phi$ maps $\pi^{-1}(W)$ onto $W{\times}F'$, where F' is the linear subspace of F spanned by $b_1, \dots, b_l$, and $W{\times}F'$ is a closed submanifold of $W{\times}F$. Both assertions of the lemma follow. □

With this background in vector bundles, we are ready to introduce the various "fields" or "bundle cross-sections" of interest on manifolds. Let (E,π,M) be a C^r vector bundle. A *field* is just another word for a cross-section, i.e., a map $\sigma{:}M{\to}E$ so that $\pi{\circ}\sigma{=}\mathbb{1}_M$. A cross-section assigns to each $x{\in}M$ an element $\sigma(x)$ of the fiber E_x lying over x. This assignment is usefully visualized as a *field* of objects of the fiber type of the bundle spread over M. For $0{\le}s{\le}r$, one may speak of a C^s cross-section if the map σ is C^s.

The set $\mathcal{F}(M)$ of real-valued functions on M is a ring under pointwise addition and multiplication of functions and has the set $C^s(M)$ of real-valued C^s functions on M as a subring. Since each fiber E_x is a vector space over $\mathbb{R}$, the set $\Gamma(E)$ of cross-sections of E becomes a module over $\mathcal{F}(M)$ under pointwise addition of sections and pointwise multiplication of sections by elements of $\mathcal{F}(M)$. We claim that the set $\Gamma^s E$ of C^s cross-sections of E is a module over $C^s(M)$. Being C^s is a local property, and this claim certainly holds for product bundles. But trivializing local diffeomorphisms are linear on fibers, hence they preserve addition of sections and multiplications by real-valued functions, therefore the claim follows.

Below we shall describe the principal examples of fields that will enter our future discussions.

8. Examples

(a) As mentioned in Example 2, a cross-section of the tangent bundle of a manifold M is a (tangent) vector field on M.

(b) A cross-section of the exterior bundle $(\Lambda^* M, \tau^*_M, M)$ of M is an ***(exterior) differential form*** on M. Specifically, a cross-section of $\Lambda^p M$ is an ***(exterior) differential p-form***.

(c) A cross-section of $L^p M$ (respectively, $L_q M$) is called a ***covariant p-tensor field*** (respectively, ***contravariant q-tensor field***) on M.

For the examples above, there is an alternative to the cross-section viewpoint that can be equally useful. This was already described for product bundles at the end of Section A of Chapter 3. We will briefly illustrate it first for C^s p-covariant tensor fields on a manifold M. Consider the set $\hat{\Gamma}^s((TM)^p)$ of C^s functions $(TM)^p \to \mathbb{R}$ the restriction of which to each fiber $(T_x M)^p$ of $(TM)^p$ is p-linear. $\hat{\Gamma}^s((TM)^p)$ is a module over $C^s(M)$ under pointwise addition of functions and multiplication by elements of $C^s(M)$:

$$f \in C^s(M), \quad \beta \in \hat{\Gamma}^s((TM)^p), \quad u_1, \ldots, u_p \in T_x M$$
$$(f\beta)(u_1, \ldots, u_p) = f(x)\, \beta(u_1, \ldots, u_p)$$

Now we define a map $\alpha \mapsto \hat{\alpha}$ from $\Gamma^s(L^p M)$ to $\hat{\Gamma}^s((TM)^p)$ as follows:

$$\hat{\alpha}|_{(T_x M)^p} = \alpha(x) \tag{6.9}$$

9. Lemma *The correspondence $\alpha \mapsto \hat{\alpha}$ gives a $C^s(M)$-module isomorphism for each s, $0 \le s \le r$.*

Proof. We need only provide an inverse for $\alpha \mapsto \hat{\alpha}$, the rest follows immediately from the definition. Suppose $\beta \in \hat{\Gamma}^s((TM)^p)$. We define $\check{\beta} \in \Gamma^s(L^p M)$ by

$$\check{\beta}(x) = \beta|_{(T_x M)^p} \tag{6.10}$$

It is a straightforward matter to check that (6.9) and (6.10) define inverse module isomorphisms at every level s, $0 \le s \le r$. □

By restriction, we obtain an alternative description of *anti-symmetric* tensor fields, i.e., exterior differential forms, the local case of which was already discussed in Chapter 3. In the case of vector fields, case (a) in Example 8, a vector field on M may be alternately viewed as a map $T^* M \to \mathbb{R}$ the restriction of which to each fiber $T^*_x M$ is linear. This is precisely the double-dual identification.

B. Vector Fields: Globalization

In this section we implant the local theory of vector fields, developed in Chapter 2, on a manifold. The reader must have a firm understanding of the material in that chapter in order to follow the sequel. In fact, most of the material in this section is little more than a restatement of the results of Chapter 2. The genuinely global study of vector fields is the subject of the theory of dynamical systems. Our general guiding principle here for globalization will be the following: Concepts that are invariant under diffeomorphisms of open subsets of $\mathbb{R}^m$ can be automatically transferred to manifolds via a chart. This is because such concepts will be well-defined under change of coordinate diffeomorphisms. These include the flow of a vector field and the Lie bracket of vector fields. In what follows, some, but not all, major concepts will be highlighted for emphasis. The reader should be able to refer back to Chapter 2 where required.

As standing hypothesis for most of this section, we work with smooth (= C^∞) manifolds and smooth vector fields. Let M be a smooth m-manifold and X a smooth vector field on M. An ***integral curve*** or a ***solution*** for X is a curve $\gamma : I \to M$ so that $\gamma'(t) = X(\gamma(t))$ for all $t \in I$. Let (U, ξ) be a chart in the smooth atlas of M. Then $\xi_*(X|_U) = T\xi \circ X|_U \circ \xi^{-1}$ is a smooth vector field on $\xi(U)$. The integral curves of $X|_U$ and $\xi_*(X|_U)$ correspond under ξ. In fact, if λ is an integral curve for $\xi_*(X|_U)$, and $\gamma = \xi^{-1} \circ \lambda$, then

$$\begin{aligned}\gamma'(t) &= (T\xi^{-1} \circ \xi_*(X|_U) \circ \lambda)(t) \\ &= (X \circ \gamma)(t)\end{aligned}$$

It follows that local propositions, such as the fundamental existence-uniqueness-uniformity theorem, carry over to manifolds, independently of coordinates. This independence can also be observed by virtue of the discussion in Subsection 10 of Chapter 2, on the correspondence of flows under coordinate change. Having established this, the stronger forms of uniqueness, and the notion of maximal solution

$$\begin{aligned}\Phi^x &: I^x \to M \\ \Phi^x(0) &= x\end{aligned}$$

also carry over verbatim to manifolds. The flow of X is defined on

$$\tilde{M} = \bigcup_{x \in M} I^x \times \{x\} \subset \mathbb{R} \times M$$

by

$$\Phi(t, x) = \Phi^x(t)$$

The group property of the flow $\Phi_t(x) = \Phi(t, x)$ also follows. We recall that a vector field (or the corresponding flow) is called ***complete*** in case $\tilde{M} = \mathbb{R} \times M$, i.e., every solution Φ^x is defined for all $t \in \mathbb{R}$.

We also include a few words about time-dependent vector fields following the approach of Subsection 11 of Chapter 2. A ***time-dependent vector field*** on a manifold M is a vector field on $\mathbb{R} \times M$ of the form $X(t, x) = \frac{d}{dt}(t) \oplus X_t(x)$, where each X_t is a vector field on M, i.e., $X_t(x) \in T_x M$. We denote the flow of X on $\mathbb{R} \times M$ by $\widetilde{\Phi}$. A family of diffeomorphisms (not generally satisfying the group property of a flow) $(\Phi_t)_t$ is defined on open subsets of M by $\Phi_t(x) = (\pi \circ \Phi)(0, x)$, where π is the projection of $\mathbb{R} \times M$ on the second component. We call the time-dependent vector field X ***complete*** in case the flow $\widetilde{\Phi}$ is complete. In this case, each Φ_t is a diffeomorphism of the entire manifold M onto itself.

The following is one of the very few global statements about vector fields in this section; the proof is essentially a repeat of the argument in 8c of Chapter 2.

10. Theorem *If M is a compact smooth manifold, then any smooth vector field on M, time-dependent or time-independent, is complete.*

Proof. We first consider the case of a time-independent vector field X. Suppose that for $x \in M$, the maximal solution is defined on $]\alpha_x, \omega_x[$ with $\omega_x < +\infty$; we derive a contradiction. A similar argument would show that $\alpha_x = -\infty$. By compactness of M, there is a sequence of points on the positive orbit of x, say $(\Phi^x(t_n))$ that converges to a point $\bar{x} \in M$, as $t_n \to \omega_x$. We may assume that $t_1 < t_2 < \cdots < \omega_x$. By the uniformity assertion of the Fundamental Theorem, there is $T > 0$ and a neighborhood V of $\bar{x}$ so that any solution starting in V is defined for at least all $0 \le t \le T$. We take N so large that $\Phi^x(t_n) \in V$ and $t_n + T > \omega_x$ for all $n > N$. It follows from the group property of flows that $\Phi^x(t_n + T) = \Phi_T(\Phi^x(t_n))$ is defined, contradicting the definition of ω_x.

In the time-dependent case, we look at the flow $\widetilde{\Phi}$ on $\mathbb{R} \times M$. Proceeding as in the proof above with the assumption $\omega_x < +\infty$, we note that because of the boundedness of the first component $\frac{d}{dt}$, the solution stays in the compact region $[0, \omega_x] \times M$ for $0 \le t < \omega_x$. Therefore, we can repeat the argument of the time-independent case to obtain a contradiction to $\omega_x < +\infty$. □

Given a chart (U, ξ) in the smooth atlas for the m-manifold M, we use local coordinates $(\xi^1, \ldots, \xi^m)$ on U, where $\xi^i = x^i \circ \xi$, with x^i denoting projection on the ith coordinate in $\mathbb{R}^m$. Vector fields $\frac{\partial}{\partial \xi^i}$ on U are defined by

$$\frac{\partial}{\partial \xi^i} = \xi_*^{-1}\left(\frac{\partial}{\partial x^i}\right)$$

Any vector field X on U, e.g., the restriction to U of a vector field on M, can be written as a linear combination of the $\frac{\partial}{\partial \xi^i}$ with functional coefficients, $X = \sum_i X^i \frac{\partial}{\partial \xi^i}$. We then have $\xi_* X = \sum_i (X^i \circ \xi^{-1}) \frac{\partial}{\partial x^i}$. Since smoothness is a local property, a vector field on M is smooth if and only if its local representations in charts as above have smooth coefficients. It follows that X is smooth if and only if all the X^i are smooth. We point out that the material of Section B of Chapter 2 can now be wholly transferred to manifolds since much of it is local in nature, and some items are based on general arguments (e.g., the classification of orbits). This includes Theorem 6 and Theorem 9, as well as the entire Subsections 8, 10 and 11 of that chapter.

We now turn to the interpretation of smooth vector fields as derivations, the local version of which was handled in Section C of Chapter 2. Let $C^\infty(M)$ be the ring of real-valued C^∞ functions on M. The set $\mathfrak{X}(M)$ of smooth vector fields on M is a module over $C^\infty(M)$. As in the local case of Chapter 2, we define a ***derivation*** of $C^\infty(M)$ to be a map $C^\infty(M) \to C^\infty(M)$, which is $\mathbb{R}$-linear and Leibnizian. The set $\mathcal{D}(M)$ of derivations is also a module over $C^\infty(M)$. The fact that every smooth vector field can be viewed as a derivation can be tied to the identification of a vector field as a map $T^*M \to \mathbb{R}$, which is linear on fibers T_x^*M. Suppose $f \in C^\infty(M)$, then df is a smooth cross-section of T^*M, and defining

$$(X \cdot f)(x) = (X(x))(df(x)) = (df(x))(X(x))$$

we obtain a derivation of $C^\infty(M)$. To show that every derivation arises from a (unique) smooth vector field, the main tool, as in Chapter 2, is the so-called *localization* technique, which we shall reproduce here for the sake of completeness.

11. Lemma (Localization) *Let M be a smooth manifold, U an open subset of M and D a derivation of $C^\infty(M)$. Then there is a (unique) derivation D^U on $C^\infty(U)$ with the property that if $f \in C^\infty(M)$, then*

$$D^U(f|_U) = (Df)|_U \tag{6.11}$$

Proof. Given $g \in C^\infty(U)$, and $a \in V$, we wish to give a definition of $(D^U g)(a)$. If g were extendable to all of M as a smooth function $\tilde{g}$, one could define $(D^U g)(a) = D\tilde{g}(a)$. However, there certainly exist $g \in C^\infty(U)$ that are not even continuously extendable to M. But since D^U is to be a derivation, $(D^U g)(a)$ will depend only on the values of g in an arbitrarily small neighborhood of a (see Lemma 13(ii) of Chapter 2), so we may try to modify g outside a small neighborhood of a so as to make it extendable to M. Taking local coordinates around a, we pick open sets V, W and a smooth function $\theta: M \to \mathbb{R}$, so that $a \in W \subset \overline{W} \subset V \subset U$, and θ has value 1 on W and value 0 outside V. Then by defining the value of g to be 0 outside U, the function $\tilde{g} = \theta g$ will be smooth on M and agree with g in a neighborhood of a. We then define $(D^U g)(a)$ equal to $D\tilde{g}(a)$. This is the unique possible value for $(D^U g)(a)$ because of (6.11). That D^U is a derivation follows from the hypothesis that D is a derivation. Finally, (6.11) holds because f itself is an extension of $f|_U$ outside a neighborhood of any $a \in U$. □

Now suppose D is a derivation of $C^\infty(M)$. Given any chart (U, ξ) in the smooth atlas of M, we obtain a derivation D^U as above. By the local correspondence proved in Chapter 2, a $C^\infty(U)$-module isomorphism exists between $\mathcal{D}(U)$ and $\mathfrak{X}(U)$. We let X^U be the smooth vector field on U corresponding to D^U. For two overlapping open subsets U and V of M, D^U and D^V localize to the same derivation for $C^\infty(U \cap V)$ because of uniqueness. Therefore, $X^U = X^V$ on $U \cap V$, hence a (unique) smooth vector field X on M corresponding to D is obtained.

The operator interpretation of vector fields enabled us to introduce the important notion of Lie bracket of vector fields in Section D of Chapter 2. Suppose X and Y are two smooth vector fields on the smooth manifold M. To define $[X, Y]$, we first take a smooth chart (U, ξ) for M, and define $[X|_U, Y|_U]$ by

$$[X|_U, Y|_U] = \xi_*^{-1}[\xi_*(X|_U), \xi_*(Y|_U)]$$

If (V, η) is another chart with $U \cap V \neq \emptyset$, then by Theorem 21 of Chapter 2, the diffeomorphism $\eta \circ \xi^{-1}$ from $\xi(U \cap U)$ to $\eta(U \cap U)$ preserves the bracket, so the definition is independent of the particular chart in the maximal atlas of M. It follows that $[X, Y]$ is well-defined on M. The properties of the bracket, being all local, carry over to the manifold M. In particular, the propositions of Subsection 18, (2.41), Theorem 23, Corollary 24 and Theorem 26 of Chapter 2 all hold on a manifold.

12. Push-forward of Vector Fields

In Subsection 10 of Chapter 2, we considered the push-forward of a vector field under a diffeomorphism. We are going to look at two distinct generalizations of this concept that often arise in practice. In each case, a smooth map of manifolds, $h{:}N\to M$ allows an association of vector fields $X\mapsto h_*X$ that is a Lie algebra homomorphism, i.e., it is $\mathbb{R}$-linear and preserves the Lie bracket.

(a) Suppose M and N are smooth manifolds, and $h{:}N\to M$ is a smooth map. Then a smooth function $Y{:}N\to TM$ is called a ***smooth vector field along*** h, if $Y(x)\in T_{h(x)}M$ for all $x\in N$. Given a smooth vector field Y along $h{:}N\to M$, suppose that a smooth tangent vector field X on N exists so that $Y=Th\circ X$, then we say that Y is ***h-related*** to X, or that Y is the ***push-forward*** of X by h, and we write $Y=h_*X$. An outstanding example is the velocity vector function $\mathbf{v}$ for a smooth curve $\gamma{:}I\to M$, where we can regard the velocity vector of γ as the push-forward of the unit vector field $\frac{d}{dt}$ by γ. We have $\mathbf{v}(t)=(T\gamma\circ\frac{d}{dt})(t)=\gamma_*(\frac{d}{dt})$. In elementary differential geometry of curves in $\mathbb{R}^3$, the acceleration vector and the normal vector fields are other examples of vector fields defined along a curve that, however, are not push-forwards.

A smooth vector field Y along $h{:}N\to M$ is called **regular** if the following holds: For every $a\in N$, there are neighborhoods U of a and V of $h(a)$ and a smooth vector field Z defined on V, so that $Y(x)=(Z\circ h)(x)$ for all $x\in U\cap h^{-1}(V)$. An important class of such regular vector fields arises from immersions. Suppose that $h{:}N\to M$ is an immersion. We know from Theorem 4b of Chapter 5 that for every $a\in N$, there is an open subset U of a in N such that the restriction of h to U is an embedding of U into M, and $h|_U$ is a diffeomorphism of U onto the submanifold $h(U)$. In this case, the concept of push-forward corresponds exactly to that considered in Subsection 10 of Chapter 2, for $h|_U$ is a diffeomorphism from U onto $h(U)$. But since h may not be globally one-to-one, the domain of definition of h_*X must be taken to be N, even though $h_*X(x)\in T_{h(x)}M$. Given $b\in M$, for every a in the pre-image of b under the immersion h, a different vector field may be defined in a neighborhood of b as the push-forward of X. We now observe that the push-forward Y defined by the immersion h is regular. For note that an immersion h possesses a smooth locally-defined left inverse k around a point $h(x)$ (see Appendix II). Then letting $Z=Th\circ X\circ k$, we obtain $Y=Z\circ h$.

If X and Y are two smooth vector fields on N, then the value of $[X, Y]$ at $a\in N$ is determined by the values of X and Y in an arbitrarily small neighborhood of a in N, it follows that the equality $h_*[X, Y]=[h_*X, h_*Y]$ remains valid for push-forwards in the context above, just as in Theorem 21 of Chapter 2 .

(b) The other major class of examples involves quotient manifolds. Let N be a smooth manifold and M be a quotient manifold of N. Thus the quotient map $q{:}N\to M$

is a smooth submersion. Let X be a smooth vector field on N that is ***q-equivariant***, i.e., it has the property that whenever $q(a_1)=q(a_2)$, then $Tq(X(a_1))=Tq(X(a_2))$. Then $Tq\circ X:N\to TM$ is a smooth map that is constant on each pre-image set $q^{-1}(b)$. It follows from the property of quotient manifolds (see Lemma b in the proof of Theorem 17 of Chapter 5) that a smooth vector field $X':M\to TM$ exists with the property that the following diagram is commutative:

$$\begin{array}{ccc} N & \xrightarrow{X} & TN \\ \downarrow{\scriptstyle q} & & \downarrow{\scriptstyle Tq} \\ M & \xrightarrow{X'} & TM \end{array}$$

In other words, $X'\circ q=Tq\circ X$. As before, we say that X' is ***q-related to*** X, and write $X'=q_*X$. The association $X\mapsto X'=q_*X$ is $\mathbb{R}$-linear, and by the remark following Theorem 21 of Chapter 2, if X and Y are q-equivariant vector fields on N, then $q_*[X,Y]=[q_*X,q_*Y]$.

C. Differential Forms: Globalization

This section follows a similar pattern as in the previous section. The material on differential forms as developed in Sections B, C and D of Chapter 3 will be transferred here to manifolds. Most of the material will consist of reminders and repetitions, so the reader is advised to be thoroughly familiar with those sections before proceeding. The concept of *orientability*, however, is a genuinely global notion, and will be discussed at some length at the end of this section.

Let M be a smooth m-manifold. We denote the module over $C^\infty(M)$ of smooth cross-sections of $\Lambda^p M$ by $\Omega^p M$. The alternate description as smooth maps $(TM)^p\to\mathbb{R}$ whose restriction to each $T_xM\times\cdots\times T_xM$ is p-linear and anti-symmetric will also be useful. The notation $\Omega^* M$ will be used for the direct sum of the Ω^p. We have $\Omega^0 M=C^\infty(M)$.

If (U,ξ) is a chart in the C^∞ atlas of M, we use local coordinates $(\xi^1,\ldots,\xi^m)$ on U, with $\xi^i=x^i\circ\xi$, and x^i projection on the ith component in $\mathbb{R}^m$. The 1-forms $d\xi^1,\ldots,d\xi^m$ are dual to $\frac{\partial}{\partial\xi^1},\ldots,\frac{\partial}{\partial\xi^m}$, in the sense that for each $a\in U$, $(d\xi^1(a),\ldots,d\xi^m(a))$ is the dual basis in T_a^*M relative to the basis $(\frac{\partial}{\partial\xi^1}(a),\ldots,\frac{\partial}{\partial\xi^m}(a))$ for T_aM. In fact,

$$\begin{aligned}(d\xi^i(a))(\frac{\partial}{\partial\xi^j}(a)) &= \big(dx^i(\xi(a))\circ T_a\xi\big)\Big(\big(T_{\xi(a)}\xi^{-1}\circ\frac{\partial}{\partial x^j}\big)(\xi(a))\Big) \\ &= dx^i(\xi(a))\big(\frac{\partial}{\partial x^j}(\xi(a))\big) \\ &= \delta^i_j\end{aligned}$$

Pointwise operations, namely addition, wedge product, interior product (contraction) and pullback, automatically transfer to vector spaces constructed from individual tangent spaces, and all the propositions in Subsections 2, 3 and 4 of Section B, Chapter 3, hold for a manifold. If α is a cross-section of $\Lambda^p M$, then $\alpha|_U$ has a local representation in the chart (U, ξ) as

$$\alpha = \sum_{i_1 < \cdots < i_p} \alpha_{i_1 \cdots i_p} d\xi^{i_1} \wedge \cdots \wedge d\xi^{i_p}$$

where the $\alpha_{i_1 \cdots i_p}$ are real-valued functions on U. $\alpha \in \Omega^p M$, i.e., α is smooth, if and only if the functional coefficients $\alpha_{i_1 \cdots i_p}$ are smooth. In general, for a cross-section α of $\Lambda^p M$, and for vector fields $X_1, \ldots, X_p$ on M, the real-valued function $\alpha(X_1, \ldots, X_p)$ on M is defined by

$$(\alpha(X_1, \ldots, X_p))(x) = \alpha(x)(X_1(x), \ldots, X_p(x)) \tag{6.12}$$

The following smoothness criterion holds here just as Lemma 1 of Chapter 3, because smoothness is a local property.

13. Lemma *α is smooth at a point a of M if and only if for every p vector fields $X_1, \ldots, X_p$, defined and smooth in a neighborhood of a, the real-valued function $\alpha(X_1, \ldots, X_p)$ is smooth at a.*

We now turn to the local concepts exterior derivative and Lie derivative. Let $\alpha \in \Omega^p M$, and suppose (U, ξ) is a chart in the C^∞ atlas of M. It was seen in Theorem 8 of Chapter 3 that the exterior derivative commutes with the pullback. We then define

$$d(\alpha|_U) = \xi^* \circ d \circ (\xi^{-1})^* (\alpha|_U)$$

This concept is invariant under smooth changes of coordinates. It follows that the definition extends globally to M. Further, Lemma 5, Proposition 6, as well as formula 3.25 of Section C, Chapter 3, being local in nature, continue to hold on the entire manifold.

The limit definition of the Lie derivative, (3.27) in Chapter 3, makes sense on a manifold. The existence of that limit was proved via Lemma 10 and Lemma 11, both of which are local and invariant under coordinate change. Therefore, the same formula for the Lie derivative of a smooth p-form α with respect to a smooth vector field X can be adopted:

$$L_X \alpha(x) = \lim_{t \to 0} \frac{\Phi_t^* \alpha(x) - \alpha(x)}{t}$$

Here Φ is the flow of X. Of course, one can also use Cartan's Formula, (12) of Chapter 3, to define the Lie derivative, as both d and i_X are well-defined on a manifold. However, in this case, one must prove the limit expression. With these two equivalent expressions for Lie derivative on hand, the same arguments as in Chapter 3, replacing the open subset U of $\mathbb{R}^m$ by a smooth manifold M yield results of the propositions 13 and 15 of Chapter 3, as well as the computational formulas in Subsection 14.

14. Volume Elements and Orientation

Let F be a real vector space of dimension m. We recall from Section D of Chapter 1 some facts about the one-dimensional vector space $\Lambda^m F$. A non-zero element ω of $\Lambda^m F$ is called a *volume element* for F and constitutes a basis for $\Lambda^m F$. Thus every volume element for F is either a positive or negative multiple of ω, dividing the set of volume elements into two disjoint classes. Each class is called an *orientation* for F. There is an equivalent approach to orientation using ordered bases. For ordered bases $\mathcal{B}=(b_1,\ldots,b_m)$ and $\mathcal{B}'=(b_1',\ldots,b_m')$, we write $\mathcal{B}'\sim\mathcal{B}$ if and only if the (unique) linear map f that maps each b_i to the corresponding b_i' has a positive determinant. $\sim$ is an equivalence relation that splits the set of ordered bases into two disjoint classes. Since $f^*\omega=(\det f)\omega$, it follows that $\omega(b_1,\ldots,b_m)$ has the same sign for all ordered bases $\mathcal{B}$ in the same equivalence class. Likewise, given an ordered basis $(b_1,\ldots,b_m)$, then $\omega'(b_1,\ldots,b_m)$ will have the same sign as $\omega(b_1,\ldots,b_m)$, if ω' is in the same class as ω. Therefore, an orientation for V may also be defined as a choice of equivalence class of ordered bases under the above equivalence relation. A vector space admits exactly two orientations. Given a volume element ω for F, an ordered basis $(b_1,\ldots,b_m)$ is called ***positively oriented*** (respectively, ***negatively oriented***), relative to ω, if $\omega(b_1,\ldots,b_m)>0$ (respectively, if $\omega(b_1,\ldots,b_m)<0$).

Let M be a C^1 manifold of dimension m. It is possible to pick a volume element $\omega(x)$ for each tangent space T_xM and consider the induced orientation for T_xM. However, one normally demands more, namely, that $\omega(x)$ depend continuously on x. We say that the m-manifold M is ***orientable*** if $\Lambda^m M$ admits a continuous non-zero cross-section ω. In this case, we provide each tangent space T_xM with the orientation induced by the volume element $\omega(x)$. We will soon see examples of both orientable and non-orientable manifolds. Note that since $\dim T_xM=m$, the vector bundle $\Lambda^m M$ has rank 1. Therefore, if M is connected and orientable, then there are precisely two possible orientations for M. By an ***oriented manifold*** we mean an orientable manifold for which a choice of orientation has been adopted.

15. Examples

(a) $\mathbb{R}^m$, and open subsets of $\mathbb{R}^m$, admit the volume element

$$\omega_0 = dx^1\wedge\cdots\wedge dx^m$$

This is called the ***standard-*** or ***Euclidean volume element*** of (subsets of) $\mathbb{R}^m$, and the induced orientation will be the ***standard orientation***. In general, an open subset of an orientable manifold is also oriented by restriction of the original volume element.

(b) Any parallelizable manifold is orientable. Suppose TM is C^0 isomorphic to the product bundle $M\times\mathbb{R}^m$, and consider the standard volume element ω_0 of $\mathbb{R}^m$. The constant cross-section, $\omega(x)=\omega_0$ provides a continuous non-zero cross-section of

$\Lambda^m(M\times\mathbb{R}^m)$. The C^0 isomorphism between TM and $M\times\mathbb{R}^m$ induces a continuous non-zero cross-section for $\Lambda^m M$.

(c) Let M be a smooth manifold and N a smooth submanifold. Then a smooth vector field X along the inclusion map $j{:}N\hookrightarrow M$ is called ***transverse to*** N if for each $x\in N$, one has $X(x)\notin T_xN$. Suppose M is orientable with volume element ω, $\dim M=m$ and N has codimension k. We assume that smooth vector fields, $\{X_1,\ldots,X_k\}$ along the inclusion map of N in M are given so that each X_i is transverse to N, and for each $x\in N$, the set $\{X_1(x),\ldots,X_k(x)\}$ is a linearly independent subset of T_xM. We claim that the $(m-k)$-form

$$j^*(i_{X_k}\circ\cdots\circ i_{X_1}\omega)$$

is a volume element for N. In particular, if the above hypothesis holds, then N is orientable.

To prove the claim, it suffices to show that for each $x\in N$, if $\{u_{k+1},\ldots,u_m\}$ is a linearly independent subset of $T_xN\subset T_xM$, then

$$(j^*(i_{X_k}\circ\cdots\circ i_{X_1}\omega)(x))(u_{k+1},\ldots,u_m)\neq 0$$

But

$$\begin{aligned}(j^*(i_{X_k}\circ\cdots\circ i_{X_1}\omega)(x))(u_{k+1},\ldots,u_m) &= ((i_{X_k}\circ\cdots\circ i_{X_1}\omega)(x))(T_xj(u_{k+1}),\ldots,T_xj(u_m))\\ &= \omega(x)(X_1(x),\ldots,X_k(x),u_{k+1},\ldots,u_m)\end{aligned}$$

The set of $\{X_1(x),\ldots,X_k(x),u_{k+1},\ldots,u_m\}$ is linearly independent since the first k elements are outside the subspace T_xN of T_xM, the last $m-k$ are in T_xN, and each of the two sets is linearly independent. Since $\omega(x)$ is a volume element for T_xM, it follows that the right-hand side of the above is non-zero.

As an important concrete example, we obtain a volume element for the unit sphere S_m, which is a submanifold of $\mathbb{R}^{m+1}$. The vector field

$$X(x)=\sum_{j=1}^{m+1}x^j\frac{\partial}{\partial x^j}(x)$$

does not lie in the tangent space to S_m at the point $x\in S_m$ (see Figure 1). If $\omega_0=dx^1\wedge\cdots\wedge dx^{m+1}$ is the standard volume element on $\mathbb{R}^{m+1}$, then $j^*(i_X\omega)$, $j{:}S_m\hookrightarrow\mathbb{R}^m$, i.e., the restriction of $i_X\omega_0$ to S_m, is called the ***standard volume element of*** S_m and is denoted by ω_{S_m}. Explicitly, using 2d and 3.19 of Chapter 3, we have

$$\begin{aligned}i_X\omega_0 &= \sum_{j=1}^{m+1}(x^j)i_{\frac{\partial}{\partial x^j}}(dx^1\wedge\cdots\wedge dx^{m+1})\\ &= \sum_{j=1}^{m+1}(-1)^{j-1}x^jdx^1\wedge\cdots\widehat{dx^j}\ldots\wedge dx^{m+1}\end{aligned}$$

Thus

$$\omega_{S_m} = \sum_{j=1}^{m+1}(-1)^{j-1}x^j dx^1\wedge\cdots\widehat{dx^j}\ldots\wedge dx^{m+1} \tag{6.13}$$

As a special case of Example (c), we have shown that for a submanifold N of

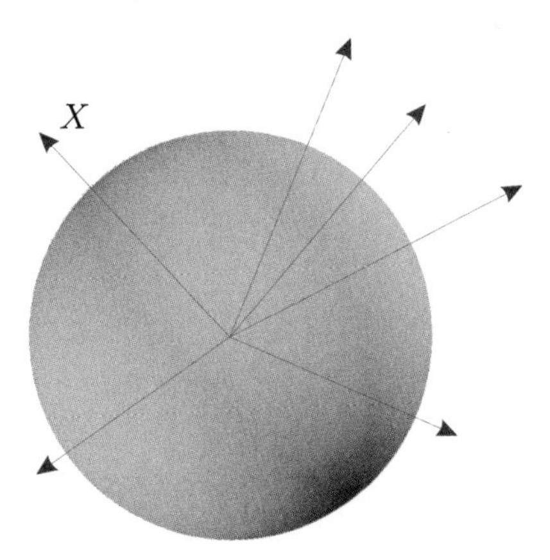

FIGURE 1. Normal vector field on the sphere

codimension one in an orientable manifold M, the existense of a smooth transverse vector field implies the orientability of N. The converse of this assertion is also true and not hard to prove. It is expedient to postpone the proof to the next section of this chapter (see Theorem 25).

(d) We show that for even integers m, the real projective space $\mathbb{RP}(m)$ is not orientable. Recall from Example 19b of Chapter 5 that there is a two-sheeted covering space $p:S_m\to\mathbb{RP}(m)$ that sends antipodal points $\{x,-x\}$ of S_m to the same point $[x,-x]$ of $\mathbb{RP}(m)$. Suppose ω is a volume element for $\mathbb{RP}(m)$. Then $\omega'=p^*\omega$ is a volume element for S_m since p is a local diffeomorphism. Therefore, we must have $\omega' = f\omega_{S_m}$, where f is a continuous non-vanishing function on S_m, and ω_{S_m} is the standard volume element of S_m described in the previous example. If $\tau:x\mapsto -x$ is the antipodal map of S_m, we have $p=p\circ\tau$, therefore

$$\omega' = p^*\omega = (\tau^*\circ p^*)\omega = \tau^*\omega'$$

Explicitly, then

$$f(x)\sum_{j=1}^{m+1}(-1)^{j-1}x^j dx^1\wedge\cdots\widehat{dx^j}\ldots\wedge dx^{m+1} = \tau^*(f(x)\sum_{j=1}^{m+1}(-1)^{j-1}x^j dx^1\wedge\cdots\widehat{dx^j}\ldots\wedge dx^{m+1})$$

$$= (-1)^{m+1}f(-x)\sum_{j=1}^{m+1}(-1)^{j-1}x^j dx^1\wedge\cdots\widehat{dx^j}\ldots\wedge dx^{m+1}$$

If m is even, it follows that $f(x)=-f(-x)$. This implies that the continuous function f changes sign on the connected space S_m, therefore it must vanish at some point of S_m, contradicting the assumption. The argument shows that for even m, $\mathbb{RP}(m)$ does not possess a volume element, i.e., it is not orientable. It turns out that this situation is quite general: Every non-orientable manifold admits an orientable, connected and two-sheeted covering space (see Exercise 6.25 at the end of this chapter). Incidentally, for odd m, since $\tau^*\omega_{S_m}=\omega_{S_m}$, and p is a local diffeomorphism, ω_{S_m} induces a volume element for $\mathbb{RP}(m)$ (see also Exercise 6.24).

We now proceed to give a characterization of orientability in terms of the maximal atlas of a manifold. A C^1 atlas $\mathcal{A}$ for a manifold M is called a ***positive atlas*** if for every $(U,\xi), (V,\eta)\in\mathcal{A}$ with $U\cap V\neq\emptyset$, and every $x\in U\cap V$, one has

$$\det\left(D(\eta\circ\xi^{-1})(x)\right) > 0$$

In other words, the derivative of the change of local coordinates must be orientation-preserving as a linear map of $\mathbb{R}^m$ to itself.

16. Theorem *A C^1 manifold M is orientable if and only if its maximal C^1 atlas possesses a positive sub-atlas.*

Proof. Let $\widetilde{\mathcal{A}}$ be the maximal C^1 atlas of M. Suppose first that a volume element ω for M is given; we produce a positive sub-atlas $\mathcal{A}$ of $\widetilde{\mathcal{A}}$. Let ω_0 be the standard volume element of $\mathbb{R}^m$, and suppose $\tau:\mathbb{R}^m\to\mathbb{R}^m$ is a fixed orientation-reversing involution of $\mathbb{R}^m$, e.g., a symmetry with respect to a hyperplane. For every chart $(U,\xi)\in\widetilde{\mathcal{A}}$ with connected U, $T_x\xi$ is either orientation-preserving for all $x\in U$, or it is orientation-reversing for all $x\in U$. This is because U is connected and $T\xi$ is continuous. In the former case, let $(U,\xi)\in\mathcal{A}$, and in the latter, we let $(U,\tau\circ\xi)\in\mathcal{A}$. The atlas $\mathcal{A}$ thus constructed is a positive atlas.

Conversely, suppose a positive sub-atlas $\mathcal{A}$ of $\widetilde{\mathcal{A}}$ is given. We may suppose by Theorem 11 of Chapter 4 that $\mathcal{A}$ is countable, $\mathcal{A}=\{(U_i,\xi_i)\}$, it is locally-finite, and that a C^1 partition of unity (θ_i) subordinate to (U_i) exists. Now $\omega_i=\xi_i^*\omega_0$ is a volume element for U_i. We claim that $\omega=\sum_i\theta_i\omega_i$ is a volume element for M. The point is that since $\theta_i(x)\geq 0$ and $\sum_i\theta_i(x)=1$ for all x, then $\omega(x)$ is everywhere non-zero. □

We note that the argument above proves a little more than the statement of the theorem. In fact, it was shown that if M is oriented, then a positive atlas can be so chosen that for every chart (U,ξ) of the atlas, the diffeomorphism $\xi:U\to\xi(U)\subset\mathbb{R}^m$ is orientation-preserving, where one uses the standard orientation of $\mathbb{R}^m$ for $\xi(U)$. Conversely, for a given positive atlas, the volume element constructed for M is such that for every chart (U,ξ) of the atlas, ξ is orientation-preserving. In such a case, the volume element (or the corresponding orientation) and the positive atlas are said to be ***concordant***.

17. Examples

(a) We show that the product of two C^1 orientable manifolds M and N is orientable (see Example 2f and Example 7c of Chapter 4 for definition and notation). Consider positive atlases $\{(U_\alpha, \xi_\alpha)_\alpha\}$ and $\{(V_\beta, \eta_\beta)_\beta\}$, respectively, for M and N. Let $(U_1{\times}V_1, \xi_1{\times}\eta_1)$ and $(U_2{\times}V_2, \xi_2{\times}\eta_2)$ be charts for $M{\times}N$ in the product atlas, with $(U_1{\times}V_1){\cap}(U_2{\times}V_2){\neq}\emptyset$. Then the derivative of the change of coordinates map has the matrix representation

$$\begin{bmatrix} \mathbf{A} & \mathbf{0} \\ \mathbf{0} & \mathbf{B} \end{bmatrix}$$

It follows from $\det \mathbf{A}{>}0$ and $\det \mathbf{B}{>}0$ that the product atlas is also positive. Inductively, the assertion holds for any finite number of oriented manifolds. Thus, for example, the m-torus is orientable. One could also show the orientability of product by exhibiting a "product volume element." If ω_i is a volume element for M_i, $i{=}1, \ldots, k$, then

$$p_1^*\omega_1 \wedge \cdots \wedge p_k^*\omega_k$$

is a volume element for $M_1{\times}\cdots{\times}M_k$, where p_i denotes the projection of $M_1{\times}\cdots{\times}M_k$ on the ith component (Exercise 6.18 at the end of this chapter).

(b) Let the manifold M be at least C^2, so that TM is a C^1 manifold. We show that TM, as a manifold, is orientable, regardless of whether M is orientable or not. Given a C^2 atlas $\mathcal{A}$ for M with charts (U, ξ), we saw in Theorem 17 of Chapter 4 that the atlas $T\mathcal{A}$, consisting of charts $(TU, T\xi)$, is a C^1 atlas for TM. Suppose that for $a{\in}M$ and a chart (U, ξ) such that $a{\in}U$, we have $\xi(a){=}(x^1, \ldots, x^m)$. If $w{\in}T_aM$, then $T\xi(w)$ will have the local representation $\sum_i u^i \frac{\partial}{\partial x^i}(x)$, where $x{=}\xi(a)$. Or, equivalently, using the original definition of the tangent space to $\mathbb{R}^m$ in Chapter 2,

$$T\xi(w) = (x^1, \ldots, x^m; u^1, \ldots, u^m)$$

Now suppose (V, η) is another chart with $U{\cap}V{\neq}\emptyset$, then the tangent map of $h{=}\eta{\circ}\xi^{-1}$ at x will be in the form

$$(x; u) \mapsto \big(h(x); (Dh(x))(u)\big)$$

Note that for fixed x, the second component is linear with respect to u. It follows that the derivative of the tangent map has the matrix representation

$$\begin{bmatrix} \mathbf{A} & \mathbf{0} \\ * & \mathbf{A} \end{bmatrix}$$

where $\mathbf{A}$ is the matrix representation of $Dh(x)$. Therefore, the determinant of the derivative of the tangent map is $\big(\det(Dh(x))\big)^2{>}0$, showing that the atlas $T\mathcal{A}$ is positive.

D. Riemannian Metrics

Riemannian metrics on open subsets of $\mathbb{R}^m$ were introduced in Section E of Chapter 3. Here we will globalize the concept on manifolds and arbitrary vector bundles, and discuss some elementary applications. More geometric applications will be encountered in later chapters, especially Chapter 9.

Let (E, π, M) be a C^r vector bundle of fiber type F, where $r \geq 0$, and F is a finite-dimensional real vector space. By a C^s ***Riemannian metric***, $0 \leq s \leq r$, on this vector bundle, we mean a C^s cross-section ρ of L^2E so that for each $x \in M$, $\rho(x)$ is an inner product on the fiber E_x. Equivalently, ρ can be viewed as a real-valued C^s function $E \oplus E \to \mathbb{R}$, the restriction of which to each $E_x \times E_x$ is bilinear, symmetric and positive-definite.

18. Lemma *Every C^r vector bundle, $0 \leq r \leq \infty$, admits a C^r Riemannian metric.*

PROOF. We fix an inner product $\bar{\rho}$ for fiber type vector space F, and consider a C^r VB-atlas for the vector bundle. We know from Theorem 11 of Chapter 4 that the VB-atlas may be taken to be a countable and locally-finite atlas (U_i, Φ_i), and that a C^r partition of unity (θ_i) subordinate to the (U_i) exists. For each i, consider the C^r trivializing diffeomorphism $\Phi_i : \pi^{-1}(U_i) \to U_i \times F$. Since the restriction of Φ_i to each fiber $E_x \subset \pi^{-1}(U_i)$ is a linear isomorphism with F, $\bar{\rho}$ induces an inner product $\rho_i(x)$ on E_x. Thus a C^r Riemannian metric ρ_i on $\pi^{-1}(U_i)$ is obtained. We define $\rho = \sum_i \theta_i \rho_i$. For each $x \in M$, the sum $\sum_i \theta_i(x) \rho_i(x)$ is a finite convex linear combination of inner products on E_x, and is hence an inner product, proving the lemma. □

A Riemannian metric for the tangent bundle of a manifold M is simply referred to as a ***Riemannian metric on*** M. Given a Riemannian metric ρ on M, one can discuss notions such as *length*, *angle* and *perpendicularity* in each tangent space T_xM. For $u \in T_xM$, the ρ-**length**, $|u|_\rho$, is defined as $(\rho(x)(u,u))^{1/2}$, and for non-zero u and v in T_xM, the ρ-**angle** between u and v is defined as

$$\cos^{-1}\left(\frac{\rho(x)(u,v)}{|u|_\rho |v|_\rho}\right)$$

In particular, we say u and v are ρ-***perpendicular*** if $\rho(x)(u,v) = 0$.

In a different vein, a Riemannian metric on M provides a very useful VB-isomorphism between TM and T^*M that we will use in this section. However, because the idea and its utility extend to certain other covariant 2-tensor fields, we will provide a more general description below. Let F be a finite-dimensional real vector space, and suppose $\beta : F \times F \to \mathbb{R}$ is a bilinear map. One defines $\beta^\flat : F \to F^*$ by

$$(\beta^\flat(u))(v) = \beta(v, u) \tag{6.14}$$

From the linearity of β in the first component, we infer that in fact $\beta^\flat(u) \in F^*$, and from the linearity in the second component, that $\beta^\flat$ is linear. β is called ***non-degenerate*** in case $\beta^\flat$ is an isomorphism. This is equivalent to $\beta^\flat(u)$ not being identically zero, unless $u=0$. Inner products are examples of non-degenerate bilinear maps. It is useful to have an explicit expression for $\beta^\flat$ in terms of β. Suppose $\mathcal{B} = (e_1, \ldots, e_k)$ is a basis for F, and $\mathcal{B}^* = (e^1, \ldots, e^k)$ is the dual basis for F^*. We know from Chapter 1 that β can be written as $\sum_{i,j} \beta_{ij} e^i \otimes e^j$. Since $\beta^\flat(e_j) \in F^*$, we may write $\beta^\flat(e_j) = \sum_i c_{ij} e^i$. Therefore,

$c_{ij} = (\beta^\flat(e_j))(e_i) = \beta(e_i, e_j) = \beta_{ij}$. It follows that the matrix of $\beta^\flat{:}F{\to}F^*$ with respect to the bases $\mathcal{B}$ and $\mathcal{B}^*$ for F and F^*, respectively, is B=$[\beta_{ij}]$. In the non-degenerate case, the inverse of $\beta^\flat$ is denoted by $\beta_\sharp$. Therefore, the matrix of $\beta_\sharp$, relative to the bases $\mathcal{B}^*$ and $\mathcal{B}$, is given by $[\beta^{ij}]$, where $[\beta^{ij}]$=B^{-1}. We summarize these results in the following lemma for future reference.

19. Lemma *Let F be a finite-dimensional vector space ove $\mathbb{R}$ with ordered basis $\mathcal{B}=(e_1,\ldots,e_k)$, and let $\mathcal{B}^*=(e^1,\ldots,e^k)$ be the dual basis for F^*. If $\beta=\sum_{ij}\beta_{ij}e^i{\otimes}e^j$ is a covariant 2-tensor on F, then the matrix of $\beta^\flat$ relative to $\mathcal{B}$ and $\mathcal{B}^*$ is given by $B=[\beta_{ij}]$. If further, β is non-degenerate, then the matrix of $\beta_\sharp$ relative to $\mathcal{B}^*$ and $\mathcal{B}$ is given by the inverse $B^{-1}=[\beta^{ij}]$.*

Now consider a C^r vector bundle (E, π, M) of fiber type F, and suppose that instead of a single bilinear map, we have a C^r cross-section of $E{\oplus}E$. For each $x{\in}M$, we obtain a linear map $\beta(x)^\flat{:}E_x{\to}E_x^*$. Thus a map $\beta^\flat{:}E{\to}E^*$ is obtained that sends the fiber E_x to the fiber E_x^*, for each x. The following commuative diagram summarizes this:

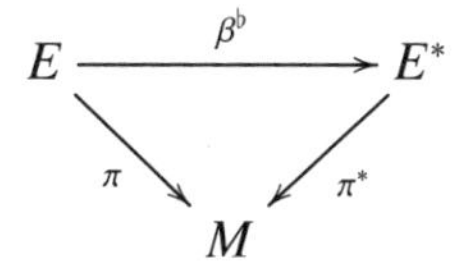

We call β ***non-degenerate*** if $\beta(x)$ is non-degenerate for all $x{\in}M$. Thus a Riemannian metric is non-degenerate.

20. Theorem *If β is a non-degenerate C^r cross-section of $E{\oplus}E$, then $\beta^\flat$ is a C^r VB-isomorphism.*

Proof. By the hypothesis of non-degeneracy, the restriction of $\beta^\flat$ to each E_x is an isomorphism onto E_x^*. It remains to show that $\beta^\flat$ is a C^r diffeomorphism. In view of the existence of global inverse, it suffices to show that $\beta^\flat$ is a local diffeomorphism. Let (U, Φ) be an arbitrary VB-chart for E, and suppose that $(e_1,\ldots,e_k)$ is a basis for F. We let $\Phi^{-1}(x, e_i) = s_i(x)$. Since Φ is C^r, so is each s_i. We denote the dual of $(s_1(x),\ldots,s_k(x))$ by $(s^1(x),\ldots,s^k(x))$; the s^i are also C^r (see the construction of the dual bundle in Example 6c). The local expression of β over U is $\beta(x)=\sum_{ij}\beta_{ij}(x)s^i(x){\otimes}s^j(x)$, with β_{ij} being C^r. It follows from Lemma 19 that the local expression of $\beta^\flat$ is given by the matrix $[\beta_{ij}(x)]$, which has C^r entries and is invertible; thus the proof is complete. □

The following consequence is immediate:

21. Corollary *Let E be a C^r vector bundle over M and suppose β is a non-degenerate C^r cross-section of $E{\oplus}E$. Then β induces a $C^r(M)$-module isomorphism of $\Gamma^r E$ and*

$\Gamma^r E^*$ *given by*

$$\Gamma^r E \ni X \mapsto \beta^\flat \circ X \in \Gamma^r E^*$$
$$\Gamma^r E^* \ni \alpha \mapsto \beta_\sharp \circ \alpha \in \Gamma^r E$$

We usually denote $\beta^\flat \circ X$ by $X^\flat$ and $\beta_\sharp \circ \alpha$ by $\alpha_\sharp$, if β is understood. As an application of this isomorphism, we can introduce the concept of "gradient," which provides a geometric interpretation for the differential of a real-valued function on M. Suppose that a smooth Riemannian metric ρ on M is given. For a smooth function $f{:}M \to \mathbb{R}$, the ***gradient of*** f ***with respect to*** ρ is defined by

$$\nabla_\rho f = (df)_\sharp \tag{6.15}$$

Thus $\nabla_\rho f$ is a vector field on M. For $w \in T_x M$, the definition of $\rho^\flat$ implies the following basic identity:

$$df(w) = \rho(w, \nabla_\rho f(x)) \tag{6.16}$$

22. Elementary Properties of the Gradient Suppose M is a smooth manifold, ρ a smooth Riemannian metric on M and $f{:}M \to \mathbb{R}$ a smooth function.
(a) If $a \in M$ is a regular point for f, then $\nabla_\rho f(a)$ is ρ-perpendicular to the level set of f that passes through a.
(b) (Lagrange) If N is a smooth submanifold of M and $a \in N$ is a critical point for the restriction of f to N, then $\nabla_\rho f(a)$ is ρ-perpendicular to $T_a N$.

Proof. (a) Let L be the level set of f that contains a. Then we know from the Regular Value Theorem of Chapter 5 that the intersection of L with a neighborhood of a in M is a submanifold of codimension one, and from Theorem 10 of Chapter 5 that $T_a L = \ker T_a f$. We must show that $\rho(w, \nabla_\rho f(a)) = 0$, if $w \in T_a L$, but this is immediate from (6.16).
(b) The hypothesis that a is a critical point for $f|_N$ means that $T_a N \subset \ker df(a)$. The claim follows again from (6.16). □

Part (b) is often described as follows. Smooth functions $g_1, \dots, g_k$ are given so that at the point $a \in M$, the set of gradients $\{\nabla_\rho g_1(a), \dots, \nabla_\rho g_k(a)\}$ is linearly independent. Then a is a regular point for the function $g = (g_1, \dots, g_k){:}M \to \mathbb{R}^k$. Let L be the level set of g that passes through a. It follows from the Regular Value Theorem that the portion of L in a neighborhood of a is a submanifold of codimension k, and by Theorem 10 of Chapter 5 that $\ker T_a g = T_a L$. Since L is contained in the level sets of all the g_i, we infer from part (a) above that every $\nabla_\rho g_i(a)$ is perpendicular to $T_a L$ at a, hence $\nabla_\rho g_1(a), \dots, \nabla_\rho g_k(a)$ form a basis for the orthogonal complement of $T_a L$ in $T_a M$. Therefore, by (b) above, if a is a critical point for $f|_L$, there are unique real numbers $\lambda_1, \dots, \lambda_k$ so that:

$$\nabla_\rho f(a) = \lambda_1 \nabla_\rho g_1(a) + \cdots + \lambda_k \nabla_\rho g_k(a)$$

This is the familiar form of the so-called ***Lagrange Multiplier Theorem***, with the λ_i known as ***Lagrange multipliers***.

We now turn to the question of volume element in the presence of a Riemannian metric. The local case was already treated in Section E of Chapter 3. Let (M, ρ) be a smooth Riemannian m-manifold, i.e., M is a smooth m-manifold, and ρ is a smooth Riemannian metric on M. An ordered m-tuple of tangent vectors $(u_1, \ldots, u_m)$ in a tangent space T_aM determines a ***parallelepiped***

$$P(u_1, \ldots, u_m) = \{t_1u_1 + \cdots + t_mu_m : 0 \leq t_i \leq 1, i = 1, \ldots, m\}$$

In the case $(u_1, \ldots, u_m)$, is an orthonormal set, i.e., $\rho(a)(u_i, u_j) = \delta_{ij}$, $P(u_1, \ldots, u_m)$ is referred to as a ***unit cube***. The following basic theorem shows that for an oriented Riemannian manifold, there is a unique volume element that assigns unit volume to each positively oriented unit cube.

23. Theorem *Let (M, ρ) be an oriented smooth Riemannian m-manifold. Then there is a unique smooth volume element ω_ρ on M so that for each positively oriented orthonormal m-tuple $(u_1, \ldots, u_m)$ in a tangent space of M, $\omega_\rho(u_1, \ldots, u_m) = 1$. Moreover, if the local expression of ρ in a coordinate chart (U, ξ) is $\rho = \sum_{ij} g_{ij} d\xi^i \otimes d\xi^j$, then the local expression of ω_ρ relative to the same chart is*

$$\omega_\rho = \sqrt{\det[g_{ij}]}\ d\xi^1 \wedge \cdots \wedge d\xi^m \tag{6.17}$$

Proof. The analogous result was proved for an open subset of $\mathbb{R}^m$ in Section E of Chapter 3. By taking a positive atlas, one can transplant that local result on the oriented manifold, but it is just as easy to go through essentially the same argument again. We take a positive atlas for M concordant with the orientation of M. Thus for every chart (U, ξ) of the atlas, the diffeomorphism ξ is orientation-preserving. Therefore, the ordered m-tuple of vector fields $(\frac{\partial}{\partial \xi^1}, \ldots, \frac{\partial}{\partial \xi^m})$ provides a positively oriented basis at each T_xM, $x \in U$. By Lemma 17 of Chapter 3, there is a positively oriented m-tuple $(X_1, \ldots, X_m)$ of smooth orthonormal vector fields on U. The dual one-forms $(\eta^1, \ldots, \eta^m)$, defined by $\eta^i(x)(X_j(x)) = \delta^i_j$, for all $x \in U$, are smooth by the construction of the dual bundle. We define ω_ρ on U by

$$\omega_\rho = \eta^1 \wedge \cdots \wedge \eta^m$$

Note that at every point $x \in U$,

$$(\omega_\rho(x))(X_1(x), \ldots, X_m(x)) = 1$$

Now suppose $(u_1, \ldots, u_m)$ is any positively oriented orthonormal m-tuple in T_xM. There is a (unique) orientation-preserving orthogonal linear isomorphism $f: T_xM \to T_xM$ such that $f(X_i(x)) = u_i$ for all $i = 1, \ldots, m$. We have $\det f = 1$, therefore

$$(\omega_\rho(x))(u_1, \ldots, u_m) = (\det f)(\omega_\rho(x))(X_1(x), \ldots, X_m(x)) = 1$$

Thus ω_ρ, as defined, satisfies the requirement of the theorem on U. In fact, this ω_ρ is the unique smooth volume element on U with this property since for any other volume element ω, one has $\omega = g\omega_\rho$, and by evaluating both sides on $(X_1(x), \ldots, X_m(x))$ we obtain $g=1$. Now the same construction can be carried out for each chart of the given positive atlas. It follows from the uniqueness just proved that the volume elements coincide on overlaps, therefore a unique globally defined ω_ρ results.

Now suppose that on the chart (U, ξ), we have $\rho = \sum_{ij} g_{ij} d\xi^i \otimes d\xi^j$. We therefore obtain

$$g_{ij} = \rho(\frac{\partial}{\partial \xi^i}, \frac{\partial}{\partial \xi^j})$$

With $(X_1, \ldots, X_m)$ as above, there are are smooth functions c_i^k defined on U so that $\frac{\partial}{\partial \xi^i} = \sum_k c_i^k X_k$. Therefore,

$$\begin{aligned} g_{ij} &= \rho(\sum_k c_i^k X_k, \sum_l c_j^l X_l) \\ &= \sum_{k,l} c_i^k c_j^l \delta_{k,l} \\ &= \sum_k c_i^k c_j^k \end{aligned}$$

In matrix notation, we have

$$[g_{ij}] = [c_j^i][c_j^i]^T$$

Therefore,

$$\det [c_j^i] = \sqrt{\det [g_{ij}]}$$

On the other hand, writing $\omega_\rho = V d\xi^1 \wedge \cdots \wedge d\xi^m$, we compute the function V by evaluating both sides on $(\frac{\partial}{\partial \xi^1}, \ldots, \frac{\partial}{\partial \xi^m})$, to obtain

$$\begin{aligned} V &= \omega_\rho(\sum_k c_1^k X_k, \ldots, \sum_k c_m^k X_k) \\ &= \det [c_j^i] \end{aligned}$$

and the result follows. □

The volume element ω_ρ is called the ***Riemannian volume element*** of the oriented Riemannian manifold (M, ρ). The real number $\det [g_{ij}]$ is often denoted by g.

24. Corollary *Let (M, ρ) be an oriented smooth Riemannian manifold and X a smooth vector field on M. If the local expression of X in a chart (U, ξ) is $X = \sum_i X^i \frac{\partial}{\partial \xi^i}$, then*

$$\text{(6.18)} \qquad \mathbf{div}_{\omega_\rho} X = \frac{1}{\sqrt{g}} \sum_i \frac{\partial(X^i \sqrt{g})}{\partial \xi^i}$$

PROOF. Just as the local formula (3.36) of Chapter 3, this follows from (3.30) of that chapter. □

We are now in a position to complete the discussion at the end of Example 15c regarding the orientability of codimension one submanifolds.

25. Theorem *Let M be a smooth orientable manifold, and suppose that N is a smooth codimension one submanifold of M. Then N is orientable if and only if N admits a smooth transverse vector field.*

PROOF. It was already shown in Example 15c that the existence of a smooth transverse vector field to N implies the orientability of N; we must show the converse. We consider a smooth Riemannian metric and an orientation for M. The metric induces, by restriction, a Riemannian metric on N. By assumption, N is orientable, so we fix an orientation for N. At each point $x \in N$, we can take a positively oriented basis $(u_1, \dots, u_{m-1})$ for $T_x N$. But $T_x M$ has been given an orientation as well, so there is a unique unit vector $u_m(x) \in T_x M$, orthogonal to $T_x N$, so that $(u_1, \dots, u_{m-1}, u_m(x))$ is a positively oriented basis for $T_x M$. We have thus constructed a transverse vector field u_m on N. The point is to show that u_m is smooth. Taking a concordant positive atlas for M, we use the parametrized version of Gram-Schmidt, Lemma 17 of Chapter 3, to prove the smoothness of u_m. First, given $x \in N$, there is a positively oriented $(m-1)$-tuple of orthonormal smooth vector fields, defined and tangent in a neighborhood of x in N. Since N is a submanifold of M, the vector fields $X_1, \dots, X_{m-1}$ have smooth extensions $Y_1, \dots, Y_{m-1}$ to a neighborhood of x in M. Taking, for example, a submanifold chart (U, ξ) around x, and extending $\xi_* X_i$ smoothly by parallel translation into $\mathbb{R}^m$, one can then use $(\xi^{-1})_*$ to transfer these extensions to the manifold. The extensions will remain linearly independent in a neighborhood of x. By Lemma 17 of Chapter 3, a smooth unit vector field X_m exists, defined in a neighborhood of x, which is everywhere orthogonal to the span of $\{Y_1, \dots, Y_{m-1}\}$ and such that $(Y_1, \dots, Y_{m-1}, X_m)$ is positively oriented. The uniqueness of $u_m(x)$ implies that $u_m(x) = X_m(x)$; therefore, the smoothness of u_m is established. □

E. Plane Fields

Suppose X is a nowhere-vanishing vector field on a manifold M. Then X determines a one-dimensional subspace of $T_x M$ for each $x \in M$. More generally, any assignment $x \mapsto \Delta_x$ of a k-dimensional subspace Δ_x of $T_x M$ to each $x \in M$ is called a ***k*-plane field** (or a ***k*-plane distribution**) on M. Δ is called C^r if the following condition is satisfied: For each $a \in M$, there is an open neighborhood U of a, and C^r vector fields $X_1, \dots, X_k$, defined on U, so that Δ_x is the linear span of $X_1(x), \dots, X_k(x)$

for every x in U. More succinctly, a C^r k-plane field on M is just a C^r vector subbundle of TM of rank k. For C^r vector fields to exist on a manifold M, the tangent bundle TM must be at least C^r, hence M has to be at least C^{r+1}.

By an ***integral manifold*** of Δ, we mean a pair (I, γ), where I is a connected manifold and $\gamma{:}I \to M$ is a one-to-one immersion, so that for each $t \in I$, one has $T_t\gamma(T_tI) = \Delta_{\gamma(t)}$. Thus if (I, γ) is an integral manifold for a C^r plane field, then I is at least C^{r+1}. A k-plane field Δ is called ***integrable*** if for every $a \in M$, there is an integral manifold (I, γ) with $a \in \gamma(I)$.

26. Examples

(a) A nowhere zero C^r vector field generates an integrable C^r 1-plane field (better known as a ***line field***). Locally, every line field is generated by a non-vanishing vector field, therefore the fundamental existence theorem of ordinary differential equations applies, and every C^1 line field is integrable.

(b) Not every C^r line field is *globally* generated by a C^r vector field. We construct an example on the cylinder $S_1 \times \mathbb{R}$ as follows. On $]\frac{-\pi}{2}, \frac{\pi}{2}[\times\mathbb{R}$, consider the graphs of functions $y = \sec x + C$, where the constant C ranges over all real numbers. As a line field on $]\frac{-\pi}{2}, \frac{\pi}{2}[\times\mathbb{R}$, we consider the tangent lines to these graphs. Extend this to $[\frac{-\pi}{2}, \frac{\pi}{2}]\times\mathbb{R}$ by adding the tangents to the two bounding vertical straight lines. This line field is smooth; it is in fact generated by the vector field

$$X = (\cos^2 x)\frac{\partial}{\partial x} + (\sin x)\frac{\partial}{\partial y}$$

For note that if $x = \pm\,\frac{\pi}{2}$, then a non-zero vertical tangent vector is obtained, and for other values of x, the slope of X at $(x, \sec x + C)$ is $\sin x/\cos^2 x$, which is the derivative of $\sec x$. By identifying the two vertical lines $x = \pm\,\frac{\pi}{2}$ via $(-\frac{\pi}{2}, y) \sim (\frac{\pi}{2}, y)$, we obtain a smooth line field on a cylinder. However, no globally-defined vector field on this cylinder can generate the line field since any vector field forces a direction on solutions, and any direction chosen for the graphs of functions $y = \sec x + C$ induces opposite directions for the two vertical lines $x = \pm\,\frac{\pi}{2}$ that have been identified (see Figure 2).

(c) Plane fields of dimension two and above need not be integrable, in general. We give an example of a 2-plane field in $\mathbb{R}^3$ that is not integrable. The example will be studied from three different points of view. Let $\Delta_{(x,y,z)}$ be the linear span of the following two vector fields:

$$z\frac{\partial}{\partial x} + \frac{\partial}{\partial y}, \quad \frac{\partial}{\partial z}$$

These two vector fields are everywhere linearly independent, so in fact generate a 2-plane field Δ. We will show that Δ is not integrable.

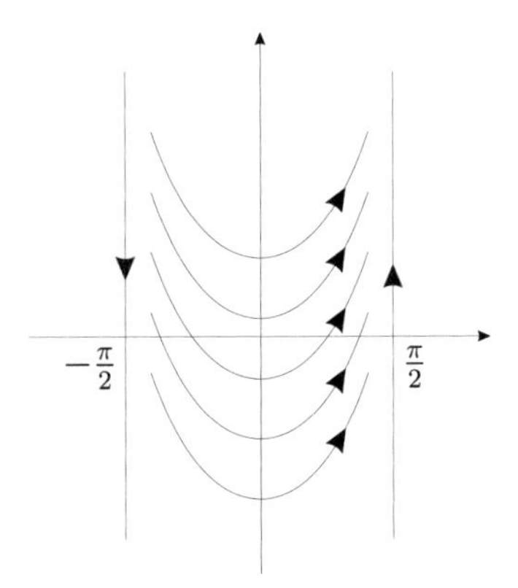

FIGURE 2. Constructing a line field on the cylinder

Method 1 (Intuitive). Note that for a given $z=z_0$, all Δ-planes at (x, y, z_0) are parallel, and they are vertical since the vertical vector $\frac{\partial}{\partial z}$ is in Δ (see Figure 3). Thus if a two-dimensional integral manifold existed, its normal vector would have to be horizontal, i.e., parallel to (x, y)-plane. This can in fact be rigorously ascertained by taking the cross-product of the two generating vectors. It follows that the image of any integral manifold must be "cylindrical," i.e., it must be obtainable by moving a horizontal curve parallel to itself in vertical direction. But this is not possible since one of the generating vectors, namely $z\frac{\partial}{\partial x}+\frac{\partial}{\partial y}$, changes with z.

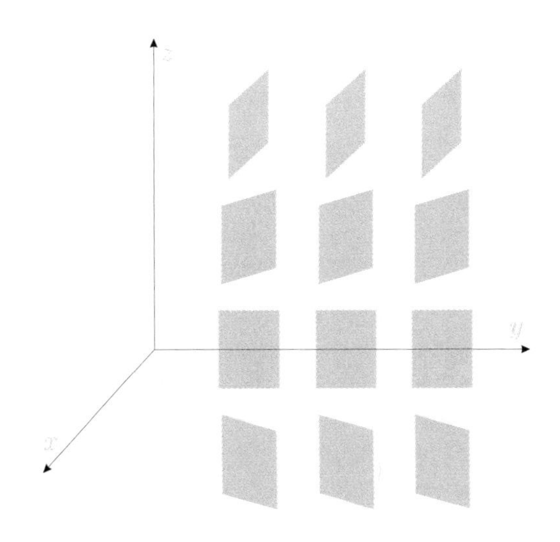

FIGURE 3. A non-integrable field in $\mathbb{R}^3$

Method 2. We use classical vector analysis in $\mathbb{R}^3$ to find a necessary condition for the integrability of 2-plane fields. Later we will generalize this and show the sufficiency of the condtion. Here "gradient" and "curl" will be used in their classical Euclidean sense. Let Δ be a smooth 2-plane field in $\mathbb{R}^3$, and suppose that two smooth vector fields X and Y locally generate Δ. Denote by F the cross-product of X and Y, $F{=}X{\times}Y$. Thus F is everywhere orthogonal to Δ. Now suppose an integral manifold (I, γ) exists for Δ. Let $a{=}\gamma(t_0)$ be a point on the image of this integral manifold. We know from Theorem 4b of Chapter 5 that if γ is restricted to a small enough neighborhood V of t_0, then $\gamma(V){=}N$ is in fact a 2-dimensional submanifold of $\mathbb{R}^3$. By the Rank Theorem, if N is taken sufficiently small, it can be realized as a regular level surface of a smooth function f defined in a neighborhood of the point a in $\mathbb{R}^3$. Therefore, $\mathbf{grad}f$ is perpendicular to N. It follows that we must have $F{=}\varphi(\mathbf{grad}f)$ for a suitable nowhere-zero smooth function φ. We now have

$$\begin{aligned}\mathbf{curl}F &= \varphi(\mathbf{curl}(\mathbf{grad}f)) + \mathbf{grad}\varphi\times\mathbf{grad}f\\ &= (\varphi^{-1})\mathbf{grad}\varphi\times F\end{aligned}$$

Therefore, we obtain the following necessary condition for integrability

$$\mathbf{curl}F \cdot F = 0 \tag{6.19}$$

where "$\cdot$" denotes the standard Euclidean inner product in $\mathbb{R}^3$. We show that in the example at hand this condition is nowhere satisfied. In fact, we can take

$$\begin{aligned}F &= (z\frac{\partial}{\partial x} + \frac{\partial}{\partial y}) \times (\frac{\partial}{\partial z})\\ &= \frac{\partial}{\partial x} - z\frac{\partial}{\partial y}\end{aligned}$$

Therefore,

$$\mathbf{curl}F \cdot F = (\frac{\partial}{\partial x}) \cdot (\frac{\partial}{\partial x} - z\frac{\partial}{\partial y}) = 1$$

and Δ has no integral manifolds.

Method 3. We assume an integral manifold (I, γ) exists and take a piece N of the image small enough to be a submanifold of $\mathbb{R}^3$. Now the vector fields $X{=}\frac{\partial}{\partial z}$ and $Y{=}z\frac{\partial}{\partial x}{+}\frac{\partial}{\partial y}$ are everywhere tangent to N, hence they can be considered as tangent vector fields to the manifold N. Therefore, the Lie bracket $[X, Y]$ must also be a tangent vector field to N. But

$$[\frac{\partial}{\partial z}, z\frac{\partial}{\partial x} + \frac{\partial}{\partial y}] = \frac{\partial}{\partial x}$$

However, it is not possible to express $\frac{\partial}{\partial x}$ as a linear combination of X and Y with smooth functional coefficients, therefore the 2-plane field is not integrable. This method will also give rise to general necessary and sufficient conditions, as we shall see.

In the remainder of this section, we will confine ourselves to smooth data to simplify the presentation. The following gives a clear-cut criterion for the integrability of plane fields. We say that a vector field X ***belongs to*** Δ, if X is defined and smooth on an open subset U of M, and $X(x){\in}\Delta(x)$ for all $x{\in}U$.

27. Theorem ***(Frobenius)*** *Let Δ be a smooth k-plane field on a smooth m-manifold M. Then the following conditions are equivalent:*
(a) Δ is integrable.
(b) Δ is closed under $[,]$, *i.e., if smooth vector fields X and Y belong to Δ, then* $[X, Y]$ *also belongs to Δ.*
(c) For any point $a{\in}M$, there is a chart (U, ξ) around a so that $\frac{\partial}{\partial\xi^1}(x), \dots, \frac{\partial}{\partial\xi^k}(x)$ span Δ_x for all $x{\in}U$.

A k-plane field for which the property (b) above holds is called ***involutive***. Thus the Frobenius Theorem is sometimes stated as the proposition that a plane field is integrable if and only if it is involutive.

Proof. The fact that (a) implies (b) was already noted in the example above. To repeat, let X and Y be as in (b), and suppose that (I, γ) is an integral manifold of Δ so that $\gamma(I)$ passes through the domain W of X and Y. We take I small enough so that $\gamma(I)$ is a submanifold of M. Then Δ_x will be the tangent space to $\gamma(I)$ at each $x{\in}\gamma(I)$, with $X(x)$ and $Y(x)$ in Δ_x. It follows that $[X, Y](x)$ is in the same tangent space Δ_x.

The main thrust of the theorem is the implication (b)$\Rightarrow$(c)[1]. By the definition of smoothness of Δ, this k-plane field is locally spanned by k smooth vector fields, and by the hypothesis, Δ is closed under Lie bracket. The idea of the proof is to show that the k spanning vector fields of Δ can be replaced by k vector fields spanning Δ having the additional property that the bracket of any pair of them vanishes. Then (c) will follow by an application of Theorem 26 of Chapter 2. Let $a{\in}M$. By the definition of smoothness of Δ, there are smooth vector fields $Y_1, \dots, Y_k$ defined in a neighborhood of a that span Δ in that neighborhood. We claim that there is a chart (V, η) around a, with V contained in the domain of the Y_i, so that $\eta(a){=}\mathbf{0}{\in}\mathbb{R}^m$, and

$$T_a\eta(Y_i(a)) = \frac{\partial}{\partial x^i}(\mathbf{0}), \quad i{=}1, \dots, k$$

Such a chart can be realized by first making a translation in $\mathbb{R}^m$ to have $\eta(a){=}\mathbf{0}{\in}\mathbb{R}^m$, and then utilizing a linear isomorphism of $\mathbb{R}^m$ to obtain $T_a\eta(Y_i(a)){=}\frac{\partial}{\partial x^i}(\mathbf{0})$, for all $i{=}1, \dots, k$. Let $q{=}p{\circ}\eta$, where $p{:}\mathbb{R}^m{\to}\mathbb{R}^k$ is the projection on the first k coordinates. By the choice of η, we have $T_aq(\Delta_a){=}\mathbb{R}^k$, and T_aq restricts to an isomorphism of Δ_a onto $\mathbb{R}^k$. By continuity of Tq, this mapping restricts to an isomorphism of Δ_x onto $\mathbb{R}^k$ for x in an open subset V' of a. Reducing the size of V, if necessary, we may assume that this property holds for all $x{\in}V$. Thus q will be a submersion from V onto some open subset of $\mathbb{R}^k$. For each $x{\in}V$, we use the induced isomorphism from Δ_x onto

[1]The argument given here seems to be due to M. Spivak, see Volume 1 of [**26**].

$T_{q(x)}\mathbb{R}^k$ to choose the unique $X_i(x)$ so that

$$T_x q(X_i(x)) = \frac{\partial}{\partial x^i}(q(x))$$

We show first that the vector fields X_i thus defined are smooth. For $x \in V$, let $Z_i(x) = Tq(Y_i(x))$. This is a smooth vector field defined *along* q. We then have, for each $i=1,\ldots,k$,

$$Z_i(x) = \sum_j a_i^j(x) \frac{\partial}{\partial x^j}(q(x))$$

with the coefficients $a_i^j(x)$ forming an invertible matrix of smooth functions. Let $[b_i^j(x)]$ be the inverse matrix (of smooth entries). We then have

$$\frac{\partial}{\partial x^j(q(x))} = \sum_i b_i^j(x) Z_i(x)$$

and the isomorphism from Δ_x to $T_{q(x)}\mathbb{R}^k$ gives $X_j(x) = \sum_i b_j^i(x) Y_i(x)$, proving the smoothness of X_j.

Now for each $x \in V$, the tangent vectors $X_1(x),\ldots,X_k(x)$ span Δ_x, and we are in the situation of Subsection 12b, where $\frac{\partial}{\partial x^i} = q_* X_i$. Therefore,

$$\begin{aligned} q_*[X_i, X_j] &= [\frac{\partial}{\partial x^i}, \frac{\partial}{\partial x^j}] \\ &= \mathbf{0} \end{aligned}$$

Since Δ is closed under the bracket operation, $[X_i, X_j](x)$ is in Δ_x, and since the restriction of $T_x q$ to Δ_x is an isomorphism, we conclude that $[X_i, X_j](x)=0$. It now follows from Theorem 26 of Chapter 2 that a chart (U, ξ) around a exists so that $X_i = \frac{\partial}{\partial \xi^i}$, $i=1,\ldots,k$.

Finally, the implication (c)$\Rightarrow$(a) is of a general nature. Each connected component I of a k-dimensional level set $\xi^{k+1}=c^{k+1},\ldots,\xi^m=c^m$, for constant $(c^{k+1},\ldots,c^m)$, is tangent to $\frac{\partial}{\partial \xi^1},\ldots,\frac{\partial}{\partial \xi^k}$, hence (I,i) is an integral manifold of Δ, where $i: I \hookrightarrow M$ is the inclusion map. □

There is a dual viewpoint on plane fields based on differential forms that is often computationally more useful than the treatment above. A k-plane field on an m-manifold M is also referred to as a ***plane field of codimension*** l, where $k+l=m$. By a ***differential system*** $\mathcal{D}$ ***of dimension*** l on an m-manifold M, we mean an assignment $x \mapsto \mathcal{D}_x$ of an l-dimensional subspace $\mathcal{D}_x$ of T_x^*M for each $x \in M$. $\mathcal{D}$ is called C^r if for every $x \in M$, there is an open neighborhood U of x and C^r one-forms $\alpha^1,\ldots,\alpha^l$ on U so that $\mathcal{D}_x$ is spanned by $\alpha^1(x),\ldots,\alpha^l(x)$ for all $x \in U$. The following describes the relation between differential systems and plane fields.

28. Lemma *Let M be a smooth manifold of dimension m. Then there is a one-to-one correspondence between smooth plane fields of codimension l and smooth differential* *systems of dimension l on M, given by*

$$\Delta_x \longleftrightarrow \mathcal{D}_x = \Delta_x{}^{\perp}$$

where, for any subset S of a vector space V, $S^{\perp}$ is defined as the subspace $\{\alpha \in V^* : \alpha(s)=0, \forall s \in V\}$.

Proof. For every fixed $x \in M$, the correspondence $\Delta_x \longleftrightarrow \Delta_x{}^{\perp}$ is standard linear algebra, we need only prove that the correspondence associates smooth plane fields with smooth differential systems. Assume Δ is smooth, then for $x \in M$, there are smooth vector fields $X_1, \ldots, X_k$, $k+l=m$, that span Δ in a neighborhood U of x. Taking U small enough to be the domain of a chart, TU is smoothly isomorphic as a vector bundle to $U \times \mathbb{R}^m$, therefore the cross-sections $X_1, \ldots, X_k$ can be extended to smooth vector fields $X_1, \ldots, X_m$ so that $(X_1(x), \ldots, X_m(x))$ is a basis for T_xM for each $x \in U$. The dual one-forms $(\alpha^1, \ldots, \alpha^m)$ are smooth, and for each $x \in U$, a basis for $\mathcal{D}_x$ is given by $(\alpha^{k+1}(x), \ldots, \alpha^m(x))$. □

Note that if W is a linear subspace of a finite-dimensional vector space V, then $(W^{\perp})^{\perp}$ is identified with W via the canonical identification of V^{**} with V. Hence a plane field Δ may alternatively be given by a differential system $\mathcal{D}$. In view of this correspondence, a differential system $\mathcal{D}$ is called *integrable* if $\mathcal{D}^{\perp}$ is integrable. The ring structure of $\Omega^* M$ provides algebraic facility in dealing with integrability. Because of the anti-commutativity of the wedge product, the concepts of right- and left-ideal coincide in the ring $\Omega^* M$, and we can simply speak of an ideal. The same considerations apply to $\Omega^* U$, where U is an open subset of M. Recall that, in general, if S is a subset of a ring A, the *ideal generated by* S, denoted by $< S >$, is the smallest ideal in A that contains S. In fact, $< S >$ consists of all finite sums

$$\beta_1 \wedge \alpha_1 + \cdots + \beta_n \wedge \alpha_n$$

where the α_i are elements of S, and the β_i are arbitrary elements of A.

We say that a 1-form α ***belongs to*** $\mathcal{D}$ if α is defined and smooth on an open subset U of M, and $\alpha(x) \in \mathcal{D}(x)$ for all $x \in U$.

29. Corollary *Let $\mathcal{D}$ be a smooth differential system on a smooth m-manifold M. Then the following conditions are equivalent:*
(a) $\mathcal{D}$ is integrable.
(b) $d\mathcal{D}$ is contained in the ideal generated by $\mathcal{D}$, in the sense that if S is a set of 1-forms belonging to $\mathcal{D}$, and $\alpha \in S$, then $d\alpha$ belongs to the ideal generated by S.
(c) For any point $a \in M$, there is a chart (U, ξ) around a so that $d\xi^{k+1}(x), \ldots, d\xi^m(x)$ span $\mathcal{D}_x$ for all $x \in U$.

Proof. The statements (a) and (c) above are equivalent to corresponding statements of the theorem. We need only check that (b) in this corollary is equivalent to any of the other statements. First assume (b) in the corollary, we show that $\Delta = \mathcal{D}^{\perp}$ is closed under brackets. Suppose that smooth vector fields X and Y belong to Δ, and let α belong to $\Delta^{\perp}$=$\mathcal{D}$. Thus $\alpha(X)$=0 and $\alpha(Y)$=0; we must show that $\alpha[X, Y]$=0. By the hypothesis, $d\alpha = \sum_{i=1}^{n} \beta_i \wedge \alpha_i$, where the α_i belong to $\mathcal{D}$. We have

$$\begin{aligned}(\beta_i \wedge \alpha_i)(X, Y) &= \beta_i(X)\alpha_i(Y) - \beta_i(Y)\alpha_i X) \\ &= 0\end{aligned}$$

Therefore, $d\alpha(X, Y)$=0. It follows then from Formula 3.26 for $d\alpha$ in Chapter 3 that $\alpha[X, Y]$=0.

Conversely, suppose that (c) holds, and α belongs to $\mathcal{D}$. On an appropriate chart (U, ξ), we may express α in the form

$$\alpha = a_{k+1} d\xi^{k+1} + \cdots + a_m d\xi^m$$

where the a_i are smooth functions. Therefore,

$$d\alpha = da_{k+1} \wedge d\xi^{k+1} + \cdots + da_m \wedge d\xi^m$$

which shows that $d\alpha$ belongs to the ideal generated by $\mathcal{D}$. If the support of α is not entirely contained in one such chart, we take a smooth partition of unity (θ_j) subordinate to a locally finite cover by charts (U_j, ξ_j) of the form described in (c), and write $\alpha = \sum_j \theta_j \alpha$. Given any point $x \in M$, there is an open neighborhood of x outside of which all but a finite number of the θ_j have constant value zero. It follows that $d\alpha$ can be globally expressed as a finite sum $\sum d(\theta_j \alpha)$. By the argument above, each $d(\theta_j \alpha)$ belongs to the ideal generated by $\mathcal{D}$, therefore $d\alpha$ belongs to that ideal. □

Just as a non-vanishing vector field generates a line field, a non-vanishing 1-form α gives rise to a codimension one plane field consisting of hyperplanes

$$\Delta_x = <\alpha(x)>^{\perp} = ker(\alpha(x))$$

The criterion of integrability in the above corollary takes on an especially simple form in this case:

30. Corollary *Let α be a smooth nowhere-vanishing* 1*-form on a manifold M. Then the corresponding codimension one plane field is integrable if and only if one of the following equivalent conditions holds.*
(a) There exists a smooth 1*-form β so that*

$$d\alpha = \beta \wedge \alpha$$

(b) $d\alpha \wedge \alpha = 0$

Proof. Condition (a) is equivalent to $d\alpha$ being in the ideal generated by α. Assuming (a), condition (b) follows since $\beta \wedge \alpha \wedge \alpha$=0. Now assume $d\alpha \wedge \alpha$=0, we show that $d\alpha$ belongs to the ideal generated by α. We first prove this locally for an open subset U of M over which TU (hence T^*U) is trivial as a vector bundle. Since α is non-vanishing, $\alpha|_U = \alpha^1$ can be extended to a basis $(\alpha^1, \ldots, \alpha^m)$ for $\Omega^1 U$. Therefore, $d\alpha^1$

has a representation as

$$d\alpha^1 = \sum_{i<j} a_{ij}\alpha^i \wedge \alpha^j$$
$$= \sum_{1=i<j} a_{1j}\alpha^1 \wedge \alpha^j + \sum_{1<i<j} a_{ij}\alpha^i \wedge \alpha^j$$

From $d\alpha \wedge \alpha = 0$, we can then conclude that

$$\sum_{1<i<j} a_{ij}\alpha^1 \wedge \alpha^i \wedge \alpha^j = 0$$

But the 3-forms $\alpha^1 \wedge \alpha^i \wedge \alpha^j$ form a linearly independent set at every point, therefore $a_{ij}=0$ for all $1<i<j$. Hence we obtain

$$d\alpha^1 = \sum_{1<j} a_{1j}\alpha^1 \wedge \alpha^j$$
$$= \beta \wedge \alpha_1$$

where $\beta = -(\sum_{1<j} a_{1j}\alpha^j)$, as claimed.

For a general 1-form α satisfying condition (b), we take a partition of unity (θ_i) subordinate to a locally-finite open cover (U_i) as above, and write $d\alpha = \sum_i \theta_i d\alpha$. Writing each $\theta_i d\alpha$ as $\beta_i \wedge \alpha$, and summing up, we obtain the desired result. □

One can now revisit Example 26c in the light of the above corollary. The condition **curl**$F \cdot F=0$ described in Method 2 of that example is equivalent to condition (b) of the corollary (see also Subsection 9 of Chapter 3). The equivalence of the two conditions in the above lemma can be generalized as follows.

31. Algebraic Criterion *Let* $\{\alpha^1, \ldots, \alpha^s\}$ *be a linearly independent set of* 1*-forms in* $\Omega^1 V$*, where V is an open subset of a smooth manifold M. For an exterior differential form* $\alpha \in \Omega^* V$ *to belong to the ideal generated by* $\{\alpha^1, \ldots, \alpha^s\}$*, it is necessary and sufficient that* $\alpha \wedge \alpha^1 \wedge \cdots \wedge \alpha^s = 0$.

Proof. If α is in the form $\sum_{i=1}^n \beta^i \wedge \alpha^i$, then $\alpha \wedge \alpha^1 \wedge \cdots \wedge \alpha^s = 0$. Conversely suppose that $\alpha \wedge \alpha^1 \wedge \cdots \wedge \alpha^s = 0$ holds for α. An arbitrary element α of $\Omega^* V$ is a finite sum of elements of various degrees. If $\alpha \wedge \alpha^1 \wedge \cdots \wedge \alpha^s = 0$, then the wedge product of the part of α in each degree with $\alpha^1 \wedge \cdots \wedge \alpha^s$ must be zero. So it suffices to prove the result for a homogeneous α, say of degree k. Just as in the proof of the corollary, if we prove the statement locally on open set U where TU is parallelizable, then by making use of a partition of unity, the result extends to V. Working on such an open set U, the linearly independent set $\{\alpha^1, \ldots, \alpha^s\}$ can be extended to a basis $(\alpha^1, \ldots, \alpha^m)$ for $\Omega^1 U$. We express α as

$$\alpha = \sum_{i_1 < \cdots < i_k} a_{i_1 \cdots i_k} \alpha^{i_1} \wedge \cdots \wedge \alpha^{i_k}$$
$$= \sum{}' a_{i_1 \cdots i_k} \alpha^{i_1} \wedge \cdots \wedge \alpha^{i_k} + \sum{}'' a_{i_1 \cdots i_k} \alpha^{i_1} \wedge \cdots \wedge \alpha^{i_k}$$

where $\sum'$ is the part of the sum where $i_1 \leq s$, and $\sum''$ denotes the rest. Now the wedge product of $\sum'$ with $\alpha^1 \wedge \cdots \wedge \alpha^s$ is zero, therefore the hypothesis implies that the following sum is zero:

$$\sum_{s<i_1<\cdots<i_k} a_{i_1\cdots i_k} \alpha^{i_1} \wedge \cdots \wedge \alpha^{i_k} \wedge \alpha^1 \wedge \cdots \wedge \alpha^s$$

Because of $s<i_1<\cdots<i_k$, the $(k+s)$-forms $\alpha^{i_1} \wedge \cdots \wedge \alpha^{i_k} \wedge \alpha^1 \wedge \cdot \wedge \alpha^s$ are linearly independent. It follows that the coefficients must all vanish leading to

$$\alpha = \sum' a_{i_1\cdots i_k} \alpha^{i_1} \wedge \cdots \wedge \alpha^{i_k}$$

Every term in this sum contains some α^i with $i = 1, \ldots, s$, therefore α belongs to the ideal generated by $\{\alpha^1, \ldots, \alpha^s\}$. □

32. Maximal Integral Manifolds and Foliations. We discussed the concept of *maximal integral curve* for vector fields in Chapter 2 and again in Section B of this chapter. We now wish to introduce the corresponding concept of *maximal integral manifold* for plane fields. This will lead to the important notion of a *foliation*, a subject we will only briefly touch on here. Let Δ be an integrable smooth k-plane field on an m-manifold M. According to Frobenius Theorem, M can be covered by charts (U, ξ) so that $\xi(U) = V \times W$, with V an open ball in $\mathbb{R}^k$, W an open ball in $\mathbb{R}^{m-k}$, and each $\xi^{-1}(V \times \{c\})$, $c \in W$, an integral manifold for Δ. Let us call each such chart (U, ξ) a **Δ-chart** and each such integral manifold a **Δ-plaque**. A relation $\sim$ on M is defined as follows. We write $A \sim B$ in case there is a continuous path $\gamma: [a, b] \to M$, and a partition $a = t_0 \leq t_1 \leq \cdots \leq t_p = b$, so that each $\gamma[t_{i-1}, t_i]$ is contained in a Δ-plaque, $\gamma(a) = A$ and $\gamma(b) = B$. $\sim$ is an equivalence relation and decomposes M into disjoint equivalence classes. For each $x \in M$, the equivalence class to which x belongs is denoted by $[x]_\Delta$, and will be called the **maximal integral manifold containing** x. The terminology will soon be justified. Note that each $[x]_\Delta$ is connected as a subspace of M since any pair of points in $[x]_\Delta$ can be joined by a continuous path. We will define a topology on $[x]_\Delta$, convert it into a smooth manifold, and show that the inclusion map $[x]_\Delta \hookrightarrow M$ is an immersion.

(a) *The topology of* $[x]_\Delta$. Fixing $x \in M$, we note that the set of Δ-plaques contained in $[x]_\Delta$ forms a base for a topology $\mathcal{F}$ on $[x]_\Delta$. Since each plaque is contained in an open subset of M, this topology is at least as fine as the subspace topology, hence Hausdorff property holds for this topology. In general, $\mathcal{F}$ could be finer than the subspace topology for $[x]_\Delta$, since the intersection of $[x]_\Delta$ with the domain U of a Δ-chart (U, ξ) may contain infinitely many connected components (plaques) that accumulate on one plaque. We show that second-countability holds for the topology on $[x]_\Delta$. The collection of Δ-charts is a smooth atlas for M, so we can extract from this atlas a countable locally-finite atlas $(U_n, \xi_n)_n$, $n = 1, 2, \ldots$. To show that $\mathcal{F}$ is second-countable, it then suffices to show that the number of connected components of the intersection of $[x]_\Delta$ with each U_n is countable. Let $\mathcal{P} \subset U_i$ be a plaque and $x \in \mathcal{P}$. Suppose $y \in [x]_\Delta \cap U_j$, then y belongs to a plaque $\mathcal{Q} \subset U_j$, and is joined to x by

a continuous path $\gamma:[a,b]\to M$, where $\gamma(a)=x$, $\gamma(b)=y$, $a=t_0\leq t_1\leq\cdots\leq t_p=b$, with each $\gamma[t_{l-1},t_l]$ contained in some U_n. Since $(U_n)_n$ is a locally-finite collection, given a natural number p, there are only a finite number of plaques that can be reached from x by a p-step path as above. It follows that the number of plaques in U_j that can be reached from x is countable, and the claim is proved.
(b) *The manifold structure of* $[x]_\Delta$. Having established the topological requirements for $[x]_\Delta$ to be a manifold, we now proceed to provide it with a smooth manifold structure of dimension k. For each Δ-chart (U,ξ) and each Δ-plaque $\mathcal{P}\subset U$ in $[x]_\Delta$, we define the chart $(\mathcal{P},\varphi)$ by $\varphi=\pi\circ\xi|_{\mathcal{P}}$, where $\xi:U\to V\times W$ is as in the definition of a Δ-chart above, and $\pi:V\times W\to V$ is projection on the first component. Let (U',ξ') be another Δ-chart with Δ-plaque $\mathcal{P}'\subset U'$ in $[x]_\Delta$ so that $\mathcal{P}\cap\mathcal{P}'\neq\emptyset$. Defining the chart $(\mathcal{P}',\varphi')$ similarly to $(\mathcal{P},\varphi)$ as above, we look at the change of variable $\varphi'\circ\varphi^{-1}$ on $p(\mathcal{P}\cap\mathcal{P}')$. First note that this function maps the open subset $\varphi(\mathcal{P}\cap\mathcal{P}')$ of $\mathbb{R}^k$ to the open subset $\varphi'(\mathcal{P}\cap\mathcal{P}')$ of $\mathbb{R}^k$ because of the form of Δ-charts. Further, the map is obtained by restricting the smooth change of variable $\xi'\circ\xi^{-1}$, so it is also smooth.
(c) *The inclusion* $[x]_\Delta\hookrightarrow M$. Using the smooth charts in (b) above, the inclusion map takes the local form $(x^1,\ldots,x^k)\mapsto(x^1,\ldots,x^k,c^{k+1},\ldots,c^m)$, with constant c^i, which is an immersion.

It follows from the above that the inclusion $[x]_\Delta\hookrightarrow M$ defines an integral manifold for the k-plane field Δ. Since connectedness and path-connectedness are equivalent for manifolds, the equivalence relation $\sim$ ensures that $[x]_\Delta$ is the maximal connected integral manifold for Δ that includes the point x. Thus, given a smooth integrable k-plane field Δ on a manifold M, the manifold is partitioned into maximally connected smooth integral manifolds of Δ. Moreover, by Frobenius Theorem, these integral manifolds are locally organized as smoothly parallel, in the sense that around any given point, a C^∞ chart (U,ξ) exists for which the connected components of maximal integral manifolds appear as the subsets $\xi^i=constant$ for $i=k+1,\ldots,m$ (see Figure 4).

The above picture leads to the following general definition. Let M be an m-

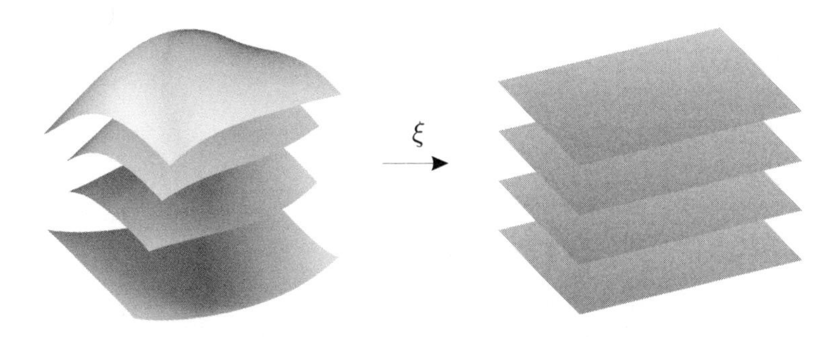

FIGURE 4. Straightening out integral manifolds

dimensional C^r manifold, $0\leq r\leq\omega$, and $0\leq s\leq r$. By a C^s ***k*-dimensional foliation** $\mathcal{F}$ of M, we mean a decomposition of M into a disjoint union of connected subsets

$(L_\alpha)_\alpha$ with the following property: For every $a \in M$, there is an open neighborhood U of a and a C^s diffeomorphism ξ from U onto an open subset $V \times W \subset \mathbb{R}^k \times \mathbb{R}^{m-k}$, so that for each L_α, every non-empty connected component of $L_\alpha \cap U$ is given by a $\xi^{-1}(V \times \{c\})$ for some $c \in W$. We also refer to this as a ***codimension (m−k) foliation***. Each connected set L_α is a ***leaf*** of the foliation.

The following assertion was demonstrated in the previous paragraphs.

33. *The maximal integral manifolds of a smooth integrable k-plane field on M constitute a smooth k-dimensional foliation of M.*

The technique of demonstration in (b) and (c) above proves the following.

34. *For a C^s k-dimensional foliation $\mathfrak{F}$ of M, each leaf L_α has the structure of a C^s manifold, and the inclusion map $L_\alpha \hookrightarrow M$ is an injective C^s immersion.*

Further examples of foliations will be found in the following.

35. Examples

(a) Let X be a C^r vector field on a manifold M (of class at least C^{r+1}). The set of singular points, $Sing(X)$, of X is a closed subspace of M, so the set $U = M - Sing(X)$ is an open subset of M, hence a submanifold. The non-singular trajectories of X constitute a C^r foliation of U by Theorem 26 of Chapter 2. Note that as we have already observed in Chapter 2, each individual leaf (=orbit=trajectory) is locally a C^{r+1} manifold. This situation, where individual leaves are smoother than the foliation, occurs quite often.

(b) Let P and Q be C^r manifolds and $f:P \to Q$ a C^r map of constant rank k, $r \geq 1$. Then by Theorem 4 of Chapter 5, P admits a C^r codimension k foliation by level sets of f. In particular, if f is a submersion, then the foliation has codimension equal to the dimension of Q.

(c) As a particular example of the above, suppose $f:M \to \mathbb{R}$ is a C^r function, $r \geq 1$. The set of regular points, $R(f)$ of f, i.e., the set of $x \in M$ where $df(x) \neq 0$ is an open submanifold of M, possibly empty. If $R(f) \neq \emptyset$, then $R(f)$ admits a codimension 1 foliation by level sets of $f|_{R(f)}$. By the Rank Theorem, the leaves are actual submanifolds of M in this case.

A p-form α that can be written as $\alpha = d\beta$ is called an ***exact p-form***. For an exact p-form α, we have $d\alpha = 0$. More generally, any p-form α that satisfies $d\alpha = 0$ is called a ***closed p-form***. Now if α is a closed nowhere-vanishing 1-form on M, then it follows from Corollary 30b that the subspaces $\ker \alpha(x)$ define an integrable codimension one plane field on M. The maximal integral manifolds form a codimension one foliation on M. Unlike the case of exact forms, the leaves need not be embedded submanifolds in this more general case (see Exercise 6.31 at the end of this chapter).

(d) (*Reeb Foliation*) This foliation plays a fundamental role in the theory of foliations. We describe a three-dimensional version. Let C be the open solid cylinder $D{\times}\mathbb{R}$, where D is the open two-dimensional disk of radius $\frac{\pi}{2}$ in $\mathbb{R}^2$. Consider a smooth surface in C such as the graph of the function $z{=}\sec r$, using cylindrical coordinates (r, θ, z). The graph has a minimum at $r{=}0$, is symmetric around the z-axis, and becomes asymptotic to the surface of the cylinder as $r{\to}1$. Translating this graph vertically along the z-axis, we obtain a foliation of the open solid cylinder given by the surfaces $z{-}\sec r{=}constant$ (see the left side of Figure 5). This foliation is equivalently given by the one-form $\alpha{=}(\cos^2 r)dz{-}(\sin r)dr$ (see Example 26b). Note that $d\alpha{\wedge}\alpha{=}0$. Now α and the foliation are invariant under vertical translations (along the z-axis), so a foliation is induced on the open solid torus $D{\times}S_1$ (see Figure 5). This is the ***Reeb foliation*** of open solid torus. The leaves of Reeb foliation are homeomorphic to $\mathbb{R}^2$. They open up while spiraling inside the solid torus toward the boundary torus. One can write the 3-sphere S_3 as the union of two closed solid tori identified along the boundary (see Figure 6). In the figure the role of meridians and parallels on the surface of the torus are switched for the two solid tori. The vertical axis that closes at ∞ to produce S_3 plays the role of the central circle for the horizontal torus. The Reeb foliation for S_3 consists of the Reeb foliations of the two solid tori together with a single compact leaf that is their common boundary.

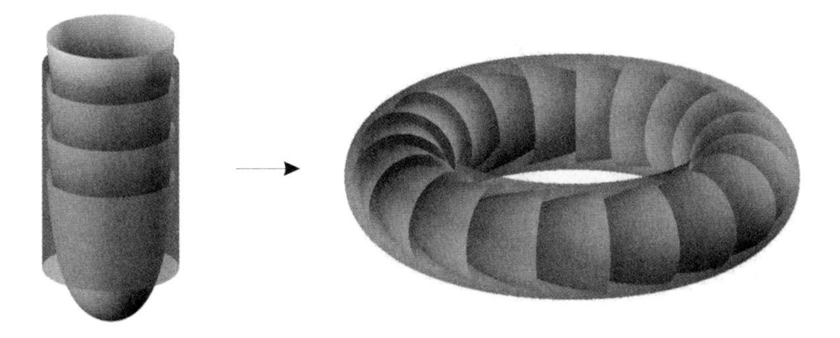

FIGURE 5. Reeb foliation

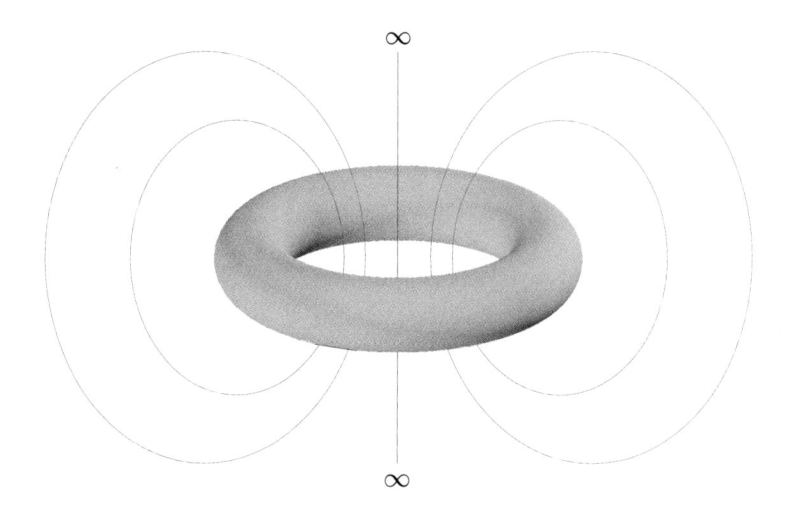

FIGURE 6. 3-sphere as the union of two solid tori

EXERCISES

6.1 Let (E,π,M) be a C^r vector bundle, $r\geq 1$. Show that M is a quotient manifold of E.

6.2 Let E and E' be two C^r vector bundles over M. A C^r map $H:E\to E'$ is called a C^r ***vector bundle homomorphism*** (in short, a C^r ***VB-homomorphism***) if the restriction of H to each E_x is a linear map from E_x to E'_x. We define $\ker(H)=\{e\in E:H(e)=\mathbf{0}\}$, $\mathrm{Im}(H)=\{H(e)\in E':e\in E\}$, and $\mathrm{Hom}(E,E')$=set of all C^r VB-homomorphisms fom E to E'.
(a) Show that $\mathrm{Hom}(E,E')$ has the structure of a C^r vector bundle over M.
(b) Give examples where $\ker(H)$ and $\mathrm{Im}(H)$ are not vector bundles over M.
(c) Show that if the rank of $H|_{E_x}$ is the same for all $x\in M$, then $\ker(H)$ and $\mathrm{Im}(H)$ admit C^r vector bundle structures over M.

6.3 Let E and E' be C^r vector bundles over M of fiber type F.
(a) Suppose that $(U_\alpha,\phi^\beta_\alpha)$ and $(U_\alpha,\psi^\beta_\alpha)$ are, respectively, the 1-cocycles of E and E'. Then E and E' are C^r-isomorphic if and only if there exist C^r maps $h^\alpha:U_\alpha\to GL(F)$ so that for all $x\in M$, $\psi^\beta_\alpha(x)\circ h^\alpha(x)=h^\beta(x)\circ\phi^\alpha_\beta(x)$.
(b) More generally, suppose that $(U_\alpha,\phi^\beta_\alpha)$ and (V_μ,ψ^ν_μ) are, respectively, the 1-cocycles of E and E'. Then E and E' are C^r-isomorphic if and only if there exist C^r maps $h^\mu_\alpha:U_\alpha\cap V_\mu\to GL(F)$ so that

$$h^\mu_\beta(x)\circ\phi^\beta_\alpha(x) = h^\mu_\alpha(x),\quad \forall x\in V_\mu\cap U_\alpha\cap U_\beta$$
$$\psi^\nu_\mu(x)\circ h^\mu_\alpha(x) = h^\nu_\alpha(x),\quad \forall x\in V_\nu\cap V_\mu\cap U_\alpha$$

6.4 Let (E,π,M) be a C^r vector bundle, M' a C^r manifold and $h:M'\to M$ a C^r map.

We define (h^*E,π',M') by

$$h^*E = \{(x,e)\in M'\times E : h(x)=\pi(e)\},\ \pi'(x,e)=x$$

(a) Show that h^*E is a C^r manifold and (h^*E,π',M') is a C^r vector bundle (called the **pullback** of E over M').
(b) Let (E,π,M) be a C^r vector bundle and $\Delta:M\to M\times M$ the *diagonal map*, $x\mapsto(x,x)$. Set up a natural isomorphism between the pullback $\Delta^*(E\times E)$ of the product $(E\times E,\pi\times\pi,M\times M)$ over M and $E\oplus E$.

6.5 (a) Show that a line bundle is trivial if and only if it admits a continuous nowhere zero cross-section.
(b) A vector bundle (E,π,M) of rank m is called ***orientable*** if the line bundle $\Lambda^m E$ over M is trivial. Give a characterization in terms of the VB-atlas.
(c) For any vector bundle E over M, show that $E\oplus E$ is orientable.

6.6 Consider the subset E of $\mathbb{RP}(m)\times\mathbb{R}^{m+1}$ consisting of pairs (x,u), where u belongs to the line determined by x. Define π as projection $(x,u)\mapsto x$.

(a) Show that $(E, \pi, \mathbb{RP}(m))$ is a smooth line bundle over $\mathbb{RP}(m)$. This is known as the ***tautological*** (or ***canonical***) ***line bundle***.
(b) For $m=1$, show that E is homeomorphic to the Möbius band (see Example 21d of Chapter 5).
(c) Show that this line bundle is not trivial for any m.

6.7 Suppose $\pi:E\to M$ is a C^r vector bundle of fiber type F. Consider an ordered basis $(b_1, \ldots, b_k)$ for F. If U is an open subset of M and $\Phi:\pi^{-1}(U)\to U\times F$ is a local-product diffeomorphism, then the ordered k-tuple of cross-sections $(\sigma_1, \ldots, \sigma_k)$ defined over U by $\sigma_i(x)=\Phi^{-1}(x, b_i)$ is called a ***local frame*** for the vector bundle. Show that any k-tuple $(\sigma_1, \ldots, \sigma_k)$ of cross-sections over U that is linearly independent at every point of U is a local frame, i.e., it can be obtained by the inverse of a local-product diffeomorphism. (Hint: Imitate the proof of Lemma 7.)

6.8 Let M be a smooth Riemannian manifold and N a smooth submanifold of M. For each $x\in N$, let $(T_xN)^\perp$ be the orthogonal complement of T_xN in T_xM with respect to the given Riemannian metric.
(a) Construct a vector bundle over N with fiber $(T_xN)^\perp$ over $x\in N$. This is called the ***normal bundle*** of N in M.
(b) Show that if M is orientable and the normal bundle of N in M is trivial, then N is also orientable. If the codimension of N in M is one, show the converse also holds.

6.9 (Based on Exercise 6.8) Let M be a compact smooth submanifold of $\mathbb{R}^N$. For each $x\in M$ and each $\delta>0$, let

$$U_\delta(x) = \{v\in T_xM^\perp : |v|<\delta\}$$

Here we are identifying T_xM and its orthogonal complement in $T_x\mathbb{R}^N$ with affine subspaces of $\mathbb{R}^N$ passing through x (see Theorem 10 of Chapter 5), and $|\cdot|$ denotes the Euclidean norm. Show that if $\delta>0$ is sufficiently small, then $U_\delta=\bigcup_{x\in M} U_\delta(x)$ is an open N-dimensional submanifold of $\mathbb{R}^N$ homeomorphic to the total space of the normal bundle of M in $\mathbb{R}^N$. U_δ is called a ***tubular neighborhood*** of M in $\mathbb{R}^N$.

6.10 Let M_1 and M_2 be smooth manifolds with X_1 and X_2 smooth vector fields on M_1 and M_2, respectively. For $x_1\in M_1$ and $x_2\in M_2$, identifying $T_{(x_1,x_2)}(M_1\times M_2)$ with $T_{x_1}M_1\times T_{x_2}M_2$, we define $X_1\oplus X_2$ on $M_1\times M_2$ by $(X_1\oplus X_2)(x_1, x_2)=(X_1(x_1), X_2(x_2))$.
(a) If X_1, Y_1 are smooth vector fields on M_1, and X_2, Y_2 are smooth vector fields on M_2, prove that

$$[X_1\oplus X_2, Y_1\oplus Y_2] = [X_1, Y_1]\oplus[X_2, Y_2]$$

(b) For X and Y smooth vector fields on $\mathbb{R}$, show that $X\oplus Y$ does not have a (non-zero) periodic orbit in $\mathbb{R}\times\mathbb{R}\cong\mathbb{R}^2$.

6.11 Let X be a smooth vector field on a smooth manifold M. Show that there exists a smooth function $f:M\to\mathbb{R}$, non-zero where X is non-zero, so that fX is a complete vector field. (Cf. Exercise 2.3, Chapter 2.)

6.12 Given a smooth non-compact manifold M, show that there exists a smooth vector field on M that is not complete.

6.13 The ***Liouville vector field*** L on the tangent bundle of a smooth manifold M is defined as follows. Since T_xM is a vector space, $T_v(T_xM)$ can be identified with $\{v\}\times T_xM$. For $v\in T_xM$, we define $L(v)=(v,v)\in T_v(T_xM)\subset T_v(TM)$. Show that L is smooth and its flow is given by $(t,v)\mapsto e^t v$.

6.14 Let $\Phi:\mathbb{R}\times M\to M$ be a smooth flow on M with generating vector field X. Show that $\Phi^+:\mathbb{R}\times TM\to TM$ defined by $\Phi^+(t,v)=T\Phi_t(v)$, is a smooth flow on TM, and denote its generating vector field by X^+. Prove that $T\tau_M\circ X^+=X\circ\tau_M$, where τ_M is the tangent bundle projection. For $f\in C^\infty(M)$, show that $X^+\cdot df=d(X\cdot f)$.

6.15 (**Ehresmann Fibration Theorem**) Let $p:E\to M$ be a smooth surjective submersion of manifolds, and assume that p is *proper*, i.e., for every compact subset K of M, the preimage $p^{-1}(K)$ is compact. Show that $p:E\to M$ is a *(local product) fibration* in the sense that for every $a\in M$, there is an open neighborhood W of a, a smooth manifold F, and a smooth diffeomorphism $h:W\times F\to p^{-1}(W)$ so that $\pi_1=p\circ h$, where π_1 is projection on the first component. (Hint: Let $F=p^{-1}(a)$, and (U,ξ) be a local chart around a. Construct vector fields X_i in a neighborhood of F in E so that $Tp(X_i)=\frac{\partial}{\partial\xi^i}$, and denote the flow of X_i by Φ^i. For $v\in F$ and $x\in W$ with $\xi(x)=(t^1,\ldots,t^m)$, define $h(x,v)=\Phi^1_{t^1}\circ\cdots\circ\Phi^m_{t^m}(v)$. You need the properness of p to ensure that the flows remain defined as long as needed.)

6.16 Let M be a smooth manifold, $\alpha\in\Omega^pM$ and N is a smooth submanifold of M.
(a) If X is a smooth vector field on M so that $X(x)\in T_xN$ for all $x\in N$, show that $(L_X\alpha)|_N$ depends only on the values of α on N.
(b) Let $X_1,\ldots,X_{p+1}$ be smooth vector fields on M so that $X_i(x)\in T_xN$ for all $x\in N$ and all $i=1,\ldots,p+1$. Is it true that $(d\alpha(X_1,\ldots,X_{p+1}))|_N$ depends only on the values of α on N?

6.17 (**Cartan Lemma**) Let M be a smooth manifold and $\alpha^i,\beta^i\in\Omega^1M$ for $i=1,\ldots,s$. Suppose further that (a) $\{\alpha^1(x),\ldots,\alpha^s(x)\}$ is linearly independent for all $x\in M$, and (b) $\sum_i\alpha^i\wedge\beta^i=0$. Show that there are smooth functions $a^i_j:M\to\mathbb{R}$, $a^i_j=a^j_i$, so that $\beta^i=\sum_j a^i_j\alpha^j$. (Cf. Exercise 1.17 of Chapter 1.)

6.18 Let $M_1,\ldots,M_k$ be oriented manifolds with volume elements $\omega_1,\ldots,\omega_k$, respectively, and denote the projection $M_1\times\cdots\times M_k\to M_i$ by p_i. Show that $p_1^*\omega_1\wedge\cdots\wedge p_k^*\omega_k$ is a volume element for $M_1\times\cdots\times M_k$.

6.19 Let ω_{S_m} be the standard volume element of the unit sphere in $\mathbb{R}^{m+1}$ (see Formula (6.13) in this chapter). Show that for each hemisphere $x^{m+1}\neq 0$, this volume element can be written as

$$\frac{(-1)^m}{x^{m+1}}dx^1\wedge\cdots\wedge dx^m$$

6.20 If a product manifold $M\times N$ is orientable, show that both M and N are also orientable.

6.21 Let M and N be smooth orientable manifolds and $f:M \to N$ a smooth submersion. Show that any non-empty level set of f is orientable.

6.22 Show that the Möbius band and Klein Bottle are not orientable. (See Example 22d,e of Chapter 5 for definitions.)

6.23 Let M be a smooth orientable m-manifold with volume element ω. Show that for any $\alpha \in \Omega^{m-1}M$, there exists a smooth vector field X on M so that $i_X\omega = \alpha$. (Hint: First do it locally.)

6.24 Let M be a smooth manifold and $\Phi: G \times M \to M$ be a free and properly discontinuous group action by diffeomorphisms on M. Thus by Theorem 21 of Chapter 5, the quotient map $q: M \to M/\Phi$ defines a covering space. Suppose $\alpha \in \Omega^p M$ is such that $\Phi_g^* \alpha = \alpha$ for all $g \in G$. Show that a unique $\beta \in \Omega^p(M/\Phi)$ exists so that $q^*\beta = \alpha$. In particular, show that if M is orientable with volume element ω, and $\Phi_g^*\omega = \omega$ for all $g \in G$, then M/Φ is orientable. Deduce that odd dimensional real projective spaces are orientable.

6.25 The goal of this exercise is to establish the existence of orientable double cover for non-orientable manifolds. Let M be a smooth, connected m-manifold and consider the line bundle $\Lambda^m M$ over M. Deleting the zero element from each $\Lambda^m(T_xM)$, we are left with an open submanifold U of $\Lambda^m M$. We define an equivalence relation $\sim$ on U in the following manner: $\alpha \sim \beta$ if and only if α and β belong to the same $\Lambda^m(T_xM)$, and there is a positive real number r so that $\beta = r\alpha$. Let us denote the resulting quotient space by $\widetilde{M}$. Prove the following:
(a) $\widetilde{M}$ is a smooth quotient manifold of U of dimension m and is orientable.
(b) The projection map of $\Lambda^m M$ on M induces a covering space projection $\widetilde{M} \to M$, and each fiber consists of two points.
(c) $\widetilde{M}$ is connected if and only if M is not orientable.

6.26 Let M be a smooth m-dimensional manifold. Show that the existence of a smooth Riemannian metric on M is equivalent to the following assertion: There is a sub-atlas $\mathcal{A}$ of the maximal atlas of M with the property that if $(U, \xi), (V, \eta) \in \mathcal{A}$, and $U \cap V \neq \emptyset$, then $D(\eta \circ \xi^{-1})(x): \mathbb{R}^m \to \mathbb{R}^m$ is an orthogonal linear map for all $x \in \xi(U \cap V)$ relative to the standard Euclidean inner product of $\mathbb{R}^m$.

6.27 Let (M, ρ) be a smooth Riemannian manifold and N a smooth submanifold of M. We denote the restriction of ρ to the tangent spaces of N by ρ'. If f is a smooth real-valued function on M, show that for all $x \in N$, $\nabla_{\rho'}(f|_N)(x)$ is the orthogonal projection, with respect to ρ, of $\nabla_\rho f(x)$ on T_xN.

6.28 For a Riemannian manifold (M, ρ), a diffeomorphism $f: M \to M$ is called an ***isometry*** if for all $u, v \in T_xM$, $x \in M$, one has $\rho(T_xf(u), T_xf(v)) = \rho(u, v)$. Suppose that M is a smooth manifold and a diffeomorphism $f: M \to M$ has the property that

$f^k = \mathbb{1}_M$ for some positive integer k. Show that there is a smooth Riemannian metric for M relative to which f is an isometry.

6.29 Let T^*M be the cotangent bundle of a smooth m-manifold M. For any chart (U, ξ) on M, we define a chart $((\tau^*_M)^{-1}(U), (q, p))$ for T^*M in the following manner. $q=(q^1, \ldots, q^m)$, where $q^i = \xi^i \circ \tau^*_M$, and $p=(p_1, \ldots, p_m)$, with $p_i = \frac{\partial}{\partial \xi^i}$. Locally, for an elcment $w = \sum_i w_i d\xi^i(x) \in T^*_x M$, $(q, p)(w) = (\xi^1(x), \ldots, \xi^m(x); w_1, \ldots, w_m)$. Then the ***Liouville one-form***, λ, on T^*M is defined as follows. For a vector field X on T^*M, and a point $w \in T^*M$,

$$(\lambda(X))(w) = w(T\tau^*_M(X))$$

Show the following:
(a) For any smooth one-form α on M, regarded as a cross-section of T^*M, one has $\alpha^*(\lambda) = \alpha$.
(b) The local expression of λ relative to the coordinate system (q, p) is $\sum_{i=1}^m p_i dq^i$.
(c) $(d\lambda) \wedge \cdots \wedge (d\lambda)$ (m times) is a volume element for T^*M. This proves that T^*M, hence also TM, is orientable (another proof was provided in Example 17b).

6.30 Give examples of smooth line fields on the 2-torus and on Möbius band that do not arise from vector fields.

6.31 Let a and b be real numbers, not both zero. The closed one-form $\theta = a dx + b dy$ induces a closed one-form on the two-torus. Give a necessary and sufficient condition on a and b for the leaves of the resulting foliation on the two-torus to be submanifolds.

6.32 Consider the one-form $\alpha = x^2 dx^1 - x^1 dx^2 + x^4 dx^3 - x^3 dx^4$ on $\mathbb{R}^4$, and let $j: S_3 \hookrightarrow \mathbb{R}^4$ be the inclusion map. Show that $j^*\alpha$ defines a codimension one plane field on S_3. Investigate whether this plane field is integrable.

6.33 Let α^1 and α^2 be smooth one-forms on a manifold M that are everywhere linearly independent. Suppose also that $d(\alpha^1 \wedge \alpha^2) = 0$. For every $x \in M$, we define $\Delta(x) = \{v \in T_x M : i_v(\alpha^1 \wedge \alpha^2) = 0\}$. Show that Δ is smooth and integrable.

6.34 Let α be a closed smooth p-form on M, and for every $x \in M$, define:

$$\Delta_x = \{v \in T_x M : i_v \alpha(x) = 0\}$$

Δ_x is a linear subspace of $T_x M$. Suppose in addition that the dimension of Δ_x is the same for all x.
(a) Prove that Δ is a smooth vector sub-bundle of TM.
(b) Prove that Δ is integrable.

6.35 Let α be a smooth nowhere zero one-form on a manifold M. Show that the codimension one plane field defined by α is integrable if and only if for every $x \in M$, there is an open set U containing x and a smooth nowhere zero function $f: U \to \mathbb{R}$ so that $f\alpha$ is closed in U (f is then called an ***integrating factor***). Equivalently, a necessary and sufficient condition for the integrability of α is that for every $x \in M$, there exist an open set U containing x and smooth functions $S, T: U \to \mathbb{R}$ so that

$\alpha = TdS$. Give an example with dim $M \geq 2$ where α itself is not closed, but appropriate f can be found.

6.36 For given smooth functions $\varphi_i : \mathbb{R}^n \times \mathbb{R} \to \mathbb{R}$, we consider the system of partial differential equations

$$\frac{\partial y}{\partial x^i} = \varphi_i(x^1, \ldots, x^n, y), \quad i = 1, \ldots, n$$

The system is called ***integrable*** if for all $(a^1, \ldots, a^n, b) \in \mathbb{R}^n \times \mathbb{R}$, an open set U around $a = (a^1, \ldots, a^n)$ and a smooth function $f : U \to \mathbb{R}$ exist so that $f(a) = b$, and $y = f(x)$ satisfies the system. For each i and j from $\{1, \ldots, n\}$, we define

$$A_{ij} = \frac{\partial \varphi_i}{\partial x^j} - \frac{\partial \varphi_j}{\partial x^i} + \varphi_j \frac{\partial \varphi_i}{\partial y} - \varphi_i \frac{\partial \varphi_j}{\partial y}$$

Show that the system is integrable if and only if $A_{ij} = 0$ for all i, j. (Hint: Look at the 1-form $\omega = dy - \sum_{i=1}^n \varphi_i(x, y) dx^i$, and use Frobenius Theorem.)

6.37 Consider the one-dimensional foliation of S_3 obtained by the circles of Hopf fibration (see Example 18d as well as Exercise 5.20 of Chapter 5). Let Δ be the plane field consisting of the orthogonal complements (with respect to the standard Riemannian metric of S_3, inherited from $\mathbb{R}^4$) of the tangent lines to the circles. Investigate whether Δ is integrable.

6.38 Let Δ be a smooth k-plane field on a manifold M. A smooth curve $\gamma : I \to M$ is an ***integral curve*** for Δ if $\gamma'(t) \in \Delta(\gamma(t))$ for all $t \in I$. Referring to the non-integrable 2-plane field of Example 26c, if P and Q are any two points of $\mathbb{R}^3$, show that there is an integral curve joining P and Q (in fact there are infinitely many). There are far-reaching generalizations of this fact due to Caratheodory, Chow and Rashevsky, see [**2**] and [**21**].

6.39 Let Δ be a smooth k-plane field on a manifold M. For $a \in M$, let $X_1, \ldots, X_k$ be smooth vector fields that span Δ in a neighborhood of a. For $u = (u^1, \ldots, u^k) \in \mathbb{R}^k$, consider the vector field $X_u = \sum_i u^i X_i$, and let γ_u be the integral curve of X_u with initial condition $\gamma_u(0) = a$. Show that the union of the images of the γ_u, $u \in \mathbb{R}^k$, form a k-dimensional submanifold of M in a neighborhood of a. (Hint: If c is a non-zero real number, how are γ_u and γ_{cu} related?)

6.40 Let U be a smooth unit vector field on $\mathbb{R}^3$. As observed in Example 26c and Corollary 30, a necessary and sufficient condition for the integrability of the plane field orthogonal to U is that $\mathbf{curl} U \cdot U = 0$. We proceed to give a geometric interpretation for the quantity $\mathbf{curl} U \cdot U$, which is a measure of deviation from integrability. Let $p \in \mathbb{R}^3$ and V, W be smooth unit vector fields defined in a neighborhood of p so that (U, V, W) is a positively oriented orthonormal frame. Denoting the image of the integral curve of V through p by N, we know from Exercise 5.8 that the union of integral curves of W emanating from N form a smooth surface M in a neighborhood of p. Let U' be the unit normal to this surface so that (U', V, W) is positively oriented, and denote by φ the angle between U' and U. Thus $\varphi(p)$=0. Prove that $\frac{d\varphi}{ds} = \mathbf{curl}U \cdot U$, where s is the arclength parameter for the integral curves of W. (Hint: Note that since the plane field orthogonal to U' is integrable, $\mathbf{curl}U' \cdot U'$=0. Write U' as a linear combination of U and V and express $\mathbf{curl}U' \cdot U'$ in terms of U, V and W.)

CHAPTER 7

Integration and Cohomology

Line integrals and surface integrals of elementary calculus are precursors of integration on manifolds. The account to be given here will include a generalized *Stokes Theorem* that unifies the integral theorems of classical vector analysis. This, in turn, can be viewed as a higher dimensional version of the Fundamental Theorem of Calculus, both in content and in the fact that the proof is little more than an application of that theorem in manifold setting. To achieve this generalization, one needs to augment the definition of a manifold to include *manifolds with boundary*. There are generally two approaches in literature to the problem of integration on manifolds. The one given here is perhaps more transparent geometrically on first encounter and leads directly to the definition of *smooth measure* on a manifold. Another approach, theoretically more powerful, is based on so-called *singular chains*, and is better suited for proving theorems especially in connection with algebraic topology. The reader is urged to consult [**26**] for this alternative. The concept of *cohomology*, more specifically here, *de Rham cohomology*, is an algebraic tool based on differential forms that reflects the topology of a manifold. Integration of differential forms helps reveal the topological utility of this concept.

A. Manifolds with Boundary

We begin by introducing the ***closed m-dimensional half-space*** as

$$\mathbb{H}_m = \{(x^1, \ldots, x^m) \in \mathbb{R}^m : x^m \geq 0\} \tag{7.1}$$

The subspace of $\mathbb{H}_m$ defined by $x^m=0$ is called the ***boundary*** of $\mathbb{H}_m$, and is denoted by $\partial\mathbb{H}_m$. We may regard $\partial\mathbb{H}_m$ as $\mathbb{R}^{m-1}$ via the canonical identification $\iota_m:(x^1, \ldots, x^{m-1}, 0) \mapsto (x^1, \ldots, x^{m-1})$. For an open subset U of $\mathbb{H}_m$, the subset ∂U defined as $U \cap \partial\mathbb{H}_m$ will be called the ***boundary*** of U. U is an open subset of $\mathbb{R}^m$ if and only if $\partial U = \emptyset$.

Just as m-dimensional manifolds are modeled on $\mathbb{R}^m$, m-dimensional manifolds with boundary are modeled on $\mathbb{H}_m$. More precisely:

1. Definition Let M be a Hausdorff and second-countable topological space. M is a ***topological m-manifold with boundary*** if every point x of M possesses an open neighborhood homeomorphic to an open subset of $\mathbb{H}_m$.

A ***chart*** will be a pair (U,ξ), where U is an open subset of M and $\xi{:}U{\to}\xi(U){\subset}\mathbb{H}_m$ a homeomorphism as described above. A point $a{\in}M$ is a ***boundary point*** if for a chart (U,ξ) with $a{\in}U$, $\xi(a)$ belongs to the boundary of $\xi(U)$. We must show this is a well-defined notion, i.e., if (V,η) is another chart with $a{\in}V$, then $\eta(a)$ is also a boundary point of $\eta(V)$. Let $W{=}U{\cap}V$, then $h{=}\xi{\circ}\eta^{-1}$ is a homeomorphism of the open subset $\eta(W)$ of $\mathbb{H}_m$ onto the open subset $\xi(W)$ of $\mathbb{H}_m$. Now if $\eta(a)$ is not a boundary point of $\eta(W)$, then by reducing the size of W, if necessary, we may assume that $\eta(W)$ is an open subset of $\mathbb{R}^m$. By the Invariance of Domain Theorem of Brouwer (see Chapter 4), $\xi(W)$ will then be an open subset of $\mathbb{R}^m$, contradicting the assumption that $\xi(a)$ is a boundary point. Thus the notion of boundary point is well-defined. We denote the set of boundary points of M by ∂M. In the case $\partial M{=}\emptyset$, we obtain the usual concept of topological manifold as defined in Chapter 4 (= *manifold without boundary*).

If U is an open subset of $\mathbb{H}_m$, a map $f{:}U{\to}\mathbb{R}^n$ is called C^r if for every $x{\in}U$, there is an open neighborhood V of x in $\mathbb{R}^m$ and a C^r map $g{:}V{\to}\mathbb{R}^n$ so that $f|_{U\cap V}{=}g|_{V\cap U}$. The chain rule holds in this context, and the concepts of C^r-compatibility of charts, C^r-atlas, maximal C^r-atlas, C^r structure and equivalence of C^r structures become available as before, and will be used.

2. Examples

(a) We show that the closed unit disk $\mathbb{D}^m{=}\{(x^1,\dots,x^m){\in}\mathbb{R}^m : \sum_i(x^i)^2{\leq}1\}$ is a smooth manifold with boundary S_{m-1}. A C^∞ atlas is constructed in the same fashion as in Example 2b of Chapter 5. The chart $(\mathbb{U}^m, \mathbb{1}_{\mathbb{U}^m})$, where $\mathbb{U}^m$ is the open unit disk, covers

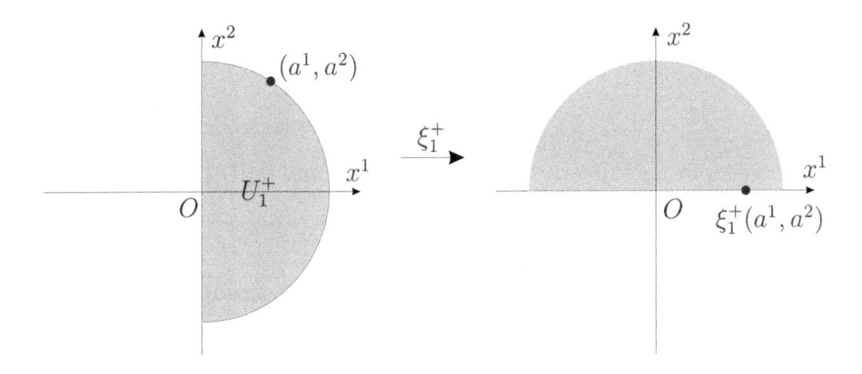

FIGURE 1. A submanifold chart

the interior points of $\mathbb{D}^m$. Charts (U_k^+,ξ_k^+) and (U_k^-,ξ_k^-), $k{=}1,\dots,m$, are given by

$$U_k^+ = \{(x^1,\dots,x^m){\in}\mathbb{D}^m : x^k{>}0\},\ \ U_k^- = \{(x^1,\dots,x^{m+1}){\in}\mathbb{D}^m : x^k{<}0\}$$

$$\xi_k^+(x^1,\dots,x^m) = (x^1,\dots,x^{k-1},x^m,x^{k+1},\dots,\sqrt{1-\sum_{i\neq k}(x^i)^2}-x^k)$$

$$\xi_k^-(x^1,\dots,x^m) = (x^1,\dots,x^{k-1},x^m,x^{k+1},\dots,\sqrt{1-\sum_{i\neq k}(x^i)^2}+x^k)$$

(see Figure 1). Each of the $\xi_k^\pm$ is smooth in its domain of definition, and so are the changes of coordinate between charts.

(b) Let M be a C^r manifold without boundary and $\gamma{:}I{\to}M$ be C^r embedding, where I is an open interval, and $r{\geq}1$. For $[a,b]{\subset}I, a{<}b$, denote $\gamma[a,b]$ by C. We show that C is a one-dimensional manifold with boundary. Using the open subsets $C^-{=}\gamma[a,b[$ and $C^+{=}\gamma]a,b]$, the two charts below establish this fact:

$$(C^-, t_a{\circ}\gamma^{-1}|C^-)\ ,\ (C^+, -t_b{\circ}\gamma^{-1}|C^+)$$

where t_r is the translation $x{\mapsto}x{-}r$ along $\mathbb{R}$.

(c) Suppose M is a C^r m-manifold without boundary and $f{:}M{\to}\mathbb{R}$ is a C^r function. We let $M^+{=}\{x{\in}M{:}f(x){\geq}0\}$, $M^-{=}\{x{\in}M{:}f(x){\leq}0\}$ and $M^0{=}\{x{\in}M{:}f(x){=}0\}$. Assume that M^0 is non-empty and that for every $a{\in}M^0$, one has $df(a){\neq}0$, i.e., 0 is a regular value for f. It follows from the Rank Theorem (see Chapter 5) that each of M^+ and M^- is an m-dimensional manifold with boundary and that $\partial M^+{=}\partial M^-{=}M^0$. Example (a) above can be cast in this framework.

To extend the theory of Chapter 6 to a C^r m-manifold M with boundary, where $r{\geq}1$, it suffices to have notions of tangent space and tangent bundle analogous to corresponding notions for manifolds without boundary. For a non-boundary point, we already have the concept of the tangent space as before. The key point for a boundary point will be that the same construction as before will yield a full linear space of the same dimension $m{=}\dim M$. This is intuitively expected, for if we imagine the tangent space at a boundary point $a{\in}\partial M$ to be the collection of all possible velocity vectors of curves passing through a, then for every tangent vector w at a, representing the velocity of a parametrized curve, the vector $-w$ will also be a tangent vector, denoting the velocity of the same curve traversed in the opposite direction (see Figure 2). More precisely, let $(U_\alpha,\xi_\alpha)_\alpha$ be the collection of all charts in the maximal C^r atlas of M so that the boundary point a belongs to U_α. As in Chapter 4, we define an equivalence relation $\sim_a$ on pairs $(\xi_\alpha(a),u)$, $u{\in}\mathbb{R}^m$, by

$$(\xi_\alpha(a),u){\sim_a}(\xi_\beta(a),v) \iff v = \Big(D(\xi_\beta{\circ}\xi_\alpha{}^{-1})(\xi_\alpha(a))\Big)(u)$$

The only point to note here is that $D(\xi_\beta{\circ}\xi_\alpha{}^{-1})$ is well-defined at $\xi_\alpha(a)$. For if h_1 and h_2 are two differentiable extensions of $\xi_\beta{\circ}\xi_\alpha{}^{-1}$ to a neighborhood of $\xi_\alpha(a)$, then the value of their derivatives at $\xi_\alpha(a)$ is uniquely determined by partial derivatives in m positive directions pointing to $\mathbb{H}_m$ and its boundary, where both functions coincide with $\xi_\beta{\circ}\xi_\alpha{}^{-1}$. It follows that the construction of tangent space and tangent bundle follow verbatim that of manifolds without boundary.

An important feature of T_aM, for $a{\in}\partial M$ is to be underlined here. Suppose that the equivalence class $[\xi_\alpha(a),u]$ represents an element of T_aM, where $a{\in}\partial M$ and $u{=}(u^1,\dots,u^m){\in}\mathbb{R}^m$. We claim that the sign of u^m, the last component of u, is independent of the particular representative. Note that the diffeomorphism $\xi_\beta \circ \xi_\alpha{}^{-1}$

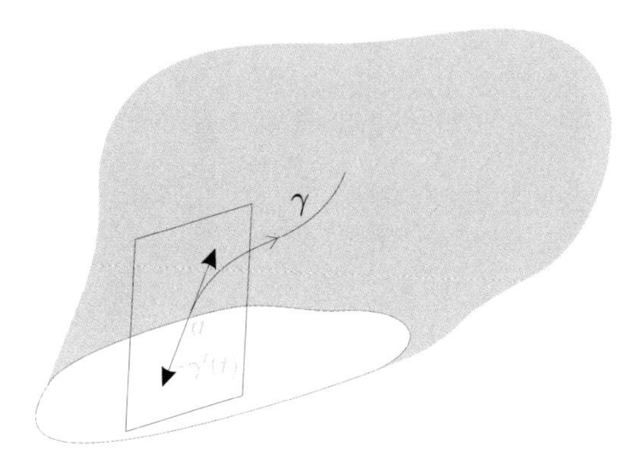

FIGURE 2. Tangent vectors at the boundary

sends $\partial\mathbb{H}_m$ to $\partial\mathbb{H}_m$ and the upper half-space to the upper half-space. Therefore, if $u^m{>}0$, then u is the velocity vector at $\xi_\alpha(a)$ of a curve pointing to the upper half-space. It follows that the image of this curve under $\xi_\beta \circ \xi_\alpha{}^{-1}$ will also point to the upper half-space, and the last component of $\big(D(\xi_\beta \circ \xi_\alpha{}^{-1})(\xi_\alpha(a))\big)(u)$ will also be positive. Of course, if $u^m{=}0$, then u is tangent to $\partial\mathbb{H}_m$, and therefore its image will also be tangent, yielding zero last coordinate. We call the tangent vector $[\xi_\alpha(a), u]$ ***inward-pointing*** (respectively, ***outward-pointing***) if $u^m{>}0$ (respectively, if $u^m{<}0$). For $u^m{=}0$, we obtain a ***tangent vector to the boundary***.

Let M be an m-manifold with boundary, $\mathcal{A}$ a C^r atlas for M, and consider ∂M with the subspace topology inherited from M. Let (U, ξ) be in $\mathcal{A}$ with $\partial U{\neq}\emptyset$. Then ∂U is a non-empty open subset of ∂M, and $\xi|_{\partial U}$, denoted by $\partial\xi$, is a homeomorphism from ∂U onto $\partial(\xi(U))$. Recalling the canonical identification $\iota_m{:}(x^1, \ldots, x^{m-1}, 0){\mapsto}(x^1, \ldots, x^{m-1})$, the composition $\iota_m \circ \partial\xi$ is then a homeomorphism from ∂U onto an open subset of $\mathbb{R}^{m-1}$. The collection of $(\partial U, \iota_m \circ \partial\xi)$ thus obtained will be denoted by $\partial\mathcal{A}$.

3. Theorem *Let M be a C^r m-manifold with non-empty boundary, $r{\geq}0$. Then ∂M is a closed subset of M. Moreover, if $\mathcal{A}$ is a C^r atlas for M, then ∂M, as a subspace of M, is a C^r manifold of dimension $m{-}1$ (without boundary) with atlas $\partial\mathcal{A}$. Further, for $r{\geq}1$, if $\mathcal{A}$ is a positive atlas for M, then $\partial\mathcal{A}$ is a positive atlas for ∂M.*

PROOF. The complement of ∂M is open since every point has a neighborhood homeomorphic to an open subset of $\mathbb{R}^m$, hence ∂M is closed. We must check C^r-compatibility of charts in $\partial\mathcal{A}$. Let $(\partial U, \iota_m \circ \partial\xi)$ and $(\partial V, \iota_m \circ \partial\eta)$ be so that $\partial U{\cap}\partial V{\neq}\emptyset$.

Then

$$\begin{aligned}(\iota_m \circ \partial\eta) \circ (\iota_m \circ \partial\xi)^{-1}(x^1, \ldots, x^{m-1}) &= \iota_m \circ \partial\eta \circ (\partial\xi)^{-1} \circ (\iota_m)^{-1}(x^1, \ldots, x^{m-1}) \\ &= \iota_m \circ \partial\eta \circ (\partial\xi)^{-1}(x^1, \ldots, x^{m-1}, 0) \\ &= \iota_m \circ \eta \circ \xi^{-1}(x^1, \ldots, x^{m-1}, 0) \\ &= \iota_m(y^1, \ldots, y^{m-1}, 0).\end{aligned}$$

Now $(y^1, \ldots, y^{m-1}, 0)$ is a C^r function of $(x^1, \ldots, x^{m-1}, 0)$ since $\eta \circ \xi^{-1}$ is C^r, hence the change of variable is C^r. Note also that as a subspace of M, ∂M is Hausdorff and second-countable, therefore the first assertion is established.

Now assume that $r \geq 1$ and that the atlas $\mathcal{A}$ is positive. Thus for two charts (U, ξ) and (V, η) in $\mathcal{A}$ with $U \cap V \neq \emptyset$, we have $\det D(\eta \circ \xi^{-1})(a) > 0$ for all $a \in \xi(U \cap V)$. In particular, if $\partial U \cap \partial V \neq \emptyset$, and $a = (a^1, \ldots, a^{m-1}, 0) \in \xi(\partial U \cap \partial V)$, the matrix of $D(\eta \circ \xi^{-1})(a)$ will have the form

$$\begin{bmatrix} & & * \\ & \mathbf{A} & \vdots \\ & & * \\ 0 \ldots 0 & & c \end{bmatrix}$$

Here $\mathbf{A}$ denotes the matrix of $D((\iota_m \circ \partial\eta) \circ (\iota_m \circ \partial\xi)^{-1})(a^1, \ldots, a^{m-1})$. We justify the form of the last row as follows. Writing $\eta \circ \xi^{-1}$ as $h = (h^1, \ldots, h^m)$, the entry at the (m, j) spot, for $j < m$, is

$$\frac{\partial h^m}{\partial x^j}(a) = \lim_{t \to 0} \frac{h^m(a^1, \ldots, a^j + t, \ldots, a^{m+1}, 0) - h^m(a^1, \ldots, a^{m+1}, 0)}{t}$$

The numerator of the fraction is zero since boundary points are mapped to boundary points, hence the (m, j) entry, for $j < m$, is zero. On the other hand, for $j = m$, and $t > 0$, one has $h^m(a^1, \ldots, a^{m-1}, t) > 0$, and $h^m(a^1, \ldots, a^{m-1}, 0) = 0$, so $c \geq 0$. But since the determinant of the matrix is non-zero (in fact positive), we have $c = \frac{\partial h^m}{\partial x^m}(a) > 0$. This implies that the determinant of $\mathbf{A}$ is also positive, proving the desired result. □

It follows of course, from Theorem 16 of Chapter 6, that the boundary of an orientable manifold is also orientable. Let M be an oriented C^1 m-manifold with boundary and suppose $\mathcal{A}$ is a concordant positive atlas. Then, *by convention*, the ***boundary orientation*** for ∂M will be $(-1)^m$ times the concordant orientation provided by the positive atlas $\partial\mathcal{A}$. In other words, if M is even-dimensional, we adopt the orientation concordant with $\partial\mathcal{A}$ for ∂M, and if M is odd-dimensional, we use the opposite orientation. The ultimate justification for this choice is that it provides for a single Stokes Formula, same for odd and even dimensions. The following useful lemma explains the geometric meaning of this choice.

4. Lemma *Let M be a C^1 manifold oriented by the volume element ω. Then for every $a \in \partial M$, the conventional boundary orientation for $T_a(\partial M)$ is the orientation provided* *by the volume element*

$$j^* i_X \omega$$

where X is any outward-pointing tangent vector at a, and $j{:}\partial M \hookrightarrow M$ is the inclusion map. Further, if $Y \in T_a(\partial M)$, then

$$j^* i_Y \omega = 0$$

Proof. Let $\mathcal{A}$ be a concordant positive atlas for M and (U, ξ) be a chart in $\mathcal{A}$ so that $a \in U$. Since the notion of "outward-pointing" is preserved by the chart, it suffices to investigate the assertion in $\mathbb{H}_m$ with boundary $\mathbb{R}^{m-1}$. Here we replace ω by $\omega_0 = dx^1 \wedge \cdots \wedge dx^m$. The orientation determined for $\partial \mathbb{H}_m = \mathbb{R}^{m-1}$ by $\partial \mathcal{A}$ will be given by $dx^1 \wedge \cdots \wedge dx^{m-1}$ (see the proof of Theorem 3). We have

$$dx^1 \wedge \cdots \wedge dx^{m-1} = (-1)^{m-1} i_{\frac{\partial}{\partial x^m}} (dx^1 \wedge \cdots \wedge dx^m)$$

Note that $\frac{\partial}{\partial x^m}$ is an *inward-pointing* vector at the boundary. If X is a tangent vector at the boundary, we can write

$$X = \sum_{k=1}^{m-1} c^k \frac{\partial}{\partial x^k} + c^m \frac{\partial}{\partial x^m}$$

where $c^m = 0$ if X is tangent to the boundary, and $c^m < 0$ if X is outward-pointing. Noting that at the boundary, $j^* i_{\frac{\partial}{\partial x^k}} (dx^m) = 0$ for $k < m$, we obtain the last claim. In general, then

$$i_X \omega_0 = (-1)^m (-c^m) j^* (dx^1 \wedge \cdots \wedge dx^{m-1})$$

where $(-c^m)$ will be positive if X is outward-pointing, and the remaining assertion is also proved. □

5. Examples

(a) We look at Example 2b of one-dimensional manifold with boundary. Here the volume element (= directed arclength element) for C may be taken to be $\gamma'(t)dt$. At the initial and terminal endpoints, we assign the signs "-" and "+," respectively, as orientations to zero-dimensional manifolds $\gamma(a)$ and $\gamma(b)$ (see Figure 3). At $\gamma(a)$, the vector $-\gamma'(a)$ is outward-pointing, therefore $(-1)^0 i_{-\gamma'(a)} \gamma'(a) dt = -|\gamma'(a)|^2 < 0$, and at $\gamma(b)$, the velocity vector $\gamma'(a)$ itself is outward-pointing and the opposite holds.

(b) In elementary calculus, one considers planar regions with boundary consisting of several closed curves, one closed curve surrounding several inner closed curves. The outer curve is oriented counter-clockwise, and the inner ones are taken in the clockwise direction. This is consistent with our orientation convention (see Figure 4).

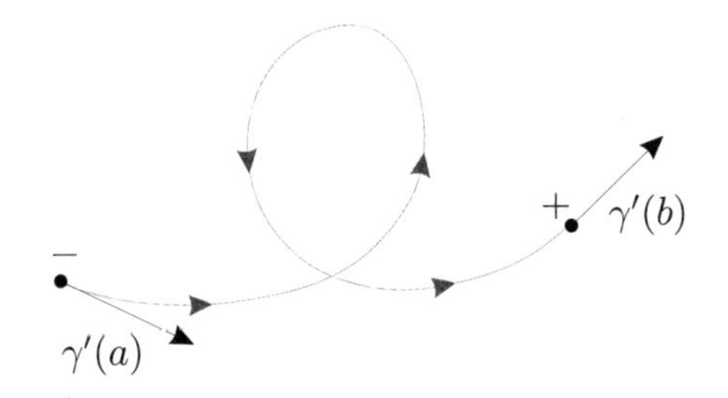

FIGURE 3. Orienting boundary points of a 1-dimensional manifold

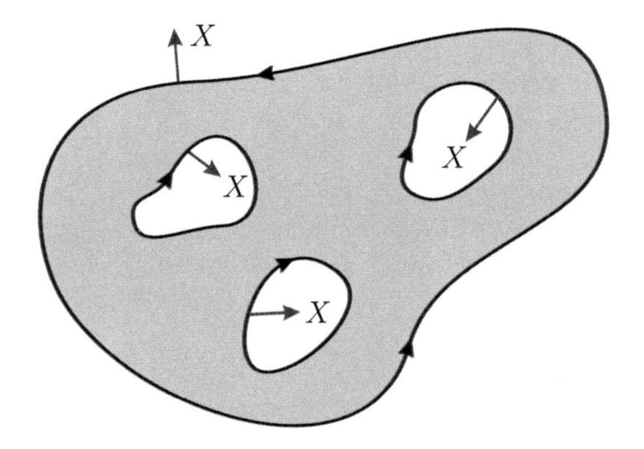

FIGURE 4. Orienting the boundary of a planar region

B. Integration on Manifolds with Boundary

In this section and throughout the rest of the chapter, integration on $\mathbb{R}^m$ will be in the sense of Riemann or Lebesgue. Functions defined on subsets of $\mathbb{R}^m$ for which the integral is considered, are assumed to be integrable without explicit mention. Further reference to integration in $\mathbb{R}^m$ and measure theory will be provided as the need arises.

Throughout this section, M will be a smooth m-dimensional manifold with boundary. A naive attempt to use charts to define the integral of a real-valued function defined on a manifold faces the same challenge of well-defined-ness as trying to define the derivative of such a function. Let us suppose that $f{:}M\to\mathbb{R}$ is continuous, and that the value of f is zero outside a compact set K that is entirely contained within the domain of a chart (U,ξ). As a first attempt, let us try the definition

$$\int_M f = \int_{\xi(U)} (f \circ \xi^{-1})$$

If (V, η) is another chart with $K \subset V$, then the above definition will make sense only if it were true that

$$\int_{\xi(U)} (f \circ \xi^{-1}) = \int_{\eta(V)} (f \circ \eta^{-1})$$

But the change of variable formula indicates that this is generally not true. In fact:

$$\int_{\eta(V)} (f \circ \eta^{-1}) = \int_{\xi(U)} (f \circ \xi^{-1}) |\det D(\eta \circ \xi^{-1})| \tag{7.2}$$

One way to interpret this is to regard each chart as inducing a measure on the chart domain relative to which f is integrated. Since the measures induced from different charts need not be the same, a rescaling factor $|\det D(\eta \circ \xi^{-1})|$ must be introduced. It turns out that by integrating m-forms instead of functions, this rescaling is automatically built into the process, exactly as in the integration of differentials in one-variable calculus. We will then switch to describing the integration of m-forms in $\mathbb{H}_m$.

In general, the ***support*** of a p-form α, $\mathrm{Supp}(\alpha)$, will be the closure of the set where α is non-zero. Suppose now that α is an m-form on $\mathbb{H}_m$, say $\alpha(x) = f(x)\omega_0(x)$, with f a continuous real-valued function of compact support, and ω_0 the standard volume element $dx^1 \wedge \cdots \wedge dx^m$. We define the ***integral of*** α over $\mathbb{H}_m$ by

$$\int_{\mathbb{H}_m} \alpha = \int_{\mathbb{H}_m} f(x^1, \ldots, x^m) dx^1 \cdots dx^m \tag{7.3}$$

with the integral on the right being in the sense of Riemann or Lebesgue. Now suppose $h: W_1 \to W_2$ is a smooth diffeomorphism of open subsets of $\mathbb{H}_m$, and $\mathrm{Supp}(\alpha) \subset W_2$. Then

$$\begin{aligned} h^*\alpha &= (f \circ h) h^* \omega_0 \\ &= (f \circ h) \det(Dh) \omega_0 \end{aligned}$$

Assuming that $\mathrm{Supp}(f)$ is connected, $|\det(Dh)| = \pm \det(Dh)$, the sign $\pm$ depending on whether h is orientation-preserving or orientation-reversing. Therefore, by (7.3) and (7.2), we have

$$\int_{W_1} h^*\alpha = \pm \int_{W_2} \alpha \tag{7.4}$$

where the sign $\pm$ is determined by whether h is orientation-preserving or orientation-reversing.

Now suppose M be a smooth oriented m-manifold with boundary, and fix a smooth volume element ω compatible with the orientation of M. For $0 \le r \le \infty$, there is a $C^r(M)$-module isomorphism, $f \mapsto f\omega$, between $C^r(M) = \Omega^0_r M$ and $\Omega^m_r M$, where $\Omega^p_r M$ denotes the module of C^r p-forms on M. This correspondence restricts to one between C^r functions of compact support, $C^r_c(M)$ and C^r m-forms of compact support, $\Omega^m_{r,c} M$. So we may equivalently discuss the integration of m-forms instead of the integration of functions. The point is that the invariant form of (7.4) makes it amenable to deployment on a manifold, as we shall now see. Let $\mathcal{A}$ be a concordant positive atlas for M, i.e., for each $(U,\xi)\in\mathcal{A}$, the diffeomorphism ξ is orientation-preserving. We define an operator

$$\int_M : \Omega^m_{0,c}M \to \mathbb{R}$$

in two steps.

Step 1 Suppose the support of $\alpha\in\Omega^m_{0,c}M$ is contained in U, where (U,ξ) is an element of $\mathcal{A}$. In this case, we define

$$\int_M \alpha = \int_{\xi(U)} (\xi^{-1})^*\alpha \tag{7.5}$$

This is now well-defined, for if $(V,\eta)\in\mathcal{A}$ is such that V also contains the support of α, letting $h=\eta\circ\xi^{-1}|_{\xi(U\cap V)}$, then

$$\begin{aligned}\int_{\eta(V)} (\eta^{-1})^*\alpha &= \int_{\eta(U\cap V)} (\eta^{-1})^*\alpha \\ &= \int_{\xi(U\cap V)} h^*((\eta^{-1})^*\alpha) \quad \text{by (7.4)} \\ &= \int_{\xi(U)} (\xi^{-1})^*\alpha\end{aligned}$$

Under the linear map $(\xi^{-1})^*$, the linear properties of integration in $\mathbb{H}_m$ are transferred to integration of m-forms with support in U. For example, the integral of the sum is the sum of integrals.

Step 2 In general, the support of $\alpha\in\Omega^m_{0,c}M$ need not be contained in a single chart domain. We take a countable, locally-finite sub-atlas $\mathcal{B}=\{(U_i,\xi_i):i=1,2,\ldots\}$ with subordinate smooth partition of unity (θ_i), $i=1,2,\ldots$, and define

$$\int_M \alpha = \sum_i \int_M \theta_i\alpha \tag{7.6}$$

Since the support of α is compact and (U_i) is locally-finite, the sum above is finite, and each summand is well-defined by Step 1. We must now show that this definition is independent of the particular locally-finite sub-atlas of $\mathcal{A}$ and the particular partition of unity. Let $\mathcal{C}=\{(V_j,\eta_j):j=1,2,\ldots\}$ be another locally-finite sub-atlas of $\mathcal{A}$ with a subordinate partition of unity (ϕ_j), $j=1,2,\ldots$. Consider the common refinement, i.e., the family $(W_{ij}=U_i\cap V_j)$, which is itself a countable and locally-finite sub-atlas of $\mathcal{A}$. The indexed set of smooth functions ψ_{ij}, where $\psi_{ij}=\theta_i\phi_j$ is a smooth partition of unity with the support of ψ_{ij} contained in W_{ij}. For a fixed i, we have $\theta_i=\sum_j\psi_{ij}$, and this sum is finite. Likewise, for fixed j, the equality $\phi_j=\sum_i\psi_{ij}$ holds.

Therefore, by the linearity mentioned at the end of Step 1, we have

$$\sum_i \int_M \theta_i\alpha = \sum_i \int_M (\sum_j \psi_{ij}\alpha)$$
$$= \sum_i \sum_j \int_M \psi_{ij}\alpha$$
$$= \sum_j \int_M (\sum_i \psi_{ij}\alpha)$$
$$= \sum_j \int_M \phi_j\alpha$$

which proves the desired independence from the locally-finite atlas and the particular partition of unity. The following theorem summarizes what has been proved.

6. Theorem *Let M be a smooth oriented m-manifold with boundary. Then there is a unique $\mathbb{R}$-linear map $\int_M : \Omega^m_{0,c}M \to \mathbb{R}$ with the following property. (*) For every open subset W of M, every orientation-preserving diffeomorphism $\xi: W \to W'$, where W' is an open subset of $\mathbb{H}_m$, and each $\alpha \in \Omega^m_{0,c}M$ with* $\mathrm{Supp}(\alpha) \subset W$, *one has*

$$\int_M \alpha = \int_{\mathbb{H}_m} (\xi^{-1})^*\alpha$$

7. Corollary *Let M_1 and M_2 be smooth, connected and oriented m-manifolds and $h: M_1 \to M_2$ a smooth diffeomorphism. Then for each $\alpha \in \Omega^m_{0,c}M_2$, one has*

$$\int_{M_1} h^*\alpha = \pm \int_{M_2} \alpha$$

with $\pm$ depending on whether h is orientation-preserving or orientation-reversing.

Proof. First suppose that h is orientation-preserving. We define

$$\int'_{M_2} : \Omega^m_{0,c}M \to \mathbb{R}$$

by $\int'_{M_2} \alpha = \int_{M_1} h^*\alpha$. If we show that operator $\int'_{M_2}$ satisfies property (*) in Theorem 6 for $\int_{M_2}$, then the result will follow from the uniqueness statement of that theorem. Now $\int'_{M_2}$ is certainly $\mathbb{R}$-linear since h^* and $\int_{M_1}$ are. Let W be an open subset of M_2, $\xi: W \to W' \subset \mathbb{H}_m$ an orientation-preserving diffeomorphism, and $\alpha \in \Omega^m_c M_2$ so that $\mathrm{Supp}(\alpha) \subset W$. Then $W'' = h^{-1}W$ is an open subset of M_1, and $\eta = \xi \circ h|_{W''}$ is an orientation-preserving diffeomorphism of W'' onto W', therefore

$$\begin{aligned}\int'_{M_2} \alpha &= \int_{M_1} h^*\alpha \\ &= \int_{\mathbb{H}_m} (\eta^{-1})^*(h^*\alpha) \\ &= \int_{\mathbb{H}_m} (\xi^{-1})^*\alpha \\ &= \int_{M_2} \alpha\end{aligned}$$

The assertion of the corollary for the orientation-preserving case then follows from the uniqueness statement of the theorem. Now consider the orientation-reversing case. Let ω be an m-form that determines the orientation of M_1. Denoting the integral of $\alpha \in \Omega^m_{0,c} M_1$ with respect to $-\omega$ on M_1 by $\widehat{\int_{M1}}\alpha$, we claim that

$$\widehat{\int_{M1}} \alpha = -\int_{M_1} \alpha$$

This follows from (7.4) and the method of construction of the integral. Now if $h{:}M_1 \to M_2$ is orientation-reversing, replacing ω by $-\omega$ for M_1, h becomes orientation-preserving, and the assertion follows from the formula above. □

We have thus arrived at the situation where, fixing a smooth volume element ω for the orientable manifold with boundary M, we can define the integral $\int_M f$ of a continuous real-valued function f with compact support as $\int_M f\omega$, and this concept of integral satisfies the usual $\mathbb{R}$-linearity property. Note further that the use of a concordant positive atlas in the definition of integral implies the *positivity property*, i.e., if $f \geq 0$, then $\int_M f \geq 0$. General representation theorems in measure theory, usually ascribed to F. Riesz, about the extension of positive linear functionals on the space of continuous functions with compact support on a Hausdorff, second-countable and locally compact space[1] then imply the following.

8. Theorem *Let ω be a smooth volume element on a manifold with boundary M. Then there is a unique Borel measure μ_ω on M with the following properties:*
(a) For every continuous real-valued function f with compact support on M

$$\int_M f\omega = \int_M f d\mu_\omega$$

(b) For every compact subset K of M, $\mu_\omega(K) < \infty$, and for every open subset U of M, $\mu_\omega(U) > 0$.

[1]See, e.g., Rudin, W. *Real and Complex Analysis*, or Lang, S. *Real and Functional Analysis* for the background material on measure theory.

(c) For every Borel subset B of M

$$\mu_\omega(B) = \inf\{\mu_\omega(U): U \text{open}, B\subset U\subset M\}$$

(d) For every Borel subset B of M with finite measure

$$\mu_\omega(B) = \sup\{\mu_\omega(K): K \text{compact}, K\subset B\}$$

We recall that Borel subsets of M are subsets obtainable from open subsets by operations of complementation, countable union and countable intersection. A measure of the type μ_ω is called a ***smooth measure*** on M. If ω and ω' are two smooth volume elements of the same orientation, then $\omega'=\delta\omega$ where δ is a smooth positive function. It follows that $\mu_{\omega'}=\delta\mu_\omega$. This implies, among other things, that *sets of measure zero* are the same with respect to all smooth measures. In fact, this can be independently proved. Recall that in $\mathbb{R}^m$ with Lebesgue measure, a set Z is of ***measure zero*** if and only if for every $\varepsilon>0$, there is a countable cover of Z by rectangles (R_i) so that $\sum_i \text{Vol}(R_i)<\varepsilon$. Let W be an open set in $\mathbb{R}^m$, Z a set of measure zero, $Z\subset W$, and $g: W\rightarrow\mathbb{R}^m$ a C^1 map. Then $g(Z)$ is a set of measure zero (Exercise 7.11 at the end of this chapter). It follows that for a C^1 manifold, one can invariantly define measure zero sets via charts.

As usual, an important application of integration is to the computation of volumes. If M is an oriented smooth manifold with boundary, and ω is a volume element compatible with the orientation, then for every Borel subset B of M, the volume of B with respect to ω (or the ***ω-volume***) is defined as

$$\text{Vol}_\omega(B) = \int_M \chi_B d\mu_\omega \tag{7.7}$$

where χ_B, the ***characteristic function*** of B, is the function that takes value 1 at points of B and value 0 outside B. By virtue of the definition of integral, where the coordinate maps were taken to be orientation preserving, the value of volume is always non-negative. For a smooth measure, if the set B contains interior points, then by part (b) of Theorem 8, the volume is strictly positive. Thus:

9. Corollary *If M is a compact, orientable smooth manifold, then the volume of M with respect to any smooth measure on M is a positive real number.*

In fact, as we will see in Subsection 11, the hypothesis of orientability is not needed here. Below we give some examples and straightforward consequences.

10. Examples

(a) Suppose M_1 and M_2 are compact manifolds (without boundary) with respective volume elements ω_1 and ω_2. We know from Example 17a of Chapter 6 that

$\omega = p_1{}^*\omega_1 \wedge p_2{}^*\omega_2$ is a volume element for $M_1 \times M_2$, where $p_i : M_1 \times M_2 \to M_i$ is projection on the ith component. We claim that

$$\mathrm{Vol}_\omega(M_1 \times M_2) = \mathrm{Vol}_{\omega_1} M_1 \times \mathrm{Vol}_{\omega_2} M_2$$

Let $\mathcal{A} = \{U_i : i = 1, 2, \ldots\}$ and $\mathcal{B} = \{V_j : j = 1, 2, \ldots\}$ be countable, locally finite, concordant positive atlases for (M_1, ω_1) and (M_2, ω_2), respectively, and suppose that (θ_i) and (ϕ_j) are respective partitions of unity subordinate to these. Then $\{W_{ij} = U_i \times V_j : i, j = 1, 2, \ldots\}$ is a concordant positive atlas for $(M_1 \times M_2, \omega)$, and $(\psi_{ij} = \theta_i \phi_j)$ is a partition of unity with the support of ψ_{ij} contained in W_{ij}. Note that if $(U, \xi) \in \mathcal{A}$ and $(V, \eta) \in \mathcal{B}$, then $\mathrm{Vol}_\omega(U \times V) = \mathrm{Vol}_{\omega_1} U \times \mathrm{Vol}_{\omega_2} V$ by virtue of Step 1 in the definition of the integral and Fubini's Theorem. The general case follows from the procedure of Step 2 in the definition of integral:

$$\begin{aligned}
\int_{M_1 \times M_2} \omega &= \sum_{i,j} \int_{W_{ij}} \psi_{ij} \omega \\
&= \sum_{i,j} \int_{W_{ij}} (\theta_i p_1{}^* \omega_1) \wedge (\phi_j p_2{}^* \omega_2) \\
&= (\sum_i \int_{U_i} \theta_i \omega_1)(\sum_j \int_{V_j} \phi_j \omega_2) \\
&= (\int_{M_1} \omega_1)(\int_{M_2} \omega_2)
\end{aligned}$$

In particular, the volume of k-torus $\mathbb{T}^k = S_1 \times \cdots \times S_1 \subset \mathbb{R}^{2k}$ is $(2\pi)^k$. This is not to be confused, e.g., with the geometric 2-tori of various sizes sitting in $\mathbb{R}^3$ (see Exercise 7.13). We point out that the same general argument holds if all, except possibly one, of the factors in $M_1 \times \cdots \times M_k$ are without boundary. The product of two smooth manifolds with boundary is not a smooth manifold with boundary in the natural way (see Exercises 7.2 and 7.3), but it can be treated in the setting of measure theory.

(b) (Volume of Spheres) The standard volume element of the unit m-sphere was defined by Formula (6.13) of Chapter 6 as

$$\omega_{S_m} = \sum_{i=1}^{m+1} (-1)^{i-1} x^i dx^1 \wedge \cdots \wedge \widehat{dx^i} \wedge \cdots \wedge dx^{m+1}$$

We claim that

$$\mathrm{Vol}_{\omega_{S_m}}(S_m) = \frac{2\pi^{\frac{m+1}{2}}}{\Gamma(\frac{m+1}{2})} \tag{7.8}$$

where Γ is the Euler gamma function, Γ, given for real values $x > 0$ by the convergent integral

$$\Gamma(x) = \int_0^\infty t^{x-1} e^{-t} dt$$

It is well-known, and elementary to verify, that $\Gamma(x+1)=x\Gamma(x)$ for all x, $\Gamma(n)=(n-1)!$ for every natural number n, and that $\Gamma(\frac{1}{2})=\sqrt{\pi}$. For the last claim, the integral is seen to equal the well-known integral $\int_0^\infty e^{-u^2}du=\sqrt{\pi}$ by a change of variable. Therefore, the explicit formulas for the volume of m-sphere are

$$\mathrm{Vol}_{\omega_{S_m}}(S_m) = \begin{cases} \frac{2\pi^k}{(k-1)!} & \text{if } m = 2k-1 \\ \frac{2^{k+1}\pi^k}{(1)(3)\cdots(2k-1)} & \text{if } m = 2k \end{cases} \tag{7.9}$$

We now turn to the proof of (7.8). Consider the diffeomorphism $H:\mathbb{R}^+\times S_m\to\mathbb{R}^{m+1}-\{\mathbf{0}\}$ given by $(t,u)\mapsto tu$. Computing the pull-back of the standard volume element on $\mathbb{R}^{m+1}$:

$$\begin{aligned} H^*(\omega_0) &= (tdu^1+u^1dt)\wedge\cdots\wedge(tdu^{m+1}+u^{m+1}dt) \\ &= t^m dt\wedge(\sum_{k=1}^{m+1}(-1)^{k-1}u^k du^1\wedge\cdots\widehat{du^k}\cdots\wedge du^{m+1}) \end{aligned}$$

since the $(m+1)$-form $u^1\wedge\cdots\wedge u^{m+1}$ necessarily vanishes on the m-dimensional S_m. Thus $H^*(\omega_0) = t^m dt\wedge\omega_{S_m}$. The vector field $\frac{d}{dt}$ is outward-pointing on the boundary S_m of the unit disk, and $i_{\frac{d}{dt}}\omega_0=t^m\omega_{S_m}$ is a positive multiple of ω_{S_m}, so H is orientation preserving. It follows from Corollary 7 that for any real-valued function f on $\mathbb{R}^{m+1}-\{\mathbf{0}\}$,

$$\int_{\mathbb{R}^+\times S_m} H^*(f\omega_0) = \int_{\mathbb{R}^{m+1}-\{\mathbf{0}\}} f\omega_0$$

We apply this to $f(x)=e^{-|x|^2}$, where $|.|$ is the Euclidean norm. The left-hand side gives

$$\begin{aligned} (\int_{\mathbb{R}^+} e^{-t^2}t^m dt)(\int_{S_m}\omega_{S_m}) &= (\frac{1}{2})(\int_{\mathbb{R}^+} e^{-s}s^{\frac{m-1}{2}}ds)\mathrm{Vol}_{S_m}(S_m) \\ &= (\frac{1}{2})\Gamma(\frac{m+1}{2})\mathrm{Vol}_{S_m}(S_m) \end{aligned}$$

For the right-hand side, we note that the addition of the single point $\{\mathbf{0}\}$ does not affect the integral of the bounded function, so the right-hand side can be written as

$$\begin{aligned} \int_{\mathbb{R}^{m+1}} e^{-|x|^2}dx^1\cdots dx^{m+1} &= (\int_{\mathbb{R}} e^{-s^2}ds)^{m+1} \\ &= \pi^{\frac{m+1}{2}} \end{aligned}$$

Comparing the two results, we obtain (7.8).

11. Integration on Non-Orientable Manifolds Our discussion of integration in this section was confined to orientable manifolds, but the theory can be extended to a non-orientable manifolds in two ways. One is to consider a volume element ω on the orientable double cover M' of the non-orientable manifold M (see Exercise 6.25), and define the measure of a Borel set B in M to be $(1/2)\mu_\omega(B')$, where B' is the pre-image of B under the covering map $M'\to M$. Another method is through the introduction of *densities*. A C^r **density** on an m-manifold M is a C^r function $(TM)^m\to\mathbb{R}$ that assigns to each m-tuple $(w_1,\ldots,w_m)$, $w_i \in T_xM$, a non-negative real number $\delta(w_1,\ldots,w_m)$, where

$$\delta(w_1,\ldots,w_m) = |\omega_x(w_1,\ldots,w_m)|$$

for some volume element ω_x for T_xM. Given a C^r m-manifold, orientable or not, and a chart (U,ξ), a C^r density is induced on U from $\xi(U) \subset \mathbb{R}^m$. Using a partition of unity, we can piece together these local densities to define a C^r density on M. A theory of integration, based on densities, in place of volume elements, can then be carried through. Finally, note that whichever approach one uses for non-orientable manifolds, Corollary 9 continues to hold.

C. Stokes Theorem

As we indicated in the introduction to this chapter, the general Stokes Theorem may be regarded as the Fundamental Theorem of Calculus on manifolds; it is the key result of this chapter. The following local lemma embodies the main ingredient of the proof.

12. Lemma *Let U be an open subset of $\mathbb{H}_m$ and $\alpha \in \Omega_c^{m-1}U$. Then*

$$\int_U d\alpha = \begin{cases} 0 & \text{if } \partial U = \emptyset \\ \int_{\partial U} j^*\alpha & \text{if } \partial U \neq \emptyset \end{cases}$$

where $j{:}\partial U \hookrightarrow U$ is the inclusion map, and ∂U is oriented according to convention.

We note that in general $\mathrm{Supp}(d\alpha) \subset \mathrm{Supp}(\alpha)$, so that if $\alpha \in \Omega_c^{m-1}M$, then $d\alpha \in \Omega_c^m M$.

Proof. We can express α in the following form:

$$\alpha(x) = \sum_{i=1}^{m} a_i(x) dx^1 \wedge \cdots \wedge \widehat{dx^i} \wedge \cdots \wedge dx^m$$

Therefore,

$$\begin{aligned} d\alpha(x) &= \sum_{i=1}^{m} \Big(\sum_{j=1}^{m} \frac{\partial a_i}{\partial x^j} dx^j \Big) \wedge dx^1 \wedge \cdots \wedge \widehat{dx^i} \wedge \cdots \wedge dx^m \\ &= \Big(\sum_{i=1}^{m} (-1)^{i-1} \frac{\partial a_i}{\partial x^i} \Big) dx^1 \wedge \cdots \wedge dx^m \end{aligned}$$

We consider a rectangle $R = [s^1,t^1] \times \cdots \times [s^m,t^m]$ with $s^m = 0$ so that $\mathrm{Supp}(\alpha) \subset R$. By extending α to be zero outside U in $\mathbb{H}_m$, we may suppose that $\alpha \in \Omega_c^{m-1}\mathbb{H}_m$ and $d\alpha \in \Omega_c^m \mathbb{H}_m$. Note that α and $d\alpha$ vanish on the sides of the rectangle R except possibly on the side $s^m=0$. Using the expression for $d\alpha$ above, we obtain

$$\int_U d\alpha = \int_R d\alpha = \sum_{i=1}^{m}(-1)^{i-1}\Big(\int_R \frac{\partial a_i}{\partial x^i} dx^1 \wedge \cdots \wedge dx^m\Big)$$

Using (7.3) and Fubini's Theorem for iterated integrals, we have

$$\int_R \frac{\partial a_i}{\partial x^i} dx^1 \wedge \cdots \wedge dx^m = \int \cdots \int \Big[\int_{s^i}^{t^i} \frac{\partial a_i}{\partial x^i} dx^i\Big] dx^1 \cdots \widehat{dx^i} \cdots dx^m$$
$$= \int \cdots \int [a_i(x^1, \ldots, x^{i-1}, t^i, x^{i+1}, \ldots, x^m) - a_i(x^1, \ldots, x^{i-1}, s^i, x^{i+1}, \ldots, x^m)] dx^1 \cdots \widehat{dx^i} \cdots dx^m$$

Because of the vanishing of α on the sides of R except possibly on $s^m=0$, we obtain

$$\int_R d\alpha = (-1)^m \int \cdots \int a_m(x^1, \ldots, x^{m-1}, 0) dx^1 \cdots dx^{m-1}$$

In the case $\partial U=\emptyset$, $a_m(x^1, \ldots, x^{m-1}, 0)=0$, and the above vanishes. Otherwise, noting that $j^*\alpha(x^1, \ldots, x^{m-1}, 0)=a_m(x^1, \ldots, x^{m-1}, 0)dx^1 \wedge \cdots \wedge dx^{m-1}$, and by virtue of the convention for the orientation of the boundary, we obtain the desired result. □

The general theorem is now an exercise in using the partition of unity.

13. Theorem (**Stokes Formula**) *Let M be an oriented smooth m-manifold with the boundary oriented according to convention (if non-empty), and let $j{:}\partial M \hookrightarrow M$ be the inclusion map. Then for every $\alpha \in \Omega_c^{m-1} M$*

$$\int_M d\alpha = \begin{cases} 0 & \text{if } \partial M=\emptyset \\ \int_{\partial M} j^*\alpha & \text{if } \partial M \neq \emptyset \end{cases}$$

Proof. We consider a countable, locally finite concordant positive atlas (U_i, ξ_i), and take a subordinate partition of unity (θ_i). We can then proceed to calculate

$$\int_M d\alpha = \int_M d\Big(\sum_i \theta_i \alpha\Big)$$

Since the support of α is compact and the atlas is locally finite, the sum on the right is finite, therefore we can write

$$\begin{aligned}\int_M d\alpha &= \sum_i \int_M d(\theta_i\alpha)\\ &= \sum_i \int_{U_i} d(\theta_i\alpha)\\ &= \sum_i \int_{\xi_i(U_i)} (\xi_i^{-1})^*(d(\theta_i\alpha))\\ &= \sum_i \int_{\xi_i(U_i)} d((\xi_i^{-1})^*(\theta_i\alpha))\end{aligned}$$

Denoting the inclusion mapping $\partial\mathbb{H}_m \hookrightarrow \mathbb{H}_m$ by j_0, we can use the previous lemma to write

$$\int_{\xi_i(U_i)} d((\xi_i^{-1})^*(\theta_i\alpha)) = \begin{cases} 0 & \text{if } \partial\xi_i(U_i)=\emptyset \\ \int_{\partial\xi_i(U_i)} j_0{}^*(\xi_i^{-1})^*(\theta_i\alpha) & \text{if } \partial\xi_i(U_i)\neq\emptyset \end{cases}$$

Now if $\partial M=\emptyset$, then all $\partial(\xi_i(U_i))$ are empty, and we obtain $\int_M d\alpha=0$. Otherwise, we can rewrite the second alternative above as

$$\int_{\xi_i(U_i)} d((\xi_i^{-1})^*(\theta_i\alpha)) = \int_{\partial\xi_i(U_i)} (\xi_i^{-1}\circ j_0)^*(\theta_i\alpha)$$

Note that $\xi_i^{-1}\circ j_0 = j\circ(\partial\xi_i)^{-1}$, therefore

$$\begin{aligned}\int_M d\alpha &= \sum_i \int_{\partial\xi_i(U_i)} (j\circ(\partial\xi_i)^{-1})^*(\theta_i\alpha)\\ &= \sum_i \int_{\partial\xi_i(U_i)} ((\partial\xi_i)^{-1})^*(j^*(\theta_i\alpha))\end{aligned}$$

But the sum on the right is precisely the definition of $\int_{\partial M} j^*\alpha$, and the proof is complete. □

Applications of Stokes Formula will be ubiquitous throughout the rest of this chapter. For a starter, we state and prove the general form of a classical theorem of vector analysis known as the *Divergence Theorem.* We first note the following Riemannian version of Lemma 4.

14. Lemma *Let M be an oriented smooth m-manifold with boundary oriented according to convention, and consider a smooth Riemannian metric ρ on M. Suppose that ω_ρ and ω'_ρ are, respectively, the Riemannian volume elements of M and ∂M compatible with the orientations, and* **n** *is the unit outward normal at the boundary. Then*

$$\omega'_\rho = i_{\mathbf{n}}\omega_\rho$$

PROOF. Working at a boundary point $a \in \partial M$, let $(e_1, \ldots, e_m)$ be a positively oriented orthonormal basis for T_aM so that $(e_1, \ldots, e_{m-1})$ span $T_a(\partial M)$ and e_m is the inward-pointing unit normal. Then $\mathbf{n} = -e_m$. If $(e^1, \ldots, e^m)$ is the dual basis, we have $\omega_\rho = e^1 \wedge \cdots \wedge e^m$, and by Lemma 4, the boundary convention is determined by

$$j^* i_{\mathbf{n}} \omega_\rho = (-1)^m e^1 \wedge \cdots \wedge e^{m-1}$$

Since the e_k form an orthonormal set, the assertion is proved. □

15. (Divergence Theorem) *Let M be a compact oriented smooth m-manifold with boundary oriented according to convention, and consider a smooth Riemannian metric ρ on M. If X is a smooth vector field on M, then*

$$\int_M (\mathbf{div}_{\omega_\rho} X)\omega_\rho = \int_{\partial M} \rho(X, \mathbf{n})\omega'_\rho$$

where ω_ρ and ω'_ρ are, respectively, the Riemannian volume elements of M and ∂M compatible with the orientations, and $\mathbf{n}$ is the unit outward normal on the boundary.

PROOF. We consider the $(m-1)$-form $\alpha = i_X\omega_\rho$. Since $d\omega_\rho = 0$, Cartan's Formula for the Lie derivative gives

$$\begin{aligned} d\alpha &= L_X\omega_\rho \\ &= (\mathbf{div}_{\omega_\rho} X)\omega_\rho \end{aligned}$$

On the other hand, the restriction of X to the boundary can be written as

$$X = Y + \rho(X, \mathbf{n})\mathbf{n}$$

where Y is the component tangent to the boundary. Therefore, using Lemma 4 and Lemma 14, we obtain

$$j^* i_X \omega = \rho(X, \mathbf{n})\omega'_\rho$$

and Stokes Formula implies the result. □

16. Examples

(a) (Volumes of Disks) The unit $m+1$-dimensional disk, $\mathbb{D}_{m+1}$, has S_m as its boundary. It follows from

$$d\Big(\sum_{k=1}^{m+1} (-1)^{k-1} x^k dx^1 \wedge \cdots \widehat{dx^k} \cdots \wedge dx^{m+1}\Big) = (m+1) dx^1 \wedge \cdots \wedge dx^{m+1}$$

and Stokes Theorem that

$$(m+1)\mathrm{Vol}_{\omega_0}(\mathbb{D}_{m+1}) = \mathrm{Vol}_{\omega_{S_m}}(S_m)$$

Therefore, the following result is obtained from (7.9):

$$\mathrm{Vol}_{\omega_0}(\mathbb{D}_m) = \begin{cases} \frac{\pi^k}{k!} & \text{if } m = 2k \\ \frac{2^{k+1}\pi^k}{(1)(3)\cdots(2k+1)} & \text{if } m = 2k+1 \end{cases} \tag{7.10}$$

(b) The identity $d \circ d=0$ shows that every exact form is closed. We will now give an example of a closed m-form on $\mathbb{R}^{m+1}-\{\mathbf{0}\}, m \geq 1$ which is not exact. Let $\nu : \mathbb{R}^{m+1}-\{\mathbf{0}\} \to S_m$ be the map $x \mapsto x/|x|$, and define

$$\Theta_m = \nu^* \omega_{S_m} \tag{7.11}$$

Θ_m is closed since $d(\Theta_m)=d(\nu^*\omega_{S_m})=\nu^*(d\omega_{S_m})$, and $d\omega_{S_m}=0$. Suppose Θ_m is exact, say $\Theta_m=d\Phi$, we will derive a contradiction. Letting i be the inclusion map of S_m in $\mathbb{R}^{m+1}-\{\mathbf{0}\}$, we obtain

$$\omega_{S_m}=i^*\Theta_m=i^*(d\Phi)=d(i^*\Phi)$$

But ω_{S_m} is a volume element for S_m and therefore has non-zero integral on S_m by Corollary 9. On the other hand, if ω_{S_m} were exact, then the integral over the manifold without boundary S_m would have to be zero by Stokes Formula.

For future reference, we give the following explicit expression for Θ_m.

$$\Theta_m = |x|^{-(m+1)} \sum_{k=1}^{m+1} (-1)^{k-1} x^k dx^1 \wedge \cdots \widehat{dx^k} \cdots \wedge dx^{m+1} \tag{7.12}$$

Here $|.|$ denotes the Euclidean norm. To prove this formula, we argue as follows. The map ν sends every ray through the origin to a single point on S_m, hence vectors tangent along rays are sent to zero under $T\nu$. Therefore, to compute the effect of $\nu^*\omega_{S_m}$ at a point x of $\mathbb{R}^{m+1}-\{\mathbf{0}\}$, one need only evaluate it on tangent vectors to the sphere of radius $|x|$. However, the effect of ν on this sphere is precisely the same as the effect of the linear homothety that sends z to $z/|x|$ with constant $|x|$. Therefore, (7.12) follows from (7.11) and the definition of ω_{S_m}.

As in (b) above, Corollary 9 and Stokes Formula show that the volume element of a compact orientable manifold without boundary, in particular the volume element of the boundary of a compact orientable manifold, cannot be exact. Recall from Theorem 3 that the boundary is always a closed subset, hence if M is compact, then so is ∂M. This simple fact can lead to some striking results. The following is a well-known instance.

17. No-Extension Theorem *Let W be a compact oriented smooth manifold with boundary M, and suppose that $f : M \to M$ is a smooth diffeomorphism. The f has no extension to a smooth map $F : W \to M$.*

Proof. Suppose such an extension F exists, thus $f=F \circ j$, where j is the inclusion map of the boundary. If ω is a volume element for M, then $F^*\omega$ is a smooth m-form on W, where $m{=}\dim M$. Then

$$\begin{aligned}\int_M \omega &= \pm\int_M f^*\omega \\ &= \pm\int_M j^*(F^*\omega) \\ &= \pm\int_W d(F^*\omega) \\ &= \pm\int_W F^*(d\omega)\end{aligned}$$

But $d\omega{=}0$ being an $(m{+}1)$-form on the m-manifold M, which contradicts the non-vanishing of volume. □

The special case, where f is the identity mapping of M, is known as the *No-Retraction Theorem*. This theorem implies the celebrated:

18. Brouwer Fixed Point Theorem *Every smooth map $f{:}\mathbb{D}_m{\to}\mathbb{D}_m$ has a fixed point, i.e., there exists a point $x{\in}\mathbb{D}_m$ with $f(x){=}x$.*

Actually, Brouwer Fixed Point Theorem holds true even for continuous maps of the disk. This more general result can be deduced from the smooth version by a simple argument based on the Weierstrass Approximation Theorem (see [**20**], [**24**]).

Proof. Suppose a smooth map $f{:}\mathbb{D}_m{\to}\mathbb{D}_m$ without a fixed point exists. We prolong the unique straight segment from $f(x)$ to x until it intersects $\partial\mathbb{D}_m{=}S_{m-1}$ at the well-defined point $F(x)$. Note that the restriction of F to the boundary is the identity map. We show that $F{:}\mathbb{D}_m{\to}S_{m-1}$ is smooth. Writing $F(x){=}x{+}tu$, where

$$u = \frac{x-g(x)}{|x-g(x)|},\quad t = -x{\cdot}u{+}\sqrt{|x{\cdot}u|^2{+}1{-}|x|^2},$$

the expression under the square-root sign is always non-zero, for if $|x|{=}1$, then x is on the boundary, and the inner product $x{\cdot}u$ is non-zero. This contradicts the No-Retraction Theorem. □

D. De Rham Cohomology

Example 16b above and the applications of Stokes Formula that followed set the stage for the introduction of de Rham cohomology groups. As we indicated in Example 16b, the identity $d{\circ}d{=}0$ implies that every exact form is closed. It turns out that the manner and degree by which the converse proposition may fail for a space, reflect certain topological properties of that space. In algebraic topology, a host of groups and other algebraic structures are devised to formulate and study topological properties of spaces. Among these are various "cohomology groups," of which "de Rham cohomology" can be realized by classes of differential forms. In this case, the groups are actually real vector spaces, with the group operation being addition in the vector space.

We begin by presenting the simplest topological case, which is the subject of the so-called *Poincaré Lemma*. This states that for a "contractible space"(to be defined), every closed form is exact. To motivate the proof of this proposition, and for future reference, two examples of independent interest will be initially studied in detail. The first actually appears in elementary calculus, often in the guise of "conservative vector fields," and the second, the case of m-forms on $\mathbb{R}^m$, will be needed in the next section. The crucial common thread of the two examples is the use of "integration by parts," which will later manifest itself under the cover of "chain homotopy."

19. Examples

(a) A subset U of $\mathbb{R}^m$ is called ***star-like*** if there is a point x_0 of U so that the line segment joining any point x of U to x_0 lies entirely in U. We show that for a C^1 one-form α defined on an open star-like subset U of $\mathbb{R}^m$, if $d\alpha=0$, then a real-valued function ("potential") f defined on U exists so that $df=\alpha$. Let $\alpha=\sum_{k=1}^m a_i dx^i$. Now

$$d\alpha=0 \iff \frac{\partial a_k}{\partial x^l}=\frac{\partial a_l}{\partial x^k}, \forall k,l=1,\dots,m \tag{7.13}$$

We look for a function $f{:}U\to\mathbb{R}$ so that $df=\alpha$. We may assume, by performing a translation in $\mathbb{R}^m$, that U contains the origin of $\mathbb{R}^m$ and that $x_0=\mathbf{0}$. Take an arbitrary point $x=(x^1,\dots,x^m)\in U$ and parametrize the line segment from $\mathbf{0}$ to x as $\sigma(t)=tx$, $0\leq t\leq 1$ (see Figure 5). If the desired function f exists, then it must satisfy $\frac{\partial f}{\partial x^i}=a_i$, and

$$\begin{aligned} f(x)-f(\mathbf{0}) &= \int_\sigma df \\ &= \int_0^1 \sum_k \frac{\partial f}{\partial x^k}\frac{d\sigma^k}{dt}dt \\ &= \int_0^1 \sum_k a_k(tx)x^k dt \\ &= \sum_k x^k\Big(\int_0^1 a_k(tx)dt\Big) \end{aligned}$$

We now verify that f defined this way actually satisfies $df=\alpha$.

$$\begin{aligned} \frac{\partial f}{\partial x^l} &= \int_0^1 a_l(tx)dt + \sum_k x^k\Big(\int_0^1 \frac{\partial}{\partial x^l}(a_k(tx))dt\Big) \\ &= \int_0^1 a_l(tx)dt + \sum_k x^k\Big(\int_0^1 \frac{\partial a_k}{\partial \sigma^l}tdt\Big) \end{aligned}$$

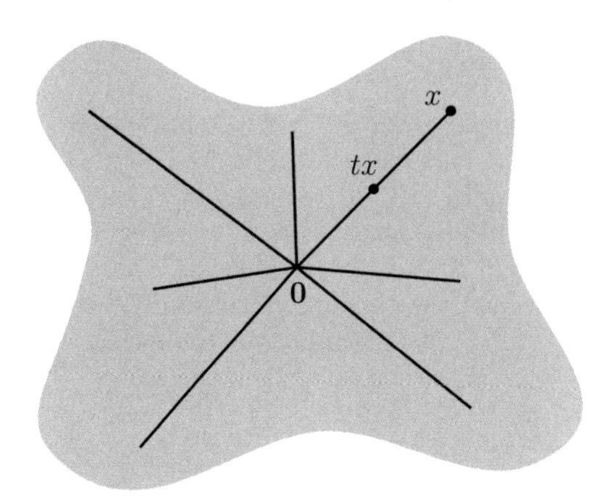

FIGURE 5. A star-like region in $\mathbb{R}^m$

Here we use the hypothesis $d\alpha$=0 via (7.13) to write

$$\begin{aligned}
\sum_k x^k\Big(\int_0^1 \frac{\partial a_k}{\partial \sigma^l} t dt\Big) &= \sum_k x^k\Big(\int_0^1 \frac{\partial a_l}{\partial \sigma^k} t dt\Big) \\
&= \int_0^1 \Big(\sum_k \frac{\partial a_l}{\partial \sigma^k} x^k\Big) t dt \\
&= \int_0^1 \frac{da_l}{dt} t dt \\
&= t a_l(tx)\big|_0^1 - \int_0^1 a_l(tx) dt \\
&= a_l(x) - \int_0^1 a_l(tx) dt
\end{aligned}$$

(note the use of integration by parts in the line before last), and we conclude that $\frac{\partial f}{\partial x^l}=a_l$ as claimed.

(b) Consider an arbitrary smooth m-form $\omega = f\omega_0$ on $\mathbb{R}^m$. Thus

$$\omega(x) = f(x) dx^1 \wedge \cdots \wedge dx^m$$

where $f{:}\mathbb{R}^m \to \mathbb{R}$ is a smooth function. $\mathbb{R}^m$ is star-like relative to every point, including the origin. Consider the $(m-1)$-form

$$\eta(x) = \Big(\int_0^1 t^{m-1} f(tx) dt\Big) \sum_{k=1}^m (-1)^{k-1} x^k dx^1 \wedge \cdots \widehat{dx^k} \cdots \wedge dx^m \tag{7.14}$$

We claim that $d\eta=\omega$. Vaguely speaking, η is constructed by splitting ω into its radial and spherical components, and performing a weighted integration along the radial segment joining the origin to the point x. Let us compute $d\eta$.

$$d\eta = \sum_{k=1}^{m}(-1)^{k-1}\Big\{\Big(\int_0^1 t^{m-1}f(tx)dt\Big)dx^k + x^k\sum_{l=1}^{m}\Big(\int_0^1 t^m\frac{\partial f}{\partial x^l}(tx)dt\Big)dx^l\Big\}\wedge dx^1\wedge\cdots\widehat{dx^k}\cdots\wedge dx^m$$

$$= \sum_{k=1}^{m}\Big(\int_0^1 t^{m-1}f(tx)dt\Big)\omega_0 + \sum_{k=1}^{m}\Big(\int_0^1 t^m\frac{\partial f}{\partial x^k}(tx)x^k dt\Big)\omega_0$$

$$= \Big(\int_0^1 mt^{m-1}f(tx)dt\Big)\omega_0 + \Big(\int_0^1 t^m\Big(\sum_{k=1}^{m}\frac{\partial f}{\partial x^k}(tx)x^k\Big)dt\Big)\omega_0$$

We now apply integration by parts to the first integral to obtain

$$\int_0^1 mt^{m-1}f(tx)dt = t^m f(tx)|_0^1 - \int_0^1 t^m\Big(\sum_{k=1}^{m}\frac{\partial f}{\partial x^k}(tx)x^k\Big)dt$$

$$= f(x) - \int_0^1 t^m\Big(\sum_{k=1}^{m}\frac{\partial f}{\partial x^k}(tx)x^k\Big)dt$$

This proves the desired result.

20. Remark We note here the relation between the $(m-1)$-form η above and Θ_{m-1} introduced earlier in (7.11) and (7.12). On $\mathbb{R}^m - \{\mathbf{0}\}$, i.e., for $x\neq\mathbf{0}$, we can rewrite η as

$$\eta(x) = |x|^m\Big(\int_0^1 t^{m-1}f(tx)dt\Big)\Theta_{m-1}(x)$$

For given x, making the change of variable $s=|x|t$, we obtain

$$\eta(x) = \Big(\int_0^{|x|} s^{m-1}f(s\nu(x))ds\Big)\Theta_{m-1}(x) \tag{7.15}$$

Going back to the main discussion, we begin by generalizing the notion of a star-like region. A manifold (with or without boundary) is ***(smoothly) contractible*** if there is a smooth map $H{:}[0,1]\times M\to M$ so that $H(0,x)=x$ for all $x\in M$, and $H(1,x)=x_0$, where x_0 is any given point in M. We may think of this as a (smooth) family H_t of maps from M to M, $H_t{:}x\mapsto(t,x)\mapsto H(t,x)$, parametrized by $t\in[0,1]$, starting with identity and smoothly deforming into a constant map sending every point to x_0. It is hard to fathom such a situation unless M is homeomorphic to a disk, but in fact non-disk examples can be constructed, the first given by J. H. C.Whitehead in 1935[2]. We note that if M is without boundary, then $[0,1]\times M$ is a manifold with boundary, and smoothness has been defined. If M itself has non-empty boundary, then smoothness with respect to t at t=0, 1 is taken as one-sided smoothness. Smooth p-forms on $[0,1]\times M$ are also defined to mean tensor fields of the form

$$f(t,x)p_1^*dt\wedge p_2^*\alpha + g(t,x)p_2^*\beta$$

[2]See the book *The Topology of 4-manifolds* (1989) by Robion Kirby.

where f and g are smooth functions, $\alpha\in\Omega^{p-1}M$, $\beta\in\Omega^p M$, and p_1, p_2 are projections on first and second components. For each $t\in[0,1]$, we have an embedding $j_t{:}M\to[0,1]\times M$ by $x\mapsto(t,x)$. (If $0<t<1$ or M is without boundary, this is a true embedding, otherwise we use one-sided derivatives to extend the meaning.) We will show that in a (smoothly) contractible space, every closed form is exact.

The following analogue of "integration by parts" will be a very powerful tool in our discussion.

21. Theorem *There is a family of linear maps $I_p{:}\Omega^p([0,1]\times M)\to\Omega^{p-1}M$ with the property that*

$$j_1^* - j_0^* = d\circ I + I\circ d$$

Here we really mean a sequence of identities, namely

$$d_{p-1}\circ I_p + I_{p+1}\circ d_p = j_1^* - j_0^* : \Omega^p([0,1]\times M) \to \Omega^p M$$

but the subscripts have been dropped. Recall also that $\Omega^p M=\{\mathbf{0}\}$ for $p<0$ by convention. The mappings I_p, which are some kind of integration operators, are said to define a ***chain homotopy*** from $\Omega^*([0,1]\times M)$ to $\Omega^* M$.

Assuming this result for the moment, we prove the following theorem.

22. Theorem **(Poincaré Lemma)** *Let M be (smoothly) contractible. If $\phi\in\Omega^p M$, $p\geq1$, is closed, then ϕ is exact.*

Proof. Since $H\circ j_0=\mathbb{1}_M$, then $(H\circ j_0)^*\phi=\phi$. On the other hand, $\{x_0\}$ is a zero-dimensional manifold and $p\geq1$, so $(H\circ j_1)^*\phi=0$. Therefore, by Theorem 21, we obtain

$$\begin{aligned}\phi &= (H\circ j_0)^*\phi - (H\circ j_1)^*\phi\\ &= (j_0^* - j_1^*)(H^*\phi)\\ &= -d\circ I(H^*\phi) - I\circ d(H^*\phi)\\ &= d(-I(H^*\phi)) - I\circ H^*(d\phi)\\ &= d(-I(H^*\phi))\end{aligned}$$

since by hypothesis, $d\phi=0$. □

Proof. (of Theorem 21) To begin, we claim that every element $\varphi\in\Omega^p([0,1]\times M)$ has a unique representation in the form

$$\varphi = (\tau\wedge\alpha) + \beta \tag{7.16}$$

where $i_{\frac{\partial}{\partial t}}\alpha=0$, $i_{\frac{\partial}{\partial t}}\beta=0$ and $\tau=p_1^*(dt)$. For uniqueness, suppose that we have two such representations $\tau\wedge\alpha_1+\beta_1=\tau\wedge\alpha_2+\beta_2$, then $\tau\wedge(\alpha_1-\alpha_2)=\beta_2-\beta_1$. Applying $i_{\frac{\partial}{\partial t}}$ to both sides, we get $\alpha_1-\alpha_2=0$, hence also $\beta_2-\beta_1=0$. To prove the existence of such a representation, simply define $\alpha = i_{\frac{\partial}{\partial t}}\varphi$ and $\beta = \varphi - \tau \wedge \alpha$.

Now we define I by giving the value of $I\varphi$ at a point $x \in M$ on a $(p-1)$-tuple of vectors $(w_2, \ldots, w_p)$, $w_i \in T_x M$.

$$I\varphi(w_2, \ldots, w_p) = \int_0^1 (j_t^* \alpha)(w_2, \ldots, w_p) dt$$

I is $\mathbb{R}$-linear. Taking a chart (U, ξ) around x in M, and suppressing the explicit mention of projections p_1 and p_2, we may express α, β and τ as

$$\alpha = f(t, \xi) d\xi^{i_2} \wedge \cdots \wedge d\xi^{i_p}, \ \beta = g(t, \xi) d\xi^{i_1} \wedge \cdots \wedge d\xi^{i_p}, \ \tau = dt$$

By additivity of I, we may prove the assertion of the theorem by verifying it on the two types of terms, $dt \wedge \alpha$ and β. For β, $I\beta = 0$, but

$$d\beta = \frac{\partial g}{\partial t} dt \wedge d\xi^{i_1} \wedge \cdots \wedge d\xi^{i_p} + \sum_k \frac{\partial g}{\partial \xi^k} d\xi^k \wedge d\xi^{i_1} \wedge \cdots \wedge d\xi^{i_p}$$

Therefore,

$$\begin{aligned} (d \circ I + I \circ d)(\beta) &= (\int_0^1 \frac{\partial g}{\partial t} dt)\ d\xi^{i_1} \wedge \cdots \wedge d\xi^{i_p} \\ &= (g(1, \xi) - g(0, x))\ d\xi^{i_1} \wedge \cdots \wedge d\xi^{i_p} \\ &= (j_1^* - j_0^*)(\beta) \end{aligned}$$

For the terms of the type $dt \wedge \alpha = f(t, \xi) dt \wedge d\xi^{i_2} \wedge \cdots \wedge d\xi^{i_p}$,

$$\begin{aligned} (d \circ I)(dt \wedge \alpha) &= d\Big((\int_0^1 f(t, \xi) dt)\ d\xi^{i_2} \wedge \cdots \wedge d\xi^{i_p} \Big) \\ &= \sum_k \frac{\partial}{\partial \xi^k} (\int_0^1 f(t, \xi) dt)\ d\xi^k \wedge d\xi^{i_2} \wedge \cdots \wedge d\xi^{i_p} \\ &= \sum_k (\int_0^1 \frac{\partial}{\partial \xi^k} f(t, \xi) dt)\ d\xi^k \wedge d\xi^{i_2} \wedge \cdots \wedge d\xi^{i_p} \end{aligned}$$

On the other hand,

$$\begin{aligned} (I \circ d)(dt \wedge \alpha) &= -I(dt \wedge d\alpha) \\ &= -I(\sum_k \frac{\partial f}{\partial \xi^k} dt \wedge d\xi^k \wedge d\xi^{i_2} \wedge \cdots \wedge d\xi^{i_p}) \\ &= -\sum_k (\int_0^1 \frac{\partial f}{\partial \xi^k} dt)\ d\xi^k \wedge d\xi^{i_2} \wedge \cdots \wedge d\xi^{i_p} \end{aligned}$$

Summing up, we obtain $(d \circ I + I \circ d)(dt \wedge \alpha) = 0$. But in this case, we also have

$$\begin{aligned} j_0^*(dt) &= d(0) = 0 \\ j_1^*(dt) &= d(1) = 0 \end{aligned}$$

and the proof is complete. □

Knowing whether a closed form defined on a manifold is exact turns out to be of importance in many problems of mathematics and applications. We have just observed that this is always the case when the manifold is contractible. In Example 16(b), we encountered the example, for $m \geq 1$, of the m-form Θ_m on $\mathbb{R}^{m+1} - \{\mathbf{0}\}$ which is closed, but not exact. For $m=1$, that example appears in elementary calculus in the equivalent form of the vector field

$$\frac{-y}{x^2+y^2}\frac{\partial}{\partial x} + \frac{x}{x^2+y^2}\frac{\partial}{\partial y}$$

on $\mathbb{R}^2 - \{\mathbf{0}\}$. Here the equality $\frac{\partial}{\partial y}(\frac{-y}{x^2+y^2}) = \frac{\partial}{\partial x}(\frac{x}{x^2+y^2})$ holds, which is equivalent to $d\Theta_1 = 0$ by (7.13), but no "potential function" f on $\mathbb{R}^2 - \{\mathbf{0}\}$ exists the gradient of which is the above vector field. It turns out, however, that the manner of this failure is itself quite systematic. In fact, the existence of potential is equivalent to the vanishing of the line integral of the corresponding 1-form (here, Θ_1) on closed curves. In the present case, the line integral turns out to be $2\pi k$, where the integer k counts the number of times that the closed curve winds around the origin in counter-clockwise direction. Another interpretation is that the "potential" is actually not a function in the usual sense, but the "multi-valued function" θ, the polar angle, which increases by 2π after each turn around the origin. This and other instances make this problem a highly interesting object of study. The "de Rham cohomology groups," which we are about to define, provide a framework for such an investigation.

Let M be a smooth manifold, with or without boundary. The set of closed p-forms in $\Omega^p M$ is an $\mathbb{R}$-linear subspace, which we denote by $Z^p M$. We let $B^p M$ be the linear subspace of exact forms in $Z^p M$. The quotient subspace $Z^p M / B^p M$ will be denoted by $H^p M$, and will be called the ***p*-th de Rham cohomology group** of M. $H^p M$ then measures the deviation of closed forms from being exact. Poincaré Lemma implies that if M is a contractible manifold, then $H^p M = \{0\}$ for $p \geq 1$. For $\alpha \in Z^p M$, we denote its equivalence class in the quotient $H^p M$ by $[\alpha]$. Now suppose $h: M \to N$ is a smooth map. It follows from the fundamental identity $h^* \circ d = d \circ h^*$ that

(7.17) $$h^*(Z^p N) \subset Z^p M$$

(7.18) $$h^*(B^p N) \subset B^p M)$$

Therefore, a linear map of the quotients is induced, which we also denote by h^p or h^*. The functorial property of $h^*: H^* N \to H^* M$ implies the following at the cohomology level.

(7.19) $$(\mathbb{1}_M)^* = \mathbb{1}_{H^p M}$$

(7.20) $$(f \circ g)^* = g^* \circ f^*$$

It follows from the above that if $f: M \to N$ is a diffeomorphism, then $f^*: H^p N \to H^p M$ is an isomorphism. Less obvious equalities arise in connection with the notion of "homotopy" that we shall now describe. For two smooth maps f and g from M

to N, we say f is ***(smoothly) homotopic*** to g, and write $f \simeq g$, if there is a smooth map $H{:}[0,1]\times M \to N$ so that $H(0,x)=f(x)$ and $H(1,x)=g(x)$, for all $x \in M$. Regarding each $\{t\}\times M$ as a copy of M, this says that f can be smoothly deformed to g. In this parlance, a contractible space is one for which the identity map is homotopic to a constant map.

23. Lemma *$\simeq$ is an equivalence relation for smooth maps $M \to N$.*

PROOF. By taking $H(t,x)=f(x)$ for all t and x, we have $f \simeq f$. If H provides a homotopy $f \simeq g$, then H' defined by $H'(t,x)=H(1-t,x)$ provides a homotopy $g \simeq f$. Now suppose H and K, respectively, establish the homotopy relations $f \simeq g$ and $g \simeq h$, then a first obvious attempt is to define $L{:}[0,1]\times M \to N$ by

$$L(t,x) = \begin{cases} H(2t,x) & \text{if } 0 \le t \le \frac{1}{2} \\ K(2t-1,x) & \text{if } \frac{1}{2} \le t \le 1 \end{cases}$$

This fails because the reparametrized H and K may not match smoothly at $t=\frac{1}{2}$. But the attempt can be salvaged by the use of a smoothing bump function as follows. Recalling the auxiliary functions constructed in the course of the proof of Lemma 12 of Chapter 2, there is a non-decreasing smooth function $\gamma{:}\mathbb{R}\to\mathbb{R}$ such that $\gamma(t)=0$ for $t \le 0$, $\gamma(t)=1$ for $t \ge 1$, and the derivatives of all orders of γ vanish at $t=0,1$. We then modify L as

$$L(t,x) = \begin{cases} H(\gamma(2t),x) & \text{if } 0 \le t \le \frac{1}{2} \\ K(\gamma(2t-1),x) & \text{if } \frac{1}{2} \le t \le 1 \end{cases}$$

This establishes $f \simeq h$. □

A smooth map $f{:}M \to N$ is called a ***(smooth) homotopy equivalence*** if there is a smooth map $g{:}N \to M$ so that $g \circ f \simeq \mathbb{1}_M$ and $f \circ g \simeq \mathbb{1}_N$. Two manifolds are ***(smoothly) homotopy equivalent*** if there is a homotopy equivalence between them. Homotopy equivalence is, in general, weaker than diffeomorphism, yet it also gives rise to an isomorphism of cohomology groups, as we shall soon see.

24. Examples

(a) A contractible manifold is homotopy equivalent to a single point $\{x_0\}$.

(b) Suppose $\pi{:}E \to M$ is a smooth vector bundle. The π is a smooth homotopy equivalence, since if $\zeta{:}x \mapsto \mathbf{0}_x$ is the zero-section, then $\pi \circ \zeta = \mathbb{1}_M$ and $\zeta \circ \pi = \pi$ is smoothly homotopic to $\mathbb{1}_E$ through the homotopy $H(t,x)=tx$.

(c) $\mathbb{R}^{m+1}-\{\mathbf{0}\}$ is homotopy equivalent to S_m. Let $j{:}S_m \to \mathbb{R}^{m+1}-\{\mathbf{0}\}$ be the inclusion map, and $\nu{:}\mathbb{R}^{m+1}-\{\mathbf{0}\} \to S_m$ be the map $x \mapsto \frac{x}{|x|}$. Then $\nu \circ j = \mathbb{1}_{S_m}$, and $j \circ \nu = \nu$ is homotopic to $\mathbb{1}_{\mathbb{R}^{m+1}-\{\mathbf{0}\}}$ through the homotopy $K(t,x)=(1-t)x+t\frac{x}{|x|}$.

(d) The m-sphere with two points removed, say $S_m-\{N,S\}$ with N and S the north and south poles $(0,\ldots,0,\pm 1)$, is homotopy equivalent to S_{m-1}. In fact, $S_m-\{N\}$ is diffeomorphic under stereographic projection to $\mathbb{R}^m$ (Example 2d and Example 7d, Chapter 4). So, further removing S, puts us into the situation of (c) above. It is useful to visualize S_{m-1} as the "equator" of S_m for $m \geq 2$. For $m=1$, S_0 is just a pair of points.

The following simple theorem establishes the basic relation of homotopy to cohomology.

25. Theorem *Let M and N be smooth manifolds.*
(a) If $f, g: M \to N$ are homotopic, then $f^* = g^*: H^p N \to H^p M$, for all p.
(b) If $h: M \to N$ is a homotopy equivalence, then $h^*: H^p N \to H^p M$ is an isomorphism, for all p.

Proof. (a) Let $H: [0,1] \times M \to N$ be a homotopy between f and g. Then $f = H \circ j_0$ and $g = H \circ j_1$ in the notation of Theorem 21, and by that theorem, $f^* - g^* = (j_0^* - j_1^*) \circ H^*$ at the level of Ω^p. For a cohomology class $[\alpha] \in H^p N$, we have $d\alpha = 0$, so

$$\begin{aligned} f^*(\alpha) - g^*(\alpha) &= (d \circ I + I \circ d)(H^* \alpha) \\ &= d(I(H^* \alpha)) + (I \circ H^*(d\alpha)) \\ &= d\beta \end{aligned}$$

where $\beta = I(H^* \alpha)$. So

$$\begin{aligned} f^*[\alpha] - g^*[\alpha] &= [f^*(\alpha)] - [g^*(\alpha)] \\ &= [g^*(\alpha) + d\beta] - [g^*(\alpha)] \\ &= 0 \end{aligned}$$

proving (a). For (b), suppose that $k: N \to M$ is so that $k \circ h \simeq \mathbb{1}_M$ and $h \circ k \simeq \mathbb{1}_N$. Then by (a) and 7.19

$$\begin{aligned} h^* \circ k^* &= (k \circ h)^* = \mathbb{1}_{H^p M} \\ k^* \circ h^* &= (h \circ k)^* = \mathbb{1}_{H^p N} \end{aligned}$$

proving (b). □

26. Examples
(a) Since a single point $\{x_0\}$ is a zero-dimensional manifold, then for $p \geq 1$, $\Omega^p\{x_0\} = \{0\}$ and $H^p\{x_0\} = \{0\}$. Hence it follows from the theorem above that for a contractible manifold M, $H^p M = \{0\}$, for all $p \geq 1$.

(b) We show that for any manifold M, $H^0 M$ is isomorphic to $\mathbb{R}^s$, where s is the number of connected components of M. The condition $f \in Z^0 M$, i.e., $df = 0$ is equivalent to f being constant on each connected component of M. On the other hand, $B^0 M = \{0\}$, therefore the claim follows.

(c) The proof of No-Extension Theorem given earlier can be cast in the framework of cohomology. For if ω is a volume element for a compact manifold $M = \partial W$, $d\omega = 0$ but

$\int_M \omega \neq 0$. On the other hand, ω cannot be exact, since otherwise, by Stokes Formula $\int_M \omega = 0$. This can be expressed as $H^m M \neq \{0\}$, where $m = \dim M$. In fact, we will see in the next section that for any compact, connected, orientable m-manifold M without boundary, $H^m M \cong \mathbb{R}$.

(d) We compute the first de Rham cohomology group of S_m. The result will be

$$H^1 S_m \cong \begin{cases} \mathbb{R} & \text{if } m = 1 \\ \{0\} & \text{if } m \neq 1 \end{cases} \tag{7.21}$$

First suppose $m>1$ and $\alpha \in Z^1 S_m$. As usual, let N and S be, respectively, the North Pole and the South Pole. We consider the restriction of α to $S^+ = S_m - \{S\}$ and to $S^- = S_m - \{N\}$, which we denote, respectively, by α^+ and α^-. Since each of S^+ and S^- is diffeomorphic to $\mathbb{R}^m$, a contractible space with $H^1 \mathbb{R}^m = \{0\}$, there are smooth real-valued functions f^+ and f^- on S^+ and S^-, respectively, so that $\alpha^+ = df^+$ and $\alpha^- = df^-$. The intersection $S^+ \cap S^-$ is connected (here we are using $m>1$), and $df^+ = df^-$ on this intersection, so there is a constant $c \in \mathbb{R}$ so that $f^+ = f^- + c$ on $S^+ \cap S^-$. Then $f: S_m \to \mathbb{R}$ given by

$$f(x) = \begin{cases} f^+(x) & \text{if } x \in S^+ \\ f^-(x) + c & \text{if } x \in S^- \end{cases}$$

is well-defined, smooth and $df = \alpha$. This shows that $H^1 S_m = \{0\}$ for $m>1$. For $m=1$, $Z^1 S_1 = \Omega^1 S_1$. Therefore, integration of 1-forms on S_1 will be a linear map $Z^1 S_1 \to \mathbb{R}$. This linear map is surjective since S_1 possesses volume elements and volume elements have non-zero integral. It therefore suffices to show that the kernel of the map is precisely $B^1 S_1$. By Stokes Formula, we have $\int_{S_1} df = 0$, hence $B^1 S_1$ is contained in the kernel. Now suppose $\int_{S_1} \alpha = 0$, we must show a smooth function $f: S_1 \to \mathbb{R}$ exists so that $df = \alpha$. Recall from Example 19d, Chapter 5, that there is a covering map $p: \mathbb{R} \to S_1$ defined by $p(t) = e^{2\pi i t}$. We will show that there is a smooth periodic map $F: \mathbb{R} \to \mathbb{R}$ of period 1 so that $dF = p^* \alpha$. Once this is proved, we finish the argument as follows. By the property of quotient manifolds (see Lemma b in the proof of Theorem 17, Chapter 5), F will induce a smooth function $f: S_1 \to \mathbb{R}$ so that $f \circ p = F$. Then

$$p^* \alpha = dF = d(f \circ p) = p^*(df)$$

Now since p is a local diffeomorphism, it follows that $\alpha = df$. Explicitly, for any tangent vector Y to S_1, there is a tangent vector X to $\mathbb{R}$ so that $Y = Tp(X)$. Therefore

$$\alpha(Y) = \alpha(Tp(X)) = p^* \alpha(X) = dF(X) = df(Tp(X)) = df(Y)$$

It remains to prove the existence of F as required. Now $p^* \alpha$ is a smooth 1-form on $\mathbb{R}$, so $p^* \alpha = u_\alpha(t) dt$ for some smooth function u_α on $\mathbb{R}$. We define $F(t) = \int_0^t u_\alpha$. Then F is smooth, and it is periodic of period 1 since $F(1) = \int_0^1 u_\alpha(t) dt = \int_{S_1} \alpha = 0$ by the hypothesis on α. This finishes the proof that $H^1 S_1 \cong \mathbb{R}$.

E. Top-dimensional Cohomology and Applications

It turns out that for a compact and connected m-dimensional manifold M without boundary, the top cohomology group, $H^m M$, is either isomorphic to $\mathbb{R}$ or is trivial, depending precisely on whether M is orientable or not. In the orientable case, the integration operator on $H^m M$ provides an isomorphism onto $\mathbb{R}$. This can be viewed as a converse to Stokes Theorem in the following sense: if $\omega \in \Omega^m M$ is such that the integral $\int_M \omega$ vanishes, then $\omega = d\varphi$ for some $\varphi \in \Omega^{m-1} M$. We will also discuss two applications in the orientable case. The first will be to the important topological notion of degree, and the second to the proof of a remarkable theorem of Moser on volume elements.

It is helpful and illuminating to discuss, on the way to proving the main result, a variant of de Rham cohomology, known as *(de Rham) cohomology with compact support*. Recall from Section B that for a smooth manifold M, the set of smooth p-forms of compact support, $\Omega_c^p M$, is a submodule of $\Omega^p M$, and that $d(\Omega_c^p M) \subset \Omega_c^{p+1} M$. We let $Z_c^p M = Z^p M \cap \Omega_c^p M$ and $B_c^p M = d(\Omega_c^{p-1} M)$. Therefore $B_c^p M$ is a subspace of $Z_c^p M$, and we can form the quotient vector space $H_c^p M$, known as the **p-th (de Rham) cohomology group with compact support** of M. If $f: M \to N$ is a smooth map of manifolds, then $f^*(Z_c^p N) \subset Z_c^p M$ and $f^*(B_c^p N) \subset B_c^p M$, so an induced map of the cohomology groups $f^*: H_c^p N \to H_c^p M$ is defined, which enjoys the same functorial properties as 7.18 and 7.19. We preface the main theorem by a useful lemma.

27. Lemma *Let M be a connected smooth m-manifold and $\mathcal{A} = \{(U_\alpha, \xi_\alpha) : \alpha \in I\}$ a subatlas of the maximal atlas of M for which each ξ_α is a diffeomorphism of U_α onto $\mathbb{R}^m$. Then for all $x, y \in M$, there is a finite sequence $U_1, \ldots, U_k$ of the chart domains in $\mathcal{A}$ with the following properties:*
(a) $x \in U_1$, $y \in U_k$, and $U_i \cap U_{i+1} \neq \emptyset$, for all $i = 1, \ldots, k-1$.
(b) The coordinate changes $\xi_{i+1} \circ \xi_i^{-1}$ are orientation-preserving in at least one connected component of each $U_i \cap U_{i+1}$, for all $i = 1, \ldots, k-1$.
Further, if M is not orientable, there is a sequence $U_1, \ldots, U_k$, satisfying the following:
(c) $U_i \cap U_{i+1} \neq \emptyset$, for all $i = 1, \ldots, k-1$, and $U_k \cap U_1 \neq \emptyset$.
(d) For $i < k$, the coordinate change $\xi_{i+1} \circ \xi_i^{-1}$ is orientation-preserving in at least one connected component of each $U_i \cap U_{i+1}$, but on $U_k \cap U_1$, the coordinate-change $\xi_1 \circ \xi_k^{-1}$ is orientation-reversing in at least one connected component.

Proof. For a given $x \in M$, we let S be the set of all $y \in M$ for which a finite chain as described in (a) and (b) exists. Certainly $x \in S$ with $k=2$, and S is open by its definition. We show that S is also closed. Since M is metrizable by Remark 4 of Chapter 4, it suffices to show that the limit y of a convergent sequence (y_n) of points in S is also in S. We take a chart $(U, \xi) \in \mathcal{A}$ around y, and choose N so that $y_N \in U$, and y_N is related to x by a chain $U_1, \ldots, U_k$, $y_N \in U_k$. Now $U_k \cap U \neq \emptyset$, and if

$\xi\circ\xi_k^{-1}$ is orientation-reversing in all components of $U_k\cap U$, then we simply compose ξ with an orientation-reversing diffeomorphism of $\mathbb{R}^m$, modifying (U,ξ) to make the coordinate-change orientation-preserving. Thus S is also closed, and it follows from the connectedness of M that $S=M$, proving the first assertion. Now suppose M is not orientable, yet no closed chain as specified in (c) and (d) exists. Fixing a point $x\in M$, we consider all chains of elements of $\mathcal{A}$ with initial point x satisfying (a) and (b) above. The U_α used in these chains cover M, since by the first part, any point y can be joined by a chain of the above form to x. We claim that this set $\mathcal{B}$ of (U_α,ξ_α) will constitute a positive atlas for M. For if not, we let (V,η) and (W,ζ) be two charts in $\mathcal{B}$ with $V\cap W\neq\emptyset$, and $\zeta\circ\eta^{-1}$ orientation-reversing in some component of $V\cap W$. Now if $\zeta\circ\eta^{-1}$ is orientation-preserving in some other component of $V\cap W$, we obtain a closed chain $U_1=V$, $U_2=W$ contradicting the non-existence of chains satisfying (c) and (d). Otherwise, $\zeta\circ\eta^{-1}$ is orientation-reversing in all of $V\cap W$. In this case, we join $y\in V\cap W$ to x, and x to y, in the manner of (a) and (b) to obtain a closed chain starting at $U_1=V$ and ending at $U_k=W$ that satisfies (c) and (d). The contradiction establishes the second assertion. □

28. Theorem *Let M be a connected and orientable m-manifold without boundary. Then $H_c^m M\cong\mathbb{R}$. If in addition M is compact, then $H^m M\cong\mathbb{R}$.*

Proof. We consider the linear map

$$\int_M : \Omega_c^m M \to \mathbb{R}$$

Since $m=\dim M$, $\Omega_c^m M=Z_c^m M$. For compact orientable manifolds, we have already seen that this linear map is surjective since the integral of a volume element is a non-zero real number. If M is non-compact and ω is a volume element, take a smooth bump function $f{:}M\to[0,1]$ with non-empty compact support. Then the integral of $f\omega$ is non-zero, which implies that integration on M is surjective onto $\mathbb{R}$. On the other hand, by Stokes Formula, the kernel contains $B_c^m M$. Therefore, our task is to prove that the kernel is exactly equal to $B_c^m M$, i.e., for $\alpha\in Z_c^m M=\Omega_c^m M$, if the integral of α over M is zero, then there exists $\varphi\in\Omega_c^{m-1}M$ with $\alpha=d\varphi$. We proceed to prove this in two steps. The first step is to prove the result for $\mathbb{R}^m$ and S_m by induction on m. Then we will generalize the result to other manifolds.

We have already proved $H^1 S_1\cong\mathbb{R}$ in Example 26d above. To prove $H_c^1\mathbb{R}\cong\mathbb{R}$, we take $\alpha\in\Omega_c^1\mathbb{R}$ with $\int_{\mathbb{R}}\alpha=0$. Then $\alpha(t)=u(t)dt$, for a smooth function u which vanishes outside an interval $[a,b]$. We define $f{:}\mathbb{R}\to\mathbb{R}$ by $f(t)=\int_a^t u$. Then $df(t)=\alpha(t)$, and f has compact support; in fact, it vanishes outside $[a,b]$. For if $x>b$, then $f(x)=\int_a^b u(t)dt$, which is zero by hypothesis.

Now suppose the theorem has been proved through dimension $k=m-1$, $m\geq 2$, for $\mathbb{R}^k$ and S_k. We prove the result first for $\mathbb{R}^m$. Given $\alpha=f\omega_0\in\Omega_c^m\mathbb{R}^m$ with $\int_{\mathbb{R}^m}\alpha=0$, we have $\alpha = d\eta$, where η is defined as in (7.14). But this η may not have compact support. We wish to find $\varphi \in \Omega_c^{m-1}\mathbb{R}^m$ so that $\alpha = d\varphi$. Let us suppose, multiplying by a constant, if necessary, that the support of α is contained in the open ball of radius one around the origin in $\mathbb{R}^m$. Recall formula (7.15) for η, where ν is the map $x \mapsto x/|x|$ from $\mathbb{R}^m - \{\mathbf{0}\}$ to S_{m-1}. Since the support of α, and hence the support of f, lies in the unit disk, we have

$$\Big(\int_0^{|x|} s^{m-1} f(s\nu(x))ds\Big) = \Big(\int_0^1 s^{m-1} f(s\nu(x))ds\Big), \text{ for } |x| \geq 1$$

i.e., the left-hand side remains constant on rays from the points of S_{m-1} outward. Therefore, it follows from (7.15), and the definition of Θ_m, that for $|x| \geq 1$

$$\begin{aligned} \eta(x) &= \nu^*\Big(\big(\int_0^1 s^{m-1} f(s\nu(x))ds\big)\omega_{S_{m-1}}\Big) \\ &= \nu^*(j^*\eta)(x) \end{aligned}$$

where j is the inclusion map of S_{m-1} in $\mathbb{R}^m$. Using Stokes Formula, we have $\int_{S_{m-1}} j^*\eta = \int_{\mathbb{D}_m} \alpha = 0$. By the induction hypothesis for the $(m-1)$-dimensional manifold S_{m-1}, there exists $\gamma \in \Omega^{p-2} S_{m-1}$ so that $d\gamma = j^*\eta$. We thus obtain

$$\eta(x) = (\nu^* \circ d\gamma)(x) = d(\nu^*\gamma)(x), \text{ for } |x| \geq 1$$

We now take a smooth function $\delta : \mathbb{R}^m \to \mathbb{R}$ that has constant value 1 for $|x| \geq 1$ and is 0 in a neighborhood of the origin. The $(m-1)$-form

$$\varphi = \eta - d(\delta\nu^*\gamma)$$

has its support in the unit disk, and $d\varphi = \alpha$, proving the claim for $\mathbb{R}^m$.

Having proved the proposition for $\mathbb{R}^m$, the next step is to prove it for an arbitrary connected and orientable m-manifold M without boundary (including S_m). We take an open subset V of M diffeomorphic to $\mathbb{R}^m$. By what was just proved, there is $\bar{\omega} \in \Omega_c^m M$ with the support of $\bar{\omega}$ contained in V so that $\int_M \bar{\omega} \neq 0$. Strictly speaking, we have $\bar{\omega} \in \Omega_c^m V$ and the integration is over V, but by extending $\bar{\omega}$ to be zero outside V, the claim is justified. Consider an arbitrary $\omega \in \Omega_c^m M$. We must show there is a real number a and $\zeta \in \Omega_c^{m-1} M$ so that $\omega = a\bar{\omega} + d\zeta$. First we assume that the support of ω is contained in an open subset U_1 of M that is diffeomorphic to $\mathbb{R}^m$. Since M is connected, there exists by Lemma 27 a finite sequence of open sets (U_i), $i = 1, \ldots, k$, $U_k = V$, so that each U_i is diffeomorphic to $\mathbb{R}^m$ and $U_i \cap U_{i+1} \neq \emptyset$, $i = 1, \ldots, k-1$ (see Figure 6). Since each $U_i \cap U_{i+1}$ is open and non-empty, it contains an open subset diffeomorphic to $\mathbb{R}^m$, hence there is $\omega_i \in \Omega_c^m M$ with support in $U_i \cap U_{i+1}$

so that $\int_M \omega_i \neq 0$. Then there are $a_i \in \mathbb{R}$ and $\zeta_i \in \Omega_c^{m-1} M$ so that

$$\begin{aligned} \omega &= a_1\omega_1 + d\zeta_1 \\ \omega_1 &= a_2\omega_2 + d\zeta_2 \\ &\vdots \\ \omega_{k-1} &= a_k\bar{\omega} \ + d\zeta_k \end{aligned}$$

By successive substitution, we obtain $\omega = a\bar{\omega} + d\zeta$, where $a = a_1 \ldots a_k$. Now consider

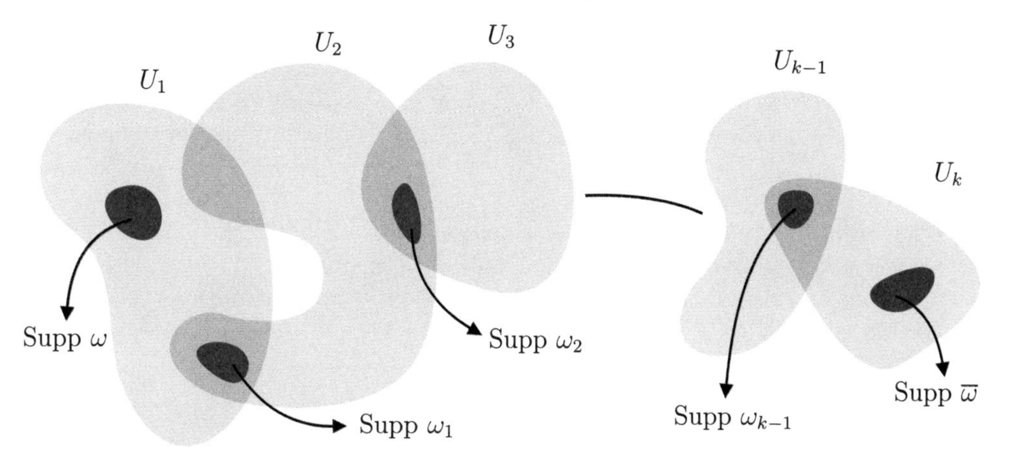

FIGURE 6. A chain

the general case, where the support of ω may not be contained in an open set diffeomorphic to $\mathbb{R}^m$. Taking a locally-finite open cover of M by sets diffeomorphic to $\mathbb{R}^m$, and a subordinate partition of unity (θ_i), it follows from the compactness of the support of ω that a finite number of these open sets cover the support of ω, and rearranging indices, if necessary, we may write

$$\omega = \theta_1\omega + \cdots + \theta_n\omega$$

Finally, writing each $\theta_i\omega$ in the above form, and adding up, the inductive proof is complete. □

29. Non-Orientable Case In the non-orientable case, we modify the argument in the last paragraph of the above proof, to prove that $H_c^m M = \{\mathbf{0}\}$. Let ω be an m-form with compact support on M. As in the above proof, we can use a smooth partition of unity to write ω as a finite sum of m-forms each with support in a chart domain diffeomorphic to $\mathbb{R}^m$. It then suffices to prove that for such m-forms ω, we have $\omega = d\eta$, where $\eta \in \Omega_c^{m-1} M$. Let $U_1, \ldots, U_k$ be a closed chain of open sets satisfying conditions (c) and (d) of Lemma 27, and suppose that the support of ω is contained in U_1. We adopt the orientation for U_1 which makes the integral of ω non-negative. By virtue of condition (d) of Lemma 27, we can pick m-forms ω_i, $i=1, \ldots, k$, with compact support in $U_i \cap U_{i+1}$, so that ω_i has positive integral on U_i, for $i \le k$. In the last step, ω_k will necessarily have negative integral with respect to the original orientation of U_1. We can then write as in the proof of the theorem above

$$\begin{aligned} \omega &= a_1\omega_1 + d\eta_1, && \text{where } a_1>0 \\ \omega_i &= a_{i+1}\omega_{i+1} + d\eta_{i+1}, && \text{where } a_{i+1}>0, \text{ for } i\le k \\ \omega_k &= a_{k+1}\omega + d\eta_{k+1}, && \text{where } a_{k+1}<0 \end{aligned}$$

Therefore, $\omega = a\omega + d\eta$, where $a = a_1 \cdots a_{k+1} < 0$. It follows that $1-a \neq 0$, $\omega = d((1-a)^{-1}\eta)$ and the claim is proved.

Theorem 28 allows us to introduce the important topological notion of "degree" for maps between compact and connected orientable manifolds of the same dimension.

30. Theorem *Let M and N be compact, connected and oriented smooth m-manifolds without boundary, and suppose that $f: M \to N$ is a smooth map. Then there is an integer deg(f), called the* **degree** *of f, with the following properties:*
(a) For every $\omega \in \Omega^m N$

$$\int_M f^*\omega = deg(f) \int_N \omega \tag{7.22}$$

(b) If $g: M \to N$ is (smoothly) homotopic to f, then deg(g)=deg(f).
(c) If $y \in N$ is a regular value for f, then

$$deg(f) = \sum_{x \in f^{-1}(y)} sgn(f, x) \tag{7.23}$$

where $sgn(f, x) = \pm 1$ depending on whether $T_x f$ is orientation-preserving or orientation-reversing.

By the celebrated Morse-Sard Theorem[3], if f is a C^1 function from M to N, manifolds of the same dimension, then almost all points of N, in the sense of smooth measure, are regular values for f, so the set of regular values is far from empty even if there are no regular points. It is understood in (7.23) that the sum on the right is zero if $f^{-1}(y)$ is empty.

Proof. Let L be a compact, connected and oriented smooth m-manifold without boundary. We established in Theorem 28 that integration of m-forms provides an isomorphism from $H^m L$ to $\mathbb{R}$. Let us denote this linear isomorphism by I_L. In the situation of the present theorem, if $f^*: H^m N \to H^m M$ is the induced linear map of cohomology vector spaces, then $I_M \circ f^* \circ I_N^{-1}$ is a linear map from $\mathbb{R}$ to $\mathbb{R}$, and is hence given as multiplication by a *real number*, which we denote by $deg(f)$. Thus

[3]see, e.g., **[13]** **[20]**.

for every $\omega \in \Omega^m N = Z^m N$

$$\int_M f^*\omega = I_M \circ f^* \circ I_N^{-1}(\int_N \omega) = deg(f) \times \int_N \omega$$

We can represent this with the following commutative diagram.

$$\begin{array}{ccc} H^*N & \xrightarrow{I_N} & \mathbb{R} \\ \Big\downarrow f^* & & \Big\downarrow \times deg(f) \\ H^*M & \xrightarrow{I_M} & \mathbb{R} \end{array}$$

This proves (a) except that we do not yet know that $deg(f)$ is an integer, a fact that will be established once (c) is proved. (b) holds since by Theorem 25(a), $g^*=f^*$. For (c), we make use of Example 19e of Chapter 5. Since M is compact, the set of R of regular values of f is open in N, in addition to being non-empty by Morse-Sard Theorem. We consider two cases: either $R \cap f(M) = Q$ is empty or not. In the case it is empty, we take $y \in R$, and $\omega \in \Omega^m N$ with support entirely contained in the complement of $f(M)$. Then $f^*\omega$ is zero and (7.23) holds as both sides are zero. If Q is not empty, then letting $P = f^{-1}Q$, we know from Example 19e of Chapter 5 that the restriction p of f to P defines a covering space $p: P \to Q$, and Q is open in N. We take $y \in Q$, and let V be an open evenly covered neighborhood of y in Q, diffeomorphic to $\mathbb{R}^m$. Then $f^{-1}(V)$ is the disjoint union of a finite number of open sets U_i, $i=1,\ldots,k$, each mapped diffeomorphically by p onto V. Let $\{x_1,\ldots,x_k\}$ be the inverse images of y with $x_i \in U_i$. Now consider $\omega \in \Omega^m N$ so that the support of ω is contained in V. The restriction of f to each U_i is a diffeomorphism, so we infer from Corollary 7 that

$$\int_{U_i} f^*\omega = \pm(\int_V \omega)$$

the sign depending on whether $f|_{U_i}$ is orientation-preserving or orientation-reversing. Since $\int_M f^*\omega = \sum_{i=1}^k \int_{U_i} f^*\omega$, this proves (7.23). □

31. Examples

(a) If $f: M \to N$ is not surjective, then $deg(f)=0$. One can take y in $N - f(M)$ as a regular value and apply (7.23). This was actually covered in the above proof.

(b) The functorial properties 7.19 imply that

$$deg(\mathbb{1}_M) = 1 \tag{7.24}$$

$$deg(f \circ g) = deg(g) \cdot deg(f) \tag{7.25}$$

(c) Let $p_n: S_1 \to S_1$ be the map $z \mapsto z^n$. We show $deg(p_n) = n$. Every point has n distinct geometric pre-images under p_n. Using local coordinates $t \mapsto e^{2\pi i t}$ for S^1, the map p_n is represented by $t \mapsto nt$, which is orientation-preserving, therefore the claim follows from (7.23). Likewise, the map $z \mapsto \bar{z}^n$ has degree $(-n)$.

(d) Consider the antipodal map $\delta_m{:}S_m \to S_m$ given by $x \mapsto -x$. We use the standard volume element of the unit sphere to calculate the degree:

$$\begin{aligned}\delta_m^*(\omega_{S_m}) &= \sum_{i=1}^{m+1}(-1)^{i-1}(-x^i)d(-x^1)\wedge\cdots\widehat{dx^i}\cdots\wedge d(-x^{m+1})\\ &= (-1)^{m+1}(\omega_{S_m})\end{aligned}$$

Therefore,

$$deg(\delta_m) = (-1)^{m+1} \tag{7.26}$$

As an application of the last example, we prove a famous theorem of Brouwer referred to earlier in text.

32. Theorem *Any smooth tangent vector field on a sphere of even dimension vanishes at some point.*

Proof. We show that the existence of a nowhere-vanishing vector field leads to the existence of a homotopy between δ_m and the identity mapping of S_m. In view of examples (b) and (d) above, this is not possible for even m. If $V(x){=}(x, v(x))$ is a non-vanishing tangent vector field defined at points x of S_m, then $U(x){=}(x, u(x))$ is a tangent vector field of unit length on S_m, where $u(x){=}\frac{v(x)}{|v(x)|}$, with $|.|$ denoting the Euclidean norm. We define $H{:}[0, 1]{\times}S_m{\to}S_m$ by

$$H(t, x) = (\cos\pi t)x + (\sin\pi t)u(x)$$

Then $|H(t, x)|{=}1$ since U being tangent to S_m implies that the Euclidean inner product $x{\cdot}u(x)$ is zero. Note that $H(0.x){=}x$ and $H(1, x){=} - x$; thus 1_{S_m} and δ_m are homotopic. □

Incidentally, Brouwer's Theorem holds for continuous tangent vector fields as well. Odd-dimensional spheres carry non-vanishing vector fields. One such tangent vector field on $S_{2n-1}{\subset}\mathbb{R}^{2n}$ is

$$X(x^1, \ldots, x^{2n}) = \sum_i x^{2i-1}\frac{\partial}{\partial x^{2i}} - \sum_i x^{2i}\frac{\partial}{\partial x^{2i-1}}$$

Next we turn to the statement and the proof of Moser's Theorem. This is sort of a converse to the fact that orientation-preserving diffeomorphisms preserve the integral. More precisely, Moser's Theorem states that if ω_0 and ω_1 are two volume elements for a compact, connected and orientable smooth manifold M without boundary so that $\int_M \omega_0{=}\int_M \omega_1$, then there is a diffeomorphism $\Phi{:}M{\to}M$ so that $\omega_0{=}\Phi^*\omega_1$. We need a couple of lemmas, the first a key lemma on time-dependent vector fields and forms. The reader should review Subsection 11 of Chapter 2 and Theorem 10 of Chapter 6.

Consider a time-dependent vector field $(X_t)_{t\in I}$, where I is an open interval in $\mathbb{R}$. This may be regarded as an ordinary vector field $\widetilde{X}$ on $I\times X$, the first component of which is the unit vector in $\mathbb{R}$ direction. Denoting the projection of $I\times X$ on the second component by π, the restriction π_t of π to $\{t\}\times X$ is a diffeomorphism with inverse $j_t{:}x\mapsto(t,x)$. We denote the push-forward $(j_t)_*X_t$ of X_t by $\widetilde{X}_t$. Thus $\widetilde{X}$ can be written as $\widetilde{X}(t,x)=\frac{\partial}{\partial t}(t,x)+\widetilde{X}_t(t,x)$. $\widetilde{X}$ has a local flow $\widetilde{\Phi}$. For $x\in M$, we define $\Phi_t(x)$ to be $\pi\circ\widetilde{\Phi}_t(0,x)$. Although this family Φ_t is (in general) not a flow, it was shown in Subsection 11 of Chapter 2 that Φ_t is a diffeomorphism of its domain (if non-empty) onto its image. Similar to a time-dependent vector field, a ***time-dependent p-form*** $(\alpha_t)_{t\in I}$ will be a smooth function $\widetilde{\alpha}{:}I\times M\to\Lambda^pM$ that assigns to each (t,x) an element of $\Lambda^p(T_xM)$. We write $\widetilde{\alpha}(t,x)=\alpha_t(x)$, and denote the restriction of $\widetilde{\alpha}$ to $\{t\}\times M$ by $\widetilde{\alpha}_t$. Thus $j_t^*\widetilde{\alpha}=\alpha_t$, and $\pi_t^*\alpha_t=\widetilde{\alpha}_t$.

33. Lemma *Let $(X_t)_{t\in I}$ and $(\alpha_t)_{t\in I}$ be, respectively, a time-dependent vector field and a time-dependent p-form on M. Then*

$$\frac{d}{dt}(\Phi_t^*\alpha_t)=\Phi_t^*(\frac{\partial\alpha_t}{\partial t}+L_{X_t}\alpha_t) \tag{7.27}$$

34. Lemma *Let M be a smooth orientable m-manifold and $\alpha\in\Omega^{m-1}M$. Then for any smooth family $(\omega_t)_{t\in I}$ of volume elements for M, there is a smooth family of vector fields $(X_t)_{t\in I}$ so that $i_{X_t}\omega_t=\alpha$.*

Assuming these lemmas for the moment, we state and prove the following.

35. Theorem (Moser) *Let M be a compact, connected, orientable smooth manifold without boundary, and ω_0, ω_1 be two volume elements for M so that $\int_M\omega_0=\int_M\omega_1$. Then there is a diffeomorphism $\phi{:}M\to M$ so that $\omega_0=\phi^*\omega_1$.*

PROOF. The m-forms ω_t defined by

$$\omega_t=(1-t)\omega_0+t\omega_1,\quad 0\le t\le 1$$

are everywhere non-zero and are hence volume elements. It follows from the equality of the integrals of ω_0 and ω_1 that for all t

$$\int_M\omega_t=\int_M\omega_0$$

Since $\int_M(\omega_0-\omega_1)=0$, there is, by Theorem 28, $\alpha\in\Omega^{m-1}M$, $m=\dim M$, so that $d\alpha=\omega_0-\omega_1$. We let $(X_t)_t$ be as in Lemma 34 and let $(\Phi_t)_t$ be the associated family of diffeomorphisms. Note here that by Theorem 10 of Chapter 6, compactness of M

implies that each Φ_t is defined for all $t\in\mathbb{R}$. Applying Lemma 33, we obtain

$$\begin{aligned}\frac{d}{dt}(\Phi_t^*\omega_t) &= \Phi_t^*(\omega_1 - \omega_0 + L_{X_t}\omega_t)\\ &= \Phi_t^*(-d\alpha + di_{X_t}\omega_t)\\ &= 0\end{aligned}$$

This shows that $\Phi_t^*\omega_t$ is constant and equal to $\Phi_0^*\omega_0=\omega_0$. In particular, $\Phi_1^*\omega_1=\omega_0$. □

We now turn to the proofs of the lemmas.

PROOF. (of Lemma 33) Using the notation introduced in the previous page for describing time-dependent vector fields and forms, we can write

$$\begin{aligned}\frac{d}{dt}(\Phi_t^*\alpha_t) &= \lim_{h\to 0}\frac{1}{h}\{\Phi_{t+h}^*\alpha_{t+h} - \Phi_t^*\alpha_t\}\\ &= \lim_{h\to 0}\frac{1}{h}\{(\pi_{t+h}\circ\widetilde{\Phi}_{t+h}\circ j_0)^*\alpha_{t+h} - (\pi_t\circ\widetilde{\Phi}_t\circ j_0)^*\alpha_t\}\\ &= \lim_{h\to 0}\frac{1}{h}\{(j_0^*\circ\widetilde{\Phi}_{t+h}^*)(\pi_{t+h}^*\alpha_{t+h}) - (j_0^*\circ\widetilde{\Phi}_t^*)(\pi_t^*\alpha_t)\}\\ &= (j_0^*\circ\widetilde{\Phi}_t^*)\Big(\lim_{h\to 0}\frac{1}{h}(\widetilde{\Phi}_h^*(\pi_{t+h}^*\alpha_{t+h}) - \pi_t^*\alpha_t)\Big)\\ &= (j_0^*\circ\widetilde{\Phi}_t^*)(\lim_{h\to 0}\frac{1}{h}(\widetilde{\Phi}_h^*\widetilde{\alpha} - \widetilde{\alpha}))\\ &= (j_0^*\circ\widetilde{\Phi}_t^*)(L_{\widetilde{X}}\widetilde{\alpha})\end{aligned}$$

For an arbitrary point $x\in M$, and $Y_1,\ldots,Y_p$ all in T_xM, we then obtain

$$\begin{aligned}\Big((\frac{d}{dt}(\Phi_t^*\alpha_t))(x)\Big)(Y_1,\ldots,Y_p) &= \Big((L_{\widetilde{X}}\widetilde{\alpha})(\widetilde{\Phi}_t(x))\Big)(T\widetilde{\Phi}_t(j_{0*}Y_1),\ldots,T\widetilde{\Phi}_t(j_{0*}Y_p))\\ &= \Big((L_{\frac{\partial}{\partial t}+\widetilde{X}_t}\widetilde{\alpha}_t)(\widetilde{\Phi}_t(x))\Big)(T\widetilde{\Phi}_t(j_{0*}Y_1),\ldots,T\widetilde{\Phi}_t(j_{0*}Y_p))\\ \text{(by (3.28) of Chapter 3)}\quad &= \Big((\frac{\partial\widetilde{\alpha}_t}{\partial t}+L_{\widetilde{X}_t}\widetilde{\alpha}_t)(\widetilde{\Phi}_t(x))\Big)(T\widetilde{\Phi}_t(j_{0*}Y_1),\ldots,T\widetilde{\Phi}_t(j_{0*}Y_p))\\ (\pi_t\text{ being a diffeomorphism})\quad &= \Big((\frac{\partial\alpha_t}{\partial t}+L_{X_t}\alpha_t)(\Phi_t(x))\Big)(T\Phi_t(Y_1),\ldots,T\Phi_t(Y_p))\\ &= \Big(\Phi_t^*(\frac{\partial\alpha_t}{\partial t}+L_{X_t}\alpha_t)(x)\Big)(Y_1,\ldots,Y_p)\end{aligned}$$

and the assertion is proved. □

PROOF. (of Lemma 34) We first prove the statement in $\mathbb{R}^m$. Let

$$\alpha(x)=\sum_{i=1}^{m} a_i(x)dx^1\wedge\cdots\widehat{dx^i}\cdots\wedge dx^m$$

and $\omega(t,x)=V(t,x)\omega_0$, where ω_0 is the standard volume element of $\mathbb{R}^m$, and $V(t,x)\neq 0$. We are seeking smooth vector fields $X(t,x)=\sum_{j=1}^m f^j(t,x)\frac{\partial}{\partial x^j}$ so that

$$\sum_{i=1}^m a_i(x)dx^1\wedge\cdots\widehat{dx^i}\cdots\wedge dx^m = i_{X(t,x)}\omega(t,x)$$

$$= \sum_{j=1}^m (-1)^{j-1} f^j(t,x)V(t,x)dx^1\wedge\cdots\widehat{dx^j}\cdots\wedge dx^m$$

Since $V(t,x)$ is assumed non-zero, we obtain the unique solution $f^j=(-1)^{j-1}(V)^{-1}a_j$. Given a manifold M, we carry out this construction on each chart. Uniqueness of solution ensures that the solutions on different charts will match to give a smooth solution on the manifold. □

EXERCISES

7.1 Let M be a C^r manifold with boundary, $r\geq 0$. The ***double*** of M, $D(M)$, is obtained by considering the disjoint union of two copies of M, namely $\{0, 1\}\times M$, and making the identification $(0, x)\sim(1, x)$ for points $x\in\partial M$. Show that $D(M)$ is a topological manifold without boundary, and there is a (topological) embedding of M into $D(M)$.

7.2 Show that the product $M\times N$ of two C^r manifolds with boundary, $r\geq 0$, is a topological manifold with boundary.

7.3 Let M and N be smooth manifolds of dimension m and n, respectively, one with boundary and the other without boundary. Show how $M\times N$ can be made into a smooth manifold of dimension $m+n$ with boundary, and identify the boundary. Suppose M is the manifold with non-empty boundary, and ω_M, ω_N are, respectively, volume elements for M and N. Consider the volume element $\pi_M^*\omega_M\wedge\pi_N^*\omega_N$ for $M\times N$, and $\pi_N^*\omega_N\wedge\pi_M^*\omega_M$ for $N\times M$. Is the natural diffeomorphism $(x, y)\mapsto(y, x)$ orientation-preserving? Compare the boundary orientations that $\partial(M\times N)$ and $\partial(N\times M)$ receive by convention.

7.4 Let M and N be smooth manifolds of dimension m and n, respectively, with $m>n$, M with non-empty boundary and N without boundary. Suppose $f{:}M\to N$ is a smooth map, and $y\in N$ is a regular value for both f and $f|_{\partial M}$. Show that $f^{-1}(y)$ is a manifold of dimension $m-n$ with boundary, and $\partial(f^{-1}(y))=f^{-1}(y)\cap\partial M$.

7.5 Let M be a smooth manifold with non-empty boundary. Show there is a smooth non-negative function $f{:}M\to\mathbb{R}$ for which 0 is a regular value and $f^{-1}(0)=\partial M$. (Hint: First do it locally at a point of the boundary.)

7.6 Let ρ be a smooth Riemannian metric for a vector bundle $p{:}E\to M$. Show that $D=\{v\in E{:}|v|_\rho\leq 1\}$ is a smooth manifold with boundary $S=\{v\in E{:}|v|_\rho=1\}$. For the tangent bundle of a Riemannian manifold, D and S are respectively known as the ***unit disk-*** and the ***unit sphere bundle*** of (M, ρ).

7.7 Let M be a compact smooth manifold with boundary and suppose that the boundary of M is the disjoint union of two manifolds without boundary, M_0 and M_1. Show that there is a smooth embedding h of M into $[0, 1]\times\mathbb{R}^N$, for appropriate N, so that $h(M_j)\subset\{j\}\times\mathbb{R}^N$, $j=0, 1$.

7.8 Consider the manifold $\mathbb{D}^m$ with boundary S_{m-1}, $m\geq 1$, and define the equivalence relation $\sim$ on $\mathbb{D}^m$ as follows: For $x\in S_{m-1}$, $x\sim y$ if and only if $y=\pm x$, otherwise $x\sim y$ if and only if $y=x$. Show that the quotient manifold exists and is diffeomorphic to $\mathbb{RP}(m)$. How do we generalize this to obtain $\mathbb{CP}(m)$?

7.9 Consider S_2 with the Riemannian metric induced from $\mathbb{R}^3$. Show that the unit sphere bundle of S_2 (see Exercise 7.6 above) is homeomorphic to $\mathbb{RP}(3)$.

7.10 Let M be a compact smooth manifold and suppose that a C^∞ diffeomorphism $f:M\to M$ has the properties: (i) f is an *involution*, i.e., $f\circ f=\mathbb{1}_M$ with $f\neq\mathbb{1}_M$, and (ii) f is fixed-point free, i.e., $f(x)\neq x$ for all $x\in M$. Show that there is a compact smooth manifold W with $\partial W=M$. (Hint: Consider $[0,1]\times M$ and pass one end to the quotient under the action of $\mathbb{Z}_2$.)

7.11 Let W be an open set in $\mathbb{R}^m$, Z a set of measure zero, $Z\subset W$, and $g:W\to\mathbb{R}^m$ a C^1 map. Then $g(Z)$ is a set of measure zero. (Hint: If C is a closed cube in W, use a bound for the derivative of the restriction of g to C to estimate the size of $g(C)$.)

7.12 Let M be a compact m-dimensional manifold with boundary in $\mathbb{R}^m$. The **centroid** $\bar{x}=(\bar{x}^1,\ldots,\bar{x}^m)$ of M is defined by

$$\bar{x}^i = \frac{\int_M x^i\omega_0}{\int_M \omega_0}$$

where ω_0 is the standard volume element of $\mathbb{R}^m$. For the upper half-disk $M=\{(x^1,\ldots,x^m) : x^m\geq 0, \sum_{i=1}^m (x^i)^2\leq 1\}$, study the behavior of $\bar{x}^m$ as the dimension m goes to ∞.

7.13 Let M be a smooth m-dimensional submanifold with boundary in the interior of the upper half-space $\mathbb{H}_m$. The **manifold of revolution** $\tilde{M}$ in $\mathbb{R}^{m+1}$ is the set

$$\{(x^1,\ldots,x^{m-1},x^m\cos\theta,x^m\sin\theta) : (x^1,\ldots,x^m)\in M, \theta\in\mathbb{R}\}$$

(See also Exercise 4.11). We use the standard metrics on $\mathbb{R}^m$ and $\mathbb{R}^{m+1}$ and the induced volume elements.
(a) Show that $\tilde{M}$ is a smooth submanifold with boundary of $\mathbb{R}^{m+1}$.
(b) Prove the **Pappus-Guldin Theorem**:

$$\mathrm{Vol}_{m+1}(\tilde{M}) = (2\pi\bar{x}^m)\mathrm{Vol}_m(M)$$

where $\bar{x}$ is the centroid of M (see the previous exercise).
(c) Compute the volume of the 2-torus $(r-a)^2+z^2=b^2$ in $\mathbb{R}^3$, where $a>b>0$ (see Example 2f(ii)).
(d) Use (b), induction and the centroid of the upper closed half-disk to derive formulas for the volumes of disks. (Strictly speaking, this does not fit the setting since the closed half-disk is not contained in the *interior* of $\mathbb{H}_m$, but the point of the original assumption was to ensure the smoothness of $\tilde{M}$, which is the case here.)

7.14 Let v_m and s_m be the m-dimensional volumes of the unit disk and the unit sphere as given in this chapter by (7.10) and (7.9). Show that both $(v_m)^{\frac{1}{m}}$ and $(s_m)^{\frac{1}{m}}$ tend to zero as the dimension m tends to ∞. (The same limit holds without the mth root, but we have rescaled to make the numbers in different dimensions meaningfully comparable.)

7.15 Provide another proof of (7.10) for the volume of the m-dimensional disk using Fubini's theorem and induction.

7.16 Provide another proof of (7.9) for the volume of the m-dimensional sphere by the use of spherical coordinates on S_m as follows:

$$(\theta_1, \ldots, \theta_m, \theta_{m+1}) \in [-\frac{\pi}{2}, \frac{\pi}{2}]^m \times [-\pi, \pi] \to S_m \subset \mathbb{R}^{m+1}$$

$$\begin{aligned} x^1 &= \sin\theta_1 \\ x^2 &= \cos\theta_1 \sin\theta_2 \\ &\vdots \\ x^m &= (\cos\theta_1)\cdots(\cos\theta_{m-1})\sin\theta_m \\ x^{m+1} &= (\cos\theta_1)\cdots(\cos\theta_{m-1})\cos\theta_m \end{aligned}$$

7.17 Let D be a compact smooth m-dimensional manifold with boundary in $\mathbb{R}^m$, and suppose that $f{:}D{\to}\mathbb{R}$ is a smooth function.
(a) Show that the graph of f, $\Gamma f{=}\{(x, f(x)){:}x{\in}D\}$ is an m-dimensional manifold with boundary $\Gamma(f|_{\partial D})$.
(b) Show that Γf is orientable, and find its Riemannian volume element induced from the standard inner product of $\mathbb{R}^{m+1}$.
(c) Show that the m-dimensional volume of Γf relative to the volume element in (b) is given by

$$\int_D \sqrt{1+\sum_{i=1}^{m}(\frac{\partial f}{\partial x^i})^2}\, dx^1\cdots dx^m$$

7.18 In Exercise 3.19 of Chapter 3, Poincaré metric on the upper-half plane and its isometries were introduced.
(a) Show that the arclength element and the Riemannian volume (=area) element for this metric are given by the following:

$$(y)^{-1}\sqrt{(\frac{dx}{dt})^2+(\frac{dy}{dt})^2}\, dt, \quad (y)^{-2}dx\wedge dy$$

(b) Consider the region between two vertical straight lines $x{=}a$, $x{=}b$ and two horizontal lines $y{=}c$ and $y{=}d$. Compute the area of this region and the length of its perimeter.
(c) Consider the circle with center on the y-axis and passing through the points $(0, R)$ and $(0, R^{-1})$, where $R{>}0$. Compute the area of the disk bounded by this circle and the circumference of the circle.
(d) Consider the unbounded region between two vertical straight lines $x{=}a$, $x{=}b$ and the semicircle in the upper-half plane with center on the x-axis and passing through $(a, 0)$ and $(b, 0)$ (see figure on the left). Show that the area of this region is π. (You will be dealing with an improper integral.)
(e) Consider the region bounded by three semi-circles in the upper-half plane that have centers on the x-axis and are mutually tangent (see figure on the right). Show

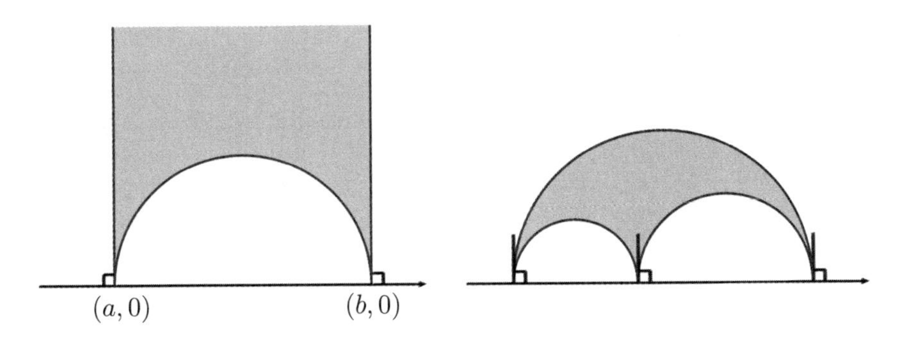

FIGURE 7

that the value of this area is also π. (Hint: Use the isometries given in Exercise 3.19 to reduce the problem to the previous part.)

7.19 Let E be a k-dimensional affine subspace of $\mathbb{R}^m$ and D a compact smooth k-dimensional submanifold with boundary of E. For each multi-index $i{=}(i_1, \ldots, i_k)$, where $i_1{<}\cdots{<}i_k$, let D_i be the orthogonal projection of D on the linear subspace $(x^{i_1}, \ldots, x^{i_k})$. Prove the following "Pythagorean theorem" for k-dimensional volumes:

$$\mathrm{Vol}(D)^2 = \sum_i \mathrm{Vol}(D_i)^2$$

(The hard part is already covered in Exercise 1.14 of Chapter 1.)

7.20 (Helpful, but not necessary, to be familiar with normal bundles in Exercise 6.8 of Chapter 6, or with *Frenet frames* for curves in $\mathbb{R}^m$, e.g., in [**17**].) Let C be a smooth simple closed curve in $\mathbb{R}^m$, i.e., the image of the unit circle under a smooth embedding into $\mathbb{R}^m$. For each $x{\in}C$, let $D_\rho(x)$ be the closed disk of radius ρ with center x perpendicular to C, and let T be the union of these disks. For $\rho{>}0$ sufficiently small, show that T is a smooth submanifold with boundary in $\mathbb{R}^m$, and that the volume of T is $\pi\rho^2 L$, where L is the length of C. (All metrics and volumes are the standard ones in $\mathbb{R}^m$.)

7.21 Consider the 1-form $\alpha{=}dx{+}xdy$ on the region $\{(x, y){\in}\mathbb{H}_2 : x^2{+}y^2{<}1\}$ with boundary $]-1, 1[\times\{0\}$. We have

$$\int_{\partial D} j^*\alpha = 2, \quad \int_D d\alpha = \frac{\pi}{2}$$

Why is this not a contradiction to Stokes Formula?

7.22 Let

$$\alpha = \sum_{i=1}^{m} x^i dx^1 \wedge \cdots \widehat{dx^i} \cdots \wedge dx^m$$

Compute $\int_{S_{m-1}} j^*\alpha$, where j is the inclusion map of S_{m-1} in $\mathbb{R}^m$.

7.23 Let M be a compact and orientable smooth m-manifold without boundary, $\omega\in\Omega^m M$, and X a smooth tangent vector field on M. Show that $\int_M L_X\omega=0$.

7.24 Let M be a compact and orientable smooth m-manifold with boundary ∂M, and suppose that ρ is a smooth Riemannian metric on M. The gradient operator is considered with respect to ρ and the divergence operator with respect to the Riemannian volume element ω_ρ. For a smooth function $f:M\to\mathbb{R}$, the ***Laplacian*** of f, $\Delta_\rho f$, is defined by

$$\Delta_\rho f = \mathbf{div}_{\omega_\rho}(\nabla_\rho f)$$

Let (U,ξ) be a chart and suppose that the Riemannian metric ρ is represented on the chart as $\sum_{ij} g_{ij} d\xi^i \otimes d\xi^j$. Denote the determinant of the matrix $[g_{ij}]$ by g, and the inverse of this matrix by $[g^{ij}]$. Then following (3.36) of Chapter 3, show that the local expression for the Laplacian is the following:

$$\Delta_\rho f = \sum_{i,j=1}^{m} g^{ij}\frac{\partial^2 f}{\partial\xi^i\partial\xi^j} + \sum_{i,j=1}^{m}\left(\frac{\partial g^{ij}}{\partial\xi^i} + \frac{1}{2}g^{ij}\frac{\partial(\log g)}{\partial\xi^i}\right)\frac{\partial f}{\partial\xi^j}$$

7.25 (Continuation of Exercise 7.24) For the smooth function $f:M\to\mathbb{R}$, the ***normal derivative*** $\frac{\partial f}{\partial n}$ of f at the boundary is defined as $\rho(\nabla_\rho f,\mathbf{n})$, where $\mathbf{n}$ is the unit outward normal at the boundary.
(a) Establish Green's identity

$$\int_M (f\Delta_\rho g - g\Delta_\rho f)\omega_\rho = \int_{\partial M}\left(f\frac{\partial g}{\partial n} - g\frac{\partial f}{\partial n}\right)\omega'_\rho$$

where ω'_ρ is the Riemannian volume element of the boundary.
(b) A function $f:M\to\mathbb{R}$ is ρ-***harmonic*** if $\Delta_\rho f$ is identically zero. Show that if a ρ-harmonic function on M vanishes on the entire boundary, or if its normal derivative vanishes on the entire boundary, then the function is identically zero.

7.26 (Continuation of Exercise 7.25) $f:M\to\mathbb{R}$ is a smooth function as before.
(a) Prove that

$$\Delta_\rho(f^2) = 2f(\Delta_\rho f) + 2|\nabla_\rho f|_\rho^2$$

(b) If M is a compact, connected and oriented manifold without boundary, prove that the only ρ-harmonic functions on M are constant functions. (Hint: Use (a) above and the Divergence Theorem.)

7.27 Let $f:\mathbb{R}^n\to\mathbb{R}$ be harmonic and suppose that D is a compact smooth n-dimensional submanifold with boundary in $\mathbb{R}^n$. Show that there is a point x on ∂D where $\nabla f(x)$ is tangent to ∂D.

7.28 Let M and N be two smooth manifolds. Show that if $H^pM\neq\{\mathbf{0}\}$, then $H^p(M\times N)\neq\{\mathbf{0}\}$.

7.29 Show that for $0\leq p\leq m$, $H^p\mathbb{T}^m$ is non-zero.

7.30 Let M be a compact smooth manifold, and suppose that $\alpha \in Z^1 M$ does not vanish anywhere on M. Show that $H^1 M \neq \{\mathbf{0}\}$. Deduce that if a compact manifold M without boundary admits a submersion onto S_1, then $H^1 M \neq \{\mathbf{0}\}$.

7.31 The aim of this exercise is to finish the computation of the cohomology groups of spheres that was started in Example 26d. The remaining cases are those of the form $H^k S_m$, where $1<k<m$. We are asking the reader to show that $H^k S_m = \{\mathbf{0}\}$ in all such cases by induction. Suppose that the claim has been proved for all pairs (l, n), where $l<k$ and $n<m$. As in Example 26d, we use the notation $S^+ = S_m - \{S\}$, $S^- = S_m - \{N\}$, and $S^0 = S^+ \cap S^-$, where S and N are, respectively, the south and north poles. Let $\alpha \in Z^k S_m$, and denote the restriction of α to $S^\pm$ by $\alpha^\pm$.
(a) Show that there exist $\beta^\pm \in \Omega^{k-1} S^\pm$ so that $\alpha^\pm = d\beta^\pm$. Letting $\beta^0 = \beta^+ - \beta^-$ on S^0, show that there exists $\gamma \in \Omega^{p-2} S^0$ with $d\gamma = \beta^0$.
(b) Consider smooth functions $\theta^\pm$ from S_m to $[0, 1]$ with support in $S^\pm$, respectively, so that $\theta^+ + \theta^- = 1$, and θ^+ (respectively, θ^-) equals 1 in a neighborhood of N (respectively, S). Define

$$\beta = \theta^+ \beta^+ + \theta^- \beta^- - d\theta^+ \wedge \gamma$$

Prove that $d\beta = \alpha$.

7.32 Let M be a smooth manifold, G a finite group and $\Phi: G \times M \to M$ an action by diffeomorphisms. Suppose that $\alpha \in B^p M$ is invariant under the action, i.e., for all $g \in G$, $\Phi_g^* \alpha = \alpha$. Show that there is $\beta \in \Omega^{p-1} M$, invariant under the action, so that $d\beta = \alpha$. Use this fact to compute all de Rham cohomology groups of real projective spaces. The result is

$$H^p(\mathbb{R}P(m)) \cong \begin{cases} \mathbb{R} & \text{if } p=0 \text{ or } p=m=2k+1 \\ \{\mathbf{0}\} & \text{otherwise} \end{cases}$$

7.33 (a) Suppose M has k compact connected components and l non-compact connected components. Compute $H_c^0(M)$.
(b) If M is smoothly contractible, show that $H_c^1(M) \cong \{0\}$.

7.34 For smooth manifolds M and N show that

$$H^1(M \times N) \cong H^1 M \oplus H^1 N$$

(For the general formula on $H^p(M \times N)$, known as Künneth Formula, see [**5**].)

7.35 Let M be a non-compact smooth m-manifold, orientable or not. Show that $H^m M = \{\mathbf{0}\}$. (Hint: Imitating the proofs of the last step of Theorem 28 or Subsection 29, take a countable locally-finite cover (U_i) of M by open sets diffeomorphic to $\mathbb{R}^m$ so that $U_i \cap U_{i+1} \neq \emptyset$, write $\omega \in \Omega^m M$ as a sum of m-forms with compact support, and push the support of each of these out of any compact set.)

7.36 Let X be a smooth tangent vector field defined on an open subset of an m-manifold M. Suppose that $X(a) = \mathbf{0}$ for some a in the domain of X, and that X does not vanish at any other point in a neighborhood of a. Taking a chart (U, ξ)

around a, we define a map $f_\epsilon: S_{m-1} \to S_{m-1}$ by

$$x \mapsto \frac{\xi_* X(\xi(a)+\epsilon x)}{|\xi_* X(\xi(a)+\epsilon x)|}$$

where $\epsilon>0$ is taken small enough so that $\xi_* X(\xi(a)+\epsilon x)$ is defined and non-zero for $0<|x|\leq 1$. The degree of f_ϵ is called the ***index of** X **at** a* and is denoted by $\mathrm{Ind}(X; a)$.
(a) Show that for $0<\epsilon'<\epsilon$, $deg(f_{\epsilon'})=deg(f_\epsilon)$.
(b) Show that the index is independent of the chart.
(c) Each of the following vector fields on $\mathbb{R}^2$ has an isolated zero at the origin. Compute the index in each case and display the integral curves around the origin.

$$x\frac{\partial}{\partial x}+y\frac{\partial}{\partial y}, \quad -x\frac{\partial}{\partial x}-y\frac{\partial}{\partial y}, \quad x\frac{\partial}{\partial x}-y\frac{\partial}{\partial y}, \quad -y\frac{\partial}{\partial x}+x\frac{\partial}{\partial y}$$

(d) Do the same as in (c) for the vector field below on $\mathbb{R}^2$, where $z=x+iy$ and $n>0$.

$$Re(z^n)\frac{\partial}{\partial x} + Im(z^n)\frac{\partial}{\partial y}$$

7.37 (Based on the previous exercise)
(a) The vector field $X = -\sum_{i=1}^n x^i \frac{\partial}{\partial x^i}$ on $\mathbb{R}^n$ has an isolated zero at $\mathbf{0}$. Show that

$$\mathrm{Ind}(X; \mathbf{0}) = (-1)^n$$

(b) Let $\mathbb{D}^n$ be the closed unit disk in $\mathbb{R}^n$ and suppose that the vector field X has only a finite number of zeros, say $a_1, \ldots, a_k$ in $\mathbb{D}^n$, all in the interior, and that X is inward-pointing on the boundary S_{n-1} of $\mathbb{D}^n$. Show that

$$\sum_{i=1}^{k} \mathrm{Ind}(X; a_i) = (-1)^n$$

7.38 Let α be a smooth nowhere vanishing integrable 1-form on a manifold M. By Corollary 30 of Chapter 6, there is a smooth 1-form η so that $d\alpha=\alpha\wedge\eta$.
(a) Show that $\alpha\wedge d\eta=0$, and $d\eta=\alpha\wedge\gamma$ for some 1-form γ.
(b) Show that $\eta\wedge d\eta$ is closed, hence a cohomology class $[\eta\wedge d\eta]\in H^3 M$ is defined.
(c) Show that the cohomology class $[\eta\wedge d\eta]$ does not depend on the particular η satisfying $d\alpha=\alpha\wedge\eta$.
(d) If f is a nowhere vanishing smooth function on M, prove that the cohomology class defined by $f\alpha$ as above is the same as $[\eta\wedge d\eta]$.

Thus by (d) and (c), the cohomology class above depends only on the foliation $\mathcal{F}$ defined by α. It is known as the ***Godbillon-Vey invariant*** of $\mathcal{F}$ and denoted by $gv(\mathcal{F})$. One can show that this invariant is zero for the Reeb foliation of S_3, but it has been shown by Thurston that for every $c\in H^3(S_3)\cong\mathbb{R}$, there is a foliation $\mathcal{F}$ of S_3 with $gv(\mathcal{F})=c$.

7.39 In this exercise, M and N are smooth manifolds, with or without boundary.
(a) Let $\alpha\in\Omega^p M$ and $\beta\in\Omega^q M$. Show that if one of α or β is exact and the other is closed, then $\alpha\wedge\beta$ is exact.

(b) We define the ***cup product*** $\smile : H^pM \times H^qM \to H^{p+q}M$ by

$$[\alpha]\smile[\beta] = [\alpha \wedge \beta]$$

Show that this is well-defined and $[\alpha]\smile[\beta]=(-1)^{pq}[\beta]\smile[\alpha]$.
(c) For a smooth function $f{:}M\to N$, show that $f^*[\alpha\smile\beta]=f^*[\alpha]\smile f^*[\beta]$.
(d) For $\alpha\in Z^pM$ and $\beta\in Z^qN$, define

$$[\alpha] \times [\beta] = [\pi_M^*\alpha \wedge \pi_N^*\beta]$$

Show that $\times$ is well-defined. If $\Delta{:}M\to M\times M$ is the diagonal map $x\mapsto(x, x)$, show that $[\alpha]\smile[\beta]=\Delta^*([\alpha] \times [\beta])$.

7.40 Let $m<n$ and M be a smooth m-manifold. Show that any smooth map $f{:}M\to S_n$ is smoothly homotopic to a constant map.

7.41 Let M and N be orientable smooth m-manifolds, $M=\partial W$, and suppose that ω is a volume element for N. Show that if a smooth map $f{:}M\to N$ has a smooth extension $W\to N$, then $\int_M f^*\omega=0$.

7.42 For $m>1$, show that any smooth map $S_m\to\mathbb{T}^m$ has degree zero.

7.43 Let $m, n\geq1$. Show that any smooth map $S_{m+n}\to S_m\times S_n$ has degree zero. (Hint: You may use the result of Exercise 7.31.)

7.44 Show that if $f{:}S_m\to S_m$ has no fixed points, then f is smoothly homotopic to the antipodal map of S_m.

7.45 Show that every smooth map $S_{2m}\to S_{2m}$ which is not of zero degree, sends a pair of antipodal points to a pair of antipodal points.

7.46 Consider the polynomial with complex coefficients $p(z)=z^n+c_{n-1}z^{n-1}+\cdots+c_0$. This defines a smooth map of S_2 to itself (see Exercise 5.27 of Chapter 5). Show that this map has degree n. Deduce the Fundamental Theorem of Algebra: If $p(z)$ is a complex polynomial of degree $n\geq1$, then $p(z)=0$ has roots in $\mathbb{C}$.

7.47 Use Lemma 33 and the idea of the proof of Moser's Theorem to give a proof of Poincaré Lemma for star-like regions in $\mathbb{R}^m$.

7.48 Let M and N be compact, connected, orientable smooth submanifolds without boundary of dimension m and n, respectively, in $\mathbb{R}^{m+n+1}$, so that $M\cap N\neq\emptyset$. Consider $\lambda{:}M\times N\to S_{m+n}$ by

$$\lambda(x, y) = \frac{x-y}{|x-y|}$$

Define the ***linking number*** $l(M, N)$ as the degree of λ. Prove the following properties.
(a) $l(N, M)=(-1)^{(m+1)(n+1)}l(M, N)$.
(b) If $M=\partial W$, and $W\cap N=\emptyset$, then $l(M, N)=0$.

(c) For m=n=1, with M and N non-intersecting smooth simple closed curves in $\mathbb{R}^3$, Gauss formula holds:

$$l(M,N) = \frac{1}{4\pi}\int_{M\times N} \frac{\mathbf{r}_M - \mathbf{r}_N}{|\mathbf{r}_M - \mathbf{r}_N|^3} \cdot (d\mathbf{r}_M \times d\mathbf{r}_N)$$

where $\mathbf{r}_M$=(x_M, y_M, z_M) and $\mathbf{r}_N$=(x_N, y_N, z_N) are the "position vectors" describing the coordinates of the points on M and N, and "·" and "×" are the usual dot product and cross product in $\mathbb{R}^3$.
(d) Let M be the (p,q) torus knot described in Example 2 f,g of Chapter 4, with parameters a=1, b=$\frac{1}{2}$ and orientation as indicated there. Suppose also that N is the unit circle in the xy-plane with counter-clockwise orientation. What is $l(M,N)$ in this case?
(e) Revisit Exercise 5.20 of Chapter 5 by computing the linking number of two circles in S_3 obtained as inverse images under Hopf fibration $S_3 \to S_2$.

7.49 For $m \geq 2$, consider a smooth map $f{:}S_{2m-1} \to S_m$, and let ω be a volume element of S_m with the standard orientation so that $\int_{S_m} \omega$=1. We know from Exercise 7.31 above that there exists $\eta \in \Omega^{m-1} S_{2m-1}$ so that $d\eta$=$f^*\omega$. Define the ***Hopf invariant***, $H(f)$ of f by

$$H(f) = \int_{S_{2m-1}} \eta \wedge d\eta$$

(a) Show that $H(f)$ is independent of the particular choices of ω and η above, with the required properties.
(b) Show that if f and g are homotopic, then $H(f)$=$H(g)$.
(c) For odd m, show that $H(f)$=0.
(d) For smooth $\phi{:}S_{2m-1} \to S_{2m-1}$, $H(f \circ \phi)$=$deg(\phi)H(f)$.
(e) For smooth $\psi{:}S_m \to S_m$, $H(\psi \circ f)$=$(deg(\psi))^2 H(f)$.

For the computation of $H(h)$, where $h{:}S_3 \to S_2$ is the Hopf fibration described in Example 18d of Chapter 5, see [**10**], pp.224-225. An alternative approach, defining the Hopf invariant as a linking number, can be found in Exercise 15 of [**20**], where it is immediate from the definition that H(f) is an integer. For a purely algebraic approach that works for continuous maps, see [**12**].

CHAPTER 8

Lie Groups and Homogeneous Spaces

The modest aim of this chapter is to expand the repertoire of manifolds at the reader's disposal to include the many examples that arise as Lie groups or Lie-group quotients known as homogeneous spaces. Ideally one would like to study this material in parallel with the previous four chapters, but the one-dimensional book format does not allow "links" or substantial detours. In the lecture/class environment, however, one can imagine various rearrangement scenarios that could serve this purpose. As standing hypothesis in this chapter, the word "manifold" always refers to a manifold without boundary.

A. Continuous Groups

Let G be a group and a topological space. We look at the group operation as a map $\mu{:}G\times G\to G$, $\mu(g,h)=gh$, and the inverse operation $g\mapsto g^{-1}$ as a map $\iota{:}G\to G$. One can hypothesize various degrees of continuity or smoothness for these operations to relate the algebraic and topological structures. If μ and ι are continuous, using the product topology for $G\times G$, we say that G is a ***topological group***. In case G is, in addition, a smooth manifold, and μ and ι are smooth, G is called a ***Lie group***. The ***dimension*** of a Lie group is the dimension of the underlying manifold. The continuity or smoothness of the pair of maps μ and ι can of course be subsumed under the continuity or smoothness of the single map $\nu{:}G\times G\to G$ by $\nu(x,y)=x^{-1}y$. It turns out that a Lie group always admits a compatible real-analytic structure relative to which the two operations are analytic. More striking is the fact that under the mere assumption that G be a topological manifold, continuity of μ and ι imply that G has a compatible analytic manifold structure, and that the two operations are analytic maps! This is the common interpretation of Hilbert's Fifth Problem that was successfully solved in 1950s.

Let G be a topological group or a Lie group. For $a\in G$, ***left translation*** by a is the bijective map $L_a{:}G\to G$ given by $x\mapsto ax$. This has as its inverse the left translation $L_{a^{-1}}$. Similarly, ***right translation*** by a, R_a, maps x to xa and has $R_{a^{-1}}$ as its inverse. Since L_a and R_a are obtained by restricting the group operation μ, they are endowed with the same degree of smoothness as μ. Thus they are homeomorphisms if G is a topological group and smooth diffeomorphisms in case G is a Lie group. Likewise, the map ι is its own inverse, hence it is a homeomorphism (respectively, a smooth diffeomorphism) if G is a topological group (respectively, a Lie group). The following notation is adopted. For subsets S and T of G,

$$ST = \{st : s\in S, t\in T\}, \quad S^{-1} = \{s^{-1} : s\in S\}$$

For $a\in G$, it is common to write aS instead of $\{a\}S$ (similarly for the right-translate). A subset S of G will be called ***symmetric*** if $S^{-1}=S$. We denote the neutral element of the group G by e except when explicitly denoted otherwise.

1. Elementary Properties *Let G and G' be topological groups.*
(a) For every neighborhood U of e, there is an open symmetric neighborhood V of e with the property $VV\subset U$.
(b) Every open subgroup of G is also closed.
(c) The connected component of e is a closed normal subgroup. Further, if G is a Lie group, then the connected component of e is also open.
(d) Every neighborhood of e generates the connected component of e in G, i.e., for every neighborhood U of e, every element of the connected component of e in G can be written as a finite product $g_1\cdots g_k$, where $g_i\in U$.
(e) If $\varphi:G\to G'$ is a group homomorphism, then φ is continuous if and only if φ is continuous at a single point. In the case G and G' are Lie groups, the group homomorphism $\varphi:G\to G'$ is smooth if and only if it is smooth at a single point.
(f) If G and G' are Lie groups and $\varphi:G\to G'$ is a smooth homomorphism, then the tangent map of φ has constant rank.

Proof. Define $\nu:G\times G\to G$ by $\nu(x,y)=x^{-1}y$. It follows from the continuity of ν at (e,e) that there are open neighborhoods V_1 and V_2 of e so that $V_1^{-1}V_2\subset U$. Then $V=V_1\cap V_2\cap V_1^{-1}\cap V_2^{-1}$ is a symmetric neighborhood satisfying (a).

Suppose the subgroup H of G is open, then each left coset of H is open because it is obtained from H by a left translation. Therefore the union of left cosets distinct from H is also open, implying that its complement, namely H, is closed, proving (b).

Let G_0 be the connected component of e in G. In a topological space, every connected component is closed, hence G_0 is closed. Moreover, for a locally-connected space, e.g., a manifold, connected components are also open. Therefore, if G is a Lie group, then the connected components are also open. To prove that G_0 is a subgroup, we first claim that

$$G_0(G_0)^{-1} = G_0 \tag{8.1}$$

Note that for every $a\in G_0$, $a(G_0^{-1})=L_a(G_0^{-1})$, hence $a(G_0^{-1})$ is connected. Moreover $e=aa^{-1}\in a(G_0^{-1})$, therefore $a(G_0^{-1})\subset G_0$. It follows that $G_0(G_0^{-1})=\bigcup_{a\in G_0} a(G_0^{-1})\subset G_0$. The inverse inclusion also holds, since for every $a\in G_0$, we have $a=ae^{-1}$. Now (8.1) implies that G_0 is a subgroup. For if $a\in G_0$, then $a^{-1}=ea^{-1}\in G_0$, and if $a,b\in G$, then $ab=a((b^{-1})^{-1})\in G_0$. But for every $x\in G$, $xG_0x^{-1}=(L_x\circ R_{x^{-1}})(G_0)$, so xG_0x^{-1} is connected. Further, this set intersects G_0 since e belongs to both, so $xG_0x^{-1}\subset G_0$, and G_0 is normal.

To prove (d), it suffices to show that an open neighborhood V as in (a) generates the connected component G_0 of e in G. Let V_k be the set of products $g_1\cdots g_k$, where $g_i \in V$. We claim that V_k is open. For $k=1$, $V_1=V$ was assumed open. If the proposition is true for V_k, then

$$V_{k+1} = V\cdot V_k = \bigcup_{g\in V} g\cdot V_k = \bigcup_{g\in V} L_g(V_k)$$

and the union of open sets is open. Therefore, $W=\bigcup_k V_k$ is an open subset of G_0. Moreover, W is a subgroup since $V=V^{-1}$ and W is closed under products. By (*b*), every open subgroup is also closed, so it follows from the connectedness of G_0 that $W=G_0$.

For (*e*), if g as an arbitrary element of G, the homomorphism property implies that

$$\varphi = L_{\varphi(g)} \circ \varphi \circ L_{g^{-1}} \tag{8.2}$$

If x and y are two elements of G, let $g=yx^{-1}$. Applying (8.2), we see that continuity (respectively, smoothness) at x implies the same property at y. Finally, (*f*) also follows from (8.2) since $L_{g^{-1}}$ and $L_{\varphi(g)}$ are diffeomorphisms. □

A group homomorphism is a called a ***topological group homomorphism*** (respectively, a ***Lie group homomorphism***) if it is continuous (respectively, smooth).

2. Examples

(a) Each of the following is a topological group with its standard topology: $(\mathbb{Z},+)$, $(\mathbb{Q},+)$, $(\mathbb{R},+)$, $(\mathbb{R}^m,+)$ and $(V,+)$, where V is a finite-dimensional vector space over $\mathbb{R}$. All except $(\mathbb{Q},+)$ are also Lie groups, but the usual topology of $\mathbb{Q}$ does not allow a manifold structure. In the case of $(V,+)$, recall that any two finite-dimensional vector spaces of the same dimension over $\mathbb{R}$ are isomorphic, so they have isomorphic group structures under addition. Further, any two norms on a finite-dimensional real vector space are equivalent, hence they induce the same topology. It follows that if $\dim V=m$, then as a Lie group, $(V,+)$ is indistinguishable from $\mathbb{R}^m$ under addition. As an example, $(\mathbb{C}^m,+)$ is canonically isomorphic, as a Lie group, with $\mathbb{R}^{2m}$ under the correspondence

$$(z^1,\ldots,z^m) \longleftrightarrow (x^1,y^1,\ldots,x^m,y^m)$$

where $z^k=x^k+iy^k$, $k=1,\ldots,m$.

(b) Each of the multiplicative groups $(\mathbb{C}^\times,\cdot)$, $(\mathbb{R}^\times,\cdot)$, $(\mathbb{Q}^\times,\cdot)$,$(\mathbb{R}^\times_+,\cdot)$ and $(\mathbb{Q}^\times_+,\cdot)$ is a topological group with its standard topology. All except $(\mathbb{Q}^\times,\cdot)$ and $(\mathbb{Q}^\times_+,\cdot)$ are also Lie groups. Here the superscript $\times$ refers to the set of non-zero elements, and the subscript $+$ indicates positive elements.

(c) (Quaternions) We introduce a (non-commutative) multiplication on $\mathbb{R}^4$ under which $\mathbb{R}^4-\{\mathbf{0}\}$ becomes a group. The division algebra thus obtained is known as the ***quaternion algebra***, and is denoted by $\mathbb{H}$ in honor of the discoverer of this multiplication[1]. It is common to denote the standard basis of $\mathbb{R}^4$ in the following way:

$$\mathbf{1} = (1,0,0,0),\ \mathbf{i} = (0,1,0,0),\ \mathbf{j} = (0,0,1,0),\ \mathbf{k} = (0,0,0,1)$$

We define the multiplication of these four basis elements and decree multiplication of arbitrary elements by bilinearity (distributive law).

(i) $\mathbf{1}$ is the identity element for multiplication.

(ii) $\mathbf{i}\cdot\mathbf{i} = \mathbf{j}\cdot\mathbf{j} = \mathbf{k}\cdot\mathbf{k} = -\mathbf{1}$.

(iii) $\mathbf{i}\cdot\mathbf{j} = -(\mathbf{j}\cdot\mathbf{i}) = \mathbf{k},\ \mathbf{j}\cdot\mathbf{k} = -(\mathbf{k}\cdot\mathbf{j}) = \mathbf{i},\ \mathbf{k}\cdot\mathbf{i} = -(\mathbf{i}\cdot\mathbf{k}) = \mathbf{j}$.

The full definition of multiplication will then be

$$\begin{aligned}(a_1\mathbf{1}+a_2\mathbf{i}+a_3\mathbf{j}+a_4\mathbf{k})(b_1\mathbf{1}+b_2\mathbf{i}+b_3\mathbf{j}+b_4\mathbf{k}) &= (a_1b_1-a_2b_2-a_3b_3-a_4b_4)\mathbf{1}\\ &+ (a_2b_1+a_3b_4-a_4b_3+a_1b_2)\mathbf{i}\\ &+ (a_3b_1+a_4b_2+a_1b_3-a_2b_4)\mathbf{j}\\ &+ (a_4b_1+a_1b_4+a_2b_3-a_3b_2)\mathbf{k}\end{aligned}$$

The associative law of multiplication can be verified by first checking it for the basis elements. To investigate the existence of inverse for non-zero elements, it is helpful to introduce the notions of *conjugate* and *norm* for quaternions. For $a=a_1\mathbf{1}+a_2\mathbf{i}+a_3\mathbf{j}+a_4\mathbf{k}\in\mathbb{H}$, the ***conjugate***, a^*, is defined by

$$a^*=a_1\mathbf{1}-a_2\mathbf{i}-a_3\mathbf{j}-a_4\mathbf{k}$$

One then verifies, by first checking on basis elements, that the following identity holds:

$$(ab)^* = b^*a^* \tag{8.3}$$

The ***norm*** of a, $N(a)\in\mathbb{R}$, is defined by

$$N(a_1\mathbf{1}+a_2\mathbf{i}+a_3\mathbf{j}+a_4\mathbf{k}) = a_1^2+a_2^2+a_3^2+a_4^2$$

It is common to identify the linear subspace of $\mathbb{H}$ spanned by $\mathbf{1}$ with $\mathbb{R}$, so that if r is a real number, then r and $r\mathbf{1}$ are used interchangeably. With this convention, one obtains

$$N(a) = aa^* \tag{8.4}$$

Using (8.3) and (8.4), we obtain

$$N(a)N(b) = N(ab) \tag{8.5}$$

Explicitly, this is the famous Euler-Lagrange identity:

$$\begin{aligned}(a_1^2+a_2^2+a_3^2+a_4^2)(b_1^2+b_2^2+b_3^2+b_4^2) &= (a_1b_1-a_2b_2-a_3b_3-a_4b_4)^2\\ &+ (a_2b_1+a_3b_4-a_4b_3+a_1b_2)^2\\ &+ (a_3b_1+a_4b_2+a_1b_3-a_2b_4)^2\\ &+ (a_4b_1+a_1b_4+a_2b_3-a_3b_2)^2\end{aligned}$$

[1]William Rowan Hamilton (1805-1865).

If $a \neq \mathbf{0}$, the element a^{-1} defined by

$$a^{-1} = \frac{1}{N(a)} a^*$$

satisfies $aa^{-1}=a^{-1}a=1$ by (8.4) and is thus the multiplicative inverse of a. The full definition of multiplication expresses each component of the product as a polynomial in the components of a and b, therefore multiplication is smooth with respect to the standard manifold structure of $\mathbb{R}^4 \times \mathbb{R}^4$. Likewise, the definition of inverse expresses the components of the inverse of a as rational functions of the components of a with non-zero denominator, therefore the inverse function is also smooth. $\mathbb{H}^\times = \mathbb{H} - \{\mathbf{0}\}$ is then a Lie group under multiplication.

(d) S_1 and S_3 are Lie groups. S_1 is the group of complex numbers z with $|z|=1$, under multiplication, and is a one-dimensional Lie group. The identity (8.5) shows that the set of quaternions of norm 1 is closed under quaternion multiplication. This is just S_3 by the definition of the norm.

(e) Let G_i, $i=1,\ldots,k$, be topological groups (respectively, Lie groups). Then $G=G_1 \times \cdots \times G_k$ is a group under component-wise multiplication, and it is a topological space under product topology. In the case of Lie groups, we also know that this topological space admits a product manifold structure. Since multiplication is defined component-wise, G becomes a topological group (or a Lie group, if the G_i are Lie groups). As an example, the m-torus $\mathbb{T}^m$ is a Lie group, which is in fact abelian since each $\mathbb{T}=S_1$ is abelian.

(f) As indicated in Example (a) above, any finite-dimensional normed vector space over $\mathbb{R}$ is a Lie group. An important example is $L(V, V)$, the space of $\mathbb{R}$-linear maps $V \to V$. This can be identified with $\mathrm{M}_n(\mathbb{R})$, the set of $n \times n$ matrices with entries from $\mathbb{R}$, once a basis is fixed for V. Since any two norms on a finite-dimensional vector space give the same topology, one can, for example, treat the entries of an $n \times n$ matrix as the components of a vector in $\mathbb{R}^{n^2}$, and use any of the norms usually used for $\mathbb{R}^N$. Most commonly, one uses the *operator norm* as follows. Let $|.|$ denote a norm on V. Then for $f \in L(V, V)$, we define $||f||$, the operator norm of f as

$$||f|| = \mathrm{Max}\{|f(x)| : |x|=1\}$$

Since a $\mathbb{C}$-linear map $\mathbb{C}^n \to \mathbb{C}^n$ may be regarded as an $\mathbb{R}$-linear map $\mathbb{R}^{2n} \to \mathbb{R}^{2n}$, the set $\mathrm{M}_n(\mathbb{C})$ of $n \times n$ matrices with complex entries is also a Lie group of dimension $2n^2$ under the operation of matrix addition[2].

(g) Let $\mathrm{GL}(n, \mathbb{R})$, respectively, $\mathrm{GL}(n, \mathbb{C})$, be the group of invertible real, respectively, complex, $n \times n$ matrices under matrix multiplication. They are known, as the real and

[2]There is a separate notion of *complex* Lie group, based on complex-analytic manifolds, which we will not discuss in this book. As a complex Lie group, $\mathrm{M}_n(\mathbb{C})$ has dimension n^2.

complex ***general linear groups***. They are, respectively, open subsets of $M_n(\mathbb{R})$ and $M_n(\mathbb{C})$, hence are also manifolds of dimension n^2 and $2n^2$, respectively. Entries of the product AB of two matrices are given as polynomials in the entries of A and B, hence the product operation is smooth. The entries of the inverse of A are rational functions of the entries of A with non-zero denominator (namely, the determinant of A), hence the inverse operation is also smooth. It follows that $GL(n,\mathbb{R})$ and $GL(n,\mathbb{C})$ are Lie groups. $GL(n,\mathbb{R})$ possesses an open subgroup of index 2 consisting of matrices with positive determinant. This is denoted by $GL^+(n,\mathbb{R})$. Certain subgroups of general linear groups, known as the *classical groups*, are the most studied of Lie groups. It follows from a theorem of Ado (see **[15]**) that these are in some sense the universal models of Lie groups. After discussing some generalities about Lie subgroups, we will introduce some of the well-known classical groups.

(h) A larger group than $GL(n,\mathbb{R})$ is obtained by including translations in $\mathbb{R}^n$. The smallest group that contains both $GL(n,\mathbb{R})$ and the translations of $\mathbb{R}^n$ is known as the ***real affine group*** and denoted by $\mathrm{Aff}(n,\mathbb{R})$. Note that translations are in one-to-one correspondence with $\mathbb{R}^n$. There are various ways to represent $\mathrm{Aff}(n,\mathbb{R})$. One is to consider $(n+1)\times(n+1)$ real matrices of the form

$$\begin{bmatrix} & & & t^1 \\ & A & & \vdots \\ & & & t^n \\ 0 & \dots & 0 & 1 \end{bmatrix} \tag{8.6}$$

where $A\in GL(n,\mathbb{R})$, and $t_i\in\mathbb{R}$. A point $x=(x^1,\dots,x^n)$ of $\mathbb{R}^n$ will then be represented by a column

$$\begin{bmatrix} x^1 \\ \vdots \\ x^n \\ 1 \end{bmatrix}$$

Then the transformation effected by the matrix sends x to $Ax+t$, where $t=(t^1,\dots,t^n)$ is the translation vector. The subgroup $GL(n,\mathbb{R})$ is represented here as the set of matrices in (8.6) with $t^1=\dots=t^n=0$. Pure translation by t is obtained by letting $A=I_n$, the identity matrix. Translation matrices also form a subgroup $T(n)$, isomorphic to $\mathbb{R}^n$. $T(n)$ is in fact a *normal* subgroup, since for $T\in T(n)$, a translation by vector t, and $M\in\mathrm{Aff}(n,\mathbb{R})$, the product MTM^{-1} is the translation by At, where A is the linear part of M. It follows that each element of $\mathrm{Aff}(n,\mathbb{R})$ has a *unique* representation as TL, where $T\in T(n)$ and $L\in GL(n,\mathbb{R})$. $\mathrm{Aff}(n,\mathbb{R})$ is a closed (n^2+n)-dimensional subgroup of $GL(n+1,\mathbb{R})$, and is hence a Lie group by a theorem we will later prove. An alternative way is to describe $\mathrm{Aff}(n,\mathbb{R})$ as the *semi-direct product* of $GL(n,\mathbb{R})$ and the normal subgroup $T(n)$ (see Exercise 8.27). If we replace the requirement $A\in GL(n,\mathbb{R})$ by $A\in GL^+(n,\mathbb{R})$, we obtain the subgroup $\mathrm{Aff}^+(n,\mathbb{R})$ of *orientation-preserving* affine transformations.

3. Definition Let G be a Lie group and H a subset of G. H is called a ***Lie subgroup*** of G if the following conditions are satisfied:
(a) H is a subgroup of G.
(b) H admits a C^∞ structure under which it is a Lie group.
(c) The inclusion map $H \hookrightarrow G$ is a smooth immersion.

We emphasize that it is *not* required for a *Lie subgroup* H of G to be a submanifold of G, i.e., carry the induced topology. For example, $(\mathbb{Q}, +)$ with the discrete topology becomes a Lie group, and a Lie subgroup of $(\mathbb{R}, +)$, although $\mathbb{Q}$ is not a submanifold of $\mathbb{R}$. An important example of a subgroup of $\mathbb{T}^2$ is obtained as follows. In $\mathbb{R}^2$, the set of points on the line $y{=}mx$ form a subgroup under addition and a submanifold, hence this line is a Lie subgroup of $\mathbb{R}^2$. The quotient projection $p{:}\mathbb{R}^2 \to \mathbb{T}^2$ (see Example 19d of Chapter 5) sends this subgroup to a Lie subgroup of $\mathbb{T}^2$. For irrational m, this Lie subgroup of $\mathbb{T}^2$ is not a submanifold (see Example 2f of Chapter 5). Of course, if H is both a subgroup and a submanifold, then it is a Lie subgroup since the inclusion $H \hookrightarrow G$ will be an embedding. The connected component of the identity, being an open subgroup by virtue of Elementary Properties 1(b) and (c), is a Lie subgroup. A theorem of Elie Cartan, to be proved in Section B, asserts that every closed subgroup of a Lie group is a Lie subgroup (with the induced topology). For most of our examples, it will not be difficult to verify the Lie subgroup property directly. The following simple lemma is often useful.

4. Lemma *Let G be a Lie group and H a subgroup of G. Suppose there is an open neighborhood U of the identity element so that $H \cap U$ is a smooth submanifold of U. Then H is a smooth submanifold of G, hence a Lie subgroup.*

Proof. Let n be the dimension of G. We will construct a submanifold chart around every point $a \in H$. Since $H \cap U$ is a smooth submanifold of U, there is a chart (V, η) around the identity element e, $V \subset U$, so that $\eta(H \cap V) = \eta(V) \cap \mathbb{R}^h$ for some $h \leq n$. Now $(aV, \eta \circ L_{a^{-1}})$ is a chart around a, and $a(H \cap V) \subset H$ since $a \in H$. We have

$$\eta \circ L_{a^{-1}}(H \cap (aV)) = \eta(H \cap V) = \eta(V) \cap \mathbb{R}^h$$

Thus $(aV, \eta \circ L_{a^{-1}})$ is a submanifold chart and H is a submanifold of dimension h. □

5. Classical Groups

In this subsection, we introduce some important Lie subgroups of $\mathrm{GL}(n, \mathbb{R})$ and $\mathrm{GL}(n, \mathbb{C})$. These are all closed subgroups of $\mathrm{GL}(n, \mathbb{R})$ or $\mathrm{GL}(n, \mathbb{C})$, hence they are Lie subgroups by virtue of Cartan's theorem mentioned earlier. In most cases, it is easy to verify that the group in question is the level set of a smooth function of constant rank, hence the submanifold nature follows from the theorems of Chapter 5, such as the Regular Value Theorem. We will carry out the proof in important sample cases and leave similar verifications to the reader.

(a) The real and complex ***special linear groups***, $\mathrm{SL}(n, \mathbb{R})$ and $\mathrm{SL}(n, \mathbb{C})$, are, respectively, the subgroups of $\mathrm{GL}(n, \mathbb{R})$ and $\mathrm{GL}(n, \mathbb{C})$ consisting of elements of determinant 1. Each is the level set $\det^{-1}\{1\}$ and is therefore closed. We prove directly that $\mathrm{SL}(n, \mathbb{R})$ is a codimension one submanifold of $\mathrm{GL}(n, \mathbb{R})$ by showing that the rank of the tangent map of the smooth function $\det : \mathrm{GL}(n, \mathbb{R}) \to \mathbb{R}$ is 1. By 1(e), it suffices to show that the rank of the derivative of det at the identity element is 1. In fact, by Example 2c of Appendix II:

$$((D\det)(\mathrm{I}_n))(\mathrm{X}) = \mathrm{tr}(\mathrm{X})$$

for $\mathrm{X} \in \mathrm{M}_n(\mathbb{R})$. Since for any real number $t \in \mathbb{R}$, there is an $n \times n$ matrix X with $\mathrm{tr}(\mathrm{X}) = t$, the rank of the derivative of det at I_n is 1. It follows from the Regular Value Theorem that $\mathrm{SL}(n, \mathbb{R})$ is a submanifold of codimension 1. A similar argument shows that $\mathrm{SL}(n, \mathbb{C})$ is a submanifold of codimension 2 of $\mathrm{GL}(n, \mathbb{C})$ (see Exercise 8.10 at the end of the chapter).

(b) The ***orthogonal group***, $\mathrm{O}(n)$, is the group of orthogonal matrices, i.e., matrices $\mathrm{A} \in \mathrm{GL}(n, \mathbb{R})$ with $\mathrm{A}^T = \mathrm{A}^{-1}$. Equivalently, these represent linear isomorphisms of $\mathbb{R}^n$ that preserve the Euclidean inner product. Since the inverse and transpose operations are continuous, $\mathrm{O}(n)$ is a closed subgroup of $\mathrm{GL}(n, \mathbb{R})$. We show directly that $\mathrm{O}(n)$ is a smooth submanifold of dimension $\frac{1}{2}n(n-1)$. Let $\mathrm{Sym}(n, \mathbb{R})$ be the set of real symmetric $n \times n$ matrices. $\mathrm{Sym}(n, \mathbb{R})$ is diffeomorphic to $\mathbb{R}^N$ with $N = \frac{1}{2}n(n+1)$. We define $\Psi : \mathrm{GL}(n, \mathbb{R}) \to \mathrm{Sym}(n, \mathbb{R})$ by $\Psi(\mathrm{A}) = \mathrm{AA}^T$. Computing the derivative of Ψ at A (see 1d, Appendix II), we obtain

$$(D\Psi(\mathrm{A}))(\mathrm{X}) = \mathrm{AX}^T + \mathrm{XA}^T$$

where $\mathrm{X} \in \mathrm{M}_n(\mathbb{R})$. This derivative is onto $\mathrm{Sym}(n, \mathbb{R})$, for if S is an arbitrary symmetric $n \times n$ matrix, then $\mathrm{X} = \frac{1}{2}\mathrm{S}(\mathrm{A}^T)^{-1}$ solves the equation

$$\mathrm{AX}^T + \mathrm{XA}^T = \mathrm{S}$$

Therefore, by the Regular Value Theorem, the set $\mathrm{O}(n) = \Psi^{-1}(\mathrm{I}_n)$ is a smooth submanifold of dimension $n^2 - \frac{1}{2}n(n+1) = \frac{1}{2}n(n-1)$. The determinant of an orthogonal matrix is ± 1, depending on whether the corresponding linear map is orientation-preserving or orientation- reversing. The set $\mathrm{SO}(n) = \mathrm{O}(n) \cap \mathrm{SL}(n, \mathbb{R})$ of orthogonal matrices with determinant $+1$ is known as the ***special orthogonal group*** and is an open and closed subgroup of $\mathrm{O}(n)$, hence a closed subgroup of $\mathrm{GL}(n, \mathbb{R})$. If M is any orthogonal $n \times n$ matrix with determinant -1, then $\mathrm{O}(n)$ can be written as the disjoint union

$$\mathrm{O}(n) = \mathrm{SO}(n) \cup \mathrm{M} \cdot \mathrm{SO}(n)$$

Thus $\mathrm{SO}(n)$ is a subgroup of index 2 in $\mathrm{O}(n)$, and a normal subgroup. In fact, we will soon see that $\mathrm{SO}(n)$ is the connected component of identity in $\mathrm{O}(n)$.

(c) Going back to the description of the real affine group in Example 2h, if we replace the matrix A∈GL(n, $\mathbb{R}$) in (8.6) by A in O(n), we obtain the ***Euclidean group***, E(n). This is the group of distance-preserving transformations of $\mathbb{R}^n$. Likewise, using SO(n), instead of O(n), we obtain the ***group of Euclidean motions***, E$^+$(n), of orientation-preserving elements of E(n).

(d) Replacing $\mathbb{R}^n$ by $\mathbb{C}^n$ and the Euclidean inner product by the standard hermitian product

$$< u|v >= u^1\bar{v}^1 + \cdots + u^n\bar{v}^n$$

for $u=(u^1,\ldots,u^n)$ and $v=(v^1,\ldots,v^n)$ in $\mathbb{C}^n$, the set of $n\times n$ complex matrices that preserve the hermitian product is the ***unitary group***, U(n). Unitary matrices are characterized by the property $\bar{A}^T = A^{-1}$. It follows that $|\det A|=1$. The subgroup SU(n)=U(n)∩SL(n, $\mathbb{C}$) is known as the ***special unitary group***. U(n) and SU(n) are closed subgroups of GL(n, $\mathbb{C}$). One can show as above that U(n) and SU(n) are submanifolds of GL(n, $\mathbb{C}$) of dimension n^2 and n^2-1, respectively (see Exercise 8.10).

The following theorem summarizes the compactness and connectedness properties of the classical groups we have introduced.

6. Theorem *(a) The following Lie groups are compact:*

$$O(n), SO(n), U(n), SU(n)$$

(b) The following Lie groups are path-connected:

$$SO(n), U(n), SU(n), SL(n,\mathbb{R}), SL(n,\mathbb{C}), GL(n,\mathbb{C}), GL^+(n,\mathbb{R}), Aff^+(n,\mathbb{R}), E^+(n)$$

Proof. (a) As subspaces of $M_n(\mathbb{R})$ or $M_n(\mathbb{C})$, these spaces inherit the topology of $\mathbb{R}^{n^2}$ or $\mathbb{C}^{n^2}\cong\mathbb{R}^{2n^2}$ and are closed subspaces. To prove they are compact, it then suffices to show that each is bounded as a subspace of $\mathbb{R}^{n^2}$ or $\mathbb{C}^{n^2}$. For this, we show that the entries of these matrix groups are uniformly bounded. Each column of a real or complex matrix can be viewed as an element of $\mathbb{R}^n$ or $\mathbb{C}^n$. If the matrix is orthogonal or unitary, then the norm (length) of each column relative to the standard inner or hermitian product is 1. Therefore, 1 is a uniform bound for the entries of the matrix, and the claim is proved.

(b) We use the canonical representations of elements of SO(n) and U(n) from linear algebra. For $g\in$SO(n), one can choose an orthonormal basis for $\mathbb{R}^n$ relative to which g has one of the following representations, depending on whether $n=2k$ or $n=2k+1$:

$$\begin{bmatrix} R_1 & 0 & \ldots & 0 \\ 0 & R_2 & \ldots & \\ \vdots & \vdots & \ddots & 0 \\ 0 & \ldots & 0 & R_k \end{bmatrix}, \quad \begin{bmatrix} 1 & 0 & \ldots & 0 \\ 0 & R_1 & \ldots & \\ \vdots & \vdots & \ddots & 0 \\ 0 & \ldots & 0 & R_k \end{bmatrix}$$

where R_j is the matrix

$$\begin{bmatrix} \cos\theta_j & -\sin\theta_j \\ \sin\theta_j & \cos\theta_j \end{bmatrix}$$

with θ_j real. Replacing each θ_j by $t\theta_j$, $0\le t\le 1$, we obtain a continuous path in SO(n) that connects the identity element I to g, establishing the path-connectedness of SO(n). Similarly for $g\in$U(n), there is an orthonormal basis relative to the standard hermitian product of $\mathbb{C}^n$, which gives the following representation for g:

$$\begin{bmatrix} e^{i\theta_1} & 0 & \dots & 0 \\ 0 & e^{i\theta_2} & \dots & \\ \vdots & \vdots & \ddots & 0 \\ 0 & \dots & 0 & e^{i\theta_n} \end{bmatrix}$$

with θ_j real. Again replacing θ_j by $t\theta_j$, $0\le t\le 1$, we obtain a continuous path in U(n) that connects the identity element to g. For an element of SU(n), the sum of the θ_j above must be an integral multiple of 2π, say $\sum_j\theta_j=2\pi m$. In fact, replacing θ_1 by $\theta_1-2\pi m$, we may assume, without affecting g, that the sum of the θ_j is zero. Now we replace each θ_j by $t\theta_j$, $0\le t\le 1$, to obtain the desired continuous path in SU(n) connecting g to the identity element.

Next we show $\mathrm{GL}^+(n,\mathbb{R})$ is path-connected. Let $g\in\mathrm{GL}^+(n,\mathbb{R})$; we connect g to the identity element in three steps (for a different proof, see Exercise 8.12 at the end of this chapter). Let $b_1,\dots,b_n$ be the columns of g viewed as elements of $\mathbb{R}^n$. Recall that the Gram-Schmidt procedure replaces $b_1,\dots,b_n$ by $a_1,\dots,a_n$, where $a_j=b_j+c_j$, with $c_1=0$ and c_j is a linear combination of $\{b_1,\dots,b_{j-1}\}$ for $j>1$. The matrix $g(t)$, $0\le t\le 1$ with columns $a_j(t)=b_j+tc_j$ has the same determinant as g and therefore belongs to $\mathrm{GL}^+(n,\mathbb{R})$. This continuous path connects g to the matrix h with columns $a_1,\dots,a_n$ that are pairwise orthogonal. We then connect h to the matrix $k\in$SO(n) that has as its jth column the n-tuple $a_j/|a_j|$. The connecting path $h(t)$, $0\le t\le 1$ has as jth column the n-tuple:

$$\frac{a_j}{1+t(|a_j|-1)}$$

The determinant of $h(t)$ is a positive multiple of the determinant of h, hence $h(t)\in\mathrm{GL}^+(n,\mathbb{R})$. Finally, k can be joined to the identity element in $\mathrm{SO}(n)\subset\mathrm{GL}^+(n,\mathbb{R})$, and SO($n$) was earlier shown to be path-connected. Thus the path-connectedness of $\mathrm{GL}^+(n,\mathbb{R})$ is established. Exactly the same argument shows the path-connectedness of $\mathrm{GL}(n,\mathbb{C})$, where no signs are involved and the determinants remain non-zero.

Now let $g\in\mathrm{SL}(n,\mathbb{R})\subset\mathrm{GL}^+(n,\mathbb{R})$. As an element of $\mathrm{GL}^+(n,\mathbb{R})$, g can be joined to the identity matrix by a continuous path $\alpha(t)$, $0\le t\le 1$, in $\mathrm{GL}^+(n,\mathbb{R})$, with $\alpha(0)=g$ and $\alpha(1)=\mathrm{I}$. One has $D(t)=\det(\alpha(t))>0$. Let $d(t)$ be the positive nth root of $D(t)$. Then dividing each entry of $\alpha(t)$ by $d(t)$, we obtain the desired continuous path in $\mathrm{SL}(n,\mathbb{R})$. For $g\in\mathrm{SL}(n,\mathbb{C})$, we take $\alpha(t)$ as above in $\mathrm{GL}(n,\mathbb{C})$; here we have $D(t)\neq 0$. Taking the continuous branch $d(t)$ of the nth root of $D(t)$ with $d(0)=1$ and dividing each entry of $\alpha(t)$ by $d(t)$, we obtain a continuous path in $\mathrm{SL}(n,\mathbb{C})$ that joins g to the diagonal matrix $d(1)\mathrm{I}$. This matrix belongs to $\mathrm{SU}(n,\mathbb{C})$, which is path-connected, so we can extend the path to reach the identity matrix I. The path-connectedness of $\mathrm{Aff}^+(n,\mathbb{R})$ and $\mathrm{E}^+(n)$ follow, respectively, from the path-connectedness of $\mathrm{GL}^+(n,\mathbb{R})$ and $\mathrm{SO}(n)$. Writing an element as TL, where T is the translation part and $\mathrm{L}\in\mathrm{GL}^+(n,\mathbb{R})$ or $\mathrm{L}\in\mathrm{SO}(n)$, we can continuously reduce the translation along a path in $T(n)\cong\mathbb{R}^n$ to zero translation, and then use the path-connectedness of $\mathrm{GL}^+(n,\mathbb{R})$ or $\mathrm{SO}(n)$. □

As corollary to the above, note that each of $\mathrm{GL}(n,\mathbb{R})$ and $\mathrm{O}(n)$ has two connected components, the connected components of the identity being, respectively, $\mathrm{GL}^+(n,\mathbb{R})$ and $\mathrm{SO}(n)$.

B. The Lie Algebra of a Lie Group

Let G be a Lie group. A vector field X on G is called ***left-invariant*** (respectively, ***right-invariant***) if for all $g\in G$, $(L_g)_*X=X$ (respectively, $(R_g)_*X=X$). Below we will generally confine the discussion to left-invariant vector fields; similar considerations hold for right-invariant ones. Let $A\in T_eG$, and consider the vector field $\tilde{A}$ defined by

$$\tilde{A}(x) = T_eL_x(A)$$

Then by the chain rule

$$((L_g)_*\tilde{A})(x) = TL_g(\tilde{A}(g^{-1}x)) = T_{g^{-1}x}L_g \circ T_eL_{g^{-1}x}(A) = T_eL_x(A) = \tilde{A}(x)$$

so that $\tilde{A}$ is left-invariant. Conversely, any left-invariant vector field $\tilde{X}$ will have to satisfy $\tilde{X}(x) = T_eL_x(\tilde{X}(e))$. We denote the set of left-invariant vector fields on G by $\mathcal{G}$. The function $A\mapsto\tilde{A}$ is an $\mathbb{R}$-linear isomorphism from T_eG onto $\mathcal{G}$ since T_eL_x is $\mathbb{R}$-linear for each $x\in G$. Thus as a real vector space, the dimension of $\mathcal{G}$ is equal to the dimension of the Lie group G.

7. Theorem *Let G be a Lie group and $\mathcal{G}$ be the space of left-invariant vector fields.*
(a) Every element of $\mathcal{G}$ is smooth.
(b) Every element of $\mathcal{G}$ is a complete vector field.
(c) For $A\in T_eG$, let $\tilde{A}$ be the associated left-invariant vector field and $a(t)$ the integral curve of $\tilde{A}$ with initial condition $a(0)=e$. Then $a(s+t)=a(s)\cdot a(t)$. The integral curve $\alpha(t)$ of $\tilde{A}$ with initial condition $\alpha(0)=g$ is $\alpha(t)=R_{a(t)}(g)=g\cdot a(t)$.
(d) $\mathcal{G}$ is a Lie algebra under the Lie bracket [,].
(e) Every Lie group is parallelizable as manifold, hence also orientable.

Proof. (a) We claim it suffices to prove smoothness in an arbitrarily small open neighborhood of e. For if x is an element of G, then $(L_x)_*$ transfers the smooth vector field to a neighborhood of x. Let (U,ξ) be a chart around e and take an open neighborhood V of e as in 1(a). Suppose $A=\sum_i a^i\frac{\partial}{\partial\xi^i}(e)$ and $\tilde{A}=\sum_i u^i\frac{\partial}{\partial\xi^i}$. Thus $u^i(e)=a^i$. We show that the u^i are smooth in V. For $g\in V$, we have

$$\sum_i u^i(g)\frac{\partial}{\partial\xi^i}(g) = T_eL_g(\sum_i a^i\frac{\partial}{\partial\xi^i}(e))$$

Evaluating both sides on the smooth function ξ^j, we obtain

$$\begin{aligned} u^j(g) &= \Big(T_e L_g\big(\sum_i a^i \frac{\partial}{\partial \xi^i}(e)\big)\Big)\cdot \xi^j \\ &= (d\xi^j(g)\circ T_e L_g)\big(\sum_i a^i \frac{\partial}{\partial \xi^i}(e)\big) \\ &= (d(\xi^j \circ L_g)(e))\big(\sum_i a^i \frac{\partial}{\partial \xi^i}(e)\big) \end{aligned}$$

Note that L_g is smooth with respect to g, being the restriction of the group operation $\mu{:}G\times G\to G$ to $\{g\}\times G$. Therefore, the right-hand side is a smooth function of g proving the smoothness of the u^j.
(b) and (c). Let $a{:}]\alpha_e,\omega_e[\to G$ be the maximal integral curve of $\tilde{A}$ with initial condition $a(0)=e$; we show that $\omega_e< +\infty$ leads to a contradiction. Pick s so that $0<s<\omega_e$, and define a curve $b{:}]\alpha_e+s,\omega_e+s[\to G$ by

$$b(u) = L_{a(s)}(a(u-s)) = a(s)\cdot a(u-s)$$

b is an integral curve for $\tilde{A}$, for

$$\begin{aligned} b'(u) &= TL_{a(s)}(a'(u-s)) \\ &= TL_{a(s)}\big(\tilde{A}(a(u-s))\big) \\ &= \tilde{A}(a(s)\cdot a(u-s)) \\ &= \tilde{A}(b(u)) \end{aligned}$$

On the other hand, $b(s)=a(s)$, so by the uniqueness of solutions, a and b match up to produce an integral curve for $\tilde{A}$, which is defined beyond $t=\omega_e$, providing a contradiction. Similarly, we can show that $\alpha_e=-\infty$. The argument above also shows that if $a(t)$ is the integral curve of $\tilde{A}$ with $a(0)=e$, and s is a real number, then $a(s)\cdot a(t)$ is an integral curve of $\tilde{A}$. Since $a(s+t)$ is also an integral curve by the chain rule and agrees with $a(s)\cdot a(t)$ for $t=0$, it follows that $a(s+t)=a(s)\cdot a(t)$. With $a(t)$ as above and $g\in G$, consider the curve $c(t)=g\cdot a(t)$. We have $c(0)=g$, and we verify that c is an integral curve for $\tilde{A}$:

$$\begin{aligned} c'(t) &= \frac{d}{dt}\big(L_g(a(t))\big) \\ &= TL_g(a'(t)) \\ &= TL_g\big(\tilde{A}(a(t))\big) \\ &= \tilde{A}(g\cdot a(t)) \\ &= \tilde{A}(c(t)) \end{aligned}$$

(d) and (e). Let $\tilde{A}$ and $\tilde{B}$ be left-invariant vector fields. Since each L_g is a diffeomorphism, and diffeomorphisms preserve the bracket, we obtain

$$(L_g)_*[\tilde{A}, \tilde{B}] = [(L_g)_*(\tilde{A}), (L_g)_*(\tilde{B})] = [\tilde{A}, \tilde{B}]$$

The vector space $\mathcal{G}$ is then closed under Lie bracket and is thus a Lie algebra. The isomorphism between T_eG and $\mathcal{G}$ shows that the dimension of $\mathcal{G}$ is the same as that of G. If $(X_1, \ldots, X_n)$ is a basis for T_eG, then the same isomorphism shows that $(\tilde{X}_1, \ldots, \tilde{X}_n)$ form a basis for the tangent space at every point of G, and G is parallelizable. Denoting the dual basis of one-forms by $(\alpha^1, \ldots, \alpha^n)$, the n-form $\alpha^1\wedge\cdots\wedge\alpha^n$ is then nowhere-vanishing and a smooth volume element for G. □

$\mathcal{G}$ is called the ***Lie algebra of*** G. Alternatively, T_eG may be regarded as the Lie algebra of G by defining

$$[A, B] = [\tilde{A}, \tilde{B}](e) \tag{8.7}$$

and using the linear isomorphism between T_eG and $\mathcal{G}$.

The ***exponential map***

$$\exp : T_eG \to G$$

is defined as follows. For $A \in T_eG$, let $a(t)$ be the integral curve of $\tilde{A}$ with $a(0)=e$. Then $\exp A = a(1)$. It follows from the completeness of $\tilde{A}$ that exp is defined on all of T_eG. Smooth dependence of solutions on parameter (Theorem 7 of Chapter 2) implies that exp is smooth. Suppose r is a real number. Consider $B=rA$, the associated left-invariant vector field $\tilde{B}$, and the integral curve $b(t)$ of $\tilde{B}$ with $b(0)=e$. We have:

$$\frac{d}{dt}a(rt) = ra'(rt) = r\tilde{A}(a(rt)) = \tilde{B}(a(rt))$$

the last step by the linearity of $A \mapsto \tilde{A}$. Therefore,

$$\exp(rA) = a(r), \text{ where } \exp A = a(1) \tag{8.8}$$

It also follows from part (c) of the theorem above that

$$\exp(s+t)A = \exp sA \cdot \exp tA \tag{8.9}$$

If Φ denotes the flow of $\tilde{A}$, then the last statement of part (c) of Theorem 7 can be expressed as

$$\Phi(t, x) = x \cdot \exp tA \tag{8.10}$$

In general, a smooth curve $a: \mathbb{R} \to G$ is called a ***one-parameter subgroup*** of the Lie group G if $a(s+t)=a(s) \cdot a(t)$. Note that for a one-parameter subgroup $a(t)$, one has $a(0)=e$, and $a(-t)$ is the inverse of $a(t)$. We know from Chapter 2, Theorem 9, that every smooth one-parameter family of diffeomorphisms Φ is generated by a smooth vector field, the flow of which is Φ. The following corollary of Theorem 7 sums up the relation between one-parameter subgroups and one-parameter groups of diffeomorphisms of a Lie group.

8. Corollary *Let $\Phi:\mathbb{R}\times G\to G$ be a one-parameter group of diffeomorphisms of the Lie group G. Then the generator of Φ is left-invariant if and only if*

$$\Phi_t \circ L_g = L_g \circ \Phi_t \tag{8.11}$$

for all $t\in\mathbb{R}$ and all $g\in G$. For every smooth one-parameter subgroup $a:\mathbb{R}\to G$, the map $\Phi:\mathbb{R}\times G\to G$ defined by

$$\Phi(t, x) = x \cdot a(t) \tag{8.12}$$

is a one-parameter family of diffeomorphisms with generator a left-invariant vector field $\tilde{X}$, and $a(t)=\exp tX$.

Proof. The first claim follows immediately from (2.26) of subsection 10 of Chapter 2. For the second claim, the definition of one-parameter subgroup and (8.12) imply that Φ is a one-parameter family of diffeomorphisms. Further, since

$$(L_g \circ \Phi_t)(x) = g \cdot x \cdot a(t) = (\Phi_t \circ L_g)(x)$$

then the vector field $\tilde{X}$ that generates Φ is left-invariant by virtue of the first claim. Finally, Φ is the flow of this vector field by Theorem 9 of Chapter 2, and therefore $a(t)$ is just $\exp tX$ by (8.10). □

The identity (8.9) should not lead the reader to believe that for arbitrary $A, B\in T_eG$, one has $\exp(A+B)=(\exp A)\cdot(\exp B)$. We shall later see that this holds if $[\tilde{A}, \tilde{B}]=0$.

9. Examples

(a) Consider the Lie group $(\mathbb{R}^n, +)$. For $a\in\mathbb{R}^n$, we have $L_a(x)=a+x$. Therefore, $DL_a(x)=\mathbb{1}_{\mathbb{R}^n}$. $\mathbb{R}^n$ being a linear space, an element of $T_x\mathbb{R}^n$ can be identified as (x, A), and we have

$$T_xL_a(x, A) = (a+x, A)$$

Writing $A=(A^1, \ldots, A^n)$, it follows that

$$\tilde{A}(a) = \sum_{i=1}^{n} A^i \frac{\partial}{\partial x^i}(a)$$

For this constant vector field we have

$$\exp tA = tA, \text{ and } \Phi(t, x) = x + tA$$

Further, if $\tilde{A}$ and $\tilde{B}$ are two left-invariant vector fields, $[\tilde{A}, \tilde{B}]=0$ and their flows commute (see Example 25a of Chapter 2). A Lie algebra for which the bracket identically vanishes is called an ***abelian (commutative) Lie algebra***. We shall later see that a connected Lie group G is abelian if and only if its associated Lie algebra is abelian. Further, in this case, $\exp(A+B)=(\exp A)\cdot(\exp B)$.

(b) The Lie group $\mathrm{GL}(n, \mathbb{R})$ is an open subset of the linear space $\mathrm{M}_n(\mathbb{R})$, hence the elements of the tangent space at the identity element can be represented as (I, A), where $A \in M_n(\mathbb{R})$. For $X \in GL(n, \mathbb{R})$, left translation L_X is the linear map $L_X Y = XY$, and, therefore,

(8.13) $$T_I L_X(I, A) = (X, XA)$$

We denote the Lie algebra of $GL(n, \mathbb{R})$ by $gl(n, \mathbb{R})$. Denoting elements A of $M_n(\mathbb{R})$ by $[A^i_j]$, we let x^i_j be the projection $M_n(\mathbb{R}) \to \mathbb{R}$ on the (i, j) entry. By (8.13), the left-invariant vector field $\tilde{A}$ determined by A (more precisely, by (I, A)), can be expressed by

$$\tilde{A}(X) = (X, XA) = \sum_{i,j} u^i_j \frac{\partial}{\partial x^i_j} \quad \text{where } u^i_j = \sum_{\mu} x^i_\mu A^\mu_j$$

Likewise for $B = [B^i_j] \in M_n(\mathbb{R})$

$$\tilde{B}(X) = (X, XB) = \sum_{i,j} v^i_j \frac{\partial}{\partial x^i_j} \quad \text{where } v^i_j = \sum_{\mu} x^i_\mu B^\mu_j$$

Now writing

$$[\tilde{A}, \tilde{B}](X) = \sum_{i,j} w^i_j \frac{\partial}{\partial x^i_j}$$

we use the formula (2.41) from Chapter 2 to calculate the w^i_j.

$$\begin{aligned} w^i_j(I) &= \sum_{k,l} \Big(\big(\sum_{\mu} x^k_\mu A^\mu_l\big) \frac{\partial v^i_j}{\partial x^k_l} - \big(\sum_{\mu} x^k_\mu B^\mu_l\big) \frac{\partial u^i_j}{\partial x^k_l} \Big)\Big|_I \\ &= \sum_{l} (A^i_l B^l_j - B^i_l A^l_j) \end{aligned}$$

The resulting expression is the (i, j) entry of the matrix $AB - BA$. This matrix is generally denoted by $[A, B]$, and is also known as the **bracket** of A and B. We have therefore demonstrated that in $GL(n, \mathbb{R})$:

(8.14) $$[\tilde{A}, \tilde{B}] = \widetilde{[A, B]}$$

Finally, we work out the exponential map for $GL(n, \mathbb{R})$. Letting $a(t)$ be the one-parameter group associated to $A \in M_n(\mathbb{R})$, then by (8.13), the left-invariance of $\tilde{A}$ is expressed by the following differential equation for $a(t) = \exp tA$:

$$a'(t) = a(t) \cdot A$$

The unique matrix solution of this differential equation with initial condition $\exp 0 = a(0) = I$ is the usual exponential e^{tA} of the matrix tA (see Appendix I):

(8.15) $$\exp tA = e^{tA} = \sum_{i=0}^{\infty} \frac{t^i A^i}{i!}$$

This fact is the source of the use of the terminology "exponential" in the lore of Lie groups. Exactly the same considerations apply to $GL(n, \mathbb{C})$ as a real Lie group, and the equations (8.14) and (8.15) hold for complex matrices. In the next examples we will describe the Lie algbras of some classical groups by identifying the corresponding subspace of the tangent space at the identity.

(c) Let us denote the Lie algebra of $G=\mathrm{SL}(n,\mathbb{R})$ by $\mathrm{sl}(n,\mathbb{R})$. Identifying this Lie algebra with T_IG, we claim that T_IG is the sub-Lie-algebra of $\mathrm{M}_n(\mathbb{R})$ consisting of matrices of trace zero. It follows from the formula

$$\exp(\mathrm{tr}\,A) = \det(\exp A) \tag{8.16}$$

of Appendix I that the exponential of a matrix of trace zero is in $\mathrm{SL}(n,\mathbb{R})$. Further, matrices of trace 0 form a linear subspace of dimension n^2-1 in $T_I\mathrm{GL}(n,\mathbb{R})\cong\mathrm{M}_n(\mathbb{R})$, and this is exactly the dimension of $\mathrm{SL}(n,\mathbb{R})$ (sec 5a), so the claim is proved. Similarly, $\mathrm{sl}(n,\mathbb{C})$, the Lie algebra of $\mathrm{SL}(n,\mathbb{C})$, can be identified with the subalgebra of $\mathrm{M}_n(\mathbb{C})$ consisting of (complex) matrices of trace zero.

(d) The Lie algebras of $\mathrm{O}(n)$ and $\mathrm{SO}(n)$, denoted respectively by $\mathrm{o}(n)$ and $\mathrm{so}(n)$, are isomorphic, each identifiable with $T_I\mathrm{SO}(n)$. We claim this is the subalgebra of $\mathrm{M}_n(\mathbb{R})$ consisting of skew-symmetric matrices, i.e., matrices A that satisfy $A^T=-A$. Taking the exponential of this relation, we obtain $(\exp A)^T=(\exp A)^{-1}$, i.e., $\exp A$ is orthogonal; in fact, $\exp A\in\mathrm{SO}(n)$ since by (8.16), $\det(\exp A)>0$. A comparison of dimensions, as in the previous example, shows that such matrices constitute the entire tangent space at the identity element. Alternatively, we can give a proof based on a useful general argument as follows. Let $A\in T_I\mathrm{SO}(n)$ and suppose that $\exp tA=a(t)=[a^i_j(t)]$. Then the orthogonality of $a(t)$ can be expressed by

$$\sum_\mu a^\mu_i(t)a^\mu_j(t)=\delta_{ij}$$

Differentiating with respect to t gives

$$\sum_\mu \{(a^\mu_i)'(t)a^\mu_j(t) + a^\mu_i(t)(a^\mu_j)'(t)\} = 0$$

At $t = 0$, substituting $a(0)=I$, we obtain

$$(a^j_i)'(0) + (a^i_j)'(0) = 0$$

or $A=a'(0)$ satisfies $A^T+A=0$, i.e., A is skew-symmetric.

(e) A similar argument to the above shows that the Lie algebra $\mathrm{u}(n)$ of $\mathrm{U}(n)$ can be identified with the subalgebra of $\mathrm{M}_n(\mathbb{C})$ consisting of skew-hermitian matrices, i.e., complex matrices A satisfying $A^T+\bar{A}=0$. For such a matrix A, the real part of the trace of A is zero. The Lie algebra $\mathrm{su}(n)$ of $\mathrm{SU}(n)$ is $\mathrm{u}(n)\cap\mathrm{sl}(n,\mathbb{C})$, i.e., the subalgebra of skew-hermitian matrices with zero trace.

(f) We work out the Lie algebras of $\mathrm{Aff}(n,\mathbb{R})$ and $\mathrm{E}(n)$. In view of Example (b) above and the representations given in Example 2h and 5c, it is easy to characterize the tangent space to identity for each group. We claim that as a linear space of dimension n^2+n, the following two sets of matrices span the tangent space at identity of $\mathrm{Aff}(n,\mathbb{R})$.

$$X = \begin{bmatrix} & M & & 0 \\ & & & \vdots \\ & & & 0 \\ 0 & \dots & 0 & 0 \end{bmatrix}, \quad Y = \begin{bmatrix} & & & t^1 \\ & \mathbf{0} & & \vdots \\ & & & t^n \\ 0 & \dots & 0 & 0 \end{bmatrix}$$

where M is an arbitrary $n \times n$ real matrix and t_j are real numbers. Then $\exp X \in \mathrm{GL}(n,\mathbb{R})$ as a subset of $\mathrm{Aff}(n,\mathbb{R})$, and $\exp Y \in \mathrm{T}(n)$. The Lie algebra structure was given in Example (b) above. Note that in general, $XY-YX \neq 0$. In fact, for $n=1$, we obtain the lowest-dimensional ($n^2+n=2$) example of a non-commutative Lie algebra (see also Exercise 8.24). For the Lie algebra of the Euclidean group $\mathrm{E}(n)$, we restrict the matrix M above to be skew-symmetric, but Y will be the same as before.

The examples of classical groups above suggest a correspondence between Lie subalgebras and Lie subgroups. This is corroborated by the following theorem.

10. Theorem *Let G be a Lie group with Lie algebra $\mathcal{G}$. Then* exp *induces a one-to-one correspondence between the Lie subalgebras of $\mathcal{G}$ and connected Lie subgroups of G. Moreover,* exp *is a local diffeomorphism from an open neighborhood of $\mathbf{0}_e$ in T_eG to an open neighborhood of e in G.*

Proof. Let H be a Lie subgroup of G with Lie algebra $\mathcal{H}$ and denote the inclusion immersion $H \hookrightarrow G$ by j. Then $j^*: \mathcal{H} \to \mathcal{G}$ is an injective homomorphism of Lie algebras by 12a of Chapter 6. The representation in the tangent space at identity is by $T_e j(T_eH) \hookrightarrow T_eG$. Note that this inclusion depends only on the connected component of e in H.

Conversely, suppose $\mathcal{H}$ is a Lie subalgebra of $\mathcal{G}$. At every point x of G, let

$$\Delta_x = \{X(x) \in T_xG : X \in \mathcal{H}\}$$

The fact that $\mathcal{H}$ is a Lie algebra says that the plane field Δ is closed under bracket and is hence integrable by Frobenius Theorem. Let H be the maximal leaf of Δ that contains e. We show that H is a connected Lie subgroup of G and has $\mathcal{H}$ as its Lie algebra. Connectedness is of course part of the definition of a leaf. By the definition of left-invariance, each $(L_g)_*$ preserves Δ, hence the diffeomorphism L_g sends leaves to leaves. If $h \in H$, since $L_h(e)=h \in H$, then $L_h(H)=H$. This implies that H is closed under group multiplication of G. Likewise for $h \in H$, $L_{h^{-1}}(h)=e \in H$, therefore $L_{h^{-1}}(H)=H$. In particular, $L_{h^{-1}}(e)=h^{-1} \in H$, implying that H is closed under inverse operation. Thus the group operations of G make H into a subgroup. These operations are smooth in G, and every point of the immersed submanifold H has a neighborhood H' (relative to the topology of H), which is a submanifold of G, therefore the group operations are also smooth as maps into H. That the Lie algebra of H is $\mathcal{H}$ follows from the fact that $T_e j(H)=\Delta_e$. Further, for $A \neq \mathbf{0}$, consider tA, the straight line through the origin of the vector space T_eG. Then exp maps tA to $a(t)$, the integral curve of $\tilde{A}$

that passes through e. Therefore, $\exp \mathcal{H} \subset H$.

Since T_eG is a linear space, its tangent space at $\mathbf{0}_e$ is canonically identified with T_eG, hence $T_{\mathbf{0}_e}\exp$ may be regarded as a linear map from T_eG to itself. We show that

$$T_{\mathbf{0}_e}\exp = \mathbb{1}_{T_eG} \tag{8.17}$$

The inverse function theorem would then imply the local diffeomorphism claim. But for $A \in T_eG$, the image of A under $T_{\mathbf{0}_e}\exp$ is precisely $a'(0)=A$ since $\exp(tA)=a(t)$ and the tangent vectors at $t=0$ correspond, therefore (8.17) holds. □

The following corollary complements Lemma 4.

11. Corollary *Let G be a Lie group with Lie algebra $\mathcal{G}$, H a subgroup of the group G and W a linear subspace of T_eG. Suppose* exp *maps an open neighborhood V of $\mathbf{0}_e$ in T_eG diffeomorphically onto an open neighborhood U of e in G so that*

$$\exp(W \cap V) = H \cap U \tag{8.18}$$

Then H is a Lie subgroup of G and its Lie algebra $\mathcal{H}$ corresponds to the linear subspace $W=T_eH$ of T_eG.

Proof. By assumption (8.18), $(U, (\exp|_V)^{-1})$ is a submanifold chart for $H \cap U$, and therefore Lemma 4 can be applied. Lemma 4 also shows that $T_eH=W$. □

To further explore the relationship between a Lie group and its Lie algebra, we will briefly discuss the important notion of *adjoint representation.* The inner automorphism (conjugation) of the group G by an element g, i.e., the automorphism $x \mapsto gxg^{-1}$ will be denoted by $ad_g: G \to G$. Thus $ad_g = L_g \circ R_{g^{-1}} = R_{g^{-1}} \circ L_g$, so ad_g is a diffeomorphism of G. We define the ***adjoint representation*** (of G on $\mathcal{G}$) by

$$Ad_g = (ad_g)_* : \mathcal{G} \to \mathcal{G}$$

We must show that if $\tilde{X}$ is left-invariant, then so is $Ad_g\tilde{X}$. It follows from the left-invariance of $\tilde{X}$ that in fact

$$Ad_g = (R_{g^{-1}})_* \tag{8.19}$$

Now since any left-translation commutes with any right-translation, we obtain

$$(L_h)_*((R_{g^{-1}})_*(\tilde{X})) = (R_{g^{-1}})_*((L_h)_*(\tilde{X})) = (R_{g^{-1}})_*(\tilde{X})$$

Thus, although a right-translation does not necessarily preserve a left-invariant vector field, it maps a left-invariant vector field to (possibly another) left-invariant vector field.

12. Elementary Properties of Adjoint Representation

(a) Ad_g is a Lie algebra automorphism.

(b) $Ad_{gh} = Ad_g \circ Ad_h$.

(c) $(Ad_g(\tilde{X}))(e) = \frac{d}{dt}(g(\exp tX)g^{-1})|_{t=0}$.

(d) $\frac{d}{dt}(Ad_{\exp tX}(\tilde{Y}))|_{t=0} = [\tilde{X}, \tilde{Y}]$.

PROOF. We know from Chapter 6 that any diffeomorphism induces a Lie algebra automorphism on the space of vector fields, hence (a). (b) follows from $ad_{gh}=ad_g \circ ad_h$. For (c), since $\exp tX$ is the integral curve of $\tilde{X}$ through e, the integral curve of $Ad_g(\tilde{X})$ through e is $ad_g(\exp tX)$, and the derivative of this curve at $t=0$ is $(Ad_g(\tilde{X}))(e)$. Finally for (d), let $x \in G$, then

$$\frac{d}{dt}(Ad_{\exp tX}(\tilde{Y})(x))|_{t=0} = \lim_{t \to 0} \frac{(R_{\exp(-tX)})_* \tilde{Y}(x) - \tilde{Y}(x)}{t}$$
$$= [\tilde{X}, \tilde{Y}](x)$$

by (8.10) and Theorem 23 of Chapter 2. □

We are now in the position to settle completely the abelian case, but first a key lemma about which we will say more following the proof.

13. Lemma *Let G be a Lie group with Lie algebra $\mathcal{G}$. Then for all $X, Y \in T_eG$*

$$(\exp tX)(\exp tY) = \exp(t(X+Y) + o(t))$$

where $o(t)$ refers to a remainder term $R(t, X, Y)$ with the property $\frac{R(t,X,Y)}{t} \to 0$ as $t \to 0$.

PROOF. The idea is to use the definition of the differentiability of group multiplication in G. For this, we have to take local coordinates in some linear space, the obvious choice being the local diffeomorphism from a neighborhood of $\mathbf{0}_e$ in $T_{\mathbf{0}_e}G$ to a neighborhood of e in G provided by exp. The function

$$f : U \times U \to \mathcal{G}, \quad f(X, Y) = \exp^{-1}((\exp X)(\exp Y))$$

is defined and differentiable in a neighborhood $U \times U$ of $(\mathbf{0}_e, \mathbf{0}_e)$ in $T_{\mathbf{0}_e}G \times T_{\mathbf{0}_e}G$, where U is a convex neighborhood of $\mathbf{0}_e$ in $T_{\mathbf{0}_e}G$. Differentiability at $(\mathbf{0}_e, \mathbf{0}_e)$ means that for $|t|<1$

$$f(tX, tY) = f(\mathbf{0}_e, \mathbf{0}_e) + L(tX, tY) + \rho(tX, tY)$$

where $L : \mathcal{G} \times \mathcal{G} \to \mathcal{G}$ is linear and $\frac{\rho(tX,tY)}{t} \to 0$ as $t \to 0$. Since $f(X, \mathbf{0}_e)=X$, and $f(\mathbf{0}_e, Y)=Y$, we obtain $L(X, Y)=X+Y$, therefore

$$f(tX, tY) = t(X+Y) + o(t)$$

Applying exp to both sides, we obtain the desired result. □

As indicated in the proof, the lemma is just an expression of the differentiability of group operation in Lie groups. In fact, the Lie group operation is C^∞ and even analytic. Higher-order differentiability and analyticity are manifested in Taylor's theorem and Taylor series. This finds expression in the so-called Baker-Campbell-Hausdorff formula (see almost any standard text on Lie groups). The higher-order terms (beyond $\exp t(X+Y)$) involve the bracket of X and Y and its iterations. For the record, the second-order approximation is

$$(8.20) \qquad (\exp tX)(\exp tY) = \exp(t(X+Y) + \frac{1}{2}t^2[X, Y] + o(t^2))$$

We only need the first-order approximation given by the above lemma for the proofs of the following theorem, as well as for Cartan's theorem, that will follow.

14. Theorem *Let G be a Lie group with Lie algebra $\mathcal{G}$, and suppose that G_0 is the connected component of the identity. Then $\mathcal{G}$ is abelian if and only if G_0 is abelian. In this case,*

$$\exp(X + Y) = \exp X \cdot \exp Y$$

for all $X, Y \in T_eG$, and exp *is onto G_0.*

PROOF. Initially note that T_eG, and hence $\mathcal{G}$, are completely determined by G_0, the rest of G having no bearing. If G_0 is abelian, then $Ad_{\exp tX} = (R_{\exp(-tX)})_* = (L_{\exp(-tX)})_*$, and by (d) of 12, $[\tilde{X}, \tilde{Y}]=0$. Now suppose that $\mathcal{G}$ is abelian. By Corollary 24 of Chapter 2,

$$(\exp sX)(\exp tY) = (\exp tY)(\exp sX), \;\; \forall X, Y \in \mathcal{G}, \forall s, t \in \mathbb{R}$$

By Theorem 10, exp maps an open neighborhood of $\mathbf{0}_e$ in $\mathcal{G}$ diffeomorphically onto an open neighborhood of e in G. Thus we can take an open neighborhood V of e in the image of exp. V generates G_0 in the sense of Proposition 1(d). For any two elements g and h of G_0, we can then write

$$g = (\exp s_1 X_1) \cdots (\exp s_k X_k), \;\; h = (\exp t_1 Y_1) \cdots (\exp t_l Y_l)$$

so that the $\exp s_i X_i$ and the $\exp t_j Y_j$ are all in V. It follows then that $gh=hg$, establishing the commutativity of G_0. In particular, we have $(\exp X)(\exp Y)=(\exp Y)(\exp X)$; we wish to show this is equal to $\exp(X+Y)$. Consider the smooth curve α defined by $\alpha(t)=(\exp tX)(\exp tY)$. It follows from commutativity that $\alpha(s+t)=\alpha(s)\cdot\alpha(t)$, i.e., α is a one-parameter subgroup. We must then have $\alpha(t)=\exp tZ$, where $Z=\alpha'(0)$. But by (8.17) and Lemma 13, it suffices to compute the derivative of the curve $t(X+Y)+o(t)$ at $t=0$, which is $X+Y$, and the claim is proved. Note that this implies that all of G_0 lies in the image of exp since G_0 is generated by V. In fact, by connectedness, $G_0 = \exp(T_eG)$. □

15. Remark

(a) Even though the image of exp includes a neighborhood of e, and any neighborhood of e generates G_0, it is not in general true that the image of exp is all of G_0 (see Exercise 8.36). In the proof above, we used commutativity to express the product of exponentials as an exponential.

(b) In Exercise 8.15, we are asking the reader to give an example of a pair of non-commuting matrices X and Y so that $\exp(X+Y)=(\exp X)(\exp Y)$. It follows, however, from (8.20), that if for matrices X and Y, $\exp t(X+Y)=(\exp tX)(\exp tY)$ for all $|t|$ sufficiently small, then X and Y commute (Exercise 8.14).

16. Theorem (E. Cartan) *Every closed subgroup of a Lie group is a Lie subgroup with the induced topology.*

PROOF. Let H be a closed subgroup of the Lie group G. We will describe a linear subspace W of T_eG for which Corollary 11 holds. Let

$$W = \{w \in T_eG : \exp(tw) \in H, \forall t \in \mathbb{R}\}$$

By the very definition, if $w \in W$ and $r \in \mathbb{R}$, then $rw \in W$. Suppose now that $w_1, w_2 \in W$. Then given a $t \in \mathbb{R}$, any integer n, it follows from Lemma 13 that

$$\begin{aligned}((\exp \frac{t}{n} w_1)(\exp \frac{t}{n} w_2))^n &= \Big(\exp(\frac{t}{n}(w_1+w_2)+o(\frac{t}{n}))\Big)^n \\ &= \exp(t(w_1+w_2)+n\cdot o(\frac{t}{n}))\end{aligned}$$

the latter by (8.9). The left-hand side is an element of H, and H is closed, therefore

$$\exp t(w_1+w_2) = \lim_{n\to\infty} (\exp t(w_1+w_2)+t\cdot\frac{n}{t}\cdot o(\frac{t}{n}))$$

is an element of H. This shows W is a linear subspace of T_eG.

If for some open neighborhood V of $\mathbf{0}_e$ in T_eG, and $U = \exp V$, the relation (8.18) holds, then the theorem is proved. Otherwise, suppose that there is an open neighborhood W' of $\mathbf{0}_e$ in W and a sequence (h_n) of points in H so that $h_n \to e$, but $h_n \notin \exp W'$. Let Z be a linear complement of W in T_eG. Consider the map E defined in a neighborhood of $\mathbf{0}_e$ from $W \oplus Z$ to G by $E(X, Y) = (\exp X)(\exp Y)$. E is smooth and T_eE is the identity map of T_eG. It follows that a rectangular open neighborhood $V_1 \times V_2$ of $\mathbf{0}_e$ in $W \oplus Z$ is mapped diffeomorphically by E onto an open neighborhood of e in G. We claim that if V_2 is sufficiently small, then

$$H \cap \exp(V_2 - \{\mathbf{0}_e\}) = \emptyset \tag{8.21}$$

Once this is proved, we will arrive at a contradiction to the existence of the set W' and the sequence h_n, as follows. Note that for n sufficiently large, $h_n \in E(V_1 \times V_2)$, hence $h_n = (\exp x_n)(\exp y_n) \in H$, where also $\exp x_n \in H$. Therefore, $\exp y_n \in H$ contradicts (8.21). It then suffices to prove (8.21). If (8.21) does not hold, then there is a sequence (z_n) in Z so that $z_n \to \mathbf{0}_e$, $z_n \neq \mathbf{0}_e$ and $\exp z_n \in H$. Let $|.|$ be a norm for T_eG. There is a subsequence (y_n) of z_n so that $y_n/|y_n| \to y$ on the unit sphere in Z. We will show that in this case, $\exp ty \in H$ for all $t \in \mathbb{R}$, i.e., $y \in W$ contradicting $W \cap Z = \mathbf{0}_e$. To see this, given any real number t and any sequence of real numbers (ϵ_n), $\epsilon_n \neq 0$, such that $\epsilon_n \to 0$, there is a sequence of integers (k_n) so that $k_n \epsilon_n \to t$. It suffices to take k_n to be the closest integer to $|t/\epsilon_n|$. In our case, $\epsilon_n = |y_n|$. We therefore obtain

$$\exp ty = \lim_{n\to\infty} \exp k_n y_n = \lim_{n\to\infty} (\exp y_n)^{k_n}$$

The right-hand side belongs to H as H is closed, and we obtain the desired contradiction to finish the proof. □

C. Homogeneous Spaces

Let G be a Lie group and suppose H is a closed (Lie) subgroup of G. We will endow the space of left cosets G/H with the structure of a quotient manifold of G in the sense of Section C of Chapter 5. More precisely, G/H, with quotient topology, will be given a smooth manifold structure so that the quotient map $G \to G/H$ is a smooth submersion. Such a space will be our model *homogeneous space*. The justification for this terminology is to be given soon, namely that under modest assumptions, spaces that admit a *transitive* group action by homeomorphisms can be represented as coset spaces. By a ***transitive*** action of a group G on a set X, we mean an action $\Phi: G \times X \to X$ so that for any $x, y \in X$, there exists $g \in G$ with $\Phi(g, x) = y$. It turns out that many spaces of interest in geometry and topology have the structure of a homogeneous space, as our examples will demonstrate.

Let us begin with a bit of terminology. If G is a topological group and X is a topological space, then an action $\Phi: G \times X \to X$ is a ***continuous action*** if Φ is continuous. In this case, Φ acts by homeomorphisms since for $g \in G$, the map $\Phi_g: X \to X$ given by $\Phi_g(x) = \Phi(g, x) = g \cdot x$ is continuous, being the restriction of Φ, and its inverse $(\Phi_g)^{-1} = \Phi_{g^{-1}}$ is also continuous. Likewise, if G is a Lie group and M is a smooth manifold, it makes sense to speak of a ***smooth action*** when the action $\Phi: G \times M \to M$ is a smooth map of manifolds. In this case, Φ acts by smooth diffeomorphisms for the same reason as above.

Let X be a set, G a group and $\Phi: G \times X \to X$ an action. For $x \in X$, the ***stabilizer*** (or ***isotropy subgroup***) of x in G is defined as $G_x = \{g \in G : g \cdot x = x\}$. G_x is a subgroup of G. A map Φ'_x from the space of left cosets G/G_x to X is defined by

$$\Phi'_x : G/G_x \to X, \quad \Phi'_x(gG_x) = g \cdot x$$

This map is well-defined since if $g^{-1}h \in G_x$, then $\Phi'_x((g^{-1}h)G_x) = x$.

17. Elementary Properties *Let $\Phi: G \times X \to X$ be a group action with G_x and Φ'_x as above. Then:*
(a) Φ'_x is one-to-one, and if Φ is transitive, then Φ'_x is onto.
(b) If X is a topological space, G is a topological group, G/G_x is given the quotient topology and Φ continuous, then Φ'_x is continuous.
(c) Under the hypothesis in (b), if X is Hausdorff, then G_x is a closed subgroup of G.
(d) Under the hypothesis in (b), if X is Hausdorff, G is compact and Φ is transitive, then Φ'_x is a homeomorphism.

Proof. (a) If $g \cdot x = h \cdot x$, then $g^{-1}h \in G_x$, therefore $gG_x = hG_x$. Now suppose Φ is transitive. For $y \in X$, there is $g \in G$ so that $g \cdot x = y$. This is equivalent to $\Phi'_x(gG_x) = y$.
(b) By a general property of quotient topology (see 14b of Chapter 5), Φ'_x is continuous if and only if $\Phi^x: G \to X$ defined by $\Phi^x(g) = \Phi(g, x)$ is continuous. But Φ^x is continuous, being a restriction of Φ.
(c) If X is Hausdorff, then $\{x\}$ is closed, and G_x, which is the inverse image of $\{x\}$ under the continuous map $\Phi^x: G \to X$, is closed.
(d) Quotient-topology projection is continuous, therefore if G is compact, then so is G/G_x, but a continuous and one-to-one map from a compact space onto a Hausdorff space is a homeomorphism. □

18. Examples

(a) We have already realized the m-torus $\mathbb{T}^m$ as the homogeneous space $\mathbb{R}^m/\mathbb{Z}^m$ in Chapter 5, but we recast it briefly in the form of the above proposition. Represent the points of $\mathbb{T}^m$ by $(e^{2\pi i\theta_1}, \ldots, e^{2\pi i\theta_m})$, $\theta_j \in \mathbb{R}$, and consider the action

$$\Phi : \mathbb{R}^m \times \mathbb{T}^m \to \mathbb{T}^m$$

$$\Phi((t_1, \ldots, t_m), (e^{2\pi i\theta_1}, \ldots, e^{2\pi i\theta_m})) = (e^{2\pi i(\theta_1+t_1)}, \ldots, e^{2\pi i(\theta_m+t_m)})$$

This is a continuous and transitive action. The stabilizer of the point $\mathbf{1}=(1, \ldots, 1)$ is $\mathbb{Z}^m$. Here the group $\mathbb{R}^m$ is not compact, but the fact that $\Phi_{\mathbf{1}}'$ is a homeomorphism was established in Example 19d of Chapter 5.

(b) Consider the action

$$\Phi : \mathrm{O}(m) \times S_{m-1} \to S_{m-1}, \quad \Phi(A, x) = Ax$$

where $A \in \mathrm{O}(m)$ is represented as an $m \times m$ matrix, and $x \in S_{m-1} \subset \mathbb{R}^m$ is represented as a column m-vector. This is a continuous action that is transitive. For any distinct pair of points $x, y \in S_{m-1}$, let E be the two-dimensional linear subspace of $\mathbb{R}^m$ determined by x and y. There is an element of $\mathrm{O}(m)$ (in fact an element of $\mathrm{SO}(m)$) that leaves $E^{\perp}$ pointwise fixed and rotates x to y in E. The stabilizer of $e_m=(0, \ldots, 0, 1)$ is the subgroup of $\mathrm{O}(m)$ consisting of matrices of the form

$$\begin{bmatrix} & & & 0 \\ & \mathrm{B} & & \vdots \\ & & & 0 \\ 0 & \ldots & 0 & 1 \end{bmatrix}$$

where $\mathrm{B} \in \mathrm{O}(m-1)$ represents an arbitrary orthogonal transformation of the subspace $(e_1, \ldots, e_{m-1})$. Identifying the subgroup of matrices of the above form with $\mathrm{O}(m-1)$, and noting that $\mathrm{O}(m)$ is compact by Theorem 6a, we obtain a homeomorphism

$$\mathrm{O}(m)/\mathrm{O}(m-1) \to S_{m-1} \tag{8.22}$$

Had we used a matrix in $\mathrm{SO}(m)$ instead of one in $\mathrm{O}(m)$, then the matrix B would have had to be picked in $\mathrm{SO}(m-1)$. We therefore also obtain a homeomorphism

$$\mathrm{SO}(m)/\mathrm{SO}(m-1) \to S_{m-1} \tag{8.23}$$

(c) Let us represent the points of the real projective space $\mathbb{RP}(m-1)$ as $[x]=\{x, -x\}$, where $x \in S_{m-1}$. As in the previous example, we have a continuous and transitive action

$$\Phi : \mathrm{SO}(m) \times \mathbb{RP}(m-1) \to \mathbb{RP}(m-1), \quad \Phi(A, [x]) = [Ax]$$

Note that this is well-defined. The stabilizer of the point $[e_m] \in \mathbb{RP}(m-1)$ consists of matrices of the form

$$\begin{bmatrix} & & & 0 \\ & \mathrm{B} & & \vdots \\ & & & 0 \\ 0 & \ldots & 0 & \pm 1 \end{bmatrix}$$

where $B \in O(m-1)$ represents an arbitrary orthogonal transformation of the subspace $(e_1, \ldots, e_{m-1})$. For $+1$ as the lower-right entry, we obtain $B \in SO(m-1)$, and for -1, B is in the complement of $SO(m-1)$ in $SO(m)$. We therefore identify $O(m-1)$ as a subgroup of $SO(m)$, and obtain a homeomorphism

$$SO(m)/O(m-1) \to \mathbb{RP}(m-1) \tag{8.24}$$

(d) A natural generalization of projective space is the *Grassmann manifold.* Recall that $\mathbb{RP}(m-1)$ may be viewed as the space of straight lines through the origin of $\mathbb{R}^m$. For integers $0 \leq k \leq m$, we define $G(k, m)$ as the set of all k-dimensional linear subspaces of $\mathbb{R}^m$. We will now endow $G(k, m)$ with a topology and later prove that $G(k, m)$ possesses a natural smooth manifold structure of dimension $k(m-k)$; this is the ***Grassmann manifold*** of k-planes in $\mathbb{R}^m$. For $k=1$, $G(1, m)=\mathbb{RP}(m-1)$. Consider the action

$$\Phi : O(m) \times G(k, m) \to G(k, m), \quad \Phi(A, E) = A(E)$$

The matrix A acting as an orthogonal transformation $\mathbb{R}^m \to \mathbb{R}^m$ maps each linear k-dimensional subspace E to a linear k-dimensional subspace $A(E)$. This action is transitive since for any pair of k-dimensional subspaces of $\mathbb{R}^m$, their chosen bases can be completed to bases for $\mathbb{R}^m$. Let E_0 be the k-dimensional linear span of $(e_1, \ldots, e_k)$. We claim that the stabilizer of E_0 consists of matrices (relative to the standard basis) of the form

$$\begin{bmatrix} M_1 & 0 \\ 0 & M_2 \end{bmatrix}$$

where M_1 and M_2 are orthogonal matrices of size, respectively, $k \times k$ and $(m-k) \times (m-k)$. Note that since E_0 is mapped to itself the first k columns have zeros as their last $(m-k)$ components. The orthogonal nature of M_1 forces the first k components of the last $(m-k)$ columns to be zero. We have therefore identified a subgroup of $O(m)$, isomorphic to $O(k) \times O(m-k)$ so that there is a bijective map

$$\frac{O(m)}{O(k) \times O(m-k)} \to G(k, m) \tag{8.25}$$

We use this bijection to topologize $G(k, m)$, homeomorphic to a quotient space of $O(m)$. Since $O(m)$ is compact, $G(k, m)$ is also compact.

(e) There is a similar generalization of a sphere in $\mathbb{R}^m$. S_{m-1} may be regarded as the set of orthonormal singletons (=unit vectors) in $\mathbb{R}^m$. For $1 \leq k \leq m$, let $S(k, m)$ be the set of ordered orthonormal k-tuples $b=(b_1, \ldots, b_k)$ in $\mathbb{R}^m$. Thus $S(1, m)=S_{m-1}$. We look at the action

$$\Phi : O(m) \times S(k, m) \to S(k, m), \quad \Phi(A, b) = (Ab_1, \ldots, Ab_k)$$

This action is transitive for the same reason as in the previous example. The stabilizer of $\bar{b}=(e_1, \ldots, e_k)$ in $O(m)$ consists of matrices of the following form:

$$\begin{bmatrix} I_k & 0 \\ 0 & M \end{bmatrix}$$

where I_k is the $k\times k$ identity matrix and $\mathrm{M}\in\mathrm{O}(m-k)$. Identifying $\mathrm{O}(m-k)$ as the subgroup $\{\mathrm{I}_k\}\times\mathrm{O}(m-k)$ of $\mathrm{O}(m)$, we obtain a bijection

$$\frac{\mathrm{O}(m)}{\mathrm{O}(m-k)} \to \mathrm{S}(k,m) \tag{8.26}$$

through which we introduce a topology on $\mathrm{S}(k,m)$. This is known as the ***Stiefel manifold***, and it is also compact.

We wish to proceed now with showing that if H is a closed subgroup of a Lie group G, then the homogeneous space G/H of left cosets can be given a (unique) smooth manifold structure as a quotient manifold of G. Note that G/H is actually a quotient topological space of G via the *right action* of H on G by

$$\Psi : G\times H \to G,\ \ \Psi(x,h) = xh$$

The same propositions as for left actions described in Chapter 5 hold here[3]. The corresponding equivalence relation $\sim$ here is

$$x\sim y \iff x^{-1}y\in H$$

Considering the continuous map $\nu{:}G\times G\to G$ by $\nu(x,y){=}x^{-1}y$, we have

$$\sim = \nu^{-1}(H)$$

Therefore, $\sim$ is closed in $G\times G$ and G/H is Hausdorff in quotient topology according to Subsection 14 of Chapter 5. Since the quotient map $G\to G/H$ is open for group actions by Lemma 16 of Chapter 5, it also follows that G/H is second-countable.

19. Theorem *Let G be a Lie group and H a closed (Lie) subgroup of G. Then G/H, with the quotient topology, possesses a unique smooth structure relative to which the quotient projection $q{:}G\to G/H$ is a smooth submersion.*

Proof. Uniqueness is a general fact proved in Theorem 17 of Chapter 5; we need only prove the existence of a smooth atlas satisfying the assertion. Let $\mathcal{H}$ be the tangent space to H at e, identified as the Lie algebra of H, and take a complementary linear subspace $\mathcal{K}$. Thus $\mathcal{K}\oplus\mathcal{H}{=}T_eG$. For the map $E{:}\mathcal{K}\oplus\mathcal{H}\to G$ by

$$E(X,Y) = (\exp X)(\exp Y)$$

we have $T_{\mathbf{0}_e}E{=}\mathbb{1}_{T_eG}$, so a symmetric open neighborhood $U_1\times V_1\subset\mathcal{K}\oplus\mathcal{H}$ of $\mathbf{0}_e$ is mapped diffeomorphically by E onto an open neighborhood W_1 of e in G. A smaller symmetric neighborhood $U\times V\subset U_1\times V_1$ of the zero element in $\mathcal{K}\oplus\mathcal{H}$ may be taken so that if $W{=}E(U\times V)$, then $W^3{=}WWW\subset W_1$. We claim that for any $g\in G$, the composition

$$\phi_g{=}q\circ L_g\circ(\exp|_U)$$

is a homeomorphism of U onto its image in G/H. First we show that by the choice of W, the map ϕ_g is one-to-one. Suppose that for $X,X'\in U$, $\phi_g(X){=}\phi_g(X')$. Then

[3]Alternatively, we may use the left action to obtain the equivalent theory for right coset space.

$(g\cdot \exp X)H=(g\cdot \exp X')H$, which implies that $(\exp X)^{-1}(\exp X')=h\in H$. Then $h\in W^2$, and we have

$$\exp X' = \exp X'\cdot \exp \mathbf{0}_e = \exp X\cdot h \in W^3\subset W_1$$

which, by the injectivity of E on $U_1\times V_1$, implies that $h=e$ and $X=X'$. ϕ_g is a composition of continuous maps and is continuous; we must show it is open. If U' is an open subset of U, then $E(U'\times V)$ is open in G, L_g is a diffeomorphism, q is open, and we have

$$\phi_g(U') = q\circ L_g\circ E(U'\times V)$$

since $E(U'\times V)\subset \exp(U')\cdot H$. This proves the claim that ϕ_g is a homeomorphism from U onto its image in G/H.

The sets $\phi_g(U)$ cover G/H since $\phi_g(\mathbf{0}_e)=gH$. They establish a C^0 atlas for G/H modeled on the linear space $\mathcal{K}$, which has dimension $\dim G-\dim H$. We let $U_g=\phi_g(U)$ and $\xi_g=(\phi_g)^{-1}$. Next we show that the collection $\{(U_g,\xi_g)\}$ is a C^∞ atlas for G/H. Suppose that $U_{g_1}\cap U_{g_2}\neq\emptyset$, we must show that $\xi_{g_2}\circ\xi_{g_1}{}^{-1}=\phi_{g_2}{}^{-1}\circ\phi_{g_1}$ is C^∞. Representing the elements of the non-empty intersection in the charts (U_{g_1},ξ_{g_1}) and (U_{g_2},ξ_{g_2}), respectively, as $(g_1\cdot \exp X_1)H$ and $(g_2\cdot \exp X_2)H$, we must show that the transition map $X_1\mapsto X_2$ is C^∞. Pick $\bar{X}_1\in\xi_{g_1}(U_{g_1}\cap U_{g_2})$. Thus,

$$(g_1\cdot \exp \bar{X}_1)H = (g_2\cdot \exp \bar{X}_2)H$$

This means there is $h\in H$ so that

$$g_1\cdot \exp \bar{X}_1\cdot h = g_2\cdot \exp \bar{X}_2$$

Now g_1W and g_2W are, respectively, open neighborhoods of $g_1\cdot \exp \bar{X}_1$ and $g_2\cdot \exp \bar{X}_2$; it therefore follows that there are open sets W_1 and W_2 so that

$$g_1\cdot \exp \bar{X}_1\in W_1\subset g_1W,\quad g_2\cdot \exp \bar{X}_2\in W_2\subset g_2W$$

and $R_hW_1=W_2$. Therefore, given X_1 and g_1, there is a unique smoothly determined (by E^{-1}) $Y\in\mathcal{H}$ so that

$$g_1\cdot \exp X_1\cdot h = g_2(\exp X_2)(\exp Y)$$

Therefore,

$$X_2 = \pi_1\circ E^{-1}\circ L_{g_2{}^{-1}g_1}\circ R_h(\exp X_1)$$

where $\pi_1{:}\mathcal{K}\oplus\mathcal{H}\to\mathcal{K}$ is projection on the first component. This shows that the atlas is C^∞. To see that q is a submersion, note that we can use the inverse mappings to $L_g\circ E|_{U\times V}$ as charts for G, therefore in local coordinates q appears as the projection $\pi_1{:}\mathcal{K}\oplus\mathcal{H}\to\mathcal{K}$, and is therefore a smooth surjective map. □

In 12b of Chapter 6 we studied the condition *q-equivariance*, under which the push-forward of a smooth vector field to a quotient manifold becomes a smooth vector field. In the case above, since G/H arises from the right action of H on G, q-equivariance of a vector field X on G is precisely the right-invariance of X. Thus the following corollary is obtained.

20. Corollary *Under the hypothesis of Theorem 19, if a smooth vector field X on G is right-invariant, then $Y{:}G/H\to T(G/H)$ defined by*

$$Y(q(x)) = T_x q(X(x)) \tag{8.27}$$

is a well-defined and smooth vector field on G/H.

As a familiar example, constant vector fields on $\mathbb{R}^m$ project to smooth vector fields on tori $\mathbb{T}^m$, the integral curves of which are either all closed curves or all dense spirals (see Example 2g of Chapter 4 and Example 2c of Chapter 5).

We can now revisit Example 18 in the light of Theorem 19. In the case of examples (d) and (e), Grassmann and Stiefel spaces, no *a priori* structure for these sets was known.[4] In fact, we defined their topologies as quotient coset spaces of the orthogonal group by appropriate subgroups. In each case, the subgroup in question is closed, therefore the quotient receives a smooth manifold structure by Theorem 19. We transfer this smooth structure to Grassmann and Stiefel spaces via the homeomorphisms given, so that we can now speak of Grassmann manifold and Stiefel manifold as homogeneous spaces. In the case of examples (a), (b) and (c), the quotient spaces were familiar examples of smooth manifolds studied earlier. Our task then is to show that the homeomorphism provided in each case is in fact a C^∞ diffeomorphism. For this we need a smooth version of the *Elementary Properties* in 17 that we will now present. We will use the notation of 17.

21. Theorem *Let G be a Lie group, M a smooth manifold and $\Phi{:}G\times M\to M$ a smooth action. Then:*
(a) For each $x\in M$, the map Φ^x has constant rank.
(b) For each $x\in M$, the stabilizer G_x of x is a closed (Lie) subgroup of G.
(c) If Φ is transitive, then for each $x\in M$, the map $\Phi^x{:}G\to M$ is a submersion.
(d) If Φ is transitive, then for each $x\in M$, the induced map $\Phi'_x{:}G/G_x\to M$ is a smooth diffeomorphism.

Proof. (a) For any $g\in G$ and any $x\in M$ we have

$$\Phi^x \circ L_g = \Phi_g \circ \Phi^x \tag{8.28}$$

If g_1, g_2 are arbitrary elements of G, then letting $g=g_2 g_1{}^{-1}$, and taking the tangent map of the two sides of (8.28) at g_1, we conclude from the fact that $L_{g_2 g_1{}^{-1}}$ and $\Phi_{g_2 g_1{}^{-1}}$ are diffeomorphisms that $T_{g_1}\Phi^x$ and $T_{g_2}\Phi^x$ have the same rank.
(b) Since G_x is a level set of a smooth function of constant rank, namely $G_x=(\Phi^x)^{-1}(\{x\})$, then it is a closed submanifold of G. It is therefore not even necessary to appeal to Theorem 16 to conclude that G_x is a Lie subgroup.
(c) By (a), Φ^x has constant rank, say k. By Theorem 4b of Chapter 5. for any $g\in G$,

[4] $G(k,m)$ was actually introduced in Exercises 4.12 and 5.22, but those accounts are not being used here. In Exercise 8.22, we are asking the reader to prove the equivalence of all these versions.

there is an open set U around g so that $\Phi^x(U)$ is diffeomorphic to an open subset of $\mathbb{R}^k$. If $k<\dim M$, then $\Phi^x(U)$ will be a set of measure zero. Now there is a countable cover of the manifold G by open sets such as U, therefore the union of the $\Phi^x(U)$ will have measure zero. This contradicts the transitivity of action by which this union should cover M. Therefore, $k=\dim M$, i.e., Φ^x is a submersion.
(d) By 17(a) and (b), Φ'_x is one-to-one, onto and continuous. It follows from Theorem 19 above and **Lemma b** following the statement of Theorem 17 of Chapter 5, that Φ'_x is smooth. Since G/G_x and M have the same dimension, Φ'_x is a diffeomorphism. □

In view of this theorem, we can conclude our visit to Example 18 by asserting the existence of the following diffeomorphisms:

$$\mathbb{R}^m/\mathbb{Z}^m \cong \mathbb{T}^m, \quad \mathrm{O}(m)/\mathrm{O}(m-1) \cong S_{m-1}, \quad \mathrm{SO}(m)/\mathrm{O}(m-1) \cong \mathbb{R}P(m-1) \tag{8.29}$$

A manifold M satisfying (d) of Theorem 21 is also referred to as a ***homogeneous manifold***.

EXERCISES

8.1 Show that a manifold with boundary does not admit a topological group structure.

8.2 Show that the closure of any subgroup of a topological group is a subgroup.

8.3 Let G be a topological group.
(a) Show that the center of G, $Z(G)=\{g\in G:gx=xg, \forall x\in G\}$, is closed.
(b) Let H be a subgroup of G. Show that the normalizer of H, namely

$$N(H)=\{g\in G:gHg^{-1}\subset H\}$$

is closed.

8.4 If G is a connected topological group, show that its commutator subgroup, i.e., the subgroup generated by elements of the form $xyx^{-1}y^{-1}$, is also connected.

8.5 Let $G\times X\to X$ be a continuous action of a compact topological group G on a compact Hausdorff space X. Show that the quotient space is Hausdorff.

8.6 Regarding $\mathbb{H}$ as a two-dimensional vector space over $\mathbb{C}$ with basis $(\mathbf{1},\mathbf{j})$, show that left translation by an element $a\mathbf{1}+b\mathbf{j}\in\mathbb{H}$ is a linear map given by the matrix

$$\begin{bmatrix} a & -\bar{b} \\ b & \bar{a} \end{bmatrix}$$

8.7 Show that SU(2) is isomorphic with the group of 2×2 complex matrices of the form above (Exercise 8.6). Describe an isomorphism between SU(2) and S_3.

8.8 Regarding S_3 as the group of unit quaternions, show that the degree of the map $a\mapsto a^m$, where m is an integer, is m. More generally, prove that for SU(n), the degree of $a\mapsto a^m$ is m^{n-1}.

8.9 Describe a two-sheeted covering space SU(2)×SU(2)→SO(4).

8.10 With reference to Subsection 5, show that SL(n, $\mathbb{C}$), U(n) and SU(n) are Lie subgroups of GL(n, $\mathbb{C}$) without appeal to Cartan's Theorem.

8.11 Describe a diffeomorphism in each case:

$$U(m)/U(m-1)\cong S_{2m-1},\quad SU(m)/SU(m-1)\cong S_{2m-1},\quad SU(m)/U(m-1)\cong \mathbb{CP}(m-1)$$

8.12 *Polar Decomposition Theorem* of linear algebra asserts the following:

(a) If A∈GL(n, $\mathbb{R}$), then A=PT, where T is orthogonal and P is symmetric with all eigenvalues of P positive.
(b) If A∈GL(n, $\mathbb{C}$), then A=PT, where T is unitary and P is hermitian with all eigenvalues of P positive.
Use polar decomposition to give another proof of the connectedness of GL$^+$(n, $\mathbb{R}$) and GL(n, $\mathbb{C}$).

8.13 Let G be a topological group and H a closed subgroup.
(a) Show that if H and G/H are compact, then so is G.
(b) Show that if H and G/H are connected, then so is G.
(c) Use the above to prove the connectedness of $\mathrm{SO}(m)$, $\mathrm{SU}(m)$ and $\mathrm{U}(m)$.

8.14 Suppose that for $n\times n$ matrices X and Y and some $t_0>0$, we have $\exp t(X+Y)=(\exp tX)(\exp tY)$ for all $|t|<|t_0|$. Show that $XY=YX$.

8.15 Give an example of a pair of non-commuting matrices X and Y so that $\exp(X+Y)=(\exp \mathrm{X})(\exp \mathrm{Y})$. (Hint: There exist 2×2 examples.)

8.16 Let G be a Lie group and $X\in T_eG$. Show that $\exp tX$ is also the integral curve of the *right-invariant* vector field determined by X that passes through e. Describe the flow of this right-invariant vector field.

8.17 Let G be a Lie group and $\iota{:}G\to G$ the inverse operation in the group. If $\tilde{X}$ is a left-invariant vector field, show that $\iota_*\tilde{X}$ is a right-invariant vector field.

8.18 (a) Let G be a topological group. If a map $f{:}G\to G$ commutes with every left-translation of G, then f is a right translation.
(b) Suppose now G is a Lie group, X is a left-invariant vector field and Y a right-invariant vector field. Show that $[X, Y]=0$.
(c) Suppose further that G is a connected Lie group and Y is a smooth vector field so that $[X, Y]=0$ for every left-invariant vector field X. Show that Y is right-invariant.

8.19 In parts (a) and (b) of this exercise we consider an important generalization of the orthogonal group. Let $\beta{:}\mathbb{R}^n\times\mathbb{R}^n\to\mathbb{R}$ be a non-degenerate bilinear function (see Section D of Chapter 6 for relevant terminology and facts). If $(e_1, \ldots, e_n)$ is the standard basis of $\mathbb{R}^n$, then β is completely characterized by the invertible matrix $\mathrm{B}=[\beta(e_i, e_j)]$, the value of $\beta(x, y)$ being given by $x^{\mathrm{T}}\mathrm{B}y$, where y is written as a column vector, and x^{T} indicates x as a row vector. Let G be the set $n\times n$ matrices M that preserve β, i.e., $\beta(Mx, My)=\beta(x, y)$, for all $x, y\in\mathbb{R}^n$.
(a) Show that G is a Lie subgroup of $\mathrm{GL}(n, \mathbb{R})$, and that the elements M of G are characterized by the property $\mathrm{M}^{\mathrm{T}}\mathrm{BM}=\mathrm{B}$.
(b) Show that the Lie algebra of G can be identified with the set of $n\times n$ matrices X that satisfy $\mathrm{X}^{\mathrm{T}}\mathrm{B}+\mathrm{BX}=0$. (Hint: Imitate the final argument in Example 9d.)
(c) Study, in particular, the non-degenerate bilinear map $\mathbb{R}^{2n}\times\mathbb{R}^{2n}\to\mathbb{R}$ given by the matrix

$$\begin{bmatrix} 0 & I_n \\ -I_n & 0 \end{bmatrix}$$

The group here is known as the ***symplectic group*** and is denoted by $\mathrm{Sp}(2n)$. Determine the form of the matrices in the symplectic group and the form of matrices in its Lie algebra. (The reader must be cautioned that this terminology and notation is also used in other contexts.)

8.20 Let G be a Lie group and H a closed normal subgroup of G. Show that G/H with the induced homogeneous structure (Theorem 19) and quotient group structure is a Lie group.

8.21 Let H and K be Lie subgroups of a Lie group G so that $H \subset K$. Prove that H is a Lie subgroup of K.

8.22 (a) Show that the Grassmann manifold $\mathrm{G}(k,m)$ can be realized as a quotient manifold of Stiefel manifold $\mathrm{S}(k,m)$ under an action of $\mathrm{O}(k)$.
(b) In addition to the definition of Grassmann manifold in Example 18d, we have encountered other definitions in Exercises 4.12, 5.22 and (a) above. Show that these are all equivalent.

8.23 By an ***n*-flag** in $\mathbb{R}^n$ we mean a finite sequence $(E_1, \ldots, E_n)$ so that E_j is a j-dimensional linear subspace of $\mathbb{R}^n$ and $E_1 \subset E_2 \subset \cdots \subset E_n = \mathbb{R}^n$. We denote the set of n-flags by $\mathrm{F}(n)$. Show that $\mathrm{F}(n)$ has the structure of a homogeneous manifold of dimension $\frac{1}{2}n(n-1)$.

8.24 Let $\mathcal{L}$ be an n-dimensional Lie algebra over $\mathbb{R}$ and consider a basis $(e_1, \ldots, e_n)$ for the linear space $\mathcal{L}$. The ***structural constants*** c_{ij}^k are defined by

$$[e_i, e_j] = \sum_{k=1}^{n} c_{ij}^k e_k$$

(a) Show that the following relations hold:

$$c_{ii}^k = 0, \quad \sum_k (c_{ij}^k c_{kl}^m + c_{jl}^k c_{ki}^m + c_{li}^k c_{kj}^m) = 0$$

for all indices i, j, k, l, m.
(b) Conversely, suppose that a set of real numbers c_{ij}^k satisfy the above relations. Show that a (unique up to isomorphism) Lie algebra with these structural constants exists.
(c) A Lie algebra is commutative if and only if all c_{ij}^k are zero. Show that there is a unique non-commutative Lie algebra of dimension 2 (necessarily) isomorphic to the Lie algebra of the affine group $\mathrm{Aff}(2, \mathbb{R})$.

8.25 The following three matrices form a basis for the Lie algebra so(3) of SO(3):

$$X = \begin{bmatrix} 0 & 0 & 0 \\ 0 & 0 & -1 \\ 0 & 1 & 0 \end{bmatrix}, Y = \begin{bmatrix} 0 & 0 & 1 \\ 0 & 0 & 0 \\ -1 & 0 & 0 \end{bmatrix}, Z = \begin{bmatrix} 0 & -1 & 0 \\ 1 & 0 & 0 \\ 0 & 0 & 0 \end{bmatrix}$$

(a) Write down the structural constants (see the previous exercise) for this Lie algebra.
(b) Show that under the linear map $\mathrm{so}(3) \to \mathbb{R}^3$ by $X \mapsto e_1$, $Y \mapsto e_2$ and $Z \mapsto e_3$, the bracket in so(3) is converted to the cross product in $\mathbb{R}^3$.

(c) Each of X, Y and Z generates a right-invariant vector field on SO(3). Describe the induced vector fields on the quotient manifolds $\mathrm{SO}(3)/\mathrm{SO}(2)\cong S_2$ and $\mathrm{SO}(3)/\mathrm{O}(2)\cong\mathbb{RP}(2)$.

8.26 Describe a diffeomorphism between SO(3) and $\mathbb{RP}(3)$.

8.27 Let H and N be Lie groups and suppose that a smooth action $\Phi{:}H\times N\to N$ is given. Consider $G=H\times N$ as a product manifold, but define an operation on pairs $(h_1,n_1),(h_2,n_2)$ as follows:

$$(h_1,n_1)*(h_2,n_2)=(h_1h_2,n_1\Phi(h_1,n_2))$$

(a) Verify that G with this product is a Lie group. This is called a ***semidirect product*** and is denoted by $H\ltimes_\Phi N$.
(b) Suppose that G and H are Lie groups, $p{:}G\to H$ a surjective homomorphism of Lie groups, and $s{:}H\to G$ a Lie group homomorphism so that $p\circ s=\mathbb{1}_H$. Show that $N=\ker p$ is a normal Lie subgroup of G, and that G is isomorphic to $H\ltimes_\Phi N$ for an action Φ of H on N that you will define.
(c) Demonstrate the following isomorphisms with semidirect products. In each case, identify the appropriate action

$$\mathrm{Aff}(n,\mathbb{R})\cong\mathrm{GL}(n,\mathbb{R})\ltimes_\Phi T(n)$$
$$\mathrm{O}(n)\cong\{1,-1\}\ltimes_\Phi\mathrm{SO}(n)$$
$$\mathrm{U}(n)\cong S_1\ltimes_\Phi\mathrm{SU}(n)$$

8.28 If $\mathcal{L}$ is a Lie algebra, by a ***derivation*** of $\mathcal{L}$ we mean a linear map $D{:}\mathcal{L}\to\mathcal{L}$ that satisfies $D[X,Y]=[D(X),Y]+[X,D(Y)]$. Let $\mathcal{H}$ and $\mathcal{N}$ be two Lie algebras and suppose that $\mathcal{H}$ acts on $\mathcal{N}$ by derivations, i.e., for each $X\in\mathcal{H}$, a linear map $\phi(X){:}\mathcal{N}\to\mathcal{N}$ is given that satisfies

$$\phi(X)[Y_1,Y_2]=[\phi(X)(Y_1),Y_2]+[Y_1,\phi(X)(Y_2)]$$

On the linear direct sum $\mathcal{H}\oplus\mathcal{N}$, we define [,] by

$$[(X,Y),(X',Y')]=([X,X'],[Y,Y']+\phi(X)(Y')-\phi(X')(Y))$$

(a) Show that under this bracket operation, $\mathcal{H}\oplus\mathcal{N}$ becomes a Lie algebra. This is called the ***semidirect sum*** of $\mathcal{H}$ and $\mathcal{N}$, and is denoted by $\mathcal{H}\oplus_\phi\mathcal{N}$.
(b) Show that $\mathcal{H}\times\{0\}$ is a sub-Lie-algebra, and $\{0\}\times\mathcal{N}$ an ideal in $\mathcal{H}\oplus_\phi\mathcal{N}$.
(c) For a semidirect product of Lie groups, show that the Lie algebra is the semidirect sum of Lie algebras in a natural way. Verify this for the examples of the previous exercise.

8.29 Show that exp is "natural," in the sense that if $f{:}G\to G'$ is a smooth homomorphism of Lie groups, then $f(\exp A)=\exp(T_ef(A))$.

8.30 Show that any continuous homomorphism f of $(\mathbb{R},+)$ into a Lie group G is a one-parameter subgroup, i.e., it is smooth. (Hint: Use the fact that exp is a local diffeomorphism at $\mathbf{0}_e$ and the previous exercise to find a one-parameter subgroup that agrees with f on a dense set, and then use continuity.)

8.31 Let $\Phi:G\times M\to M$ be a smooth action of a Lie group on a manifold. For each $A\in T_eG$ and each $x\in M$, define $\Psi_A(t,x)=\Phi(\exp tA, x)$. Show this is a smooth flow on M and denote its generator by $\hat{A}$. Show that $A\mapsto\hat{A}$ is a Lie algebra homomorphism.

8.32 Let $\alpha\in\Omega^1 S_2$ and suppose $\rho\in\mathrm{SO}(2)$ is such that $\rho^*\alpha=\alpha$. Show that $\alpha=0$. Can you generalize?

8.33 Let G be a Lie group with Lie algebra $\mathcal{G}$. A p-form $\alpha\in\Omega^p G$ is called ***left-invariant*** if $L_g^*\alpha=\alpha$ for all $g\in G$.
(a) Show that every left-invariant p-form is smooth.
(b) If $(X_1,\ldots,X_n)$ is a basis for $\mathcal{G}$, show that the dual one-forms $(\alpha^1,\ldots,\alpha^n)$ form a basis for the space of left-invariant one-forms over $\mathbb{R}$. More generally, any left-invariant p-form has a unique representation as

$$\sum_{i_1<\cdots<i_p} c_{i_1\cdots i_p}\alpha^{i_1}\wedge\cdots\wedge\alpha^{i_p}$$

where $c_{i_1\cdots i_p}\in\mathbb{R}$.

8.34 Let G be a Lie group with Lie algebra $\mathcal{G}$, and denote the inverse mapping $g\mapsto g^{-1}$ by $\iota:G\to G$. Suppose that a p-form α is both left- and right-invariant.
(a) Show that $\iota^*\alpha=(-1)^p\alpha$.
(b) Show that $d\alpha=0$.
(c) Deduce that if G is abelian, then so is $\mathcal{G}$.

8.35 Let G be a Lie group and $\alpha\in\Omega^p G$. Show that if α is left-invariant, and X is a right-invariant vector field, then $L_X\alpha=0$. Suppose G is connected, α is left-invariant and $L_X\alpha=0$. Conclude that X is right-invariant.

8.36 Show that the map exp is not onto for $\mathrm{SL}(2,\mathbb{R})$ even though this group is connected, by showing that the following matrix is not in the image of exp:

$$\begin{bmatrix} -2 & 0 \\ 0 & -1/2 \end{bmatrix}$$

8.37 Let G be a compact and connected Lie group. Show that every left-invariant volume element for G is also right-invariant.

Part IV

Geometric Structures

CHAPTER 9

An Introduction to Connections

In the manifold machinery that we have developed, the acceleration vector of a parametrized curve on a manifold M appears as a tangent vector to the tangent bundle TM of the manifold. Likewise, if X is a tangent vector field on M, then its derivative (tangent map) takes values in $T(TM)$. One can argue that differential geometry proper begins when one can realize these objects as tangent vectors to the base manifold M. Second derivatives and derivatives of vector fields are ubiquitous throughout classical differential geometry of curves and surfaces in $\mathbb{R}^3$, where the prevailing linearity of Euclidean space allows one to visualize the second derivative without resorting to the tangent bundle. The same can be said of classical mechanics in Euclidean space wherein the all-important acceleration vector seems to coexist with the velocity vector as a directed segment emanating from the position of a moving particle. The present chapter is aimed at providing a bridge between basic manifold theory of the previous chapters and the study of manifold-based differential and Riemannian geometry. The goal of the first two sections is to convince the reader of the necessity of a new structure, namely a "connection." We have tried to patiently map out a natural path to the definition of a "covariant derivative" in the third section. Logically, one can start with the axiomatic approach to covariant derivative (see Definition 17), adopt that as the definition of a "linear connection", and proceed from there, as is often done. But we feel that this leaves a gap in the intuitive preparation of the newcomer, hence our exploratory discussions of the first two sections. It is possible to start out with Section C and work around the references to Sections A and B. The highlight of Section D is Gauss' *Theorema Egregium*, the turning point that may have led to Riemann's development of modern differential geometry. In the final section, we treat the foundations of classical mechanics in a spirit akin both to the earlier sections and to the original Newtonian approach.

A. The Geography of the Double Tangent Bundle

Let M be a smooth m-manifold, TM the tangent bundle and $T^2M{=}T(TM)$ the *double tangent bundle*. The tangent-bundle projections are, respectively, $\tau_M{:}TM{\to}M$ and $\tau_{TM}{:}T^2M{\to}TM$. Since τ_M is a submersion by Example 8d of Chapter 5, the set T_aM is a smooth submanifold of TM for each $a{\in}M$, being the inverse image of $\{a\}$. TM is conveniently visualized as the disjoint union of the T_aM as a traverses M (see Figure 1). The double tangent bundle, more difficult to visualize, has a rich and interesting structure. In the Subsections below, we will describe some features of the structure.

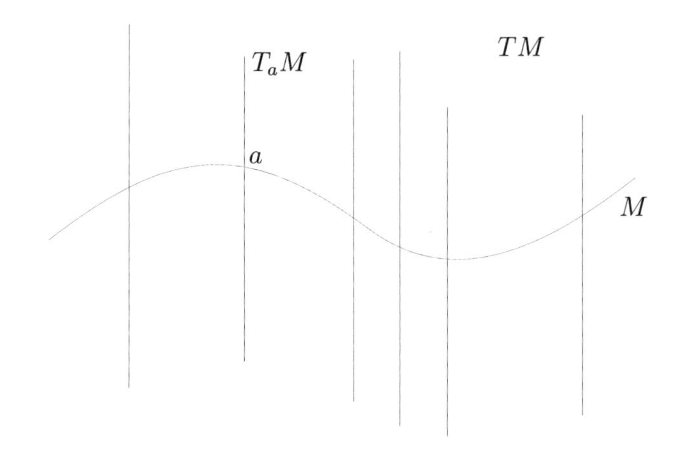

FIGURE 1. Visualizing the tangent bundle

1. The Zero-Section

Recall from Example 12c of Chapter 5 that the zero-section

$$\zeta_M : M \to TM, \quad \zeta_M(a) = \mathbf{0}_a \in T_aM$$

is a smooth embedding of M into TM. It is often useful to visualize M as a submanifold of TM by treating the image of the zero-section as a copy of M. Then at a point $\mathbf{0}_a$, $a \in M$, one can observe two copies of the m-dimensional T_aM as subspaces of the $2m$-dimensional $T_{\mathbf{0}_a}(TM)$. These are complementary and span $T_{\mathbf{0}_a}(TM)$. One is $T_{\mathbf{0}_a}(T_aM)$, and the other is $T_{\mathbf{0}_a}(\zeta_M(M))$, treating T_aM and $\zeta_M(M)$ as submanifolds of TM (see Figure 2). To see that these are actually complementary, note that the first is the kernel of $T_{\mathbf{0}_a}\tau_M$, while the restriction of $T_{\mathbf{0}_a}\tau_M$ to the latter is an isomorphism since $\tau_M \circ \zeta_M = \mathbb{1}_M$. We call $T_{\mathbf{0}_a}(T_aM)$ and $T_{\mathbf{0}_a}(\zeta_M(M))$, respectively, the ***vertical subspace*** and the ***horizontal subspace*** of $T_{\mathbf{0}_a}(TM)$.

2. The Vertical Bundle

Let $z \in T_aM \subset TM$. Since T_aM is a submanifold of TM, the tangent space $T_z(T_aM)$ is canonically identified as an m-dimensional linear subspace of the $2m$-dimensional linear space $T_z(TM)$. We call $T_z(T_aM)$ the ***vertical subspace*** (of $T_z(TM)$) at z, and denote it by V_z. The assignment $z \mapsto V_z$ is a smooth m-plane field on TM and defines a vector sub-bundle of $\tau_{TM}: T^2M \to TM$. Note that if $X \in V_z$, then $T\tau_M(X) = \mathbf{0}_a$ since τ_M maps the entire T_aM to the single point $\{a\}$. The (disjoint) union of V_z's, as z ranges over T_aM, is the tangent bundle of T_aM and will be denoted by V_aM. The (disjoint) union of the V_aM, as a ranges over M, is denoted by VM, and will be called the ***vertical bundle*** over TM. VM is the total space of the sub-bundle of

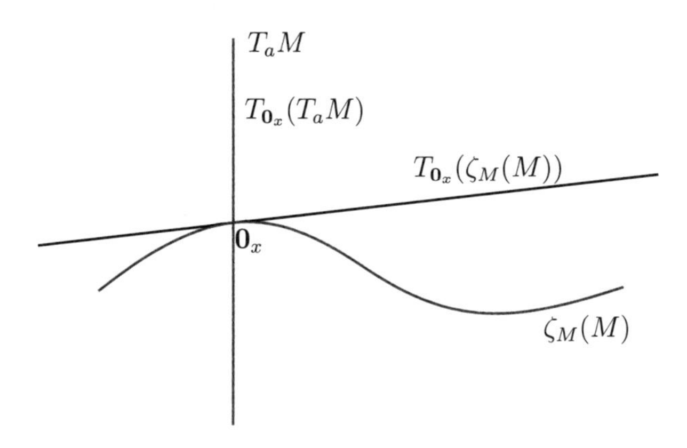

FIGURE 2. Two replicas of tangent space

$\tau_{TM}:T^2M \to TM$ defined by the V_z; it is a $3m$-dimensional smooth submanifold of T^2M by Lemma 7 of Chapter 6.

We note that the definition of VM is "natural," in the sense that if $f:M \to M'$ is a smooth map of manifolds, then

$$T^2 f(VM) \subset VM' \tag{9.1}$$

Specifically, for $z \in TM$ and $z' = Tf(z)$, one has $T^2 f(V_z) \subset V_{z'}$. This follows from the commutative diagram on the right, below, which itself follows from the commutative diagram on the left by application of the functor T.

$$\begin{array}{ccccccc} TM & \xrightarrow{Tf} & TM' & \quad & T^2M & \xrightarrow{T^2 f} & T^2M' \\ \downarrow{\tau_M} & & \downarrow{\tau_{M'}} & & \downarrow{T\tau_M} & & \downarrow{T\tau_{M'}} \\ M & \xrightarrow{f} & M' & & TM & \xrightarrow{Tf} & TM' \end{array} \tag{9.2}$$

Of course, if $f:M \to M'$ is a smooth diffeomorphism, then the inclusion in (9.1) becomes an equality.

3. Maps $VM \to TM$

There are three distinct maps from VM into TM that we now describe.

(a) The restriction of $T\tau_M:T^2M \to TM$ to VM, which maps VM onto $\zeta_M(M)$. In fact, VM was defined as the inverse image of $\zeta_M(M)$ under $T\tau_M$.

(b) The restriction of $\tau_{TM}:T^2M \to TM$ to VM, which assigns to each $X \in V_z$ the base point z. This map is onto TM, and is the bundle projection.

(c) We now describe a third map, $o_M:VM \to TM$, also onto TM, that will be of great utility in the sequel. Since each T_aM is a linear space, its tangent space at every point z (i.e., the space V_z) is canonically isomorphic to T_aM itself. In the convention of Chapter 2, V_z can be considered as the set of ordered pairs (z,z'), with z fixed and z' ranging over T_aM. o_M is the map that projects (z,z') to z'. A schematic drawing is attempted in Figure 3. One can visuaslize o_M as the map that slides each $X \in V_z = T_z(T_aM)$ parallel to itself down to the origin $\mathbf{0}_a$ of T_aM.

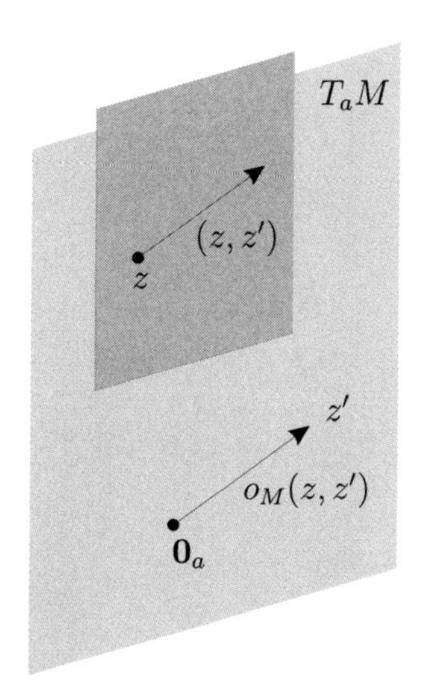

FIGURE 3. Visualizing o_M

In the subsequent discussion, we explore two distinct vector bundle structures for T^2M over TM. The following lemma will pave the way.

4. Lemma *Let $\pi{:}E\to P$ a smooth vector bundle of fiber type F over a p-manifold P. Then $T\pi{:}TE\to TP$ can be made into a smooth vector bundle of fiber type $F\times F$.*

PROOF. Suppose U is an open subset of P, and $\Phi{:}\pi^{-1}(U)\to U\times F$ is a local-product diffeomorphism. Then by applying the functor T to the commutative diagram on the left, below, we obtain the commutative diagram on the right.

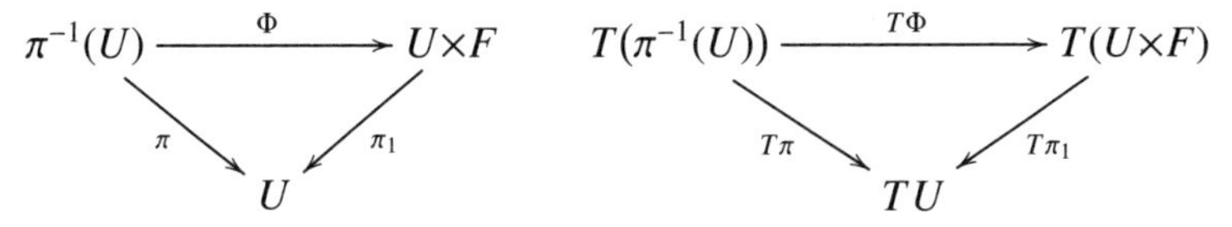

In the case of an open subset U of $\mathbb{R}^p$, $T(U\times F) = (U\times F)\times(\mathbb{R}^p\times F)$. An element of this set can be represented as $(a,e_1;u,e_2)$, and $T\pi_1(a,e_1;u,e_2) = (a,u)$. Using the canonical diffeomorphism $C{:}T(U\times F)\to TU\times TF$ (see Formula (4.18) of Chapter 4 and Example 2c of Chapter 6), and denoting projection on the first component

$TU{\times}TF{\to}TU$ by Π_1, we obtain the desired local-product structure from the following commutative diagram.

$$\begin{array}{ccccc} T(\pi^{-1}U) & \xrightarrow{T\Phi} & T(U{\times}F) & \xrightarrow{C} & TU{\times}TF \\ & {\scriptstyle T\pi}\searrow & \downarrow{\scriptstyle T\pi_1} & \swarrow{\scriptstyle \Pi_1} & \\ & & TU & & \end{array} \tag{9.3}$$

Of course, $TF{=}F{\times}F$, and the fiber over $(a,u){\in}TU$ is in bijective correspondence with $F{\times}F$. Now suppose that a VB-atlas $\{(U_\alpha,\Phi_\alpha)\}$ for E is given, and $U_\alpha{\cap}U_\beta{\neq}\emptyset$. We show that there exists a smooth map

$$\psi_\alpha^\beta : TU_\alpha{\cap}TU_\beta \to \mathrm{GL}(F{\times}F)$$

so that

$$(C{\circ}T\Phi_\beta){\circ}(C{\circ}T\Phi_\alpha)^{-1}((a,u),(e_1,e_2)) = ((a,u),\psi_\alpha^\beta(a,u)(e_1,e_2))$$

Let us compute the left-hand side, given that $\Phi_\beta{\circ}{\Phi_\alpha}^{-1}(a,e){=}(a,(\phi_\alpha^\beta(a))(e))$.

$$\begin{aligned}(C{\circ}T\Phi_\beta){\circ}(C{\circ}T\Phi_\alpha)^{-1}((a,u),(e_1,e_2)) &= C\Big(T(\Phi_\beta{\circ}{\Phi_\alpha}^{-1})((a,e_1),(u,e_2))\Big)\\ &= C\Big((a,e_1),(D(\Phi_\beta{\circ}{\Phi_\alpha}^{-1})(a,e_1))(u,e_2)\Big)\\ &= C\Big((a,e_1),\Big(u,(D\phi_\alpha^\beta(a)(e_1))(u){+}(\phi_\alpha^\beta(a))(e_2)\Big)\Big)\\ &= \Big((a,u),\Big(e_1,(D\phi_\alpha^\beta(a)(e_1))(u){+}(\phi_\alpha^\beta(a))(e_2)\Big)\Big)\end{aligned}$$

We then define ψ_α^β by

$$(\psi_\alpha^\beta(a,u))(e_1,e_2) = \Big(e_1,(D\phi_\alpha^\beta(a)(e_1))(u){+}(\phi_\alpha^\beta(a))(e_2)\Big)$$

Note that $(\psi_\alpha^\beta(a,u))(e_1,e_2)$ is linear in (e_1,e_2) and can be represented in matrix form as

$$\begin{bmatrix} I_k & \mathbf{0} \\ * & \phi_\alpha^\beta(a)\end{bmatrix}$$

where $k{=}\dim F$. It follows that indeed $\psi_\alpha^\beta(a,u){\in}\mathrm{GL}(F{\times}F)$, and the proof is complete. □

Applying the above lemma to $\tau_M{:}TM{\to}M$, we obtain a new vector bundle structure for T^2M over TM, namely $T\tau_M{:}T^2M{\to}TM$, which is entirely distinct from $\tau_{TM}{:}T^2M{\to}TM$. For example, for $\mathbf{0}_a{\in}TM$, the fiber of τ_{TM} over $\mathbf{0}_a$ consists of all vectors tangent to TM at this point, i.e., the fiber is $T_{\mathbf{0}_a}(TM)$. On the other hand, the fiber of $T\tau_M$ over $\mathbf{0}_a$ is V_aM. We thus have a picture of the typical fiber of $T\tau_M$ in case the base point z is on the zero-section. What if $z{\neq}0_a$? Perhaps the best way to visualize $(T\tau_M)^{-1}(z)$ is as follows. For $z{\in}T_aM$, first consider $\sigma(z){=}T_a\zeta_M(z)$, which is an element of the horizontal subspace at 0_a. Now the fiber of $T\tau_M$ over z can be visualized as the set of vectors tangent to TM at points of T_aM that project vertically down to $\sigma(z)$ under $T\tau_M$ (see Figure 4).

We will now describe the linear space operations on the fibers of $T\tau_M$. Let us

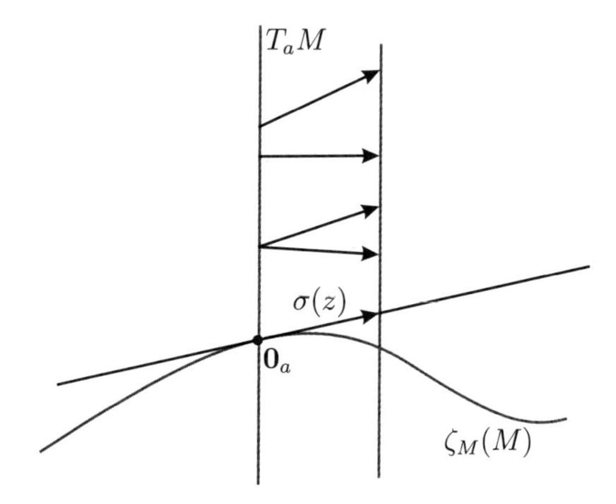

FIGURE 4. Visualizing the fiber over z

denote the vector space operations of addition and multiplication by scalars on the tangent bundle $\tau_M{:}TM{\to}M$ by A and m_r, respectively. For $z_1, z_2, z \in T_aM$ and $r{\in}\mathbb{R}$:

$$A(z_1, z_2) = z_1 + z_2$$
$$m_r(z) = rz$$

It follows from the local product structure of $T\tau_M$, (9.3), that the vector space operations in the fibers of $T\tau_M$ will be TA and Tm_r. If $X, X_1, X_2{\in}(T\tau_M)^{-1}(z)$, and $r{\in}\mathbb{R}$, then

$$Tm_r(X) \in (T\tau_M)^{-1}(z), \quad TA(X_1, X_2) \in (T\tau_M)^{-1}(z)$$

We will use the notation $r{\boxdot}X$ and $X_1{\boxplus}X_2$, respectively, for $Tm_r(X)$ and $TA(X_1, X_2)$, to describe the linear operations in the vector bundle $T\tau_M$. The symbols $\cdot$ and $+$ will continue to be used for linear operations in the fibers of τ_M and τ_{TM}.

5. Local Representation

At this point, it is useful to describe the two vector bundle structures in local coordinates. Let (U, ξ) be a chart for M, $a{\in}U$ and $z{\in}T_aM$. Thus z has the local representation $z{=}\sum_{i=1}^m z^i \frac{\partial}{\partial \xi^i}(a)$. As coordinates induced by the local chart $(TU, T\xi)$, we take $(q, \dot{q})$, where $q{=}(q^1, \ldots, q^m)$ with $q^i{=}\xi^i{\circ}\tau_M$, and $\dot{q}^i{=}d\xi^i$. Thus $(q, \dot{q})(z){=}(\xi^1(a), \ldots, \xi^m(a); z^1, \ldots, z^m)$. A tangent vector X to TU, $X{\in}T_z(TM)$, where $z{\in}T_aM$, will have the representation

$$X = \sum_{i=1}^m u^i \frac{\partial}{\partial q^i}(z) + \sum_{i=1}^m \dot{u}^i \frac{\partial}{\partial \dot{q}^i}(z) \tag{9.4}$$

where $u^i, \dot{u}^i \in \mathbb{R}$. We will then have

$$\tau_{TM}(X) = z = \sum_i z^i \frac{\partial}{\partial \xi^i}(a), \quad T\tau_M(X) = \sum_i u^i \frac{\partial}{\partial \xi^i}(a) \tag{9.5}$$

Since for $r \in \mathbb{R}$, we have $m_r(z) = \sum_i (rz^i)\frac{\partial}{\partial \xi^i}(a)$, then

$$r \boxdot X = Tm_r(X) = \sum_i u^i \frac{\partial}{\partial q^i}(rz) + \sum_i r\dot{u}^i \frac{\partial}{\partial \dot{q}^i}(rz) \tag{9.6}$$

Suppose that $z_j \in T_aM$, $j=1,2$, with $z_j = \sum_i z^i_j \frac{\partial}{\partial \xi^i}(a)$. Then $A(z_1, z_2) = \sum_i (z^i_1 + z^i_2)\frac{\partial}{\partial \xi^i}(a)$. Now let $X_j \in T_{z_j}(TM)$, $j=1,2$, so that $T\tau_M(X_1) = T\tau_M(X_2)$, i.e., X_1 and X_2 belong to the same fiber of $T\tau_M : T^2M \to TM$. Thus for $j=1,2$:

$$X_j = \sum_{i=1}^m u^i_j \frac{\partial}{\partial q^i}(z_j) + \sum_{i=1}^m \dot{u}^i_j \frac{\partial}{\partial \dot{q}^i}(z_j)$$

where $u^i_1 = u^i_2 := u^i$ because of the second equality in (9.5). It follows then from $A(z_1, z_2) = z_1 + z_2$ that

$$X_1 \boxplus X_2 = TA(X_1, X_2) = \sum_{i=1}^m u^i \frac{\partial}{\partial q^i}(z_1+z_2) + \sum_{i=1}^m (\dot{u}^i_1 + \dot{u}^i_2)\frac{\partial}{\partial \dot{q}^i}(z_1+z_2) \tag{9.7}$$

By setting $r=0$ in (9.6), we identify the zero element of the vector space $(T\tau_M)^{-1}(z)$, $z \in T_aM$, which is

$$\sum_{i=1}^m z^i \frac{\partial}{\partial q^i}(\mathbf{0}_a) = \sigma(z) \tag{9.8}$$

This representation remains valid under coordinate change since $\sigma(z)$ is canonically defined.

The following theorem describes the relation of the two vector bundle structures $T^2M \to TM$.

6. Theorem *Let M be a smooth manifold.*
(a) For $z \in TM$, $X, Y \in T_z(TM)$ and $r \in \mathbb{R}$

$$r \boxdot (X + Y) = (r \boxdot X) + (r \boxdot Y) \tag{9.9}$$

(b) For $a \in M$ and $z_1, z_2 \in T_aM$, suppose that $X_1, Y_1 \in T_{z_1}(TM)$ and $X_2, Y_2 \in T_{z_2}(TM)$ are so that $T\tau_M(X_1) = T\tau_M(X_2)$ and $T\tau_M(Y_1) = T\tau_M(Y_2)$, then

$$(X_1 + Y_1) \boxplus (X_2 + Y_2) = (X_1 \boxplus X_2) + (Y_1 \boxplus Y_2) \tag{9.10}$$

Proof. (a) This is equivalent to $Tm_r(X+Y) = Tm_r(X) + Tm_r(Y)$, which holds for the tangent map.

(b) We use the local representation as above. Let

$$X_j = \sum_{i=1}^{m} u_j^i \frac{\partial}{\partial q^i}(z_j) + \sum_{i=1}^{m} \dot{u}_j^i \frac{\partial}{\partial \dot{q}^i}(z_j)$$

$$Y_j = \sum_{i=1}^{m} v_j^i \frac{\partial}{\partial q^i}(z_j) + \sum_{i=1}^{m} \dot{v}_j^i \frac{\partial}{\partial \dot{q}^i}(z_j)$$

The hypotheses $T\tau_M(X_1)=T\tau_M(X_2)$ and $T\tau_M(Y_1)=T\tau_M(Y_2)$ imply that $u_1^i=u_2^i:=u^i$ and $v_1^i=v_2^i:=v^i$ for all i. Using (9.7), we verify that both sides of (9.10) are equal to the following:

$$\sum_i (u^i+v^i)\frac{\partial}{\partial q^i}(z_1+z_2) + \sum_i (\dot{u}_1^i+\dot{u}_2^i+\dot{v}_1^i+\dot{v}_2^i)\frac{\partial}{\partial \dot{q}^i}(z_1+z_2)$$

This finishes the proof. □

7. Second Order ODE on Manifolds

Recall from Chapter 6 that a tangent vector field on a manifold M is the analogue of a system of autonomous ordinary differential equations (ODE). A *second order* (autonomous) ODE on M will be a special type of tangent vector field on TM. Let M be a C^r manifold, $r \geq 2$. A ***second order ordinary differential equation on*** M is a vector field X on TM, which is also a cross-section of the bundle $T\tau_M : T^2M \to TM$. Thus two relations are to hold here:

$$\tau_{TM} \circ X = \mathbb{1}_{TM}, \quad T\tau_M \circ X = \mathbb{1}_{TM}$$

We will explain why this definition corresponds to a second order ODE on an open subset U of $\mathbb{R}^m$ in the usual sense. Denoting points of TU by (x, v), $x \in U$, $v \in \mathbb{R}^m$, a vector field X on TU will be in the form

$$X : (x, v) \mapsto (x, v; f(x, v), g(x, v))$$

where $f : TU \to \mathbb{R}^m$ and $g : TU \to \mathbb{R}^m$. This is equivalent to the following system of differential equations on $TU = U \times \mathbb{R}^m$:

$$\begin{cases} \frac{dx}{dt} = f(x, v) \\ \frac{dv}{dt} = g(x, v) \end{cases}$$

By the second relation in (9.5), the condition $T\tau_M \circ X = \mathbb{1}_{TM}$ is equivalent to $f(x, v) = v$. Therefore the first equation reads as $\frac{dx}{dt} = v$, and substituting this into the second equation, we obtain

$$\frac{d^2x}{dt^2} = g(x, \frac{dx}{dt}) \tag{9.11}$$

which is the familiar general form of an explicit system of second order autonomous ordinary differential equations on U.

We assume from now on that X (or, locally, the function g), is at least C^1, so that the classical existence-uniqueness-uniformity theorem of ODE holds. Classically, a solution of (9.11) is a curve $\gamma{:}I{\to}U$ so that $x{=}\gamma(t)$ satisfies (9.11). Writing $\gamma(t){=}(\gamma_1(t),\ldots,\gamma_m(t))$, then

$$\gamma'(t) = (\gamma_1(t),\ldots,\gamma_m(t); \frac{d\gamma_1}{dt},\ldots,\frac{d\gamma_m}{dt})$$

and therefore $\gamma'(t)$ is an integral curve of the vector field X. Henceforth we shall distinguish between the terms "solution" and "integral curve" for a second order ODE by calling $\gamma{:}I{\to}M$ a *solution*, and $\gamma'{:}I{\to}TM$ an *integral curve*. Notice it follows from the condition $f(x,v){=}v$ that any integral curve $\lambda{:}I{\to}TU$ is necessarily in the form $\lambda(t){=}(\gamma(t),\frac{d\gamma}{dt})$. In fact if $\lambda(t){=}(\lambda_1(t),\ldots,\lambda_{2m}(t))$, then

$$\gamma(t){=}(\tau_M{\circ}\lambda)(t){=}(\lambda_1(t),\ldots,\lambda_m(t))$$

satisfies the requirement of being a solution. The condition being local, the same applies to integral curves of a second order ODE on a manifold M. The existence-uniqueness assertions now take the following form:

(i) Given $a{\in}M$, $z{\in}T_aM$ and $t_0{\in}\mathbb{R}$, there is an open interval I containing t_0 and a solution $\alpha{:}I{\to}M$ so that $\alpha(t_0){=}a$ and $\alpha'(t_0){=}z$.

(ii) If I and J are two open intervals containing t_0 and $\alpha{:}I{\to}M, \beta{:}J{\to}M$ two solutions satisfying $\alpha(t_0){=}\beta(t_0)$ and $\alpha'(t_0){=}\beta'(t_0)$, then $\alpha(t){=}\beta(t)$ for all $t{\in}I{\cap}J$.

It follows from (ii) that $\alpha'(t){=}\beta'(t)$ for all $t{\in}I{\cap}J$. The maximal integral curve passing through z in TM projects down to a maximal solution in M. Note, however, that solutions may intersect in M, or even self-intersect. The only restriction is that at a point of intersection, the velocity vectors must be distinct.

B. Descent of the Second Derivative

With the tangent map $Tf{:}TM{\to}TN$ replacing the derivative as the linear approximation to a differentiable map $f{:}M{\to}N$, it is natural to expect that the double tangent map $T^2f{=}T(Tf){:}T^2M{\to}T^2N$ play the role of the second derivative of $f{:}M{\to}N$. However, T^2M, T^2N and T^2f are abstract entities far removed from the "ground manifolds" M and N, and we are accustomed to picturing the manifestations of the second derivative in the original spaces of discourse. This is important since the presence of the second derivative, especially the acceleration vector, is pervasive throughout the geometry of curves and surfaces, as well as in classical mechanics. In the geometry of curves, the magnitude of the acceleration vector of a parametrized curve at a point is related to the intensity of the bending of the curve at that point. If the curve is the actual path of motion of a physical particle obeying Newton's laws of motion, then the acceleration vector points in the same direction as the force acting on the particle and is displayed as a vector emanating from the position of the particle. Likewise for a parametrized surface in the three space, the second order partial derivatives of the parametric description of the surface have visual representations in $\mathbb{R}^3$ that embody the information about the bending and turning of the suface in space.

If $\gamma{:}I\to M$ is a smooth curve, we know that the velocity vector $\gamma'(t){=}T_t\gamma(\frac{d}{dt})$ is an element of the tangent space at $\gamma(t)$ since the tangent map, $T_t\gamma$, maps the tangent vectors to I at t to tangent vectors to M at $\gamma(t)$. In elementary mathematics and physics we also display the acceleration vector $\gamma''(t)$ as a vector emanating from $\gamma(t)$. What is the justification for this? On the one hand, this seems to be necessary, for example in mechanics, where the force field of a conservative system is a vector field on the space of positions (the so-called "configuration space"), and the acceleration vector is proportional to the force vector. On the other hand, it turns out that this is completely fortuitous, tied to our habit of working in the "flat" linear space $\mathbb{R}^m$. Let us look at the situation more closely. For the smooth curve $\gamma{:}I\to\mathbb{R}^m$, since each $T_{\gamma(t)}\mathbb{R}^m$ is canonically identified with $\mathbb{R}^m$ (a fact that has no analogue, replacing $\mathbb{R}^m$ by a general manifold), we may regard $\gamma'(t)$ as an element of $\mathbb{R}^m$. We thus obtain another smooth curve $\gamma'{:}I\to\mathbb{R}^m$. Taking the derivative of this curve, we obtain $\frac{d}{dt}\gamma'(t){\in}T_{\gamma'(t)}\mathbb{R}^m$. Finally, using the canonical identification of $T_{\gamma'(t)}\mathbb{R}^m$ with $\mathbb{R}^m$ and hence with $T_{\gamma(t)}\mathbb{R}^m$, we transfer the acceleration vector $\gamma''(t)$ to an element of $T_{\gamma(t)}\mathbb{R}^m$.

Algebraically, using the notation of (2.14) and (2.15) of Chapter 2, here is the sequence of identifications behind our habitual display of the acceleration vector as emanating from the point $\gamma(t)$:

$$\begin{aligned}
(9.12)\qquad \gamma''(t) &= \left(\gamma'(t), \frac{d}{dt}(\gamma'(t))\right) = (\gamma(t), \frac{d\gamma}{dt}(t); \frac{d\gamma}{dt}(t), \frac{d^2\gamma}{dt^2}(t)) \\
&\leftrightarrow (\frac{d\gamma}{dt}(t), \frac{d^2\gamma}{dt^2}(t)) \\
&\leftrightarrow \frac{d^2\gamma}{dt^2}(t) \\
&\leftrightarrow (\gamma(t), \frac{d^2\gamma}{dt^2}(t))
\end{aligned}$$

We now consider the general situation of a smooth curve on a manifold, $\gamma{:}I\to M$. Here $\gamma'(t){\in}T_{\gamma(t)}M$, i.e., $\gamma'{:}I\to TM$ is a curve in TM. There is, in general, no identification between tangent spaces to M and the manifold M itself, hence no hope of regarding γ' as a curve on M. The naive attempt to pass from TM to M by projection will only carry γ' back to γ, $\tau_M\circ\gamma'{=}\gamma$. Let us now look at the second derivative, or the acceleration. This is the velocity vector of the curve γ'. Thus

$$\gamma''(t){=}(\gamma')'(t){\in}T_{\gamma'(t)}(TM)$$

i.e., $\gamma''(t)$ is an element of $T(TM){=}T^2M$. To construct an acceleration vector for the curve γ that can be realized as a tangent vector to M, we must somehow transfer $\gamma''(t)$ to a vector in $T_{\gamma(t)}M$. We will discuss a couple of natural attempts.

Attempt 1. Using local charts, we define the acceleration vector as the final expression in (9.12), and investigate whether the definition is independent of the chart. Let U be an open subset of $\mathbb{R}^m$, $\alpha{:}I\to U$ a smooth curve and $f{:}U\to V$ a smooth diffeomorphism of U onto an open subset V of $\mathbb{R}^m$ (e.g., a coordinate change

$f = \eta \circ \xi^{-1}$). Letting $\beta = f \circ \alpha$, we should check whether the following holds:

$$Tf(\alpha(t), \frac{d^2\alpha}{dt^2}(t)) = (\beta(t), \frac{d^2\beta}{dt^2}(t))$$

or simply

$$\frac{d^2\beta}{dt^2}(t) = Df(\alpha(t))(\frac{d^2\alpha}{dt^2}(t))$$

But

$$\frac{d\beta}{dt}(t) = Df(\alpha(t))(\frac{d\alpha}{dt}(t))$$

Note that the right-hand side expresses the effect of a linear map on a vector, so it is bilinear in the two arguments, and Leibniz rule (Appendix II) gives

$$\frac{d^2\beta}{dt^2}(t) = Df(\alpha(t))(\frac{d^2\alpha}{dt^2}(t)) + D^2 f(\alpha(t))(\frac{d\alpha}{dt}(t), \frac{d\alpha}{dt}(t))$$

In view of the appearance of the second term on the right-hand side, this attempt does not succeed. In other words, defining the acceleration vector using local coordinates fails to be coordinate-independent, and is hence not transferable to manifolds. But note, incidentally, that when the velocity $\frac{d\alpha}{dt}(t)$ is zero, i.e., the velocity curve crosses the zero-section, the second term vanishes and the acceleration vector transforms correctly under coordinate change. We will later have an interpretation for this fact.

Attempt 2. The identifications in $\mathbb{R}^m$ as expressed in (9.12) lead to a more geometric approach. We wish to extract $(\gamma(t), \frac{d^2\gamma}{dt^2}(t))$ from the 4-tuple on the right-hand side of the first line of (9.12). The first two components of the 4-tuple just indicate the base point of the vector $\gamma''(t)$, and the last two constitute the directed-line vector of acceleration, with the fourth component being what we normally regard as the *real* acceleration. In $\mathbb{R}^{4m}$, to extract this fourth component, we project the vector $(\frac{d\gamma}{dt}(t), \frac{d^2\gamma}{dt^2}(t))$ on the second (i.e., the vertical) component to obtain an element of the vertical subspace $V_{\gamma'(t)}$. Then the application of o_M will bring this vector down to an element of $T_{\gamma(t)}M$:

$$(\gamma(t), \frac{d^2\gamma}{dt^2}(t)) = \downarrow(\gamma''(t))$$

Here we use $\downarrow$ as abbreviation for $o_M \circ v$, with v indicating projection on the vertical subspace at $\gamma'(t) \in TM$. But the catch is that although the vertical subspace $V_{\gamma'(t)}$ of $T_{\gamma'(t)}(TM)$ is well-defined, there is no canonically defined projection from $T_{\gamma'(t)}(TM)$ to $V_{\gamma'(t)}$! From linear algebra we know that projection on a linear subspace requires the selection of a complementary subspace. Except at the zero-section, where a complementary horizontal subspace was canonically defined, there is no distinguished choice of a complementary subspace for a vertical subspace. (This also explains or corroborates the last remark in *Attempt 1* regarding the zero-section.)

Here is where *connections* come in: We simply define a ***connection*** for the smooth m-manifold M as a smooth m-plane field H on TM with two properties:

(i) H_z is complementary to V_z for every $z \in TM$.
(ii) At a point z of the zero-section, H_z coincides with the horizontal subspace as defined in Subsection1.

Once a connection is adopted (and we do not yet know that one exists), we can transfer the acceleration vector to the manifold as a tangent vector to M. We define

$$\ddot{\gamma}(t) = \downarrow(\gamma''(t)) \tag{9.13}$$

where, as before, $\downarrow$ stands for the composition $o_M \circ v$, and v is the projection on the vertical subspace relative to the given connection (see Figure 5). The notation $\ddot{\gamma}(t)$ has been adopted to distinguish the acceleration vector in TM from the one in T^2M. $\ddot{\gamma}(t)$ is called the ***intrinsic acceleration*** (with respect to the given connection). The choice of terminology seems odd in view of the dependence on the connection, but this follows the tradition of classical differential geometry. The word "intrinsic" has the connotation of being determined intrinsically by M itself; in Section D we show how a Riemannian metric on M determines a distinguished connection, and Riemannian metrics are perceived as detectable by measurements on the manifold itself.

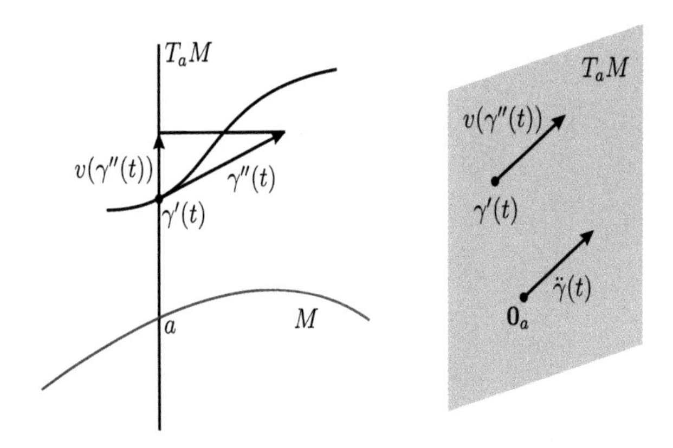

FIGURE 5. Constructing acceleration vector as a tangent to M

8. Example Let M be a parallelizable manifold, e.g., an open subset of $\mathbb{R}^m$ or a Lie group. Thus TM is VB-isomorphic to the product bundle $M \times \mathbb{R}^m$ over M. We describe an obvious *product connection* for $M \times \mathbb{R}^m$, which can then be transferred to TM by the VB-isomorphism. The tangent space to $T(M \times \mathbb{R}^m)$ at $(a, u) \in M \times \mathbb{R}^m$ is the set

$$\{(a, u; v, w) : v, w \in \mathbb{R}^m\}$$

Since $\tau_M(a, u) = a$, then for $X = (a, u; v, w)$, we obtain

$$\tau_{TM}(X) = (a, u), \quad T\tau_M(X) = (a, v)$$

It follows that

$$V_{(a,u)}=\{(a,u;0,w):w\in\mathbb{R}^m\}$$
$$V_a=\{(a,u;0,w):u,w\in\mathbb{R}^m\}$$
$$VM=\{(a,u;0,w):a\in M;u,w\in\mathbb{R}^m\}$$

The maps v, o_M, and $\downarrow$ are then given by

$$v(a,u;v,w)=(a,u;0,w),\ \ o_M(a,u;0,w)=(a,w),\ \ \downarrow(a,u;v,w)=(a,w)$$

At a point $(a,0)$ of the zero section, the set $\{(a,0;v,0):v\in\mathbb{R}^m\}$ is the horizontal subspace. The ***product connection*** is defined by specifying the ***horizontal subspace*** at an arbitrary point (a,u) of TM to be

$$H_{(a,u)} = \{(a,u;v,0) : v\in\mathbb{R}^m\}$$

This satisfies conditions (i) and (ii) of a connection. For a smooth curve $\gamma:I\rightarrow M$

$$\ddot{\gamma}(t) = \downarrow(\gamma(t),\frac{d\gamma}{dt}(t);\frac{d\gamma}{dt}(t),\frac{d^2\gamma}{dt^2}(t)) = (\gamma(t),\frac{d^2\gamma}{dt^2}(t)) \tag{9.14}$$

We now transfer the product connection to the parallelizable manifold M. Let $\Phi:TM\rightarrow M\times\mathbb{R}^m$ be a smooth diffeomorphism of vector bundles over M, so that we have $\tau_M=\pi_1\circ\Phi$, with π_1 projection on the first component. Here, instead of just one embedding of M into TM, namely the zero-section, we have a family of embeddings that fill out TM. For each $u\in\mathbb{R}^m$, we define $E_u:M\rightarrow TM$ by $E_u(x)=\Phi^{-1}(x,u)$. This is an embedding, being the composition of an embedding into the product with a diffeomorphism. For $u=0$, we obtain the zero-section. The horizontal subspace at $E_u(x)$ is then defined to be the tangent space to the submanifold $E_u(M)$ at this point.

The justification for our previous long discussion is that most manifolds are not parallelizable, so the above example has limited utility. In fact, as we shall see, even parallelizable manifolds admit useful non-product connections.

It turns out that the definition of the connection given above is too general for most concrete purposes. But before adding further conditions, we can already describe some key concepts. A smooth curve $\gamma:I\rightarrow M$ is a ***geodesic*** (with respect to the given connection) if $\ddot{\gamma}(t)=0$ for all $t\in I$. Let us look at the geodesics in the example above. For $M=\mathbb{R}^m$ and a curve $\gamma:I\rightarrow\mathbb{R}^m$, we have:

$$\gamma''(t) = (\gamma(t),\frac{d\gamma}{dt}(t);\frac{d\gamma}{dt}(t),\frac{d^2\gamma}{dt^2}(t))$$
$$\ddot{\gamma}(t) = (\gamma(t),\frac{d^2\gamma}{dt^2}(t))$$

Therefore, γ is a geodesic if and only if $\frac{d^2\gamma}{dt^2}(t)=0$, which is equivalent to $\gamma(t)=at+b$, a and b constants. Thus a geodesic is a straight line parametrized proportional to arclength (or a single point, if $a=0$). Physically, this describes *inertial motion*, i.e., movement along a straight line with constant speed, which occurs in the absence of force.

For a Lie group G with $\dim G{=}m$, we consider a product connection as above. Explicitly, pick a basis $(A_1,\dots,A_m)$ for T_eG, and transfer these via TL_g to points g of G to obtain a global basis $(\tilde{A}_1,\dots,\tilde{A}_m)$ for left-invariant vector fields on G. Defining the diffeomorphism $\Phi{:}TG{\to}G{\times}\mathbb{R}^m$ by

$$c_1\tilde{A}_1(g)+\cdots+c_m\tilde{A}_m(g)\mapsto(g;(c_1,\dots,c_m))$$

we see that the left-invariant vector fields form a basis for the horizontal subspace induced by Φ. The integral curves of a left-invariant vector field $\tilde{A}$ parametrized as $\exp tA{\cdot}g$ are therefore geodesics for this connection. Our next theorem will show that these are the *only* geodesics for this connection

Suppose that a connection H is specified for M. Given $z{\in}T_aM$, then the restriction of $T\tau_M$ to H_z is an isomorphism onto T_aM since H_z is complementary to V_z. It follows that for every $z'{\in}T_aM$, there is a unique element of H_z that is mapped to z' by $T\tau_M$. We denote this element by $\mathfrak{h}_z(z')$ and call it the ***horizontal lift of*** z' ***to base*** z. $\mathfrak{h}_z$ is then a linear isomorphism from T_aM onto H_z. Further, $\mathfrak{h}_z(z')$ is smooth in both z and z' since H is smooth. We define the ***geodesic vector field*** G (with respect to the given connection H) on TM by

$$G(z)=\mathfrak{h}_z(z) \tag{9.15}$$

The following basic theorem is rather obvious in the present framework.

9. Theorem *Let H be a connection for the smooth manifold M and suppose that G is the associated geodesic vector field. Then:*
(a) G is a second oder ODE on M.
(b) A smooth curve $\gamma{:}I{\to}M$ is a geodesic if and only if the velocity curve $\gamma'{:}I{\to}TM$ is an integral curve for G.

Proof. By the definition of horizontal lift, $T\tau_M{\circ}G{=}\mathbb{1}_{TM}$, so G is a second order ODE on M. Suppose γ' is an integral curve for G. Since $G(z)$ is horizontal, $v(\gamma''(t)){=}0$. Therefore, $\ddot{\gamma}(t){=}{\downarrow}(\gamma''(t)){=}0$. Conversely, suppose that $\ddot{\gamma}(t){=}0$. Then $v(\gamma''(t)){=}0$, i.e., the tangent vector $\gamma''(t)$ to the curve γ' at the point $\gamma'(t)$ is horizontal. But $\gamma''(t)$ has the property that its image under $T\tau_M$ is $\gamma'(t)$ since:

$$\begin{aligned}T\tau_M(\gamma''(t))&=(T\tau_M{\circ}T\gamma')(\frac{d}{dt}(t))\\&=T(\tau_M{\circ}\gamma')(\frac{d}{dt}(t))\\&=T\gamma(\frac{d}{dt}(t))\\&=\gamma'(t)\end{aligned}$$

So $\gamma''(t)$ is the horizontal lift of $\gamma'(t)$ to the base $\gamma'(t)$, i.e., $\gamma''(t){=}G(\gamma'(t))$, and the assertion is proved. □

10. Corollary *Let M be a smooth manifold and H a connection for M. Then given $a \in M$, $z \in T_aM$ and $t_0 \in \mathbb{R}$, there is an interval I containing t_0 and a geodesic (relative to H) $\alpha:I \to M$ that satisfies the initial conditions $\alpha(t_0)=a$ and $\alpha'(t_0)=z$. Moreover, if J is another interval containing t_0 and $\beta:J \to M$ another geodesic satisfying the same initial conditions, then $\alpha(t)=\beta(t)$ for all $t \in I \cap J$.*

PROOF. In view of the theorem, geodesics are just solutions of the second order ODE G on M. So the assertion follows from general considerations of Subsection 7. □

The corollary shows that we have described all the geodesics in the examples of $\mathbb{R}^m$ and Lie groups above.

11. Vertical Spread

In addition to the notion of horizontal lift, there is a notion of lifting to vertical vectors that will be of use in future sections. Let Y be a vector field on M. The ***vertical spread*** of Y to a vertical vector field $\overline{Y}$ on TM is defined in the following way. For $a \in M$, regard the tangent vector $Y(a)$ as a tangent vector to the linear space T_aM at $\mathbf{0}_a$. We can translate this vector to every point z of this linear space as a tangent vector $(z, Y(a))$, which we will denote by $\overline{Y}(z)$. $\overline{Y}$ is a smooth vertical vector field on TM, the restriction of which to each T_aM is a constant vector field (see Figure 6).

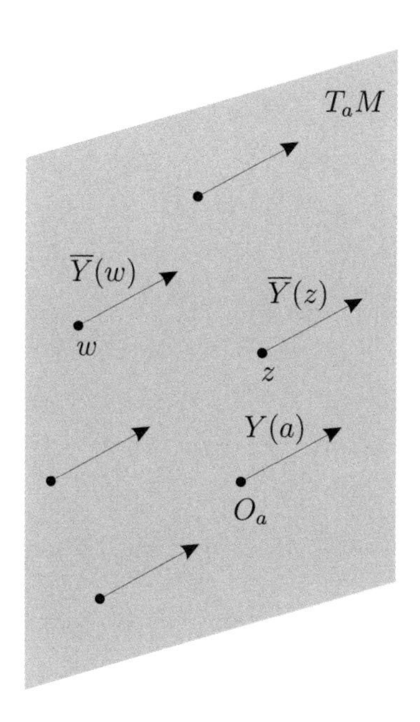

FIGURE 6. The vertical spread

So far we have not imposed any stringent requirements on the relationship of horizontal subspaces at different points of TM; the only restriction has been that H be smooth. We are now going to define *linear connections* for which a certain linear relation governs the distribution of horizontal subspaces throughout TM. Among other things, this will guarantee that a linear change of parameter will not change the geodesic nature of a geodesic. For the example of product connection that we considered, this was automatically the case, but it need not hold in general unless some restrictions on the connection are imposed.

12. Definition *Let M be a smooth manifold. A connection H for M is called a* **linear connection** *if H is invariant under linear space operations on the fibers of the vector bundle $T\tau_M{:}T^2M\to TM$. Explicitly:*
(i) For every $z\in TM$ and every $r\in\mathbb{R}$, $r\boxdot H_z=H_{rz}$.
(ii) For all z_1 and z_2 in the same tangent space T_aM, $H_{z_1+z_2}=H_{z_1}\boxplus H_{z_2}$.

The last statement is taken to mean the following: For $X_1\in H_{z_1}$ and $X_2\in H_{z_2}$ with $T\tau_M(X_1)=T\tau_M(X_2)$, one has $X_1\boxplus X_2\in H_{z_1+z_2}$, and conversely, every element of $H_{z_1+z_2}$ possesses such a representation. The following describes some useful elementary consequences of the definition. Given a connection H for M, $z\in TM$ and $X\in T_z(TM)$, we denote by $v(X)\in V_z$ and $h(X)\in H_z$, respectively, the **vertical projection** and the ***horizontal projection*** of X.

13. Elementary Properties of Linear Connections
Let M be a smooth manifold and H a linear connection for M. Then:
(a) For all $X\in T^2M$ and $r\in\mathbb{R}$

$$v(r\boxdot X) = r\boxdot v(X), \quad h(r\boxdot X) = r\boxdot h(X)$$

(b) Let $X_1\in T_{z_1}(TM)$ and $X_2\in T_{z_2}(TM)$, where $z_1,z_2\in T_xM$. If $T\tau_M(X_1)=T\tau_M(X_2)$, then

$$v(X_1\boxplus X_2)=v(X_1)\boxplus v(X_2), \quad h(X_1\boxplus X_2)=h(X_1)\boxplus h(X_2)$$

(c) If G is the geodesic vector field associated with H, $r\in\mathbb{R}$ and $z\in TM$, then

$$G(r\cdot z) = r\boxdot(r\cdot G(z))$$

(d) For any smooth curve $\gamma{:}I\to M$, define the curve λ by $\lambda(t)=\gamma(at+b)$, where $a,b\in\mathbb{R}$. Then

$$\ddot{\lambda}(t) = a^2\ddot{\gamma}(at+b)$$

indexAcceleration vector!intrinsic In particular, if γ is a geodesic, then so is λ.

Statements (a) and (b) above can be summarized by saying that v and h commute with the vector space operations in $T\tau_M$.

Proof. (a) We write $X=h(X)+v(X)$. Then by (9.9)

$$r\boxdot X = r\boxdot h(X) + r\boxdot v(X)$$

By (i) of the definition, $r\boxdot h(X)$ is horizontal. On the other hand, m_r maps each T_xM linearly into itself, so any vertical vector is mapped to a vertical vector under Tm_r, hence $r\boxdot v(X)$ is vertical. The result then follows from the uniqueness of decomposition into horizontal and vertical components.
(b) We write each of X_1 and X_2 as a sum of its horizontal and vertical parts. Since $v(X_1)$ and $v(X_2)$, being vertical, belong to the same fiber $T\tau_M{}^{-1}(\mathbf{0}_x)$ of $T\tau_M$, it makes sense to add them under $\boxplus$. Also, $T\tau_M(X_i)=T\tau_M(h(X_i))$, so one can consider $h(X_1)\boxplus h(X_2)$. Therefore, using (9.10), we obtain

$$\begin{aligned} X_1 \boxplus X_2 &= (h(X_1)+v(X_1)) \boxplus (h(X_2)+v(X_2)) \\ &= (h(X_1)\boxplus h(X_2)) + (v(X_1)\boxplus v(X_2)) \end{aligned}$$

Again the uniqueness of decomposition into horizontal and vertical components yields the result.
(c) Since $r\cdot G(z)$ is a tangent vector to TM based at z, $r\boxdot(r\cdot G(z))$ is a tangent vector based at $r\cdot z$. Moreover, this vector is horizontal by the definition of a linear connection. To prove (c), it then suffices to show that $r\boxdot(r\cdot G(z))$ projects down to $r\cdot z$ under $T\tau_M$. But

$$\begin{aligned} T\tau_M\Big(r\boxdot(r\cdot G(z))\Big) &= (T\tau_M\circ Tm_r)(r\cdot G(z)) \\ &= T\tau_M(r\cdot G(z)) \\ &= rT\tau_M(G(z)) \\ &= r\cdot z \end{aligned}$$

(d) We have $\lambda'(t)=a\gamma'(at+b)$, therefore

$$\lambda''(t) = Tm_a(a\cdot\gamma''(at+b)) = a\boxdot(a\cdot\gamma''(at+b))$$

Applying the vertical projection v, using part (a) and the linearity of projection, we obtain

$$v(\lambda''(t)) = a\boxdot\Big(a\cdot v(\gamma''(at+b))\Big)$$

Finally, applying o_M, we have

$$\downarrow(\lambda''(t)) = a^2\downarrow(\gamma''(at+b))$$

which is the desired result. □

C. Covariant Derivative

In the previous section we laid the conceptual foundation for the notion of a connection and pointed out its necessity. While our discussion revolved mainly around attempts to bring the acceleration vector down to the base manifold, connections can actually achieve much more; they in fact enable one to define the derivative of one vector field with respect to another as a vector field on the base manifold. This is the concept of the *covariant derivative* associated with a linear connection. We will actually show that there is a one-to-one correspondence between covariant-derivative operators and linear connections. The concept of covariant derivative is often taken as the definition of a (linear) connection since this is generally a highly efficient tool for computation and derivation of formulas.

14. Definition *Let M be a smooth manifold and H a linear connection for M. For $a \in M$, $e \in T_aM$ and a smooth vector field Y defined in a neighborhood of a in M, the* **covariant derivative** *of Y with respect to e is the tangent vector*

$$\nabla_e Y = \downarrow(T_aY(e)) \in T_aM$$

Note the analogy with the definition of intrinsic acceleration. The value of the tangent map of Y on e at x is $T_aY(e)$, but this object lives in the set $T_{Y(a)}(TM)$. By applying $o_M \circ v$, we transfer the vertical component of the value of the tangent map of Y on e down to the base manifold.

Let us write out the local expression for T_aY for reference. Suppose that Y is given on a chart (U, ξ) by $x \mapsto \sum_j Y^j(x) \frac{\partial}{\partial \xi^j}(x)$, or $(\xi_1, \ldots, \xi_m) \mapsto ((\xi_1, \ldots, \xi_m), (Y_1, \ldots, Y_m))$. Then for $e = \sum_i u^i \frac{\partial}{\partial \xi^i}(a) \in T_aU$

$$T_aY(e) = \sum_j u^j \frac{\partial}{\partial q^j}(Y(a)) + \sum_j \Big(\sum_i \big(\frac{\partial Y^j}{\partial \xi^i}(a) u^i \big) \Big) \frac{\partial}{\partial \dot{q}^j}(Y(a)) \tag{9.16}$$

where, as in Subsection 5, $q^i = \xi^i \circ \tau_M$ and $\dot{q}^i = d\xi^i$.

15. Elementary Properties of the Covariant Derivative
Let M be a smooth manifold, $a \in M$, $e \in T_aM$ and Y a smooth vector field defined in a neighborhood of a. Given a linear connection H for M with the associated covariant derivative ∇, the following hold.
(a) The map $e \mapsto \nabla_e Y$ is an $\mathbb{R}$-linear map of T_aM to itself.
(b) For any smooth real-valued function f, defined in a neighborhood of a,

$$\nabla_e(fY) = f(a)\nabla_e Y + (df(e))Y(a)$$

(c) If $\gamma : I \to M$ is a smooth curve with $\gamma(t_0) = a$ and $\gamma'(t_0) = e$, then $\nabla_e Y$ depends only on the values of Y along γ.

Note that (b) implies the $\mathbb{R}$-linearity of the covariant derivative with respect to Y, for if we take f to be a constant (function), then $df = 0$.

Proof. (a) follows from the $\mathbb{R}$-linearity of the maps T_aY, v and o_M. For (b), which is a local statement, we first compute $T_afY(e)$ using (9.16).

$$T_afY(e) = \sum_j u^j \frac{\partial}{\partial q^j}(f(a)Y(a)) + \sum_j \Big(f(a) \Big(\sum_i \big(\frac{\partial Y^j}{\partial \xi^i}(a) u^i \big) \Big) + (df(e)) Y^j(a) \Big) \frac{\partial}{\partial \dot{q}^j}(f(a)Y(a))$$

Therefore,

$$T_afY(e) = f(a) \boxdot T_aY(e) + df(e) \sum_j Y^j(a) \frac{\partial}{\partial \dot{q}^j}(f(a)Y(a))$$

Recalling the definition of vertical spread from Subsection 11, we have established the formula

$$T_a fY(e) = f(a) \boxdot T_a Y(e) + df(e)\overline{Y}(f(a)Y(a)) \tag{9.17}$$

Applying $o_M \circ v$ to the two sides and using 13a, we obtain (b).
For (c), we have

$$T_a Y(e) = T_a Y(\gamma'(t_0)) = (T_a Y \circ T_{t_0}\gamma)(\frac{d}{dt}(t_0))$$

Hence by the chain rule,

$$T_a Y(e) = T_{t_0}(Y \circ \gamma)(\frac{d}{dt}(t_0)) \tag{9.18}$$

which proves the contention. □

In view of (9.18) above, if $\gamma: I \rightarrow M$ is a smooth curve, $t \in I$, and Y is a vector field defined along γ (see Subsection 12 of Chapter 6), the notation $\nabla_{\gamma'(t)}\gamma'$ makes sense since γ' is defined for $t \in I$. In fact, we have the following corollary.

16. Corollary

$$\ddot{\gamma}(t) = \nabla_{\gamma'(t)}\gamma'$$

Proof. This follows from (9.13), Definition 14 and (c) above. □

We now turn to the general axiomatic treatment of covariant derivative operators.

17. Definition *Let M be a smooth manifold, $C^\infty(M)$ the algebra of smooth real-valued functions on M and $\mathfrak{X}(M)$ the module of smooth vector fields on M. A map*

$$\nabla : \mathfrak{X}(M) \times \mathfrak{X}(M) \rightarrow \mathfrak{X}(M)$$

usually denoted by $\nabla_X Y$ instead of $\nabla(X, Y)$ is called a **covariant derivative (operator)** *if it satisfies the following two properties:*
(a) $\nabla_X Y$ is $C^\infty(M)$-linear with respect to X.
(b) $\nabla_X Y$ is Leibnizian with respect to Y, i.e., for $X, Y \in \mathfrak{X}(M)$ and $f \in C^\infty(M)$,

$$\nabla_e(fY) = f(a)\nabla_e Y + (df(e))Y(a)$$

As in the case of property (b) in 15, we infer that $\nabla_X Y$ is $\mathbb{R}$-linear with respect to Y. By (a), at a point $a \in M$, $\nabla_X Y(a)$ is determined by the value of X at a, so one can write $\nabla_{X(a)} Y$ instead of $\nabla_X Y(a)$. Thus, in fact, if $e \in T_a M$, the notation $\nabla_e Y$ makes sense, as e can be extended to a smooth vector field on M, e.g., by extending locally using a chart and then multiplying by a bump function. Our immediate goal is to show that every covariant derivative (operator) is induced by a (unique) linear connection. For this, we need a couple of lemmas.

18. Lemma *Let ∇ be a covariant derivative operator. Then $\nabla_X Y(a)$ depends only on the values of Y in an arbitrarily small neighborhod of a.*

Proof. Suppose that the smooth vector field Y' agrees with Y in a neighborhood U of a. We take a smooth bump function $\theta{:}M\to\mathbb{R}$ that vanishes outside U and has value 1 on a neighborhood V of a with $V\subset\overline{V}\subset U$. Then $Y-Y'=(1-\theta)(Y-Y')$. Using (b), we can write

$$\begin{aligned}\nabla_{X(a)}(Y-Y') &= \nabla_{X(a)}((1-\theta)(Y-Y'))\\ &= (1-\theta)(a)\nabla_{X(a)}(Y-Y') + (d(1-\theta))(a)(Y-Y')(a)\end{aligned}$$

The first term on the right vanishes since $\theta(a)=1$, and $(d(1-\theta))(a)=0$ since θ is constant in a neighborhood of a. This proves the lemma. □

19. Lemma *Let ∇ be a covariant derivative operator, and suppose that Z is a smooth vector field on M with the property that $Z(a)=0$. Then for $e\in T_aM$*

$$\nabla_e Z = \downarrow(T_aZ(e))$$

Note this shows that if $Z(a)=0$ and $e\in T_aM$, then the value of $\nabla_e Z$ is independent of the particular covariant derivative operator since there is a canonical splitting of $T_{0_a}(TM)$ and v is canonically defined.

Proof. The proposition is local, so we can work in a chart (U,ξ) around a. The right-hand side can be computed from (9.16) by substituting $Z(a)=0$ for $Y(a)$, and $e=\sum_{i=1}^m u^i\frac{\partial}{\partial\xi^i}(a)$ to obtain

$$\begin{aligned}\downarrow(T_aZ(e)) &= \downarrow\Big(\sum_j u^j\frac{\partial}{\partial q^j}(\mathbf{0}_a) + \sum_j\Big(\sum_i\big(\frac{\partial Z^j}{\partial\xi^i}(a)u^i\big)\Big)\frac{\partial}{\partial\dot{q}^j}(\mathbf{0}_a)\Big)\\ &= \sum_j\sum_i u^i\big(\frac{\partial Z^j}{\partial\xi^i}(a)\big)\frac{\partial}{\partial\xi^j}(a)\end{aligned}$$

On the other hand, computing the left-hand side using properties (a) and (b)

$$\begin{aligned}\nabla_e Z &= \sum_i u^i\nabla_{\frac{\partial}{\partial\xi^i}(a)}\Big(\sum_j Z^j\frac{\partial}{\partial\xi^j}\Big)\\ &= \sum_i\sum_j u^i\Big(\frac{\partial Z^j}{\partial\xi^i}(a)\frac{\partial}{\partial\xi^j}(a) + Z^j(a)\nabla_{\frac{\partial}{\partial\xi^i}(a)}\frac{\partial}{\partial\xi^j}\Big)\\ &= \sum_j\sum_i u^i\big(\frac{\partial Z^j}{\partial\xi^i}(a)\big)\frac{\partial}{\partial\xi^j}(a)\end{aligned}$$

This proves the assertion. □

We now proceed to show that for a given covariant derivative operator ∇ on a smooth m-manifold M, a (unique) linear connection exists for which the associated covariant derivative is ∇. The idea is to reverse the process of the construction of the covariant derivative associated to a linear connection. Let $z \in T_aM$, we wish to define a horizontal subspace $H_z \subset T_z(TM)$. If $z = \mathbf{0}_a$, we are required to define $H_{\mathbf{0}_a}$ as the tangent space to the embedded zero-section at $\zeta_M(a)$. For other z, H_z will be defined as the "horizontal lifting" of $H_{\mathbf{0}_a}$ to the base z. There is a smooth vector field Y defined in a neighborhood of a so that $Y(a)=z$. One can, e.g., take a chart (U, ξ) around a and the constant cross-section Y relative to this chart with value z. For $X \in T_aM$, we define $\mathfrak{h}_z(X) \in T_z(TM)$ by

$$\mathfrak{h}_z(X) = T_aY(X) - \overline{\nabla_X Y}(z) \tag{9.19}$$

where $\overline{\nabla_X Y}$ is the vertical spread of $\nabla_X Y$. We show the following:
(i) $\mathfrak{h}_z$ is a well-defined one-to-one linear map from T_aM into $T_z(TM)$.
(ii) The image H_z of $\mathfrak{h}_z$ is complementary to the vertical subspace through z.
(iii) The assignment $z \mapsto H_z$ defines a smooth m-plane field on TM and a linear connection for M.
(iv) H is the unique linear connection with associated covariant derivative ∇.

For (i), suppose Y' is another smooth vector field defined in a neighborhood of a with $Y'(a)=Y(a)=z$. Let $Z=Y'-Y$ on the common domain, thus $Z(a)=0$. Using the notation

$$\mathfrak{h}'_z(X) = T_aY'(X) - \overline{\nabla_X Y'}(z)$$

and (9.10), we can write

$$\begin{aligned}
\mathfrak{h}'_z(X) &= (T_aY(X) \boxplus T_aZ(X)) - (\overline{\nabla_X Y}(z) \boxplus \overline{\nabla_X Z}(\mathbf{0}_a)) \\
&= (T_aY(X) - \overline{\nabla_X Y}(z)) \boxplus (T_aZ(X) - \overline{\nabla_X Z}(\mathbf{0}_a)) \\
&= \mathfrak{h}_z(X) \boxplus h(T_aZ(X)) \\
&= \mathfrak{h}_z(X) \boxplus T_a\zeta_M(X) \quad \text{(by Lemma 19)} \\
&= \mathfrak{h}_z(X)
\end{aligned}$$

the last line because $T_a\zeta_M(X)$ is the zero-element in the fiber $(T\tau_M)^{-1}(X)$. $\mathbb{R}$-linearity of $\mathfrak{h}_z$ follows from the $\mathbb{R}$-linearity of the right-hand side of (9.19) in X. To show $\mathfrak{h}_z$ is one-to-one, suppose $\mathfrak{h}_z(X)=0$, then $T_aY(X)$ must be vertical, i.e., $(T\tau_M \circ TY)(X)=X$ must be zero. The same argument shows that $\mathfrak{h}_z(X)$ cannot be vertical unless $X=0$, and this establishes (ii).

Smoothness of $z \mapsto H_z$ is a local proposition. If $(X_1, \ldots, X_m)$ is a smooth local basis for TU, where U is a chart domain in M, then since $\mathfrak{h}_z$ was proved to be one-to-one linear, $(\mathfrak{h}_z(X_1), \ldots, \mathfrak{h}_z(X_m))$ is a basis for H_z. H is smooth because the right-hand side of (9.19) is smooth in z. We now prove that H is a *linear* connection. Let $\mathfrak{h}_z(X) \in H_z$

and $r\in\mathbb{R}$, we show that $r\boxdot\mathfrak{h}_z(X)\in H_{rz}$. We have

$$\begin{aligned} r\boxdot\mathfrak{h}_z(X) &= T_z m_r(T_a Y(X) - \overline{\nabla_X Y}(z)) \\ &= T_z m_r(T_a Y(X)) - T_z m_r(\overline{\nabla_X Y}(z)) \\ &= (T_a(rY))(rX) - \overline{\nabla_{rX} Y}(rz) \\ &= \mathfrak{h}_{rz}(rX) \end{aligned}$$

which, by definition, is an element of H_{rz}. Next suppose $z_1, z_2\in T_a M$, and $X\in T_a M$; we want to show that $\mathfrak{h}_{z_1}(X)\boxplus\mathfrak{h}_{z_2}(X)\in H_{z_1+z_2}$. Let Y_1 and Y_2 be smooth vector fields defined in a neighborhood of a so that $Y_1(a)=z_1$ and $Y_2(a)=z_2$. Then using (9.10):

$$\begin{aligned} \mathfrak{h}_{z_1}(X)\boxplus\mathfrak{h}_{z_2}(X) &= (T_a Y_1(X) - \overline{\nabla_X Y_1}(z_1)) \boxplus (T_a Y_2(X) - \overline{\nabla_X Y_2}(z_2)) \\ &= (T_a Y_1(X)\boxplus T_a Y_2(X)) - (\overline{\nabla_X Y_1}(z_1)\boxplus\overline{\nabla_X Y_2}(z_2)) \\ &= (T_a(Y_1+Y_2)(X)) - (\overline{\nabla_X Y_1+\nabla_X Y_2})(z_1+z_2) \\ &= (T_a(Y_1+Y_2)(X)) - (\overline{\nabla_X(Y_1+Y_2)})(z_1+z_2) \\ &= \mathfrak{h}_{z_1+z_2}(X) \end{aligned}$$

which is an element of $H_{z_1+z_2}$.

Finally, relation (9.19) and the definition of H_z show that the covariant derivative associated to H is precisely the given ∇. Further, given ∇, (9.19) also shows that the horizontal subspace at z of any related linear connection must necessarily contain all elements of the form $\mathfrak{h}_z(X)$, and these form a subspace of dimension m, therefore uniqueness is also established.

The discussion above is summarized in the following theorem.

20. Theorem *Given a smooth manifold M, there is a one-to-one correspondence between linear connections and covariant derivative operators for M.*

We will then freely use one approach or the other as the case demands. In what follows, the covariant derivative will be the main tool.

21. Local Computation Let (U,ξ) be a chart on the smooth m-manifold M. Thus the vector fields $\frac{\partial}{\partial\xi^i}$, $i=1,\ldots,m$, form a basis for the tangent space at every point of U. We will simplify notation by writing ∂_i for $\frac{\partial}{\partial\xi^i}$. Smooth functions Γ^k_{ij}, known as ***Christoffel symbols (of the second kind)***, are then introduced by

$$\nabla_{\partial_i(x)}\partial_j = \sum_k \Gamma^k_{ij}(x)\partial_k(x) \tag{9.20}$$

By virtue of the two defining properties of a covariant derivative, given smooth functions Γ^k_{ij} on U, the above equation defines a unique covariant derivative for U.

For a vector field $Y=\sum_j Y^j\partial_j$, we have

$$\nabla_{\partial_i} Y = \sum_k \Big(\frac{\partial Y^k}{\partial \xi^i} + \sum_j \Gamma^k_{ij} Y^j\Big)\partial_k \tag{9.21}$$

If $\gamma{:}I\to U$ is a smooth curve, $\gamma^i(t)=(\xi^i\circ\gamma)(t)$ then $\gamma'(t)=\sum_{i=1}^m(\gamma^i)'(t)\partial^i(\gamma(t))$. Suppose that $\eta=Y\circ\gamma$ is a regular smooth vector field defined along γ (cf. Sub-section 12, Chapter 6). We can write Y as $Y(x)=\sum_j Y^j(x)\partial^j(x)$, so that $Y(\gamma(t))=\sum_j Y^j(\gamma(t))\partial^j(\gamma(t))$. Then using (9.21):

$$\begin{aligned}
\nabla_{\gamma'(t)}Y &= \sum_i (\gamma^i)'(t)\nabla_{\partial_i(\gamma(t))}\Big(\sum_j Y^j\partial_j\Big)\\
&= \sum_i (\gamma^i)'(t)\Big(\sum_j \frac{\partial Y^j}{\partial \xi^i}\partial_j(\gamma(t)) + \sum_{j,k} Y^j(\gamma(t))\Gamma^k_{ij}(\gamma(t))\partial_k(\gamma(t))\Big)\\
&= \sum_j \frac{d}{dt}(Y^j\circ\gamma)\partial_j(\gamma(t)) + \sum_{i,j,k}(\gamma^i)'(t)Y^j(\gamma(t))\Gamma^k_{ij}(\gamma(t))\partial_k(\gamma(t))\\
&= \sum_k \Big(\frac{d}{dt}(Y^k\circ\gamma) + \sum_{i,j}(\gamma^i)'(t)Y^j(\gamma(t))\Gamma^k_{ij}(\gamma(t))\Big)\partial_k(\gamma(t))
\end{aligned}$$

Therefore,

$$\nabla_{\gamma'(t)}Y = \sum_k \Big(\frac{d\eta^k}{dt} + \sum_{i,j}\Gamma^k_{ij}(\gamma(t))\,(\gamma^i)'(t)\,\eta^j(t)\Big)\,\partial_k(\gamma(t)) \tag{9.22}$$

Let us look at the special case where $\eta=Y\circ\gamma=\gamma'$. In this case, $\eta^j(t)=(\gamma^j)'(t)$ and $\eta'(t)=\gamma''(t)$. For a geodesic, using (16), we obtain

$$\frac{d^2\gamma^k}{dt^2} + \sum_{i,j}\Gamma^k_{ij}(\gamma(t))\,\frac{d\gamma^i}{dt}\,\frac{d\gamma^j}{dt} = 0, \quad k=1,\dots,m \tag{9.23}$$

This system of differential equations is known as the ***geodesic equation***. In general, a regular smooth vector field $\eta=Y\circ\gamma$ defined along $\gamma{:}I\to M$ is called ***parallel*** if

$$\nabla_{\gamma'(t)}Y = 0, \quad \forall t\in\mathbb{R} \tag{9.24}$$

In this case, then

$$\frac{d\eta^k}{dt} + \sum_{i,j}\Gamma^k_{ij}(\gamma(t))\,\frac{d\gamma^i}{dt}\,\eta^j(t) = 0, \quad k=1,\dots,m \tag{9.25}$$

22. Example In the case of product connection (see Example 8) on $\mathbb{R}^m$, we have globally defined constant basis vector fields ∂_i, so $T\partial_i=0$ and $\Gamma^k_{ij}=0$. Therefore, (9.25) becomes $\frac{d\eta^k}{dt}=0$, or $\eta^k=constant$. Thus a parallel vector field along a curve is parallel in the usual Euclidean sense and has constant Euclidean length. We should warn the reader that the same conclusion $\Gamma^k_{ij}=0$ does not hold, in general, for a Lie group

G, even though G is parallelizable. The reason is that the basis consisting of left-invariant vector fields $X_1, \ldots, X_m$ is not, in general, a *coordinate basis*. For a non-commutative Lie group, the bracket $[X_i, X_j]$ does not generally vanish. See also Exercise 9.18 at the end of this chapter.

The following theorem, a simple corollary of standard facts about linear differential equations, is important in several ways. Among other things, it provides a method for defining geometrically meaningful isomorphisms between tangent spaces at different points of a manifold.

23. Theorem *Let M be a smooth manifold with covariant derivative ∇, and suppose $\gamma{:}I{\to}M$ is a regular smooth curve with $\gamma(t_0){=}a$ and $\gamma'(t_0){=}Y_0{\in}T_aM$. Then there exists a unique parallel vector field $\eta{=}Y{\circ}\gamma$ defined along γ with the property that $\eta(t_0){=}Y_0$ and $\eta(t){\in}T_{\gamma(t)}M$. Further, the map $Y_0{\mapsto}\eta(t)$ is an isomorphism from T_aM to $T_{\gamma(t)}M$.*

Proof. Let $T{\in}I$ and suppose $T > t_0$. By compactness of $[t_0, T]$, there is a partition $t_0 < \cdots < t_k{=}T$ of the interval so that each $\gamma[t_{i-1}, t_i]$ is contained in a coordinate neighborhood. We take a chart (U, ξ) so that $\gamma[t_0, t_1]{\subset}U$. The sought-after vector field η along γ must satisfy the system of linear differential equations (9.25) with initial condition $\eta(t_0){=}Y_0$. The following are classical facts about linear differential equations:
(i) The unique solution with initial condition $\eta(t_0){=}Y_0$ is defined for all t for which the equation is defined.
(ii) The transition map $Y_0{\mapsto}\eta(t)$ is a linear isomorphism from T_aM to $T_{\gamma(t)}M$[1].

Proceeding similarly with the interval $[t_1, t_2]$ and continuing a finite number of steps, the theorem is established for the entire interval $[t_0, T]$. The argument for $T{<}t_0$ is similar, and the proof is complete. □

In the situation of the above theorem, one says that the vector field η is obtained by ***parallel transport*** of Y_0 along γ.

D. Curvature and Torsion

In Chapters 2 and 6, we considered the bracket $[X, Y]$ of two vector fields X and Y on a smooth manifold M. This was an anti-symmetric measure of the rate of change of one vector field along the trajectories of the other. Given a covariant derivative operator ∇ for M, one may also consider $\nabla_X Y - \nabla_Y X$ as another anti-symmetric measure of the rate of change of these vector fields relative to each other. A natural problem is to compare these two measures. It turns out that there are linear connections on M for which the two are equal, and this happens in many cases of geometric interest. However, what is *always* the case is that the difference of these two measures is a tensor field, i.e., it is $C^\infty(M)$-bilinear in X and Y. Therefore, the value of this difference at a point $a{\in}M$ is determined completely by the values $X(a)$

[1]See, e.g., **[3]** **[6]** **[14]**, or Exercises 2.23 and 2.24 of Chapter 2.

and $Y(a)$. This is unlike $\nabla_X Y$, $\nabla_X Y - \nabla_Y X$ and $[X, Y]$, individually, which depend on the values of X, Y or both, in a neighborhood of a. Explicitly, let

$$T(X, Y) = \nabla_X Y - \nabla_Y X - [X, Y] \tag{9.26}$$

T is known as the ***torsion tensor*** of ∇ or of the associated linear connection. The claim is laid out in the following lemma.

24. Lemma *For smooth vector fields X and Y on M and each $a \in M$, $T(X, Y)(a)$ is completely determined by $X(a)$ and $Y(a)$.*

Proof. Let $f \in C^\infty(M)$, then

$$\begin{aligned} T(fX, Y)(x) &= (\nabla_{fX} Y)(x) - (\nabla_Y fX)(x) - [fX, Y](x) \\ &= f(x)(\nabla_X Y)(x) - f(x)(\nabla_Y X)(x) - (Y(x)\cdot f)X(x) - f(x)[X, Y](x) + (Y(x)\cdot f)X(x) \\ &= f(x)T(X, Y)(x) \end{aligned}$$

A completely similar computation works for $T(X, fY)(x)$. □

A linear connection (or the associated covariant derivative) is called ***torsion-free*** (or ***symmetric***) if $T=0$. As an example, consider the product connection on $\mathbb{R}^m$, Example 22. Here $\Gamma^k_{ij}=0$, and (9.21) together with (2.41) of Chapter 2 show that $T=0$.

We consider a variation of the discussion that led to the torsion tensor. The definition of the bracket says that $X\cdot(Y\cdot f) - Y\cdot(X\cdot f) - [X, Y]\cdot f = 0$, where f is a smooth function from M to $\mathbb{R}$. We may consider a function $F=(f^1, \ldots, f^m)$ from M to $\mathbb{R}^m$ and extend the definition component-wise to write $X\cdot(Y\cdot F) - Y\cdot(X\cdot F) - [X, Y]\cdot F = 0$. An alternative to a function F from an m-dimensional manifold M to $\mathbb{R}^m$ is a tangent vector field Z on M. Here, replacing the vector field operation $X\cdot$, by ∇_X, we may consider

$$\nabla_X(\nabla_Y Z) - \nabla_Y(\nabla_X Z) - \nabla_{[X,Y]} Z$$

which is generally not zero. This vector field valued function of three vector fields is called the ***(Riemann) curvature tensor*** and is traditionally denoted by $R(X, Y)Z$. Just as in the case of T, this turns out to be a tensor field.

25. Lemma *For smooth vector fields X, Y and Z on M and $a \in M$, $(R(X, Y)Z)(a)$ is completely determined by $X(a)$, $Y(a)$ and $Z(a)$.*

Proof. We have to show that for $f \in C^\infty(M)$,

$$R(fX, Y)Z = R(X, fY)Z = R(X, Y)fZ = fR(X, Y)Z$$

The verification is direct as in the previous lemma, but a bit longer.

$$\begin{aligned} R(fX,Y)Z &= \nabla_{fX}(\nabla_Y Z) - \nabla_Y(\nabla_{fX}Z) - \nabla_{[fX,Y]}Z \\ &= f\nabla_X(\nabla_Y Z) - \nabla_Y(f\nabla_X Z) - f\nabla_{[X,Y]}Z + (Y\cdot f)\nabla_X Z \\ &= f\nabla_X(\nabla_Y Z) - (Y\cdot f)\nabla_X Z - f\nabla_Y(\nabla_X Z) - f\nabla_{[X,Y]}Z + (Y\cdot f)\nabla_X Z \\ &= fR(X,Y)Z \end{aligned}$$

The case of fY is similar. For fZ:

$$\begin{aligned} R(X,Y)fZ &= \nabla_X(\nabla_Y fZ) - \nabla_Y(\nabla_X fZ) - \nabla_{[X,Y]}fZ \\ &= \nabla_X((Y\cdot f)Z+f\nabla_Y Z)-\nabla_Y((X\cdot f)Z+f\nabla_X Z) - ([X,Y]\cdot f)Z-f\nabla_{[X,Y]}Z \\ &= (X\cdot(Y\cdot f))Z+(Y\cdot f)\nabla_X Z+(X\cdot f)\nabla_Y Z+f\nabla_X\nabla_Y Z-(Y\cdot(X\cdot f))Z-(X\cdot f)\nabla_Y Z \\ &\quad -(Y\cdot f)\nabla_X Z-f\nabla_Y\nabla_X Z-([X,Y]\cdot f)Z-f\nabla_{[X,Y]}Z \\ &= fR(X,Y)Z \end{aligned}$$

by virtue of cancellations and the definition of $[X,Y]$. □

26. Local Computation Let (U,ξ) be a chart for the m-manifold M with $\partial_i=\frac{\partial}{\partial\xi^i}$. We write $T(\partial_i,\partial_j)=\sum_\nu T^\nu_{ij}\partial_\nu$. Using (9.20) and $[\partial_i,\partial_j]=0$, we obtain

$$T^\nu_{ij} = \Gamma^\nu_{ij} - \Gamma^\nu_{ji} \tag{9.27}$$

Therefore, the connection is torsion-free if and only if $\Gamma^\nu_{ij}=\Gamma^\nu_{ji}$, which is the reason for the nomenclature "symmetric."

Likewise, we write $R(\partial_i,\partial_j)\partial_k=\sum_\nu R^\nu_{kij}\partial_\nu$. A straightforward computation using the definition of R, (9.20) and $[\partial_i,\partial_j]=0$ gives

$$R^\nu_{kij} = (\partial_i\Gamma^\nu_{jk}-\partial_j\Gamma^\nu_{ik}) + \sum_l(\Gamma^l_{jk}\Gamma^\nu_{il}-\Gamma^l_{ik}\Gamma^\nu_{jl}) \tag{9.28}$$

Let (M,ρ) be a smooth Riemannian manifold. A linear connection (or the associated ∇) for M is ***ρ-compatible*** if parallel transport along curves preserves inner product by ρ. Explicitly, suppose $X_0, Y_0\in T_aM$, $\gamma:I\to M$ is a smooth curve with $\gamma(t_0)=a$, and ξ, η are the parallel transports, respectively, of X_0, Y_0 along γ. Then it is required that for all $t\in I$,

$$\rho(\xi(t),\eta(t)) = \rho(X_0,Y_0) \tag{9.29}$$

27. Lemma *Let (M,ρ) be a smooth Riemannian manifold and consider a covariant derivative ∇ on M.*
(a) For any two smooth vector fields $\xi=X\circ\gamma$ and $\eta=Y\circ\gamma$ along a smooth curve γ,

$$\frac{d}{dt}\rho(\xi(t),\eta(t)) = \rho(\nabla_{\gamma'(t)}X,\eta(t)) + \rho(\xi(t),\nabla_{\gamma'(t)}Y)$$

(b) ∇ is ρ-compatible if and only if for all smooth vector fields X and Y and all $e\in T_aM$,

$$e\cdot\rho(X,Y) = \rho(\nabla_e X, Y(a)) + \rho(X(a),\nabla_e Y)$$

Proof. (a) Let $a=\gamma(t_0)$ be a point on the image of the curve γ. Pick an orthonormal basis (relative to ρ) for T_aM, say $(\overline{e}_1,\ldots,\overline{e}_m)$, and parallel transport the $\overline{e}_i$ to $(e_1(t),\ldots,e_m(t))$ along γ. Let $\xi(t)=\sum_i a^i(t)e_i(t)$ and $\eta(t)=\sum_i b^i(t)e_i(t)$, so that $\rho(a(t),b(t))=\sum_i a^i(t)b^i(t)$, and

$$\frac{d}{dt}\rho(\xi(t),\eta(t)) = \sum_i (a^i)'(t)b^i(t) + \sum_i a^i(t)(b^i)'(t)$$

On the other hand,

$$\begin{aligned}\nabla_{\gamma'(t)}X &= \nabla_{\gamma'(t)}\sum_i (a^i(t)e_i(t))\\ &= \sum_i da^i(\gamma'(t))e_i(t) + \sum_i a^i(t)\nabla_{\gamma'(t)}e_i\\ &= \sum_i (a^i)'(t)e_i(t)\end{aligned}$$

the last line by the chain rule, and the fact that e_i is parallel along γ. Therefore, $\rho(\nabla_{\gamma'(t)}X,\eta(t))=\sum_i(a^i)'(t)b^i(t)$. A similar computation for $\rho(\xi(t),\nabla_{\gamma'(t)}Y)$ proves (a).

(b) If H is ρ-compatible, then we take a smooth curve $\gamma{:}I\to M$ with $\gamma(t_0)=a$ and $\gamma'(t_0)=e$ and apply part (a). Conversely, suppose ξ and η are parallel along γ, we must show $\frac{d}{dt}\rho(\xi(t),\eta(t))=0$. But

$$\begin{aligned}\frac{d}{dt}\rho(\xi(t),\eta(t)) &= \gamma'(t)\cdot\rho(X,Y) \quad \text{(chain rule)}\\ &= \rho\big(\nabla_{\gamma'(t)}X, Y(\gamma(t))\big) + \rho\big(X(\gamma(t)),\nabla_{\gamma'(t)}Y\big)\\ &= 0\end{aligned}$$

the last line since ξ and η are parallel. □

So far we have not demonstrated the existence of connections for arbitrary smooth manifolds. But we saw in Chapter 6 that a smooth manifold admits infinitely many smooth Riemannian metrics. The following theorem then implies that an arbitrary smooth manifold admits infinitely many linear connections.

28. Theorem *Let (M,ρ) be a smooth Riemannian manifold. There is a unique torsion-free ρ-compatible linear connection for M.*

Proof. We take an arbitrary chart (U,ξ) of M and restrict ρ to U. We then show that a *unique* torsion-free ρ-compatible linear connection exists for U. The uniqueness will show immediately that the connections obtained on charts glue together smoothly to produce a connection for M. As usual, we adopt the short-hand notation ∂_i for $\frac{\partial}{\partial\xi^i}$; further we denote the smooth function $\rho(\partial_i,\partial_j)$ by g_{ij}. First we establish uniqueness assuming existence. From Lemma 27b we obtain the following:

$$\begin{cases} \partial_i g_{jk} = \rho(\nabla_{\partial_i}\partial_j, \partial_k) + \rho(\partial_j, \nabla_{\partial_i}\partial_k) \\ \partial_j g_{ki} = \rho(\nabla_{\partial_j}\partial_k, \partial_i) + \rho(\partial_k, \nabla_{\partial_j}\partial_i) \\ \partial_k g_{ij} = \rho(\nabla_{\partial_k}\partial_i, \partial_j) + \rho(\partial_i, \nabla_{\partial_k}\partial_j) \end{cases} \tag{9.30}$$

Since the connection is assumed to be symmetric (torsion-free), we have $\nabla_{\partial_\mu}\partial_\nu = \nabla_{\partial_\nu}\partial_\mu$ for all μ and ν. Applying this fact to the right-hand side and solving the three equations, we obtain

$$\rho(\nabla_{\partial_i}\partial_j, \partial_k) = \frac{1}{2}(\partial_i g_{jk} + \partial_j g_{ki} - \partial_k g_{ij})$$

or

$$\sum_\nu \Gamma^\nu_{ij} g_{k\nu} = \frac{1}{2}(\partial_i g_{jk} + \partial_j g_{ki} - \partial_k g_{ij})$$

Fixing i and j, and letting $k=1,\ldots,m$, we obtain m linear equations for unknowns Γ^k_{ij}. The matrix $[g_{ij}]$ is invertible since inner products are non-degenerate 2-tensors, so we can solve the system for Γ^k_{ij}. In fact, denoting the inverse of $[g_{ij}]$ by $[g^{ij}]$, we have

$$\Gamma^k_{ij} = \frac{1}{2}\sum_\nu g^{k\nu}(\partial_i g_{j\nu} + \partial_j g_{\nu i} - \partial_\nu g_{ij}) \tag{9.31}$$

This is known as ***Christoffel's First Identity***. Since the Γ^k_{ij} determine the connection, uniqueness is hereby proved. In fact, defining Γ^k_{ij} as above, a covariant derivative for U is obtained that is torsion-free since the expression is symmetric in i and j. We thus obtain a torsion-free linear connection on the domain of every chart. Further, uniqueness implies that these connections coincide on intersections, therefore a global linear connection is defined. It remains to check that this connection is ρ-compatible. Invertibility of $[g_{ij}]$ implies that (9.30) can be derived from (9.31). But (9.30) expresses ρ-compatibility for basis vector fields by Lemma 27b. Expanding and using the properties of covariant derivative, ρ-compatibility is deduced. □

The connection defined in Theorem 28 will be referred to as the ***Levi-Civita connection*** of the Riemannian manifold (M,ρ). The name *Riemannian connection* is also used for this, while some authors use this term to indicate a ρ-compatible connection for a Riemannian metric ρ.

29. Example For a Euclidean space, i.e., $\mathbb{R}^m$ with a fixed inner product $g_{ij}=constant$, we obtain $\Gamma^k_{ij}=0$, hence by (9.28), $R=0$ for this Levi-Civita connection.

Let (M,ρ) be a Riemannian manifold, $a\in M$, and suppose that Π is a two-dimensional linear subspace of T_aM. A key fact of Riemannian geometry is that Π is intrinsically endowed with a certain real number, called the *sectional curvature*. In the case of a two-dimensional submanifold M of $\mathbb{R}^3$ and its tangent plane $\Pi=T_aM$, the sectional curvature turns out to be the classical *Gaussian curvature*, a concept originally devised to describe the bending of the surface M in $\mathbb{R}^3$ at the point a. The following theorem will furnish the derivation of this invariant. Recall that given an inner product ρ on a vector space V, a linear map $T: V \to V$ is *skew-adjoint* if the relation $\rho(Tv, w) + \rho(v, Tw)$ holds for all $v, w \in V$.

30. Theorem *Let (M, ρ) be a Riemannian manifold and R be the curvature tensor of the Levi-Civita connection for (M, ρ). Then for any $a \in M$, any two-dimensional subspace Π of $T_a M$ and any basis (X, Y) for Π, the linear endomorphism $\rho_{a,X,Y}$ of Π given by*

$$\rho_{a,X,Y} : Z \mapsto R(X, Y)Z$$

is skew-adjoint. Further, if (X, Y) is an orthonormal basis for Π, then the matrix representation of $\rho_{a,X,Y}$ with respect to (X, Y) is in the form

$$\begin{bmatrix} 0 & K(a) \\ -K(a) & 0 \end{bmatrix} \tag{9.32}$$

where $K(a)$ is independent of the particular orthonormal basis (X, Y).

Proof. To prove the skew-adjoint property, we must show that for all $Z, W \in \Pi$

$$\rho(R(X, Y)Z, W) + \rho(Z, R(X, Y)W) = 0$$

This will be a straightforward computation using the definition of R, the definition of $[X, Y]$ as $X \cdot Y - Y \cdot X$, and Lemma 27b. We have

$$\begin{aligned} \rho(R(X, Y)Z, W) &= \rho(\nabla_X \nabla_Y Z - \nabla_Y \nabla_X Z - \nabla_{[X,Y]} Z, W) \\ &= X \cdot \rho(\nabla_Y Z, W) - \rho(\nabla_Y Z, \nabla_X W) - Y \cdot \rho(\nabla_X Z, W) + \rho(\nabla_X Z, \nabla_Y W) \\ &\quad - [X, Y] \cdot \rho(Z, W) + \rho(Z, \nabla_{[X,Y]} W) \end{aligned}$$

Note that the second and fourth terms of the above together are anti-symmetric in (Z, W), therefore they will cancel out with the corresponding terms of $\rho(Z, R(X, Y)W)$. On the other hand, the sum of the first and third terms is

$$\begin{aligned} &X \cdot [Y \cdot \rho(Z, W) - \rho(Z, \nabla_Y W)] - Y \cdot [X \cdot \rho(Z, W) - \rho(Z, \nabla_X W)] \\ &= [X, Y] \cdot \rho(Z, W) - X \cdot \rho(Z, \nabla_Y W) + Y \cdot \rho(Z, \nabla_X W) \end{aligned}$$

It follows that

$$\rho(R(X, Y)Z, W) = \rho(Z, \nabla_{[X,Y]} W) - X \cdot \rho(Z, \nabla_Y W) + Y \cdot \rho(Z, \nabla_X W) + s(Z, W)$$

where $s(Z, W)$ is anti-symmetric in (Z, W). Therefore,

$$\begin{aligned} \rho(R(X, Y)Z, W) + \rho(Z, R(X, Y)W) &= [X, Y] \cdot \rho(W, Z) - X \cdot (Y \cdot \rho(Z, W)) + Y \cdot (X \cdot \rho(Z, W)) \\ &= 0 \end{aligned}$$

The matrix of any skew-adjoint linear map T with respect to an orthonormal basis (X, Y) is skew-symmetric, i.e., it is in the form

$$\begin{bmatrix} 0 & c \\ -c & 0 \end{bmatrix} \tag{9.33}$$

where $c=\rho(T(Y), X)=-\rho(T(X), Y)$. In our case, suppose (X, Y) and (X', Y') are orthonormal bases for Π. Thus $X'=a_{11}X+a_{21}Y$ and $Y'=a_{12}X+a_{22}Y$, where the matrix $\mathrm{A}=[a_{ij}]$ is orthogonal, hence it has determinant ± 1. Using successively the facts that $R(X, Y)Z$ is anti-symmetric bilinear in (X, Y) and $\rho(R(X, Y)Z, W)$ is anti-symmetric bilinear in (Z, W), we obtain

$$\begin{aligned} \rho(\rho_{a,X',Y'}(Y'), X') &= \rho((\det \mathrm{A})R(X, Y)Y', X') \\ &= (\det \mathrm{A})\rho(R(X, Y)Y', X') \\ &= (\det \mathrm{A})^2\rho(R(X, Y)Y, X) \\ &= \rho(\rho_{a,X,Y}(Y), X) \end{aligned}$$

orthogonal Therefore, the entry $K(a)$ of the matrix (9.33) is independent of the particular orthonormal basis. □

The real number

$$K(a) = \rho(r_{a,X,Y}(Y), X) = -\rho(r_{a,X,Y}(X), Y) \tag{9.34}$$

is called the ***sectional curvature*** of Π. In the case of a two-dimensional Riemannian manifold (M, ρ), the sectional curvature of $\Pi=T_aM$ is better known as the ***Gaussian curvature*** of (M, ρ) at the point a. For a surface embedded in Euclidean $\mathbb{R}^3$, $K(a)$ is normally defined with reference to the way the surface curves in the three-dimensional Euclidean space. A pinnacle of classical differential geometry was Gauss' *Theorema Egregium* to the effect that K depends only on the inner products induced from $\mathbb{R}^3$ on the tangent spaces to the surface. This is tantamount to deducing the formula above for Gaussian curvature as a theorem from the classical definition of the curvature of a surface. The startling theorem became the starting point for Riemann's introduction of the curvature tensor and the founding of modern differential geometry. Below we proceed to give an account of Gauss' Theorem by reverting the historical process to obtain the original extrinsic interpretation of $K(a)$.

Let $(\tilde{M}, \tilde{\rho})$ be a Riemannian manifold and M a submanifold of $\tilde{M}$. Then $\tilde{\rho}$ induces a Riemannian metric ρ on M. (M, ρ) will be referred to as a ***Riemannian submanifold*** of $(\tilde{M}, \tilde{\rho})$. We consider the Levi-Civita connections for $(\tilde{M}, \tilde{\rho})$ and (M, ρ), with associated covariant derivative operators $\tilde{\nabla}$ and ∇, respectively. Suppose X and Y are smooth vector fields on M, then $\nabla_X Y(a)$ is defined for every $a \in M$. By taking local extensions $\tilde{X}$ and $\tilde{Y}$ of X and Y, respectively, to a neighborhood of a, we may also consider $\tilde{\nabla}_{\tilde{X}}\tilde{Y}(a)$. But, by 15c, this depends only on the values of $\tilde{Y}$ along the trajectory of $\tilde{X}$ passing through a, so it is independent of the particular extensions.

We thus write $\tilde{\nabla}_X Y(a)$ instead of $\tilde{\nabla}_{\tilde{X}} \tilde{Y}(a)$. The following expresses the relation of the two covariant derivatives.

31. Lemma *$\nabla_X Y(a)$ is the orthogonal projection, with respect to $\tilde{\rho}$, of $\tilde{\nabla}_X Y(a) \in T_a\tilde{M}$ on the subspace T_aM.*

Proof. Let us denote the orthogonal projection of $\tilde{\nabla}_X Y(a)$ on T_aM by $\nabla'_X Y(a)$. We show that ∇' satisfies the definition of Levi-Civita connection for the Riemannian manifold (M, ρ). Then the uniqueness statement of Theorem 28 would imply the desired result. Now properties (a) and (b) of Definition 17 follow from the linearity of projection from $T_a\tilde{M}$ onto T_aM, as does the torsion-free property. The orthogonality of projection will be used in verifying ρ-compatibility, using (b) of Lemma 27. Suppose $e \in T_aM$, then we write

$$\tilde{\nabla}_e X = \nabla'_e X(a) + X', \quad \tilde{\nabla}_e Y = \nabla'_e Y(a) + Y'$$

where $X', Y' \in T_a\tilde{M}$ are ρ-perpendicular to T_aM. Then

$$\begin{aligned} e \cdot \rho(X, Y) &= \rho(\tilde{\nabla}_e X, Y(a)) + \rho(X(a), \tilde{\nabla}_e Y) \\ &= \rho(\nabla'_e X + X', Y(a)) + \rho(X(a), \nabla'_e Y + Y') \\ &= \rho(\nabla'_e X, Y(a)) + \rho(X(a), \nabla'_e Y) \end{aligned}$$

proving the lemma. □

In fact, there is more to be said about the relationship of the two covariant derivatives. Consider the case where the submanifold M is a **hypersurface** in $\tilde{M}$, i.e., $\dim \tilde{M} = \dim M + 1$. Given $a \in M$, for a sufficiently small connected neighborhood of a in M, precisely two unit normal vector fields to M exist on that neighborhood. Let us fix one of these unit normal vector fields and denote it by $\mathbf{n}$. For a tangent vector field Z to M, defined in a neighborhood of a in M, it makes sense to consider $\tilde{\nabla}_Z \mathbf{n}$, for one can extend Z and $\mathbf{n}$ smoothly to a neighborhood of a in $\tilde{M}$, and the value of $\tilde{\nabla}_Z \mathbf{n}$ depends only on the values of $\mathbf{n}$ along the trajectory of Z that passes through a.

32. Lemma *Let (M, ρ) be a Riemannian hypersurface of the Riemannian manifold $(\tilde{M}, \tilde{\rho})$ and $a \in M$. If X and Y are smooth tangent vector fields defined in a neighborhood of a in M, then*

$$\tilde{\nabla}_X Y = \nabla_X Y + l(X, Y)\boldsymbol{n} \tag{9.35}$$

where

$$l(X, Y) = -\tilde{\rho}(\tilde{\nabla}_X \boldsymbol{n}, Y), \text{ and } l(X, Y) = l(Y, X) \tag{9.36}$$

Proof. Taking the inner product of the two sides of (9.35) with $\mathbf{n}$, we obtain

$$l(X, Y) = \tilde{\rho}(\tilde{\nabla}_X Y, \mathbf{n})$$

Now $\tilde{\rho}(Y, \mathbf{n})$ is identically zero, hence using Lemma 27b, we obtain

$$0 = X \cdot \tilde{\rho}(Y, \mathbf{n}) = \tilde{\rho}(\tilde{\nabla}_X Y, \mathbf{n}) + \tilde{\rho}(Y, \tilde{\nabla}_X \mathbf{n})$$

from which the left-side relation of (9.36) follows. The right-side equation is a consequence of the torsion-free property of Levi-Civita connections:

$$\begin{aligned} l(X, Y)\mathbf{n} &= \tilde{\nabla}_X Y - \nabla_X Y \\ &= \tilde{\nabla}_Y X + [X, Y] - (\nabla_Y X + [X, Y]) \\ &= l(Y, X)\mathbf{n} \end{aligned}$$

as claimed. □

Note, in addition, that the first relation in (9.36) implies that l is *tensorial*, i.e., $l(X, Y)(a)$ is completely determined by the values $X(a)$ and $Y(a)$, and l is bilinear. It is classically referred to as the ***Second Fundamental Form*** of the hypersurface, with the ***First Fundamental Form*** being the Riemannian metric on the hypersurface induced from the ambient manifold.

The unit normal map $\mathbf{n}$ induces a linear map $S_a : T_a M \to T_a M$, known as the ***shape operator***[2] or ***Weingarten map***, which we define as

$$S_a(X) = \tilde{\nabla}_X \mathbf{n}, \quad \text{for } X \in T_a M \tag{9.37}$$

We first note that S_a is well-defined. This follows from the pointwise dependence of any covariant derivative $\nabla_X Y$ on X. We then claim that indeed $S_a(X) \in T_a M$. Since $\tilde{\rho}(\mathbf{n}, \mathbf{n})$ is identically equal to 1, one has $X \cdot \tilde{\rho}(\mathbf{n}, \mathbf{n}) = 0$, so using Lemma 27b, we obtain $\tilde{\rho}(\tilde{\nabla}_X \mathbf{n}, \mathbf{n}) = 0$, hence $\tilde{\nabla}_X \mathbf{n} \in T_a M$. The symmetry relation $l(X, Y) = l(Y, X)$ implies that S_a is self-adjoint, so its eigenvalues are real. These are known (up to sign) as ***principal curvatures*** of (M, ρ) at a. In the case $\tilde{M} = \mathbb{R}^3$ with Euclidean metric, the determinant of shape operator is the classical definition of Gaussian curvature. In general, the determinant of S_a is also known as *Gauss-Kronecker* or *Killing-Lipschitz curvature*.

33. Remark In case the ambient manifold $\tilde{M}$ is $\mathbb{R}^N$ with the Euclidean metric, the Weingarten map can be regarded as the derivative of $\mathbf{n}$ in the following sense. $\mathbf{n}$ takes its values in the unit sphere $S_{N-1} \subset \mathbb{R}^N$, and for a point $a \in M$, the tangent space to M at a and the tangent space to S_{N-1} at $\mathbf{n}$ are parallel. By canonically identifying these two linear spaces, the derivative of $\mathbf{n}$ at a can be regarded as a linear endomorphism of $T_a M$. However, the Levi-Civita connection associated to the Euclidean metric of $\mathbb{R}^N$ is just the product connection for the parallelizable manifold $\mathbb{R}^N$ (see Examples 8 and 29), so the covariant derivative in (9.37) coincides with the usual derivative. The unit normal field $\mathbf{n}$ defined (locally) on a hypersurface is also referred to as the ***Gauss map*** of the hypersurface, thus the common statement that "the Weingarten map is the derivative of the Gauss map" for hypersurfaces in $\mathbb{R}^N$.

---

[2] Often $-S$ is taken as the shape operator.

We work out the relationship of curvature tensors $\tilde{R}$ and R of $(\tilde{M},\tilde{\rho})$ and (M,ρ), where (M,ρ) is a Riemannian hypersurface of $(\tilde{M},\tilde{\rho})$. Let X, Y and Z be tangent vector fields defined in a neighborhood of a in M. By the repeated application of Lemma 32, we can write

$$\begin{aligned}\tilde{\nabla}_X\tilde{\nabla}_Y Z &= \tilde{\nabla}_X(\nabla_Y Z + l(Y,Z)\mathbf{n})\\ &= \nabla_X\nabla_Y Z + l(X,\nabla_Y Z)\mathbf{n} + l(Y,Z)\tilde{\nabla}_X\mathbf{n} + (X\cdot l(Y,Z))\mathbf{n}\\ &= \nabla_X\nabla_Y Z + l(Y,Z)\tilde{\nabla}_X\mathbf{n} + (l(X,\nabla_Y Z) + X\cdot l(Y,Z))\mathbf{n}\end{aligned}$$

By symmetry

$$\tilde{\nabla}_Y\tilde{\nabla}_X Z = \nabla_Y\nabla_X Z + l(X,Z)\tilde{\nabla}_Y\mathbf{n} + (l(Y,\nabla_X Z) + Y\cdot l(X,Z))\mathbf{n}$$

Finally,

$$\tilde{\nabla}_{[X,Y]}Z = \nabla_{[X,Y]}Z + l([X,Y],Z)\mathbf{n}$$

Therefore,

$$\tilde{R}(X,Y)Z = R(X,Y)Z + l(Y,Z)\tilde{\nabla}_X\mathbf{n} - l(X,Z)\tilde{\nabla}_Y\mathbf{n} + A\mathbf{n} \tag{9.38}$$

where

$$A = l(X,\nabla_Y Z) + X\cdot l(Y,Z) - l(Y,\nabla_X Z) - Y\cdot l(X,Z) - l([X,Y],Z) \tag{9.39}$$

34. Lemma ***(Gauss Equation)***
Let (M,ρ) be a Riemannian hypersurface of the Riemannian manifold $(\tilde{M},\tilde{\rho})$, and suppose that X, Y, Z and W are smooth tangent vector fields on M. Then

$$\tilde{\rho}(W,\tilde{R}(X,Y)Z) - \rho(W,R(X,Y)Z) = l(X,Z)l(Y,W) - l(X,W)l(Y,Z) \tag{9.40}$$

Proof. This is obtained by taking the inner product of the two sides of (9.38) with W. □

35. Theorem ***(Theorema Egregium)***
Let M be a two-dimensional smooth manifold in $\mathbb{R}^3$ with induced Riemannian metric from the Euclidean metric on $\mathbb{R}^3$. Then the Gaussian curvature of M at each point of M is equal to the determinant of the shape operator at that point.

Proof. It was shown in Subsection 29 that for the Euclidean metric on $\mathbb{R}^3$, $\tilde{R}=0$. For any point $a\in M$, we may take an orthonormal pair of smooth vector fields (X,Y) defined in a neighborhood of a, e.g., by applying the parametrized Gram-Schmidt procedure (Lemma 17, Chapter 3). Then the determinant of S will be

$$\rho(S(X),X)\rho(S(Y),Y) - \rho(S(X),Y)\rho(S(Y),X)$$

By substituting $Z=Y$ and $W=X$ in Gauss' Equation and using (9.37) and (9.36), we obtain the desired result. □

We note that the choice of the unit normal has no effect on the sign of the determinant since we are dealing with a 2×2 matrix. As noted earlier, the original definition of Gaussian curvature was essentially as the determinant of Weingarten map, and Theorema Egregium stated that this can be expressed purely in terms of the First Fundamental Form. The adoption of the definition in terms of the First Fundamental Form opens the way for the study of geometries with properties not inherited from the ambient Euclidean space. One of the most famous examples is the so-called *Poincaré Half Plane*, which was one of the early models of non-Euclidean geometry.

36. Poincaré Half Plane

We consider the open upper half plane $\mathbb{U}=\{(x,y)\in\mathbb{R}^2 : y>0\}$ with the Riemannian metric

$$\rho(x,y) = \frac{1}{y^2}(dx\otimes dx + dy\otimes dy)$$

or equivalently $g_{11}=g_{22}=y^{-2}$ and $g_{12}=g_{21}=0$ (see also Exercises 3.19 and 7.18). Using (9.20) we obtain

$$\Gamma^2_{11} = -\Gamma^1_{12} = -\Gamma^1_{21} = -\Gamma^2_{22} = y^{-1},\ \Gamma^1_{11}=\Gamma^1_{22}=\Gamma^2_{12}=\Gamma^2_{21}=0 \tag{9.41}$$

At an arbitrary point $(x,y)\in\mathbb{U}$, we consider the orthonormal basis $(X=y\frac{\partial}{\partial x}(z), Y=y\frac{\partial}{\partial y}(z))$ for the tangent space and use (9.34) to compute the Gaussian curvature. But first, (9.28) and (9.41) give

$$R(\frac{\partial}{\partial x}, \frac{\partial}{\partial y})\frac{\partial}{\partial y} = \frac{-1}{y^2}\frac{\partial}{\partial x}$$

Therefore,

$$\begin{aligned} K(z) &= \rho(R(X,Y)Y,X) \\ &= y^4\rho(R(\frac{\partial}{\partial x}, \frac{\partial}{\partial y})\frac{\partial}{\partial y}, \frac{\partial}{\partial x}) \\ &= y^4\rho(\frac{-1}{y^2}\frac{\partial}{\partial x}, \frac{\partial}{\partial x}) \\ &= -1 \end{aligned}$$

Next we give a brief account of the geodesics in $\mathbb{U}$. Substituting the values of the Γ^k_{ij} obtained above in (9.23), we obtain

$$\begin{cases} \ddot{x} - \frac{2}{y}\dot{x}\dot{y} = 0 \\ \ddot{y} + \frac{1}{y}\dot{x}^2 - \frac{1}{y}\dot{y}^2 = 0 \end{cases} \tag{9.42}$$

where we have used the short-hand notation $\dot{x}$ for $\frac{dx}{dt}$, $\ddot{x}$ for $\frac{d^2x}{dt^2}$, etc. The vertical straight half-line $x=x_0(constant)$ satisfies the first equation. Substituting in the second, we obtain $y\ddot{y}-\dot{y}^2=0$, therefore $\frac{\dot{y}}{y}$ is a constant k, from which we get $y=y_0e^{kt}$. Thus the vertical half-line parametrized as (x_0, y_0e^{kt}) is a geodesic. In fact, given a vertical tangent vector $b\frac{\partial}{\partial y}(x_0,y_0)$ at (x_0,y_0), the vertical half-line with $k=\frac{b}{y_0}$ is the unique geodesic with that initial condition. The other solutions of this system of differential equations can be obtained by computational artistry (see, e.g., [**17**]), but we will give a more insightful account in the next section using a theorem of E. Nöther. It turns out that the other geodesics are half-circles in $\mathbb{U}$ with center on the x-axis, appropriately parametrized as $\gamma(t)$, so that the open half-circle is traversed once as t goes from $-\infty$ to $+\infty$. Geodesics are the natural analogues, in Riemannian geometry, of straight lines in Euclidean geometry in that they represent paths of acceleration-free motion. Another standard property is that they are the shortest paths between nearby points[3]. In the model of non-Euclidean geometry represented by the upper half plane $\mathbb{U}$, given a completely extended geodesic α and a point P not on the image of the geodesic, there are precisely two geodesics that pass through P and are parallel to α in the following sense: two geodesics whose extensions become tangent on the limit line y=0 are considered parallel, as are two vertical straight half-lines (see Figure 7). There are an infinite number of geodesics that pass through P and do not intersect α.

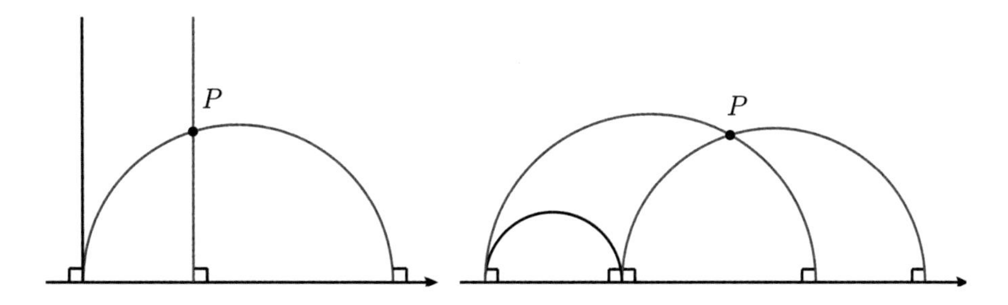

FIGURE 7. Constructing two parallels

E. Newtonian Mechanics

We will present here a brief account of Newtonian mechanics in the framework of Riemannian manifolds. Contrary to popular belief, Newtonian mechanics need not be predicated on a Euclidean space structure. Newton's own cosmological and philosophical speculations were probably shaped by the then prevalent Euclidean outlook on geometry, but these have no bearing on the actual physical predictions and mathematical derivations of Newtonian theory in the scale to which Newtonian theory is deemed to apply. Here are the two basic principles on which the present treatment of a *Newtonian system* will be based.

Principle 1. The possible physical *positions* of a Newtonian system constitute a smooth manifold M, known as the ***configuration space***. The tangent bundle TM, known as the ***state space***, is the manifold on which the *Newtonian vector field*

[3]See any standard text on Riemannian geometry, e.g., [**8**].

will be defined. A Riemannian metric ρ and its induced Levi-Civita connection are hypothesized as given data for the Newtonian system.

Principle 2. A ***Newtonian vector field*** is a smooth vector field X on TM of the form $X{=}G{+}\overline{F}$, where G is the geodesic vector field of the Levi-Civita connection and $\overline{F}$ is a vertical vector field, i.e., $T\tau_M(\overline{F}(z)){=}\mathbf{0}_x$ for all $z{\in}T_xM$. G is known as the ***inertial system*** and $\overline{F}$ as the ***force system*** of X.

Elaborating on the above, the time evolution of a Newtonian system is represented by the integral curves of X and is hence deterministic on account of the fundamental existence-uniqueness theorem of ordinary differential equations. Thus a given state of the system determines its complete future and past trajectory. Since G is a second order ODE on M and $\overline{F}$ is vertical, $X{=}G{+}\overline{F}$ is also a second order ODE on M. The solutions $\gamma{:}I{\to}M$ of X are the ***physical paths***, while the integral curves γ' are referred to as ***state paths***. For $\overline{F}{=}0$, we have an inertial system with physical paths being geodesics, i.e., curves of zero intrinsic acceleration. We are using the notation $\overline{F}$ for the force system because in an important case that we will later explore, $\overline{F}$ is the vertical spread of a *force field* F, which is defined on M (see Subsection 11 for the definition of vertical spread and notation). In actual examples, the Riemannian metric is determined by certain positive parameters known as *mass* that are incorporated into the Riemannian metric. Put succinctly, a Newtonian system can be defined as a triple $(M,\rho,\overline{F})$, where M is a smooth manifold, ρ is a Riemannian metric for M and $\overline{F}$ is a smooth vertical vector field on TM. All else can be defined in terms of the given triple.

Given the Riemannian manifold (M,ρ), the ***kinetic energy*** is the function $E_k{:}TM{\to}\mathbb{R}$ defined by

$$E_k(z) = \frac{1}{2}\rho(z,z) = \frac{1}{2}|z|_\rho^2 \tag{9.43}$$

where $|z|_\rho$ denotes the vector norm of z relative to the inner product given by ρ. The following special case of the *conservation of energy* (for inertial systems) may be construed as the ***Principle of Inertia***.

37. Lemma *For any geodesic γ of the Levi-Civita connection of a Riemannian manifold (M,ρ), $|\gamma'(t)|_\rho$ is independent of t.*

Proof. By Lemma 27

$$\frac{d}{dt}E_k(\gamma'(t)) = \rho(\nabla_{\gamma'(t)}\gamma'(t),\gamma'(t))$$

But for a geodesic γ, we have $\nabla_{\gamma'(t)}\gamma'(t){=}\ddot{\gamma}(t){=}0$, and the assertion follows. □

An important special case of a Newtonian vector field is a ***(classical) conservative vector field***, in which $\overline{F}$ is the vertical spread of a vector field F on M of the form $F=-\nabla_\rho V$, where $V:M\to\mathbb{R}$ is a smooth function. V is called the ***potential (energy)*** and F the ***force field***. The function $E:TM\to\mathbb{R}$ defined by

$$E = E_k + V\circ\tau_M$$

is called the ***(total) energy*** of the system.

38. Theorem ***(Conservation of Energy)***
Let γ be a physical path for a conservative Newtonian vector field. Then $E(\gamma'(t))$ is independent of t.

PROOF.

$$\begin{aligned}\frac{d}{dt}((E\circ\gamma')(t)) &= dE(\gamma''(t))\\ &= (d(V\circ\tau_M))(\gamma''(t)) + dE_k(\gamma''(t))\\ &= dV(\gamma'(t)) + dE_k\big(G(\gamma'(t))\big) + dE_k\big(\overline{F}(\gamma'(t))\big)\end{aligned}$$

The middle term on the right is zero by Lemma 37. The left-most term on the right is equal to

$$\rho\big(\nabla_\rho V(\gamma(t)), \gamma'(t)\big)$$

by the definition of the gradient. We show that the last term equals the negative of the above. The vector field $\overline{F}$ is tangent to the fibers of TM and the restriction to each fiber is a constant vector field $F(\gamma(t))= -\nabla_\rho V(\gamma(t))$. Therefore,

$$\begin{aligned}dE_k\big(\overline{F}(\gamma'(t))\big) &= (\overline{F}\cdot E_k)(\gamma'(t))\\ &= \frac{d}{ds}\Big|_{s=0}\Big\{E_k\big(\gamma'(t)-s\nabla_\rho V(\gamma(t))\big)\Big\} \text{ (by Theorem 17 of Chapter 2)}\\ &= \frac{d}{ds}\Big|_{s=0}\Big\{(\frac{1}{2})\rho\big(\gamma'(t)-s\nabla_\rho V(\gamma(t)), \gamma'(t)-s\nabla_\rho V(\gamma(t))\big)\Big\}\\ &= -\rho\big(\nabla_\rho V(\gamma(t)), \gamma'(t)\big)\end{aligned}$$

which proves the result. □

The above is the most basic of conservation laws in classical mechanics. Other conservation laws prevail in various categories of classical systems. We are going to present a version of an important theorem of E. Nöther that describes how symmetries (=group actions) give rise to conservation laws. By an ***isometry*** of a Riemannian manifold (M,ρ) we mean a smooth diffeomorphism $f:M\to M$ with the property that $f^*\rho=\rho$. Explicitly, given $z_1, z_2\in T_aM$, the requirement is that

$$\rho(T_af(z_1), T_af(z_2)) = \rho(z_1, z_2) \tag{9.44}$$

39. Lemma *Isometries of Riemannian manifolds preserve Levi-Civita connection and hence map geodesics to geodesics.*

Proof. The coefficients Γ_{ij}^k are determined by the g_{ij} according to (9.31), and the Γ_{ij}^k determine the linear connection as well as the geodesic equation. □

Recall from Chapter 6, and especially Theorem 9 of Chapter 2, that if (Ψ_s) is a smooth local one-parameter group of diffeomorphisms of a manifold M, then (Ψ_s) is the local flow of the vector field X defined by

$$X(x) = \frac{d}{ds}\Big|_{s=0} \Psi_s(x) = (\Psi^x)'(0)$$

X was called the generator of the local flow (Ψ_s). In the case each Ψ_s is an isometry, X is called an ***infinitesimal isometry*** or a **Killing vector field**.

40. Lemma *Let (M, ρ) be a Riemannian manifold and (Ψ_s) a local one-parameter group of isometries with generator A. Then for every geodesic γ of Levi-Civita connection, $\rho\big(A(\gamma(t)), \gamma'(t)\big)$ is independent of t.*

Proof. We define a smooth map Z from a neighborhood of $(0, 0)$ in $\mathbb{R}^2$ to M by

$$Z(s, t) = (\Psi_s \circ \gamma)(t)$$

and consider vector fields $Z_*(\frac{\partial}{\partial s})$ and $Z_*(\frac{\partial}{\partial t})$ defined along Z. Note that $Z_*(\frac{\partial}{\partial s})=A$, and we denote $Z_*(\frac{\partial}{\partial t})$ by B. Then by Lemma 27

$$\frac{d}{dt}\rho\big(A(\gamma(t)), \gamma'(t)\big) = \rho\big(\nabla_{B(0,t)}A(\gamma(t)), \gamma'(t)\big) + \rho\big(A(\gamma(t)), \nabla_{B(0,t)}\gamma'(t)\big)$$

Now $\gamma'(t)=B(0, t)$, and $\nabla_{B(0,t)}\gamma'(t)=0$ since γ is a geodesic. It therefore suffices to show that $\rho(\nabla_B A, B)=0$. For a given t, $\rho(B(s, t), B(s, t))$ is independent of s since Ψ_s is an isometry. It follows by taking derivative with respect to s that $\rho(\nabla_A B, B)=0$. Now $[A, B] = 0$ (see Subsection 12 of Chapter 6), and Levi-Civita connection is torsion-free, therefore $\nabla_B A = \nabla_A B$, and $\rho(\nabla_B A, B)=0$ as claimed. □

By incorporating the contribution of potential energy we obtain the following:

41. Theorem *(E. Nöther)*
Let (M, ρ) be a Riemannian manifold, $V{:}M\to\mathbb{R}$ a smooth function, $F=-\nabla_\rho V$ and $X=\overline{F}+G$ be the corresponding Newtonian vector field. Suppose that (Ψ_s) is a local one-parameter group of isometries of M with generator A and the property that $V\circ\Psi_s=V$ for all s. Then the function $J{:}TM\to\mathbb{R}$ defined by

$$J(z) = \rho(z, (A\circ\tau_M)(z))$$

is constant on the integral curves of X.

The function J is called the ***momentum*** associated to the local one-parameter group (Ψ_s), and the assertion of the theorem is known as the ***conservation of momentum J***.

PROOF. Let $\gamma:I\to M$ be a solution of the second order ODE X. We must show that $J(\gamma'(t))$ is independent of t, or that $(J\circ\gamma')'(t)=0$. But

$$\begin{aligned}(J\circ\gamma')'(t) &= dJ(\gamma''(t))\\ &= dJ\big(\overline{F}(\gamma'(t))\big) + dJ\big(G(\gamma'(t))\big)\end{aligned}$$

We prove that, in fact,

$$dJ\big(\overline{F}(\gamma'(t))\big) = 0$$
$$dJ\big(G(\gamma'(t))\big) = 0$$

The second equation is just the content of Lemma 40. For the first, let z be an arbitrary point of TM, and $\tau_M(z)=a$. Denoting the flow of $\overline{F}$ by Φ_t, we have $\Phi_t(z)=z+tF(a)$. Therefore,

$$\begin{aligned}dJ(\overline{F}(z)) &= (\overline{F}\cdot J)(z)\\ &= \lim_{h\to 0}\frac{J(\Phi_h(z)) - J(z)}{h}\\ &= \lim_{h\to 0}\frac{\rho(z+hF(a), A(a)) - \rho(z, A(a))}{h}\\ &= \rho(F(a), A(a))\end{aligned}$$

Now $A(a)$ is tangent to the level set of V that passes through a since for the flow (Ψ_s) of A, we have $V\circ\Psi_s=V$ by the hypothesis. On the other hand, $F(a)=-\nabla_\rho V(a)$ is perpendicular to the level sets of V, so $\rho(F(a), A(a))=0$, and the proof is complete. □

42. Examples

(a) (**Angular momentum**) In many problems of mechanics, the configuration space M is $\mathbb{R}^m$ or $\mathbb{R}^m$ with a set of measure zero removed, and the Riemannian structure is given by a Euclidean metric $\sum_i m_i dx^i\otimes dx^i$, where the m_i are given positive constants (masses). In the general *central force problem*, for example, the configuration space M is $\mathbb{R}^m$ or $\mathbb{R}^m-\{\mathbf{0}\}$, and a potential $V:M\to\mathbb{R}$ is given with the property that $V(x)$ depends only on $|x|$, i.e., on the distance of x to the center of force. The Riemannian metric is $\sum_i m dx^i\otimes dx^i$. In this case, a one-parameter group of planar rotations preserves both V and the metric, and a momentum function arises that is the *angular momentum* related to the plane of rotation. Explicitly, consider the one-parameter group of rotations in (x^i, x^j)-plane given by

$$\Psi_s(x^i, x^j) = (x^i\cos s - x^j\sin s, x^i\sin s + x^j\cos s),\quad s\in\mathbb{R}$$

The infinitesimal generator here is the vector field

$$-x^j \frac{\partial}{\partial x^i} + x^i \frac{\partial}{\partial x^j}$$

Therefore, the corresponding momentum according to Nöther's theorem is $m(-x^j v^i + x^i v^j)$, where the v^k are the components of the velocity vector $\mathbf{v}$. In $\mathbb{R}^3$, we can take three independent one-parameter rotation groups, one for each coordinate plane, and obtain three such momentum functions. These are usually treated as the components of a vector angular momentum $\mathbf{J}=m\mathbf{v} \times \mathbf{x}$, where $\times$ is the cross product in $\mathbb{R}^3$.

(b) (**Linear momentum**) We show that the center of mass of n bodies under mutual gravitational attraction moves with constant velocity. This will be a case of the conservation of *linear momentum*. Consider n mass particles in $\mathbb{R}^3$, the ith particle with position vector $x_i=(x_i^1, x_i^2, x_i^3)$, velocity vector $v_i=(v_i^1, v_i^2, v_i^3)$ and mass m^i. The center of mass is the point $\bar{x}=\frac{1}{m}(m^1 x_1 + \cdots + m^n x_n)$, where $m=\sum_i m^i$. $\bar{x}$ will then have the velocity $\bar{v}=\frac{1}{m}(m^1 v_1 + \cdots + m^n v_n)$. The gravitational potential for this system is taken to be

$$-k \sum_{i \neq j} \frac{m^i m^j}{|x_i - x_j|}$$

where $k>0$ is a constant and $|.|$ denotes the Euclidean distance. The relevant point here is the fact that the potential depends only on the relative distances of mass particles. The configuration space is $M=\mathbb{R}^{3n} - \Delta$, where Δ is the closed subset of $\mathbb{R}^3$ corresponding to particle collisions $x_i=x_j$, and the Riemannian metric is

$$\rho = \sum_{i=1}^{n} m^i (\sum_{j=1}^{3} dx_i^j \otimes dx_i^j)$$

For $U \in \mathbb{R}^3$, consider the one-parameter group of translations in the configuration space

$$\Psi_s(x) = x + s\tilde{U}$$

where $\tilde{U}$ is the n-tuple $(U, \ldots, U)$. This is a one-parameter group of isometries that preserves the potential and has generator $\tilde{U}$. Therefore, a momentum function J_U is obtained as

$$J_U(x, v) = \rho(\tilde{U}, v) = \sum_{i=1}^{n} m^i (U \cdot v_i)$$

where $U \cdot v_i$ denotes the standard Euclidean inner product in $\mathbb{R}^3$. In particular, setting $U=e_j$, $j=1,2,3$, the standard basis for $\mathbb{R}^3$, we obtain the following conserved momenta:

$$J^j = \sum_{i=1}^{n} m^i v_i^j$$

It follows that $\bar{v}=\frac{1}{m}(m^1 v_1 + \cdots + m^n v_n)$ is constant, as claimed.

(c) (**Clairaut function**) In this and the next example, we deal with purely geometric applications of Nöther's theorem, where the potential V is taken to be zero, so the vector field in question is just the geodesic vector field G. Thus, in fact, Lemma 40 is sufficient. Let I be an open interval in $\mathbb{R}$ and $f{:}I{\to}\mathbb{R}$ be an everywhere positive smooth function. We define $F{:}I{\times}\mathbb{R}{\to}\mathbb{R}^3$ by

$$F(x, \theta) = (x, f(x)\cos\theta, f(x)\sin\theta)$$

Thus the image M of F may be regarded as the result of rotating the graph of f in (x, y, z)-space with $y{=}f(x)\cos\theta$ and $z{=}f(x)\sin\theta$ (see Figure 8). It is straightforward to check that F is an immersion and M is a two-dimensional submanifold of $\mathbb{R}^3$, called a ***surface of revolution*** (see also Exercise 4.11 of Chapter 4). We endow M with the Riemannian metric induced from the Euclidean metric of $\mathbb{R}^3$ and the corresponding Levi-Civita connection (see Lemma 31 and the discussion preceding it). There is a one-parameter group of isometries (Ψ_s) (rotations) defined by

$$\Psi_s(x, y, z) = (x, y\cos s - z\sin s, y\sin s + z\cos s)$$

which maps $F(x, \theta)$ to $F(x, \theta{+}s)$. The infinitesimal generator of (Ψ_s) at a point (x, y, z) of M is then

$$A(x, y, z) = -f(x)(\sin\theta)\frac{\partial}{\partial y} + f(x)(\cos\theta)\frac{\partial}{\partial z}$$

It follows from Theorem 41, or Lemma 40, that the following function is constant along a geodesic $\gamma(t)$:

$$\begin{aligned} J(t) &= \rho\big(A(\gamma(t)), \gamma'(t)\big) \\ &= f(x)|\gamma'(t)|\cos\alpha \end{aligned}$$

where α is the angle between $\gamma'(t)$ and the *meridian x=constant* passing through $\gamma(t)$. By Lemma 37, $|\gamma'(t)|$ is constant along the geodesic $\gamma(t)$, therefore the function $f(x)\cos\alpha$ remains constant along $\gamma(t)$. This latter function is known as the ***Clairaut function***.

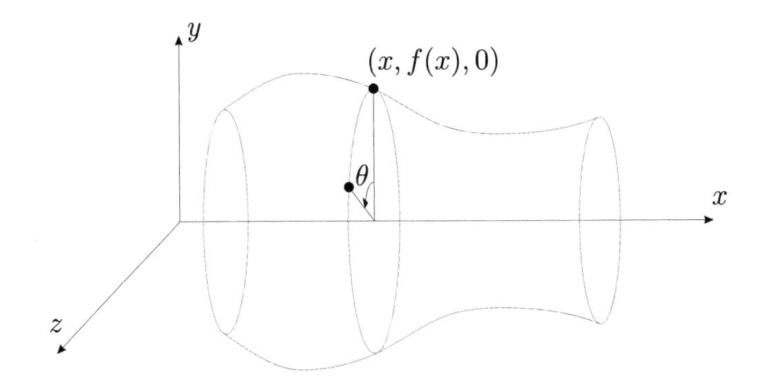

FIGURE 8. Surface of revolution

(d) (**Poincaré half plane**) Using Nöther's theorem, we complete the determination of the geodesics of Poincaré half plane that was started in Subsection 36. If γ is a geodesic, we continue to denote $\frac{d\gamma}{dt}$ by $(\dot{x}, \dot{y})$. For any $s \in \mathbb{R}$, the map $\Psi_s : (x, y) \mapsto (x+s, y)$ of $\mathbb{U}$ to itself is an isometry with respect to Poincaré metric $y^{-2}(dx \otimes dx + dy \otimes dy)$ (see also Exercise 3.19 of Chapter 3). We therefore obtain a one-parameter group of isometries, and the generator for this flow is $\frac{\partial}{\partial x}$. By Theorem 41, or Lemma 40, the following function is then constant on geodesics:

$$J_1((x, y), (\dot{x}, \dot{y})) = \rho(\frac{\partial}{\partial x}, \gamma'(t)) = \frac{\dot{x}}{y^2}$$

Another one-parameter group of isometries of $\mathbb{U}$ is provided by the homotheties:

$$\Theta_s(x, y) = (e^s x, e^s y)$$

Letting $\xi = e^s x$ and $\eta = e^s y$, we observe that

$$\eta^{-2}(d\xi \otimes d\xi + d\eta \otimes d\eta) = y^{-2}(dx \otimes dx + dy \otimes dy)$$

so that Θ_s is an isometry. The generator of this flow is the vector field $x\frac{\partial}{\partial x} + y\frac{\partial}{\partial y}$, which gives rise to the following momentum function:

$$J_2((x, y), (\dot{x}, \dot{y})) = \rho(x\frac{\partial}{\partial x} + y\frac{\partial}{\partial y}, \gamma'(t)) = \frac{x\dot{x} + y\dot{y}}{y^2}$$

At a point $z = ((x, y), (\dot{x}, \dot{y}))$ of the tangent bundle of $\mathbb{U}$, where $J_1(z) \neq 0$ (equivalently, $\dot{x} \neq 0$), we can set $J_2(z) = aJ_1(z)$, for appropriate constant a, and obtain

$$x\dot{x} + y\dot{y} = a\dot{x}$$

$$(x-a)\dot{x} + y\dot{y} = 0$$

Integrating, we have

$$(x-a)^2 + y^2 = c, \quad c : \text{constant}$$

It follows that the images of geodesics, that are not vertical lines, are half-circles in $\mathbb{U}$, with the center of the circle on the x-axis. This proves the claim in Subsection 36. To find the geodesic parametrization for the half-circles, it is convenient to use another isometry of $\mathbb{U}$, namely the map $u \mapsto (\bar{u})^{-1}$, where $\bar{u}$ is the complex conjugate of $u \in \mathbb{U}$, to transfer the parametrization given for vertical half-lines in Subsection 36 to half-circles (see Exercise 9.20 at the end of the chapter).

It is important to put in perspective what was accomplished in the last example above. Conservation of energy (or constancy of the length velocity vectors of geodesics, in the case of Poincaré half plane) confines the non-zero integral curves of G to a three-dimensional submanifold ("energy level") in the tangent bundle of $\mathbb{U}$. We then found two more apparently "independent" real-valued functions J_1 and J_2 that are constant on the integral curves of G. The common level set of the three functions, expectedly one-dimensional, is then an integral curve for G. In general, there may not exist enough momentum functions (also known as *first integrals*) to separate all integral curves (e.g., if integral curves become dense in an open subset), but knowledge of existing momenta provides valuable information about the behavior of integral curves. For an application of these ideas to the classical N-body problem of classical mechanics, see [**25**]. Regarding the notion of "independence" of first integrals, see Exercise 9.21 at the end of the chapter.

EXERCISES

9.1 Let M be a smooth manifold. Show that there is a smooth diffeomorphism $\iota{:}T^2M{\to}T^2M$ with the following properties:
(a) $\iota{\circ}\iota{=}\mathbb{1}_{T^2M}$.
(b) $\tau_{TM}{=}T\tau_M{\circ}\iota$, and $T\tau_M{=}\tau_{TM}{\circ}\iota$.
(c) ι is a VB-isomorphism between $\tau_{TM}{:}T^2M{\to}TM$ and $T\tau_M{:}T^2M{\to}TM$.
(Hint: Using (9.4), define ι locally by

$$\sum_{i=1}^{m} u^i \frac{\partial}{\partial q^i}(z) + \sum_{i=1}^{m} \dot{u}^i \frac{\partial}{\partial \dot{q}^i}(z) \mapsto \sum_{i=1}^{m} z^i \frac{\partial}{\partial q^i}(u) + \sum_{i=1}^{m} \dot{u}^i \frac{\partial}{\partial \dot{q}^i}(u)$$

Show that this map is independent of coordinates and satisfies the requirements.)

9.2 A second order ODE X on M is called a ***spray*** if for all $z{\in}TM$ and all $r{\in}\mathbb{R}$, one has

$$X(r{\cdot}z) = r\boxdot(r{\cdot}X(z))$$

Thus by 13c, the geodesic vector field is a spray.
(a) Show that for a second order ODE X, being a spray is equivalent to the following property: If $\alpha(t)$ is a solution of X, then so is $\alpha(rt)$ for all real numbers r.
(b) Show that a differential equation $\frac{d^2x}{dt^2}{=}g(x, \frac{dx}{dt})$ on an open subset U of $\mathbb{R}^m$ is a spray if and only if g is homogeneous of degree 2 in $\frac{dx}{dt}$.

9.3 Consider a spray on a smooth manifold M, e.g., a geodesic vector field relative to a linear connection. For $v{\in}T_aM$, let $\gamma^v{:}I{\to}M$ be the solution with initial condition $\gamma^v(0){=}a, (\gamma^v)'(0){=}v$.
(a) Show that the set V consisting of $v{\in}TM$ for which γ^v is defined throughout $[0, 1]$ is an open neighborhood of the image of the zero section in TM. We will denote $\gamma^v(1)$ by $\exp(v)$. This "exponential" generalizes the exponential functions defined for Lie groups.
(b) Prove that the map $(\tau_M, \exp){:}V{\to}M{\times}M$ by $v{\mapsto}(\tau_M(v), \exp(v))$ restricts to a diffeomorphism from a neighborhood of $\zeta_M(M)$ in TM to a neighborhood of the diagonal $\Delta{=}\{(x, x){:}x{\in}M\}$ in $M{\times}M$.

9.4 Show that the product connection is a linear connection.

9.5 Give an example of a product connection that is not a Levi-Civita connection.

9.6 Let M be a smooth m-manifold and H a linear connection for M. H is called a ***flat connection*** if the associated curvature tensor identically vanishes.
(a) Show that H is a flat connection if and only if H is integrable as an m-plane field on TM.
(b) We observed in Example 29 that the product connection on $\mathbb{R}^m$ is flat. Show that in fact any product connection is flat.
(c) Conversely, prove that any flat connection is *locally* a product, i.e., for every $z{\in}TM$, there is a neighborhood W of z in TM, and a smooth diffeomorphism

$h{:}W \to U \times V$, where U and V are open subsets of $\mathbb{R}^m$, so that Th maps horizontal and vertical subspaces of W, respectively, to horizontal and vertical subspaces of the product $U \times V$.

9.7 Using local coordinate system $(q, \dot{q})$ as in Subsection 5, suppose $z = \sum_j \dot{q}^j \frac{\partial}{\partial q^j}(a)$ is in $T_a M$, and $X \in T_z(TM)$ is represented by

$$X = \sum_k v^k \frac{\partial}{\partial q^k} + \sum_k w^k \frac{\partial}{\partial \dot{q}^k}$$

If h and v are, respectively, horizontal and vertical projections with respect to a linear connection, show that

$$h(X) = \sum_k v^k \frac{\partial}{\partial q^k} - \sum_k \Big(\sum_{i,j} \Gamma^k_{ij}(a) v^i \dot{q}^j \Big) \frac{\partial}{\partial \dot{q}^k}$$

$$v(X) = \sum_k \Big(w^k + \big(\sum_{i,j} \Gamma^k_{ij}(a) v^i \dot{q}^j \big) \Big) \frac{\partial}{\partial \dot{q}^k}$$

9.8 Let (M, ρ) be a Riemannian manifold.

(a) If ∇ is the covariant derivative operator of Levi-Civita connection for (M, ρ), and X, Y, Z are smooth vector fields on M, prove the following formula:

$$2\rho(\nabla_X Y, Z) = X{\cdot}\rho(Y, Z) + Y{\cdot}\rho(Z, X) - Z{\cdot}\rho(X, Y) - \rho(X, [Y, Z]) - \rho(Y, [Z, X]) + \rho(Z, [X, Y])$$

(Hint: Imitate the proof of the uniqueness of the Levi-Civita connection.)
(b) Using (a), give a coordinate-free proof of the existence and uniqueness of the Levi-Civita connection.

9.9 Let M be a hypersurface of a Riemannian manifold $(\tilde{M}, \tilde{\rho})$ with the induced Riemannian metric and Levi-Civita connection. If $\gamma{:}I \to M$ is a smooth curve on M, show that $\ddot{\gamma}(t)$ is the orthogonal projection of $\mathbf{a}(t)$ on $T_{\gamma(t)}M$, where $\mathbf{a}(t)$ is the intrinsic acceleration of γ regarded as a curve in $(\tilde{M}, \tilde{\rho})$ with the Levi-Civita connection. Deduce that γ is a geodesic if and only if the acceleration vector $\mathbf{a}(t)$ is perpendicular to M. Find all the geodesics of the m-sphere S_m as a hypersurface in $\mathbb{R}^{m+1}$.

9.10 Let (M, ρ) be a Riemannian manifold equipped with the Levi-Civita connection, and suppose $\gamma{:}I \to M$ is a regular smooth curve. We call $\mathbf{T}(t) = \frac{1}{|\gamma'(t)|_\rho} \gamma'(t)$ the ***unit tangent*** (in the direction of motion).
(a) Show that $\rho(\nabla_{\mathbf{T}} \mathbf{T}, \mathbf{T}) = 0$. The ***geodesic curvature*** of γ is defined to be $|\nabla_{\mathbf{T}} \mathbf{T}|_\rho$. Show that γ is a geodesic if and only if $|\gamma'(t)|_\rho$ is constant and the geodesic curvature identically vanishes.
(b) Show that the intrinsic acceleration $\ddot{\gamma}(t)$ lies in the span of $\mathbf{T}(t)$ and $\nabla_{\mathbf{T}(t)} \mathbf{T}$.
(c) Consider the case $\dim M = 2$ and M is orientable. For each $t \in I$, let $\mathbf{N}(t)$ be the unique element of $T_{\gamma(t)}M$ so that $(\mathbf{T}(t), \mathbf{N}(t))$ is a right-handed orthonormal basis for

$T_{\gamma(t)}M$. Then $\nabla_{\mathbf{T}}\mathbf{T}=\kappa_g\mathbf{N}$, and κ_g is called the ***signed geodesic curvature*** of γ. In this case, show that $\nabla_{\mathbf{T}}\mathbf{N}=-\kappa_g\mathbf{T}$.

9.11 Show that the torsion and curvature tensors vanish for one-dimensional manifolds. Give an example of a connection on a two-dimensional manifold with non-vanishing torsion.

9.12 In this exercise, we consider the 2-torus $\mathbb{T}^2$ with various connections.
(a) In $\mathbb{R}^3$, the 2-torus is commonly represented as the image of the following immersion:

$$\Psi_{a,b}(\phi,\theta) = ((a+b\cos\phi)\cos\theta, (a+b\cos\phi)\sin\theta, b\sin\phi)$$

where $a>b>0$ are constants. Compute the Gaussian curvature at $\Psi_{a,b}(\phi,\theta)$. Determine for which values of the constants c and c', the circles $\phi=c$ and $\theta=c'$ can be parametrized as geodesics.
(b) Represent $\mathbb{T}^2$ in $\mathbb{R}^4$ as the set

$$\{(\cos\alpha, \sin\alpha; \cos\beta, \sin\beta) : \alpha,\beta \in \mathbb{R}\}$$

with the Riemannian metric and Levi-Civita connection induced from $\mathbb{R}^4$. Show that the Gaussian curvature is everywhere zero and that all circles $\alpha=c$ and $\beta=c'$ can be parametrized as geodesics. What are the other geodesics? This is known as the ***flat torus***.
(c) Consider $\mathbb{T}^2$ as an abelian Lie group with the product connection arising from (left) invariant vector fields. Show there is an isometry between this and the flat torus.

9.13 Let M be a smooth m-manifold, U an open subset of M and $e=(e_1,\dots,e_m)$ an m-tuple of tangent vector fields defined on U so that at each $x\in U$, $(e_1(x),\dots,e_m(x))$ is a basis for T_xM; then (U,e) is called a ***moving frame*** on M. We let $(\theta^1,\dots,\theta^m)$ be the dual basis of one-forms on U. Suppose a linear connection on M is given. We define ω_j^k by

$$\nabla_{e_i}e_j = \sum_{k=1}^{m}\omega_j^k(e_i)e_k$$

It follows from the $C^\infty(M)$-linearity of ∇ with respect to e_i that ω_j^k are C^∞ one-forms. If T and R are, respectively, the torsion and curvature tensors of ∇, define T^i and R_j^i by

$$T(X,Y) = \sum_i T^i(X,Y)e_i$$

$$R(X,Y)e_j = \sum_i R_j^i(X,Y)e_i$$

(a) Show that T^i and R_j^i are smooth 2-forms on U.

(b) Prove the following:

$$T^i = d\theta^i + \sum_j \omega^i_j \theta^j \qquad \textbf{(Cartan's first structural equation)}$$

$$R^i_j = d\omega^i_j + \sum_k \omega^i_k \wedge \omega^k_j \qquad \textbf{(Cartan's second structural equation)}$$

9.14 (Continuation of Exercise 9.13) Let (M,ρ) be a Riemannian manifold. The moving frame (U,e) is called ***orthonormal*** if $(e_1(x),\ldots,e_m(x))$ is an orthonormal basis for T_xM for each $x \in U$. Show that ∇ arises from the Levi-Civita connection for (M,ρ) if and only if for each orthonormal moving frame the following are satisfied:

$$\omega^i_j + \omega^j_i = 0$$

$$d\theta^i + \sum_j \omega^i_j \wedge \theta^j = 0$$

9.15 (Continuation of Exercise 9.13) In the situation of Exercise 9.13, suppose that $\dim M = 2$.

(a) Show the following:

$$d\theta^1 = \theta^2 \wedge \omega^1_2, \quad d\theta^2 = \omega^1_2 \wedge \theta^1$$

(b) If K denotes the Gaussian curvature of M, then

$$d\omega^1_2 = k\, \theta^1 \wedge \theta^2$$

(c) Use (b) above to compute the Gaussian curvature of Poincaré upper half plane.
(d) Use (b) above to compute the Gaussian curvature of a surface of revolution. (See Example 42c.)
(e) Use (b) above to compute the Gaussian curvature of the 2-torus in each of the cases of Exercise 9.12.

9.16 For a Riemannian manifold (M,ρ) with the Levi-Civita connection and associated curvature tensor R, prove the following identities:
(a) $R(X,Y)Z + R(Y,Z)X + R(Z,X)Y = 0$ (First Bianchi Identity)
(b) $\rho(X, R(Z,W)Y) = \rho(Y, R(W,Z)X)$

9.17 Let ∇ and ∇' be two covariant derivative operators for a smooth manifold M, and denote $\nabla_X Y - \nabla'_X Y$ by $\beta(X,Y)$.
(a) Show that β is tensorial, i.e., $\beta(X,Y)(a)$ is determined bilinearly by the values $X(a)$ and $Y(a)$.
(b) Show that ∇ and ∇' possess the same geodesics if and only if $\beta(X,X)=0$ for every smooth vector field X.
(c) Show that if ∇ and ∇' possess the same geodesics and equal torsions, then $\nabla=\nabla'$.
(d) For every covariant derivative operator ∇ on M, prove that there is a torsion-free covariant derivative operator on M with the same geodesics.

9.18 A ***bi-invariant metric*** on a Lie group G is a Riemannian metric that is both left- and right-invariant. Let G be a Lie group that admits such a bi-invariant metric ρ, and consider the ∇ associated to the Levi-Civita connection of (G,ρ). Show that for left-invariant vector fields X, Y and Z, one has

$$\nabla_X Y = \frac{1}{2}[X, Y]$$
$$R(X, Y)Z = \frac{1}{4}[Z, [X, Y]]$$

(Hint: You may use Exercise 9.8.) Identify the geodesics.

9.19 Consider $\mathbb{R}^{m+1}$ with the Euclidean metric and suppose M is a hypersurface with the induced Levi-Civita connection and covariant derivative ∇. If S is the Weingarten map, show that for any two smooth tangent vector fields X and Y on M

$$S[X, Y] = \nabla_X(SY) - \nabla_Y(SX) \quad \textbf{(Codazzi-Mainardi Equation)}$$

(Hint: Use (9.39).)

9.20 Representing the points of the Poincaré upper half plane $\mathbb{U}$ by $z{=}x{+}iy$, show that the map $z{\mapsto}\bar{z}^{-1}$ is an isometry. Obtain the parametrization of geodesic half-circles in $\mathbb{U}$ using this isometry.

9.21 A collection $\{f_1, \ldots, f_k\}$ of first integrals for a vector field X on a manifold is called ***independent*** if the map $(f_1, \ldots, f_k)$ to $\mathbb{R}^k$ is a submersion. Referring to Example 42d, show that J_1, J_2 and the kinetic energy constitute three independent first integrals for the geodesic flow on $T\mathbb{U}$.

9.22 Let (M,ρ) be a Riemannian manifold. If $R{:}TM{\to}\mathbb{R}$ is a smooth function, then the ***fiber gradient*** $\check{\nabla}_\rho R$ of R is defined as follows. For $z{\in}T_aM$, we let R_a be the restriction of R to T_aM, and we define $\check{\nabla}_\rho R(z)$ to be $\nabla_\rho R_a(z)$. The fiber gradient is then a vertical vector field on M. Thus if X is a second order ODE on M, then so is $X{+}\check{\nabla}_\rho R$.
(a) Show that the fiber gradient of the kinetic energy is the Liouville vector field L. (For the definition of Liouville vector field, see Exercise 6.13 of Chapter 6.)
(b) Let X be a conservative mechanical system. For a smooth function $R{:}TM{\to}\mathbb{R}$, we call $X_R{=}X{+}\check{\nabla}_\rho R$ a ***dissipative system*** with ***Rayleigh dissipation function*** R. If E is the energy function of X, show that $dE(X_R){=}dR(L)$.

9.23 Let (M,ρ) be a Riemannian manifold. The ***Liouville 1-form*** λ_ρ is defined by

$$\lambda_\rho(X) = \rho(z, T\tau_M(X)), \quad \text{where } X{\in}T_z(TM)$$

(a) If ρ has the local expression $\sum_{ij} g_{ij} d\xi^i{\otimes}d\xi^j$ relative to the chart (U,ξ), and $(TU,(q,\dot{q}))$ is the corresponding chart on TM, show that the local expression for λ_ρ is

$$\lambda_\rho = \sum_{ij} g_{ij}\dot{q}^j dq^i$$

(b) If G is the geodesic vector field of (M, g) and E_k is the kinetic energy, show the following:

$$i_G\lambda_\rho = 2E_k, \quad L_G\lambda_\rho = dE_k, \quad i_G(d\lambda_\rho) = -dE_k$$

(c) If $V{:}M\to\mathbb{R}$ is smooth, $F= -\nabla_\rho V$ and $\overline{F}$ is the vertical spread of F, then

$$i_{\overline{F}}(d\lambda_\rho) = -d(V\circ\tau_M)$$

(d) If $X = G+\overline{F}$, and E is the energy function, deduce that

$$i_X(-d\lambda_\rho) = dE$$

$(-d\lambda_\rho)$ is called the ***ρ*-symplectic form** of TM.
(e) Show that $\Omega_\rho=\lambda_\rho\wedge d\lambda_\rho{}^{m-1}$ is a volume element for the unit sphere bundle of M. (See Exercise 7.6 of Chapter 7 for the definition of the unit sphere bundle.)
(f) By Lemma 37, the unit sphere bundle is invariant under the geodesic flow. Show that $\mathbf{div}_{\Omega_\rho}G=0$.

9.24 (Continuation of Exercise 9.23) Let (M, ρ) be a smooth Riemannian manifold. By Theorem 20 of Chapter 6, ρ induces a vector bundle isomorphism $\rho^\flat$ between TM and T^*M.
(a) Show that under $\rho^\flat$, λ_ρ corresponds to Liouville 1-form λ on T^*M (see Exercise 6.29 of Chapter 6), hence $(\rho_\sharp)^*\lambda_\rho$ is independent of the Riemannian metric ρ.
(b) Letting $\sigma= -d\lambda$, $H=E\circ\rho_\sharp$ and $X_H=(\rho^\flat)_*X$, show that $i_{X_H}\sigma=dH$.

APPENDIX I

The Exponential of a Matrix

Let $A=[a_{ij}]$ be an $m\times m$ real or complex matrix. The k-fold product of A by itself will be denoted by $A^k=[a_{ij}^{(k)}]$, and A^0 will be, by convention, the $m\times m$ identity matrix. We then claim that the following series of matrix sums converges:

$$\sum_{k=0}^{\infty} \frac{A^k}{k!} \tag{I.1}$$

To see this, let M be an upper bound for the absolute values $|a_{ij}|$ of matrix entries. We observe inductively then that

$$|a_{ij}^{(k)}| \leq m^{k-1} M^k$$

Therefore,

$$|\sum_{k=0}^{p} \frac{a_{ij}^{(k)}}{k!}| \leq \frac{1}{m} \sum_{k=0}^{p} \frac{(Mm)^k}{k!}$$

It follows that the matrix series (I.1) converges, and the ij entry of the sum has uniform bound $\frac{1}{m}\exp(Mm)$. The convergent sum (I.1) will be denoted by $\exp A$ or e^A and will be called the ***exponential*** of A.

1. Examples

(a) Consider the diagonal matrix

$$D = \begin{bmatrix} \lambda_1 & 0 & \dots & 0 \\ 0 & \lambda_2 & \dots & \\ \vdots & \vdots & \ddots & 0 \\ 0 & \dots & 0 & \lambda_m \end{bmatrix}$$

We then obtain

$$D^k = \begin{bmatrix} \lambda_1^k & 0 & \dots & 0 \\ 0 & \lambda_2^k & \dots & \\ \vdots & \vdots & \ddots & 0 \\ 0 & \dots & 0 & \lambda_m^k \end{bmatrix}, \quad \exp D = \begin{bmatrix} e^{\lambda_1} & 0 & \dots & 0 \\ 0 & e^{\lambda_2} & \dots & \\ \vdots & \vdots & \ddots & 0 \\ 0 & \dots & 0 & e^{\lambda_m} \end{bmatrix}$$

(b) Consider the $m \times m$ nilpotent matrix

$$N = \begin{bmatrix} 0 & 1 & \dots & \dots & 0 \\ 0 & 0 & 1 & \dots & 0 \\ \vdots & \vdots & \ddots & \ddots & 0 \\ 0 & \dots & \dots & 0 & 1 \\ 0 & \dots & \dots & \dots & 0 \end{bmatrix}$$

By taking successive powers of N, the diagonal of 1s recedes to the right, so that we obtain $N^m = 0$, and the entries of $\exp N = [e_{ij}]$ will be as follows:

$$e_{ij} = \begin{cases} 0 & \text{if } i > j \\ \frac{1}{n!} & \text{if } j - i = n \end{cases}$$

(c) Let A be the matrix

$$A = \begin{bmatrix} 0 & -1 \\ 1 & 0 \end{bmatrix}$$

We verify that $A^2 = -I$, $A^3 = -A$, $A^4 = I$, and, in general, $A^{k+4} = A^k$. It follows that for any number t, we have

$$\exp tA = \begin{bmatrix} \cos t & -\sin t \\ \sin t & \cos t \end{bmatrix}$$

We will later relate this to the flows of linear differential equations.

2. Theorem *Let A and B be $m \times m$ matrices so that $AB = BA$. Then*

$$\exp(A+B) = (\exp A)(\exp B)$$

Proof. Using a variable t, we show that

$$\exp[t(A+B)] = (\exp tA)(\exp tB) \tag{I.2}$$

Then the desired result is obtained by setting $t = 1$. Each side of the above is a power series in t with matrix coefficients. We compare the coefficient of t^k on the two sides. For the left-hand side, the kth degree term is $\frac{1}{k!} t^k (A+B)^k$. It follows from $AB = BA$ that $(A+B)^k$ obeys the binomial expansion

$$(A+B)^k = \sum_{i=0}^{k} \binom{k}{i} A^i B^{k-i}$$

therefore

$$\frac{1}{k!} t^k (A+B)^k = \sum_{i=0}^{k} \frac{(tA)^i}{i!} \cdot \frac{(tB)^{k-i}}{(k-i)!}$$

The right-hand side of the above is the kth degree term of the right-hand side of (I.2), and the result follows. □

3. Corollary *For complex numbers s and t and any n×n matrix A*

$$e^{(s+t)A}=e^{sA}\cdot e^{tA}$$

4. Corollary *If* $\lambda_1,\ldots,\lambda_m$ *are the eigenvalues of the m×m matrix A, then* $e^{\lambda_1},\ldots,e^{\lambda_m}$ *are the eigenvalues of* e^A. *In particular,* $\exp A$ *is always non-singular, and*

$$\exp(\operatorname{tr} A) = \det(\exp A) \tag{I.3}$$

PROOF. Suppose A is in Jordan canonical form $A=D+N$, where D and N are in the forms of Example 1(a) and (b), and $ND=DN$. It then follows from Theorem 2 and the above examples that $\exp A$ is an upper triangular matrix with diagonal entries $e^{\lambda_1},\ldots,e^{\lambda_m}$. In general, there is a non-singular complex matrix P so that $PAP^{-1}=B$ is in Jordan canonical form. Since $PA^kP^{-1}=B^k$ for all k, then $P(\exp A)P^{-1}=\exp B$, and the first assertion of the corollary holds in general. The remaining assertions follow from the fact that the determinant is the product of eigenvalues and the trace is the sum. □

The following theorem is referred to in Chapters 2 and 8. Below we represent points of $\mathbb{R}^m$ as column m-vectors and linear endomorphisms of $\mathbb{R}^m$ by $m\times m$ matrices relative to the standard basis of $\mathbb{R}^m$.

5. Theorem *Consider the linear differential equation on* $\mathbb{R}^m$ *given by*

$$\frac{dx}{dt} = A\cdot x, \quad x\in\mathbb{R}^m \tag{I.4}$$

If $\bar{x}\in\mathbb{R}^m$, *then the solution* $x(t)$ *of the differential equation with initial condition* $x(0)=\bar{x}$ *is defined for all* $t\in\mathbb{R}$, *and in fact* $x(t)=\exp tA\cdot\bar{x}$.

PROOF. The proposed solution satisfies the initial condition, we must verify that it satisfies the differential equation (I.4). We have

$$\begin{aligned}\frac{d}{dt}(\exp tA\cdot\bar{x}) &= \lim_{h\to 0}\frac{e^{(t+h)A}-e^{tA}}{h}\cdot\bar{x}\\ &= \lim_{h\to 0}\frac{e^{tA}\cdot e^{hA}-e^{tA}}{h}\cdot\bar{x}\\ &= e^{tA}\cdot\lim_{h\to 0}\frac{e^{hA}-I}{h}\cdot\bar{x}\\ &= e^{tA}\cdot A\cdot\bar{x}\\ &= A\cdot x(t)\end{aligned}$$

and the theorem is proved. □

APPENDIX II

Differential Calculus in Normed Space

In this Appendix we will collect and review, mostly without proof, some results from differential calculus on finite-dimensional real normed spaces that are used in the text. Proofs can be found in standard references such as [**6**], [**19**] and [**24**].

Any two norms on a finite-dimensional vector space are equivalent in the sense that if $|.|$ and $|.|'$ are two norms on a finite-dimensional real vector space V, then there are positive real numbers m and M so that

$$m|x| \leq |x|' \leq M|x|, \quad \forall x \in V$$

It follows from these inequalities that the concepts of limit, convergence, continuity and Cauchy sequence are independent of the particular norm. Finite-dimensional normed spaces are Banach spaces, i.e., they are complete. Linear and multilinear maps between finite-dimensional normed spaces are continuous. In the sequel, all vector spaces will be finite-dimensional vector spaces over $\mathbb{R}$.

Let V and V' be vector spaces, U an open subset of V and $a \in U$. A map $f{:}U \to V'$ is said to be ***differentiable*** at a if there is a linear map $L{:}V \to V'$ so that

$$f(x) - [f(a) + L(x-a)] = o(|x-a|)$$

This means that if we denote the left-hand side by $\rho(x, a)$, then

$$\lim_{x \to a} \frac{\rho(x, a)}{|x-a|} = 0$$

If f is differentiable at a, then the linear map L is uniquely determined, it is called the ***derivative*** of f at a and is denoted by $Df(a)$.

1. Elementary Properties

(a) If f is differentiable at a, then f is continuous at a.
(b) (Chain Rule) Suppose V_1, V_2 and V_3 are vector spaces with U_1 and U_2, respectively, open subsets of V_1 and V_2. If $f{:}U_1 \to V_2$ is differentiable at a_1 and $g{:}U_2 \to V_3$ is differentiable at $a_2 = f(a_1)$, then $g \circ f$ is defined in an open neighborhood of a_1, is differentiable at a_1, and

$$D(g \circ f)(a_1) = Dg(a_2) \circ Df(a_1)$$

(c) If $f, g{:}U\to V'$ are differentiable at $a\in U$, and $r\in\mathbb{R}$, then $rf+g$ is differentiable at a and

$$D(rf+g)(a) = rDf(a) + Dg(a)$$

(d) (Leibniz Rule) Any p-linear map $f{:}V_1\times\cdots\times V_p\to V'$ is differentiable at every point $a=(a_1,\ldots,a_p)\in V_1\times\cdots\times V_p$, and the linear map $Df(a){:}V_1\times\cdots\times V_p\to V'$ is given by

$$(Df(a))(h_1,\ldots,h_p) = \sum_{i=1}^{p} f(a_1,\ldots,a_{i-1},h_i,a_{i+1},\ldots,a_p)$$

2. Examples (a) Let $A{:}V\to V'$ be an affine map, i.e., $A(x)=L(x)+B$ for all $x\in V$, where $L{:}V\to V'$ is linear and $B\in V'$. Then A is differentiable at every point x and $DA(x)=L$.

(b) Let V and V' be finite-dimensional normed vector spaces, then $\mathcal{L}(V,V')$, the space of linear maps $V\to V'$, is a finite-dimensional normed space (see, e.g., Example 2f of Chapter 8). Consider the ***evaluation map***

$$e : \mathcal{L}(V,V')\times V \to V'$$

given by

$$e(f,v) = f(v)$$

e is bilinear, therefore by Leibniz Rule:

$$(De(f,v))(g,w) = g(v) + f(w)$$

(c) Suppose V is an n-dimensional normed space. Choosing a basis for V, $\mathcal{L}(V,V)$ can be identified with the space $M_n(\mathbb{R})$ of real $n\times n$ matrices. Regarding the columns of a matrix $A\in M_n(\mathbb{R})$ as elements of $\mathbb{R}^n$, the determinant function is an n-linear map

$$\det : \mathbb{R}^n\times\cdots\times\mathbb{R}^n \to \mathbb{R}$$

Let I be the $n\times n$ identity matrix and $H=[h_{ij}]$ be an arbitrary element of $M_n(\mathbb{R})$. Then by the Leibniz Rule:

$$\begin{aligned}(D\det(I))(H) &= h_{11}+\cdots+h_{nn}\\ &= \operatorname{tr} H\end{aligned}$$

It turns out that some major theorems of differential calculus require a hypothesis stronger than simple differentiability of a function, e.g., some hypothesis on the derivative itself. Suppose that the function $f{:}U\to V'$ is differentiable at every point of its domain, then we obtain a function $Df{:}U\to\mathcal{L}(V,V')$ that assigns to every $x\in U$, the derivative $Df(x)$ of f at x. We say that f is a C^1 (or a ***continuously differentiable***) function (at a) if Df is continuous (at a). If Df, as a map to a normed space, is itself differentiable at a point $a\in U$, we say that f is ***twice differentiable*** at a, and we call the derivative of Df at a, the ***second derivative*** of f at a, and denote it by $D^2f(a)$. Thus we have the following linear map:

$$D^2f(a)=(D(Df))(a) : V \to \mathcal{L}(V,V')$$

If $D^2 f$ is defined in a neighborhood of a and is continuous at a, we say that f is C^2 at a. One can discuss inductively still higher degrees of differentiability. Suppose that f is an r times differentiable function from a neighborhood U of a in V to V'. Thus

$$D^r f : U \to \underbrace{\mathcal{L}(V, \mathcal{L}(\cdots, \mathcal{L}}_{r \text{ times}}(V, V')\cdots))$$

If $D^r f$ is differentiable at $a \in U$, then we obtain a linear map

$$D^{r+1} f(a) = (D(D^r f))(a) : V \to \underbrace{\mathcal{L}(V, \mathcal{L}(\cdots, \mathcal{L}}_{r \text{ times}}(V, V')\cdots)),$$

or

$$D^{r+1} f(a) \in \underbrace{\mathcal{L}(V, \mathcal{L}(\cdots, \mathcal{L}}_{r+1 \text{ times}}(V, V')\cdots))$$

These look rather unwieldy, but the following general algebraic fact provides a more manageable way of looking at higher derivatives. Let us denote the space of p-linear maps

$$V_1 \times \cdots \times V_p \to V'$$

by $\mathcal{L}^p(V_1 \times \cdots \times V_p, V')$.

3. Lemma *There is a canonical isomorphism*

$$l : \mathcal{L}(V_1, \mathcal{L}(\cdots, \mathcal{L}(V_p, V')\cdots)) \to \mathcal{L}^p(V_1 \times \cdots \times V_p, V')$$

given by

$$(l(\varphi))(x_1, \ldots, x_p) = \Big(\cdots\big((\varphi(x_1))(x_2)\big)\ \cdots\Big)(x_p)$$

Proof. The fact that $l(\varphi) \in \mathcal{L}^p(V_1 \times \cdots \times V_p, V')$ follows from the linearity of each application of the iteration on the left-hand side, and the linearity of l is just the definition of the vector-space structure in spaces of linear maps. Finally, the inverse

$$k : \mathcal{L}^p(V_1 \times \cdots \times V_p, V') \to \mathcal{L}(V_1, \mathcal{L}(\cdots, \mathcal{L}(V_p, V')\cdots))$$

is given by

$$\Big(\cdots\big((k(\psi)(x_1))(x_2)\big)\ \cdots\Big)(x_p) = \psi(x_1, \ldots, x_p)$$

□

We now go back to the definition of higher derivatives. Let us denote the r-fold Cartesian product of V by itself by V^r, so that $\mathcal{L}^r(V^r, V')$ is the space of r-linear maps from V^r to V'. Suppose that $f : U \to V'$ is r times differentiable at $a \in U$, then we write

$$d^r f(a) = l \circ D^r f(a)$$

and call $d^r f(a)$ the rth (order) ***differential*** of f at a. Thus the rth differential is an r-linear map from V^r to V', and $df(a) = Df(a)$. The following is the first basic theorem on higher derivatives.

4. Theorem *Let V and V' be finite-dimensional normed spaces, U an open subset of V, $a{\in}U$, and $f{:}U{\to}V'$ an r times differentiable function at a. Then $d^r f(a)$ is a symmetric r-linear map, i.e., for any permutation σ of $\{1,\ldots,r\}$*

$$d^r f(a)(v_{\sigma(1)},\ldots,v_{\sigma(r)}) = d^r f(a)(v_1,\ldots,v_r)$$

The symmetry of higher derivatives is traditionally expressed by the interchange of the order of *partial differentiation*. This will result from the discussions that follow.

If $v{\in}V$, the **derivative of f at a with respect to v**, if it exists, is defined as the following limit:

$$D_v f(a) = \lim_{h\to 0} \frac{f(a{+}hv){-}f(a)}{h}$$

In particular, if $(e_1,\ldots,e_n)$ is a basis for V, and points of V are represented by $x{=}\sum_{j=1}^{n} x^j e_j$, then $D_{e_j} f(a)$ is also written as $D_j f(a)$ or $\frac{\partial f}{\partial x^j}(a)$, and is called the ***partial derivative*** of f with respect to x^j. Note that the target space of each partial derivative function is V' itself, not a more involved linear space. If $D_j f(x)$ exists for all $x{\in}U$, then

$$D_j f = \frac{\partial f}{\partial x^j} : U \to V'$$

This makes the iteration of partial derivatives conceptually simpler than the iteration of derivatives. We can speak of n^r partial derivatives of order r by considering differentiation with respect to various x^j. The following basic assertion is an easy consequence of the definition of differentiability.

5. Theorem *If $f{:}U{\to}V'$ is differentiable at $a{\in}U$, then the derivative of f at a with respect to any $v{\in}V$ exists and*

$$D_v f(a) = (Df(a))(v)$$

It follows inductively from this theorem and Theorem 4 that if f is r times differentiable (at a), then the value of a partial derivative of order r (at a) is independent of the order of variables used in differentiation. Theorem 5 also explains the usual representation of the derivative as the "Jacobian matrix." As above, let us write points of U as $x{=}\sum_{j=1}^{n} x^j e_j$. Taking a basis $(e'_1,\ldots,e'_m)$ for V', and writing the points of V' as $y{=}\sum_{i=1}^{m} y^i e'_i$, the partial derivative can be represented as:

$$D_j f(a) = \sum_{i=1}^{m} \frac{\partial y^i}{\partial x^j}(a)\, e'_i$$

It follows from Theorem 5, taking $v{=}e_j$, that the matrix representation of the linear map $Df(a)$ with respect to the above bases is the $m{\times}n$ matrix $[\frac{\partial y^i}{\partial x^j}(a)]$.

If the function f is r times differentiable in a neighborhood U of the point a and the rth order differential

$$d^r f : U \to \mathcal{L}^r(V^r, V')$$

is continuous at a, we say f is C^r at a. A continuous function is sometimes called a C^0 function. By a C^∞ (or ***smooth***) function we mean a function that has derivatives of all orders (necessarily all continuous). The following is another very useful and basic theorem of differential calculus.

6. Theorem *Let V and V' be finite-dimensional normed spaces, U an open subset of V, $a \in U$, and $f{:}U \to V'$. Then f is C^r at a if and only if all partial derivatives of f up to and including order r exist and are continuous at a.*

Throughout the text, we are making use of a number of existence theorems centered around the *Inverse Function Theorem.* These will occupy the rest of this Appendix.

7. Inverse Function Theorem *Let V and V' be finite-dimensional normed spaces, U an open subset of V and $f{:}U \to V'$ a C^r map, $1 \leq r \leq \infty$, so that $Df(x)$ is invertible for every $x \in U$. Then:*
(a) f is an open map.
(b) For every $a \in U$, there is an open neighborhood W of a in V and an open neighborhood W' of $f(a)$ in V' so that f maps W in one-to-one manner onto W', and the inverse $g = (f|_W)^{-1}{:}W' \to W$ is also C^r. Further, for every $x \in W$, one has

$$Dg(f(x)) = (Df(x))^{-1}$$

Of course, the last formula is a consequence of the chain rule, once the differentiability of g has been established. Note that for the hypothesis of invertability of the linear map $Df(x)$ to hold, the dimensions of V and V' must necessarily be equal. An invertible C^r map $\varphi{:}W \to W'$ with C^r inverse is called a C^r ***diffeomorphism***.

One can generalize the notion of partial derivative to derivative with respect to components of a subspace decomposition. Let U be an open subset of a product $V = V_1 \times V_2$ and $f{:}U \to V'$ a mapping. At a point $(a_1, a_2) \in V_1 \times V_2$, we will consider the notion of the derivatives of f with respect to subspaces V_1 and V_2, which we will denote, respectively, by $D_{V_1} f(a_1, a_2)$ and $D_{V_2} f(a_1, a_2)$. Let U_1 and U_2 be open neighborhoods, respectively, of a_1 in V_1 and of a_2 in V_2 so that

$$U_1 \times \{v_2\} \subset U, \quad \{v_1\} \times U_2 \subset U$$

Then consider $g_1{:}U_1 \to V'$ and $g_2{:}U_2 \to V'$ given by

$$g_1(x) = f(x, a_2), \quad g_2(y) = f(a_1, y)$$

We then define

$$D_{V_1}f(a_1,a_2) = Dg_1(a_1)$$
$$D_{V_2}f(a_1,a_2) = Dg_2(a_2)$$

If f is differentiable at $a=(a_1,a_2)$, then $D_{V_1}f(a_1,a_2)$ and $D_{V_2}f(a_1,a_2)$ exist. For matrix representation, let $(e_1,\ldots,e_n)$, $(e_{n+1},\ldots,e_{n+m})$ and $(e'_1,\ldots,e'_p)$ be bases for V_1, V_2 and V', respectively, and represent the points of these spaces, respectively, by $x=\sum_{j=1}^{n} x^j e_j$, $y=\sum_{k=1}^{m} y^k e_{n+k}$ and $z=\sum_{i=1}^{p} z^i e_i$. Then the matrix of $Df(a)$ is a $p\times(n+m)$ matrix, the first n columns of which represent $D_{V_1}f(a)$ in the form $[\frac{\partial z^i}{\partial x^j}(a)]$, and the last m columns represent $D_{V_2}f(a)$ in the form $[\frac{\partial z^i}{\partial y^k}(a)]$.

The *Implicit Function Theorem* below describes sufficient conditions under which the level set of a C^r function can be represented as the graph of a C^r function.

8. Implicit Function Theorem *Let V_1, V_2 and V' be finite-dimensional normed spaces with* $\dim V_2 = \dim V'$, *U an open subset of $V_1\times V_2$, and $F{:}U\to V'$ a C^r function, $r\geq 1$. Suppose that at $a=(a_1,a_2)\in U$, $F(a)=0$ and $D_{V_2}F(a)$ is invertible. Then there are open neighborhoods U_1 and U_2 of a_1 and a_2, respectively, with $U_1\times U_2\subset U$, and a C^r function $f{:}U_1\to U_2$ with the following properties:*
(a) $f(a_1)=a_2$
(b) The intersection of the level set $F^{-1}(0)$ with $U_1\times U_2$ is the graph of f:

$$F^{-1}(0)\cap U_1\times U_2 = \{(x,f(x)) : x\in U_1\}$$

(c) For U_1 sufficiently small and $x\in U_1$, $D_{V_2}F(x,f(x))$ is invertible and

$$Df(x) = -\big(D_{V_2}F(x,f(x))\big)^{-1}\circ D_{V_1}F(x,f(x))$$

The final result to be presented here is the *Rank Theorem*, which is the main tool of Chapter 5.

9. Rank Theorem *Let V and V' be normed spaces of dimension p and q, respectively, with U an open subset of V. Suppose that a C^r function $f{:}U\to V'$, $r\geq 1$, is given with the property that $Df(x)$ has the same rank for all $x\in U$. Then given $a\in U$, there are open neighborhoods $W_1\subset U$ of a, W'_1 of $f(a)$, W_2 of $\mathbf{0}\in\mathbb{R}^p$ and W'_2 of $\mathbf{0}\in\mathbb{R}^q$, and C^r diffeomorphisms $\phi{:}W_1\to W_2$ and $\psi{:}W'_1\to W'_2$ so that $\phi(a)=\mathbf{0}\in\mathbb{R}^p$, $\psi(f(a))=\mathbf{0}\in\mathbb{R}^q$, and $\psi\circ f\circ\phi^{-1}{:}\phi(W_1\cap f^{-1}(W'_1))\to\mathbb{R}^q$ is the restriction of a linear map from $\mathbb{R}^p$ to $\mathbb{R}^q$.*

The theorem is often expressed by saying that a C^r map, $r\geq 1$ of constant rank can be locally linearized by C^r change of coordinates. Note that since ϕ and ψ are diffeomorphisms, the linear map $\psi\circ f\circ\phi^{-1}$ has the same rank as $Df(a)$. Further, it is a simple fact of linear algebra that if g is a linear map $\mathbb{R}^p\to\mathbb{R}^q$ of rank k, then one can choose bases for $\mathbb{R}^p$ and for $\mathbb{R}^q$ relative to which g can be represented as a projection:

$$(x^1,\ldots,x^p)\mapsto(x^1,\ldots,x^k,\underbrace{0,\ldots,0}_{q-k})$$

Another elementary fact of linear algebra is that if $g{:}\mathbb{R}^p\to\mathbb{R}^q$ is a surjective (respectively, injective) linear map, then g possesses a linear right inverse (respectively, a linear left inverse), i.e., a linear map $h{:}\mathbb{R}^q\to\mathbb{R}^p$ exists so that $g\circ h=\mathbb{1}_{\mathbb{R}^q}$ (respectively, $h\circ g=\mathbb{1}_{\mathbb{R}^p}$). The corollary below, used in the text, then follows.

10. Corollary *Under the hypothesis of Theorem 9, suppose that the constant rank of the derivative is k.*
(a) If $k=q\leq p$, then for every $a\in U$, there is an open neighborhood W of $f(a)$ in V' and a C^r map $s{:}W\to U$ so that $(f\circ s)(y)=y$ for all $y\in W$.
(b) If $k=p\leq q$, then for every $a\in U$, there is an open neighborhood W of $f(a)$ in V' and a C^r map $t{:}W\to U$ so that $(t\circ f)(x)=x$ for all $x\in f^{-1}(W)$.

Bibliography

[1] J. F. Adams, *Lectures on Lie Groups*, W. A. Benjamin, New York 1969.
[2] Yu. Aminov, *The Geometry of Vector Fields*, Gordon and Breach Science Publishers, Amsterdam 2000.
[3] V. I. Arnold, *Ordinary Differential Equations*, MIT Press, Cambridge MA 1973.
[4] M. Berger and B. Gostiaux, *Differential Geometry: Manifolds, Curves, and Surfaces*, Springer-Verlag 1988.
[5] R. Bott and L. W. Tu, *Differential Forms in Algebraic Topology*, Springer-Verlag 1982.
[6] H. Cartan, *Cours de calcul différentiell*, Hermann, Paris 1977.
[7] C. Chevalley, *Theory of Lie Groups I*, Princeton University Press, Princeton, NJ 1946.
[8] S. Gallot, D. Hulin, and J. Lafontaine, *Riemannian Geometry* (3rd. ed.), Springer-Verlag 2004.
[9] C. Godbillon, *Géometrie différentielle et mécanique analytique*, Hermann, Paris 1969.
[10] C. Godbillon, *Éléments de topologie algébrique*, Hermann, Paris 1971.
[11] P. Hartman, *Ordinary Differential Equations* (2nd. ed.), Birkhäuser, Boston 1982.
[12] A. Hatcher, *Algebraic Topology*, Cambridge University Press, Cambridge, UK 2002.
[13] M. W. Hirsch, *Differential Topology*, Springer-Verlag 1976.
[14] M. W. Hirsch, S. Smale, and R. L. Devaney, *Differential Equations, Dynamical Systems, and an Introduction to Chaos*, Academic Press, Boston, MA 2004.
[15] N. Jacobson, *Lie Algebras*, Dover, New York 1979.
[16] J. L. Kelley, *General Topology*, Van Nostrand, Princeton 1955.
[17] W. Klingenberg, *A Course in Differential Geometry*, Springer-Verlag 1978.
[18] J. Lafontaine, *An Introduction to Differential Manifolds*, Springer-Verlag 2015.
[19] S. Lang, *Real and Functional Analysis*, Springer-Verlag 1993.
[20] J. W. Milnor, *Topology from a Differentiable Viewpoint* (Revised ed.), Princeton University Press, Princeton 1997.
[21] R. Montgomery, *A Tour of Subriemannian Geometries, Their Geodesics and Applications*, Amer. Math. Soc., Providence 2002.
[22] J. Munkres, *Topology*, 2nd. ed., Prentice Hall, Upper Saddle River, NJ 2000.
[23] W. A. Poor, *Differential Geometric Structures*, Dover, New York 2007.
[24] C. C. Pugh, *Real Mathematical Analysis*, Springer-Verlag 2010.
[25] S. Smale, "Topology and Mechanics I and II," *Inventiones math.*, **10**, 305-331 and **11**, 45-64 (1970).
[26] M. Spivak, *A Comprehensive Introduction to Differential Geometry*, Volumes 1-5, Houston, Publish or Perish 1994.
[27] N. Steenrod, *The Topology of Fibre Bundles*, Princeton University Press, Princeton 1951.
[28] S. Sternberg, *Curvature in Mathematics and Physics*, Dover, New York 2012.

List of symbols

Index